P9-CAS-286

Nobel Laureate	Country*	Year of Award	Accomplishment
Julius Wagner-Jauregg	Austria	1927	Discovery of the therapeutic value of malaria inoculation in the treatment of dementia paralytica
Charles J. H. Nicolle	France	1928	Studies on the cause and transmission of epidemic typhus
Christiaan Eijkman	Netherlands	1929	Discovery of the antineuritic vitamin
Frederick Gowland Hopkins	England		Discovery of the growth-stimulating vitamins
Karl Landsteiner	United States (Aust.)	1930	Discovery of the human blood groups
Otto Warburg	Germany	1931	Discovery of the nature and mode of action of the respiratory enzyme
Charles Scott Sherrington	England	1932	Functions of neurons
Edgar Douglas Adrian	England		
Thomas Hunt Morgan	United States	1933	Function of chromosomes in the transmission of heredity
George Hoyt Whipple	United States	1934	Discoveries concerning liver therapy against anemias
George Richards Minot	United States		
William Parry Murphy	United States		
Hans Spemann	Germany	1935	Discovery of the organizer effect in embryonic development

Nobel Laureate	Country*	Year of Award	Accomplishment
Henry Dale	England	1936	Chemical transmission of nerve impulses
Otto Loewi	Germany		
Albert Szent-Györgyi von Nagyrapolt	Hungary	1937	Indentification of vitamin C and discoveries in cell metabolism
Corneille Heymans	Belgium	1938	Discovery of the role played by the sinus and aortic mechanisms in the regulation of respiration
Gerhard Domagk	Germany	1939	Discovery of the chemotherapeutic effects of prontosil
No Award		1940–1942	
Henrik Dam	Denmark	1943	Discovery of vitamin K
Edward A. Doisy	United States		
Joseph Erlanger	United States	1944	Functions of nerve fibers
Herbert Spencer Gasser	United States		
Sir Alexander Fleming	England	1945	Discovery and development of penicillin
Ernst Boris Chain	England Germany		
Sir Howard W. Florey	England (Australia)		
Hermann Joseph Muller	United States	1946	Discovery of the production of mutations by means of X-ray irradiation

Continued on inside back cover

BIOLOGY

Cover The leopard (*Panthera pardus*) is surely among the most beautiful and graceful of the big cats. The characteristic black spots of a leopard are arranged in rosettes, typically against a background of straw-colored fur. However, leopards display a strong tendency toward melanism, an increase in the amount of pigment in the fur around the spots. Melanistic forms, often called "black panthers," have glossy dark brown or black fur. The spots of a black leopard, such as the one shown here, can be seen only when light strikes them from a particular angle. Unlike most species of cats, whose pupils contract in bright light to form ovals or vertical slits, the pupils of leopards and a few other cat species remain round when contracted.

Second Edition

BIOLOGY

Leland G. Johnson
Augustana College

Wm. C. Brown Publishers
Dubuque, Iowa

Book Team

John D. Stout
Executive Editor

Kevin Kane
Editor

Karen Slaght
Production Editor

Teresa E. Webb
Designer

Jeanne M. Rhomberg
Design Layout Assistant

Shirley M. Charley
Photo Research Editor

Mavis M. Oeth
Permissions Editor

Matt Shaughnessy
Product Manager

wcb
Wm. C. Brown Publishers, College Division

G. Franklin Lewis
Executive Vice-President, General Manager

E. F. Jogerst
Vice-President, Cost Analyst

George Wm. Bergquist
Editor in Chief

Beverly Kolz
Director of Production

Chris C. Guzzardo
Vice-President, Director of Marketing

Bob McLaughlin
National Sales Manager

Craig S. Marty
Manager, Marketing Research

Colleen A. Yonda
Production Editorial Manager

Marilyn A. Phelps
Manager of Design

Faye M. Schilling
Photo Research Manager

wcb group

Wm. C. Brown
Chairman of the Board

Mark C. Falb
President and Chief Executive Officer

Cover image © Morton Beebe/The Image Bank

The credits section for this book begins on page 1087, and is considered an extension of the copyright page.

Copyright © 1983, 1987 by Wm. C. Brown Publishers. All rights reserved

Library of Congress Catalog Card Number: 86-70548

ISBN 0-697-04972-8

No part of this publication may be reproduced, stored in a retrieval system, or transmitted, in any form or by any means, electronic, mechanical, photocopying, recording, or otherwise, without the prior written permission of the publisher.

Printed in the United States of America
10 9 8 7 6 5 4 3 2 1

To Rebecca whose creative work is evident throughout this book and to Michael, Julie, and Franklin who grew up with "the book."

Brief Contents

Expanded
Contents

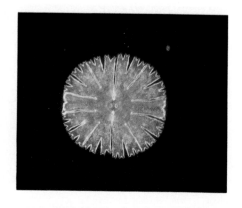

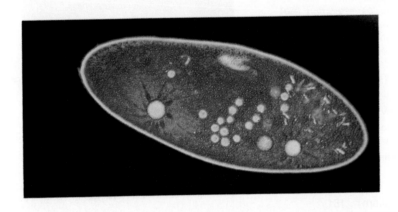

Part 2

Plants and Animals
Function and Structure of Organisms

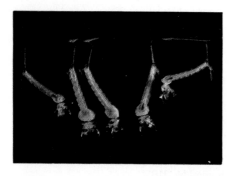

Part 3

Regulation in Organisms

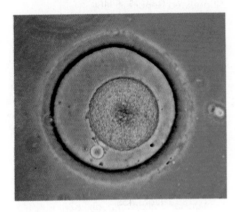

Part 4

Genetics and Development
The Continuity of Life

20
Cell Division 487

21
Fundamentals of Genetics 507

22
Chromosomes and Genes 527

23
Molecular Genetics 551

24
Genetic Expression 573

Part 5

Evolution

Part 6

Behavior and Ecology

Part 7

The Diversity of Life

Preface *continued*

Approach

Biology reflects a thorough and complete balance of plant and animal biology. Fundamental biological unity is recognized where it is scientifically and educationally sound to do so. Thus, plants and animals are considered together when discussing basic cell biology, genetics, reproduction, development, evolution, and ecology.

But plants and animals also differ in basic ways. For example, while water and mineral ion absorption by roots of vascular plants and the digestive and absorptive processes by which animals obtain nutrients are processes that might be broadly categorized as nutrient procurement, they both are worthy of careful, separate examination. Likewise, plant transport processes and animal circulation are fundamentally different sets of processes. Separate consideration also is given to plant hormonal regulation, which is primarily involved in growth control, and animal hormonal regulation, which is involved much more in short-term modulation of physiological processes.

A special organizational approach has been taken in discussing chemical reaction mechanisms, photosynthesis, and cell respiration. Chapter 6 deals with energy relationships and chemical reaction mechanisms, and it includes information relevant to both chapter 7 (photosynthesis) and chapter 8 (cellular metabolism). Chapters 7 and 8 are specifically designed to be interchangeable in sequence, however, so that instructors are free to treat either photosynthesis or cell metabolism first.

Another special approach has been taken to the study of one of the most rapidly expanding fields in modern biology: the study of resistance and immune mechanisms. Because expressions of immune responses involve developmental events that occur throughout the lives of organisms, these topics have been included in a chapter on lifelong development (chapter 29). Chapter 29 also deals with wound healing and regeneration, cancer and other abnormal growths, and aging.

Organization

The organization of this book is straightforward to allow for direct pursuit of its goals and principal themes. Chemistry and cell biology are introduced early in Part 1 to provide the foundation for understanding the more complex organismic functions and processes presented in Parts 2 and 3.

Genetics is presented in Part 4 with an eye toward its role in reproduction and development, which are presented in the later chapters of Part 4.

After the fundamental processes have been presented, the broader topics of evolution and behavior and ecology are discussed in Parts 5 and 6 respectively. Surveys of organisms are presented at the end of the book, Part 7, for purposes of flexibility, since there is tremendous variety in the placement of the surveys in courses at various institutions. The survey material can be inserted at any point that suits local course requirements. Within this framework, sections and individual chapters are designed to be sufficiently autonomous to make the text functional in virtually any course organization.

Pedagogy

Student learning aids, outlined in detail in Aids to the Reader, include part introductions, chapter outlines, chapter concepts, chapter introductions, key terms, boxed essays, chapter summaries, chapter questions for review and for analysis and discussion, suggested readings, the glossary, appendixes, and the index. The International System of Units (SI units) has been used for all physical measurements, except for heat energy, where the more familiar calories that have a strong reference framework in everyday life have been used. All learning aids are designed to facilitate the understanding and retention of biology for the student, and to make the learning process enjoyable and interesting.

biology, and it is a key theme in this text. Homeostasis permits organisms to exploit more variable external environments. Environmental changes may swirl about an organism, but its body cells live and function in a sheltered, stable internal environment.

Diversity

While we recognize the fundamental unity of life on earth, we also recognize that living things are very diverse. They come in a great variety of shapes and sizes, and they are adapted to life in a broad range of habitats. There are organisms that live in fresh water, in seawater, in freezing cold, in blistering heat, and in all environments in between.

There are literally millions of kinds of living things. How, you may wonder, do biologists make sense of this almost bewildering array of life on earth?

Biologists describe, name, and classify living things on the basis of similarities and differences. Such a system of **taxonomy** (naming and grouping of living things) is necessary in order to investigate, describe, and discuss various organisms. In this text, we will use a taxonomic scheme known as the **five-kingdom system.** The five kingdoms are **Monerans, Protists, Fungi, Plants,** and **Animals.** The following brief survey will introduce you to these kingdoms.

Monerans

This kingdom includes bacteria (singular: bacterium) and cyanobacteria (also called blue-green algae). The key characteristic of monerans is that they are **prokaryotic**—their cells do not have a membrane-enclosed nucleus or any of several other membrane-enclosed cellular structures found in the cells of members of the other four kingdoms.

Bacteria are microscopic single-celled organisms that often occur in colonies containing millions of cells. A great many bacteria cause infection and disease in humans and other animals and plants, but not all bacteria are harmful or dangerous. Bacteria are used in many industrial processes including the manufacture of cheeses and other foods. Bacteria play vital roles in natural decay processes, which release important nutrients from dead organisms and make them available for use by other organisms. Bacteria also are critically important in sewage treatment.

Cyanobacteria are prokaryotic organisms that carry on photosynthesis (figure 1.10). **Photosynthesis** is the use of light energy to synthesize nutrients from carbon dioxide and water. Cyanobacteria are found in a variety of environments, including very harsh ones such as hot springs. They usually play a positive role as a nutrient source for organisms that eat them, but occasionally they grow so numerous in a lake or pond that they make the water unpleasant or even dangerous for other living things.

Figure 1.10 The cyanobacterium (blue-green alga) *Oscillatoria* grows in many freshwater environments. Note that *Oscillatoria* grows in filaments, end-to-end chains of cells. Cyanobacteria are important food sources for many animals.

Protists

All protists are single **eukaryotic** cells or colonies of eukaryotic cells. A eukaryotic cell possesses a nucleus that is enclosed by a membrane, as well as several other kinds of membrane-enclosed structures. This kingdom includes protozoa and unicellular algae.

You can often see hundreds of protozoa in a single drop of pond water (figure 1.11a). Some whip through the water with amazing speed, while others slowly glide along. Most protozoa live independently, but some are disease-causing parasites. For example, malaria is caused by a protozoan.

Many kinds of single-celled algae are included in the protists (figure 1.11b). They carry out photosynthesis and, small though they are, are very important as major producers of organic nutrients, especially in the oceans of the world. The nutrients they produce support all of life in the oceans, ranging from tiny organisms to the largest of whales.

Fungi

The fungi (singular, fungus) include many familiar organisms such as molds, mildews, mushrooms, and puffballs (figure 1.12). The cells of fungi have cell walls and characteristically are arranged end-to-end in tubular filaments. Each fungus consists of a mass of these filaments. Fungi obtain nutrients by breaking down and absorbing chemical substances from their environment, and are very important participants in natural decay. A few fungi cause diseases.

Figure 1.11 Protists. (a) *Paramecium*, a single-celled organism common in many freshwater environments. These protists obtain nutrients by engulfing small particles and smaller organisms in the water around them. (b) A diatom. Diatoms are single-celled algae that have beautiful, glassy cell walls. Diatoms carry out photosynthesis and are so numerous in the world's waters that they provide food for many animals and supply much of the atmospheric oxygen used by all living things.

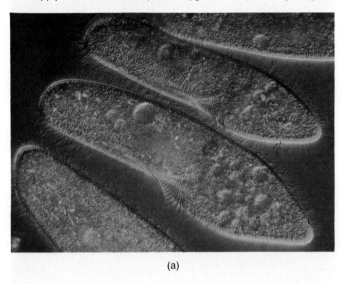

(a)

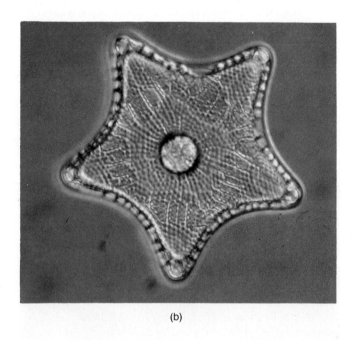

(b)

Figure 1.12 Mushrooms, such as this red-capped *Amanita* (a poisonous mushroom), are members of the kingdom Fungi. Most of the complex tubular network that makes up this fungus is underground. The familiar mushroom is a specialized reproductive structure that is produced only occasionally.

Plants

Plants are multicellular organisms that have eukaryotic cells enclosed by rigid cell walls. Plants possess the green pigment chlorophyll and carry out photosynthesis. Members of the plant kingdom range from microscopic multicellular algae that form delicate spheres or chains of cells, to enormous trees that are the largest of all living things (figure 1.13).

Animals

Animals are distinguished from members of other kingdoms by several key characteristics. They are multicellular and their cells do not have cell walls. Animals cannot manufacture their own organic nutrients by photosynthesis. But unlike fungi, which absorb organic nutrients produced by other organisms, animals ingest (eat) other organisms or material produced by other organisms.

Animals range in diversity from simply organized animals such as sponges and jellyfish to complex animals that can withstand rigorous challenges of life in the terrestrial environment (figure 1.14).

Figure 1.13 The plant kingdom. (a) *Volvox,* a multicellular green alga that is a colony of cells arranged in a sphere. Note the daughter colonies developing inside the sphere. They will be released later and become new independent colonies. (b) Kelp (brown algae) exposed on a seashore at low tide. (c) *Marchantia,* a liverwort. These small plants grow in moist environments. At times, upright stalks with egg-producing structures develop on some plants and sperm-producing structures develop on others. (d) Ferns. Most ferns grow in moist, sheltered environments, but a few hardy ferns grow in open meadows. (e) Flowers and buds of a marigold. There are a great many flowering plants, ranging from grasses to large trees. (f) A giant sequoia tree, which is one of the largest living things and an example of a coniferous (cone-bearing) plant. Many common trees such as pines and spruces are also conifers.

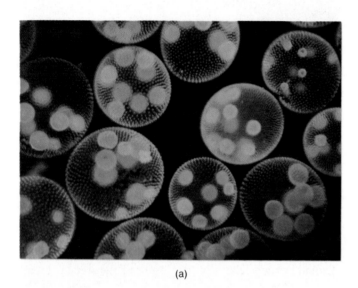

(a)

(b)

(c)

(d)

(e)

(f)

Figure 1.14 The animal kingdom. (a) Sponges on the ocean floor. (b) The Portuguese man-of-war. This is a colony of individuals living together. One is specialized as a gas-filled float. Others are specialized for defense and can deliver painful stings when touched. (c) Muscle tissue with three cysts containing *Trichinella,* a parasitic roundworm. (d) Snails belong to a group of animals called molluscs, most of which have shells. The molluscs include clams, oysters, octopuses, squids, and slugs. (e) A reef lobster. Lobsters, crabs, and crayfish are crustaceans, a subdivision of the arthropods (''joint-footed animals''). Other major groups of arthropods are the arachnids (scorpions, spiders, ticks, and mites) and the insects. (f) The Viceroy butterfly, an insect. Insects are the largest group of arthropods. (g) A variety of kinds of fish in a coral reef off the coast of Florida. Fish are the most numerous vertebrate animals (animals with backbones). The other major groups of vertebrate animals are amphibians, reptiles, birds, and mammals. (h) A poison arrow frog, an amphibian. Toads and salamanders are also amphibians. (i) A female lion with her cubs. Lions and other cats are mammals, as are humans and many other familiar vertebrate animals.

(a)

(b)

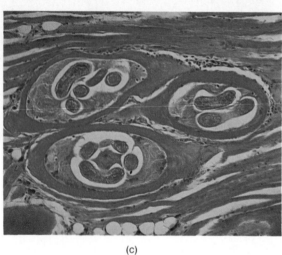

(c)

(d)

The Science of Biology

(e)

(f)

(g)

(h)

(i)

Human Beings in the Living World

Having briefly explored the diversity of life, we should add a word about ourselves. Biology is a very personal science because we humans are an integral part of the living world that we study. We learn about ourselves not only through direct consideration of the human body, but also through study of all the living things with which we share this planet.

In developing our theme of human beings in the living world, we will focus on the human organism at many points in this book, but we will always do so clearly recognizing that human biology is part of biology in general. We humans do not stand apart from other living things—we share numerous problems and processes with many of them and our activities often affect not only our own environment, but that of countless other organisms.

Biology Today

Literally thousands of intriguing questions remain unanswered concerning many aspects of the lives of the vast array of living things that populate the earth. Past discoveries and the development of new research techniques have opened the way for investigating many of these questions. But how does a young person become prepared to join this fascinating search?

Earlier we discussed the methods of investigation used by scientists and we saw that biologists actually take a variety of approaches to the study of life. Some biologists simply make observations and then systematize those observations. Others devise and test hypotheses by experimentation. Some biologists use combinations of these methods, while still others use methods that defy precise description or classification. Regardless of the methods a biologist uses, however, he or she must meet several basic requirements in order to work successfully and derive satisfaction from the work. First, biologists must have adequate background information on the processes or topics they are studying. Second, they must be able to use the "tools of the trade"—in modern biology, these include knowledge of the physical sciences and skills in mathematics as well as basic education in biology. Third, biologists must recognize that what is thought and known today will be corrected and modified in the future, just as the viewpoints of earlier biologists have been corrected and modified.

Figure 1.15 Albert Szent-Györgyi. "Discovery consists in seeing what everybody else has seen and thinking what nobody else has thought."

Present knowledge is only a set of ideas to work with, not unalterable truth. Sometimes emotional involvement in one's own scientific work makes it difficult to keep this perspective, but realizing that scientific knowledge itself is continually evolving helps to promote objectivity.

Finally, and perhaps most importantly, science requires an open and inquiring mind. Perhaps Albert Szent-Györgyi (figure 1.15) put it best in his succinct analysis of scientific work: "Discovery consists in seeing what everybody else has seen and thinking what nobody else has thought."

Summary

Biology is the study of life. Humans have always been aware of the world around them. Motivated by curiosity, people throughout history have added to biological knowledge by direct observation. Biologists currently employ a variety of methods in their attempts to answer questions about the world of life.

Within the framework of the scientific method, biologists develop hypotheses that can be tested through experimentation; results of experiments are used to draw conclusions. A large body of sound conclusions forms a theory.

Darwin and Wallace traveled extensively and developed the theory of natural selection on the basis of their observations of many biological phenomena. Gregor Mendel's experimental work on garden peas laid the groundwork for the modern science of genetics. Watson and Crick applied data gathered by other investigators, doing no actual experiments themselves, to successfully construct a model of DNA.

There is a fundamental unity among all forms of life on earth. The shared characteristics of all living things are evolutionary adaptation, complex organization, metabolism, irritability, growth and development, and reproduction.

Unicellular organisms meet all of their needs by making exchanges directly with the external environment. Multicellular organisms are characterized by a "division of labor" that contributes to maintenance of a stable internal environ-that contributes to maintenance of a stable internal environment. Homeostasis is a dynamic steady state maintained in the internal body environment of organisms.

We can organize the diverse life on our planet into five kingdoms: Monerans, Protists, Fungi, Plants, and Animals.

Whatever the outcome of research, biologists must recognize that present scientific knowledge is only a set of ideas with which to work, not unalterable truth.

Questions for Review

1. What is biology? How has the study of biology changed from its beginnings to the present?
2. Define these terms: experiment, hypothesis, control group.
3. How did Darwin and Wallace develop the theory of natural selection?
4. How did Darwin and Wallace's methods of scientific investigation differ from those of Gregor Mendel?
5. Explain how Watson and Crick discovered the structure of DNA without conducting any experiments.
6. List and briefly describe six characteristics of life.
7. What is homeostasis?
8. What are the five kingdoms of life on earth? Give at least one example of an organism found in each kingdom and explain why you included it in that group.

Questions for Analysis and Discussion

1. Contrast the use of the word *theory* in everyday conversation with its use in science.
2. Why is it important for biologists to publish the results of their investigations, thus making their work available to other scientists?

Suggested Readings

Books

Bronowski, J. 1973. *The ascent of man.* Boston: Little, Brown and Co.

Gardner, E. J. 1972. *History of biology.* 3d ed. Minneapolis: Burgess Publishing.

Kormondy, E. J. 1966. *General biology: Molecules and cells.* Dubuque, Iowa: Wm. C. Brown Publishers.

Sayre, A. 1975. *Rosalind Franklin and DNA.* New York: Norton.

Watson, J. D. 1968. *The double helix.* New York: Signet Books.

Articles

Eiseley, L. C. February 1956. Charles Darwin. *Scientific American.*

Eiseley, L. C. February 1959. Alfred Russel Wallace. *Scientific American.*

Gould, S. J. January 1980. Wallace's fatal flaw. *Natural History.*

Marshack, A. 1975. Exploring the mind of Ice Age man. *National Geographic* 147:64.

Straus, L. G. 1985. Stone age prehistory of northern Spain. *Science* 230:501.

Figure 2.14 Relationships of pH, hydrogen ion concentration [H⁺], and hydroxide ion concentration [OH⁻].

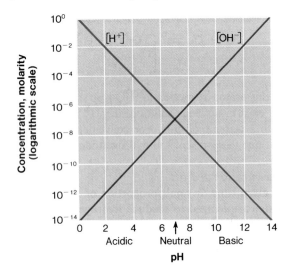

Figure 2.15 The pH scale showing the pH of various common fluids.

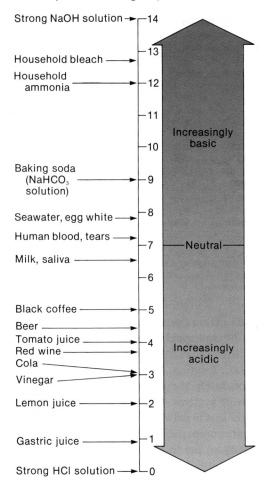

A solution is **neutral** if its H⁺ and OH⁻ concentrations are equal. For water, $[H^+] = [OH^-]$ when both exist in concentrations of 10^{-7} M; that is, when the pH is 7. A solution is **acidic** if it has a higher H⁺ concentration than OH⁻ concentration, and **basic** if its H⁺ concentration is lower than its OH⁻ concentration (figure 2.14). Therefore, acidic solutions have pH values below 7, and basic solutions have pH values above 7 (figure 2.15).

Acids and Bases

Substances that change the concentrations of H⁺ or OH⁻ ions when they dissolve in water are known as *acids* and *bases*. An **acid** is a *proton (hydrogen ion) donor;* it causes an increase in H⁺ ion concentration when it ionizes in solution. For example, hydrochloric acid (HCl) is ionized quite completely by the reaction:

$$HCl \longrightarrow H^+ + Cl^-$$

Adding hydrochloric acid to a solution increases the relative number of H⁺ ions and, therefore, lowers the pH of the solution. The solution of gastric juices in the human stomach, for example, contains a great deal of hydrochloric acid and sometimes may have a pH as low as 1.0. This is such an acidic solution that stomach contents coming in contact with skin could cause a serious acid burn.

A **base** is a *proton acceptor* and has the opposite effect on a solution. For example, sodium hydroxide (NaOH) ionizes almost completely in solution and strongly increases the relative concentration of OH⁻ ions. Hydroxide ions (OH⁻) combine with H⁺ ions to form water, and thereby decrease the H⁺ ion concentration (thus raising the pH of the solution).

Buffers

Even small pH changes may have great effects on biological processes. Because many processes are sensitive to pH changes, it is important that the pH of body fluids remains stable. Cell interiors have an average pH of 7.0 to 7.3; and in humans, for example, blood and tissue fluids have a pH of 7.4 to 7.5. These values remain very stable despite the fact that many biochemical reactions either release or incorporate H⁺ ions.

How does pH remain so constant in living things? Such pH stability is possible because organisms have built-in mechanisms to prevent pH change. The most important of these mechanisms are **buffers.** A buffer resists pH change by removing H^+ ions when the H^+ ion concentration rises, and releasing H^+ ions when the H^+ ion concentration falls.

An example of a buffer is the carbonic acid–bicarbonate ion buffer system, which is involved in the buffering of human blood:

$$H_2CO_3 \rightleftarrows H^+ + HCO_3^-$$

Carbonic acid   Bicarbonate ion

In this example, if H^+ ions are added to the system, they combine with HCO_3^- to form H_2CO_3. This reaction removes extra H^+ ions and keeps the pH from changing. If H^+ ions are removed from the system (for example, if OH^- ions are added and H_2O is formed), more H_2CO_3 will ionize and replace the H^+ ions that were used. Again, pH stability is maintained. This and other buffer systems prevent potentially harmful pH changes in living things.

Organic Molecules

Most of the thousands of kinds of chemical compounds in living organisms are **organic compounds,** compounds that contain carbon. The most important characteristics of organic compounds depend on properties of this key element, carbon.

The carbon atom has an atomic number of six, with six protons and six electrons. Since such an atom has two electrons in the first shell and four electrons in the second shell, it can form covalent bonds with as many as four other atoms. A carbon atom can bond to a variety of elements, but it most commonly bonds to hydrogen, oxygen, nitrogen, and carbon.

The carbon-to-carbon bonding ability of carbon atoms makes possible carbon chains (sometimes called carbon skeletons) of various lengths and shapes. The carbon skeleton establishes the overall framework of an organic molecule. Carbon atoms can form single, double, or triple covalent bonds with one another, and such bonding patterns determine the characteristics of organic compounds (figure 2.16).

Many organic compounds consist of just carbon and hydrogen, but many others contain **functional groups** as well. Functional groups are characteristic patterns of atoms other than those involved in carbon-carbon and carbon-hydrogen bonds. These functional groups are the parts of organic molecules that are most often involved in chemical reactions (figure 2.17).

Figure 2.16 Possible arrangements of bonds around the carbon atom. Single and double bonds are common in organic compounds found in living things, but triple bonds are not. In the methane molecule, the dashed line indicates that the carbon-hydrogen bond goes away from you, and the extended triangle indicates that the carbon-hydrogen bond comes toward you, out of the plane of the page.

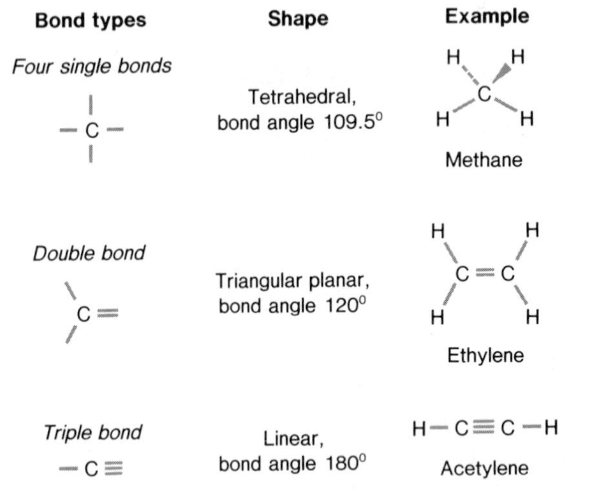

Bond types	Shape	Example
Four single bonds	Tetrahedral, bond angle 109.5°	Methane
Double bond	Triangular planar, bond angle 120°	Ethylene
Triple bond	Linear, bond angle 180°	Acetylene

One reason there are so many organic compounds is that carbon forms chains of atoms better than any other element. An unusually long noncarbon chain might contain ten atoms. On the other hand, organic compounds with chains of more than fifty carbon-carbon bonded atoms can be found in living systems, and chains containing carbon along with other elements such as nitrogen and oxygen are sometimes thousands of atoms long.

Another factor contributing to the diversity of organic compounds is the presence of numerous **isomers.** Isomers are compounds having the same molecular formulas but different atom arrangements. *Structural isomers* are completely different chemical compounds; they share the same molecular formula but have different pairs of atoms connected by bonds (figure 2.18a). *Geometric isomers* differ from each other only in the orientations of bonds between particular pairs of atoms (figure 2.18b). These differences are possible in the presence of certain rings or C=C double bonds. Although rotation around C— C single bonds occurs readily, rotation around C=C double bonds does not. *Optical isomers* occur as pairs of nonsuperimposible mirror images, similar to the relationship between your right and left hands. Optical isomers have at least one asymmetric carbon atom, a carbon atom attached to four different groups. This makes it possible for the four different groups to occupy two different arrangements in space. These mirror-image forms are designated L– and D–. Such differences can be biologically important. For example, it is interesting and a little curious that all of the amino acids that occur in proteins in living things are of the L– form (figure 2.18c).

Figure 2.17 Some examples of functional groups found in organic molecules. The functional group is enclosed in a colored box in each case. (a) Neutral functional groups. (b) Acidic and basic functional groups.

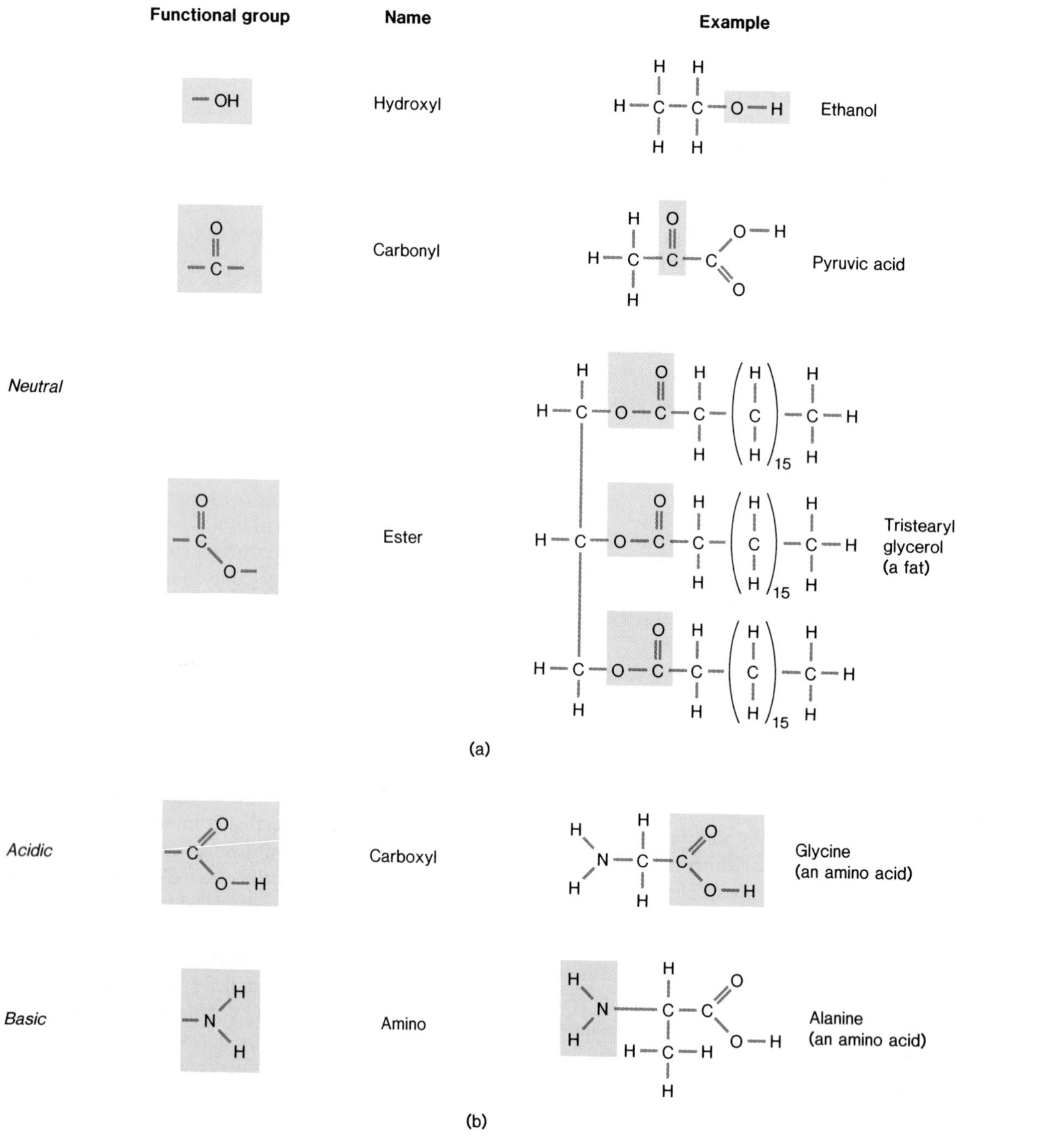

(a)

(b)

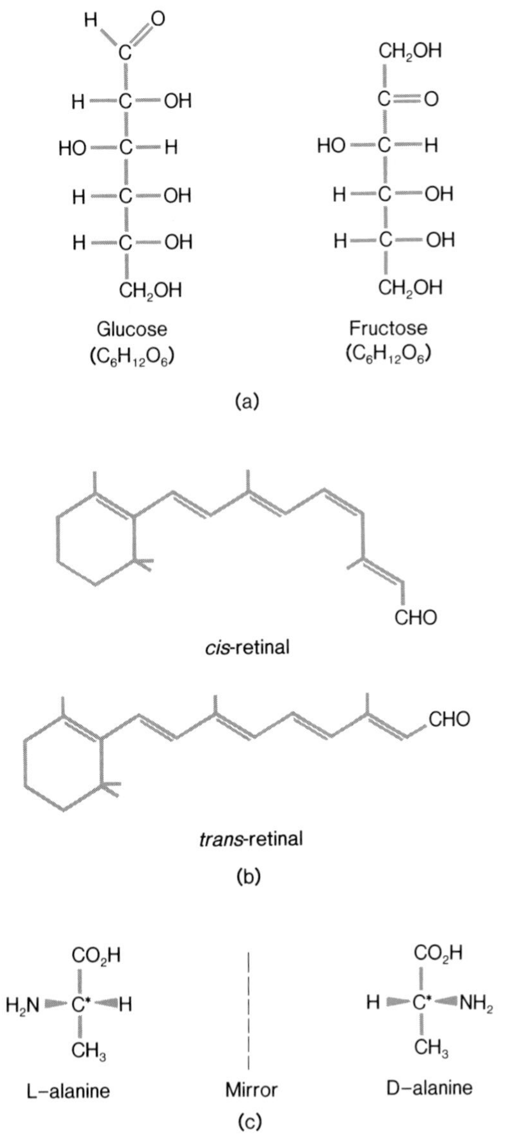

Figure 2.18 Some examples of isomers. (a) Structural isomers are molecules with identical molecular formulas, but different pairs of atoms connected by bonds. The sugars glucose and fructose both have the molecular formula ($C_6H_{12}O_6$), but their atoms are arranged differently. (b) Geometric isomers are molecules that have different orientations of bonds in space, but that are not mirror images of each other. Each line indicates a C–C bond in retinal. Hydrogens are not shown. In the retina of the eye, *cis*-retinal is present (combined with a protein) as rhodopsin. Light energy is absorbed by *cis*-retinal, converting it to *trans*-retinal. Thus, geometric isomers are important in the initial step in vision. (c) Optical isomers are pairs of molecules that are nonsuperimposable mirror images of each other. Most optical isomers contain one or more asymmetric carbon atoms (indicated by an asterisk). An asymmetric carbon atom is a carbon bonded to four different groups in a molecule. L-alanine occurs commonly in organisms; D-alanine does not.

Glucose
($C_6H_{12}O_6$)

Fructose
($C_6H_{12}O_6$)

(a)

cis-retinal

trans-retinal

(b)

L–alanine

Mirror

D–alanine

(c)

Summary

The study of modern biology requires a background in the physical sciences, especially chemistry.

The basic substances composing all forms of matter are called elements. Atoms are the smallest units having all the characteristic properties of the elements, but atoms are made up of protons (positively charged particles), neutrons, and electrons (negatively charged particles). Protons and neutrons are located in the nucleus of an atom, while electrons circle the nucleus in orbitals in a series of shells.

An element's atomic number is equal to the number of protons in its nucleus. Normally, the number of electrons in shells around the nucleus is the same as the number of protons. Different isotopes of an element have different numbers of neutrons. Some isotopes are radioactive and spontaneously emit particles.

The numbers and locations of electrons in the orbital shells around different atoms determine the chemical properties and bonding characteristics of those atoms.

Ionic bonding involves the transfer of electrons between atoms. Covalent bonds form when atoms share pairs of electrons in their outer orbital shells. Single, double, and triple covalent bonds result from the sharing of one, two, or three pairs of electrons.

Weaker bonds also hold molecules together. Hydrogen bonding occurs both between molecules and within molecules, where it plays a role in establishing complex three-dimensional molecular structures.

Water is the most abundant material in living things. Water's polarity explains its solvent properties; polar substances are soluble in water and nonpolar substances are not. Water's physical properties, such as its high heat capacity, high heat of vaporization, and high surface tension, also serve useful functions in living systems.

One of the most useful units for measuring and expressing concentration is molarity. Molarity is the number of moles of solute per liter of solution.

Hydrogen ion concentration can be expressed through the concept of pH. Acidic solutions have high H^+ concentrations and low pH numbers, while basic solutions have high OH^- concentrations and high pH numbers. Buffer systems that resist pH changes in cells and in body fluids are very important for the maintenance of normal biochemical functions.

Figure 3.11 The relationship of primary, secondary, and tertiary protein structure. Expanded diagram of a small part of the polypeptide chain shows peptide bonds connecting several amino acids.

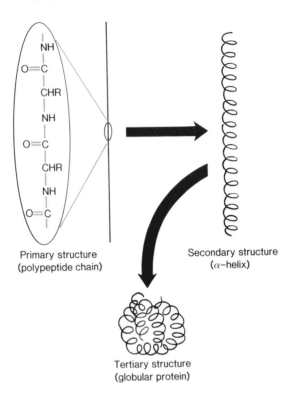

Primary structure
(polypeptide chain)

Secondary structure
(α–helix)

Tertiary structure
(globular protein)

Figure 3.12 X-ray crystallography of the protein myoglobin, which reversibly stores oxygen in muscle tissue. The three-dimensional structure of myoglobin was the first protein structure to be determined by this technique. (a) Diagram of the X-ray diffraction technique. (b) X-ray diffraction picture of myoglobin. (c) Three-dimensional representation of myoglobin as it was determined by John Kendrew using X-ray diffraction techniques and complex mathematical analysis of the X-ray diffraction pictures he obtained. Note the α-helix secondary structure and the folded tertiary structure. The heme group (involved in oxygen binding) is shown in black.

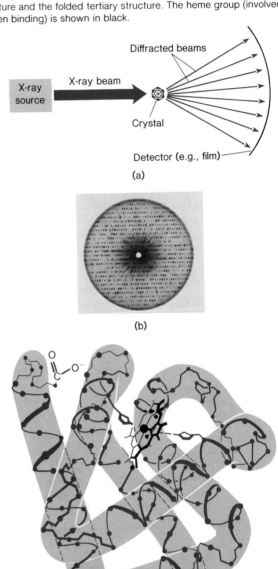

The **tertiary structure** of a protein, which also is stabilized by weak bonds between amino acids in different parts of the molecule, is formed when the polypeptide chain, with its orderly secondary structure, is bent and folded into an even more complex three-dimensional arrangement (figure 3.11).

One of the most powerful and useful techniques for determining the three-dimensional structure of a substance such as a protein is **X-ray crystallography.** In this technique, an X-ray beam is passed through a crystal of the substance being studied. Part of the beam is scattered (diffracted) as it passes through the crystal; the way it scatters depends upon the molecule's structure. On the other side of the crystal, a photographic plate records a pattern of spots representing the intensity of the emergent X rays (figure 3.12). Scientists can then use this pattern to determine the three-dimensional shapes of molecules.

The form of a protein's amino acid chain in a crystal seems to be similar to its form in solution. Therefore, the information provided by X-ray crystallography is relevant to the study of protein structure in solution within living cells.

Forces Determining Protein Structure

Secondary and tertiary structures are formed by weak, noncovalent bonds. Because their bond energies are so small, however, noncovalent bonds constitute a significant factor in determining protein structure only when they are formed in relatively large numbers. Furthermore, the strength of these types of bonds decreases as their length increases. As a result, a protein molecule folds in such a way that the number of weak bonds is maximized and their bond lengths are optimized (shortened). Thus, a protein chain, which may already contain stretches of α-helix, folds upon itself to form its tertiary structure, a more complex, often globular, shape.

Because proteins interact with other molecules, they must fold very precisely in assuming their tertiary structures. A given protein will interact strongly with another molecule only if the surface shapes of the two molecules are complementary. That is, the two molecules must fit together precisely—much like two puzzle pieces—for a stable attachment to occur. This need for a precise fit between molecules is the basis for the high degree of specificity in the interactions between proteins and other molecules.

The studies of Christian Anfinsen on the enzyme ribonuclease revealed much about how a protein's complex structure comes about. Ribonuclease, a digestive enzyme produced by the pancreas, hydrolyzes ribonucleic acid (RNA) in the digestive tract. It is a relatively small protein consisting of a single polypeptide chain made up of 124 amino acids and cross-linked by four disulfide bonds. Anfinsen discovered that when ribonuclease is treated with urea and β-mercaptoethanol (a substance that breaks disulfide bonds), the peptide chain unfolds. When this happens, the protein is no longer able to act as an enzyme because its normal functioning depends upon its specific three-dimensional shape (figure 3.13). Such a disruption of a protein's structure with an accompanying decrease in biological activity is called **denaturation.**

Anfinsen also found that if he slowly and carefully removed the urea and β-mercaptoethanol from the denatured ribonuclease, the disulfide bonds reformed and the polypeptide chains refolded into their proper secondary and tertiary structures—thus regaining their enzymatic ability. As a result of Anfinsen's experiments, we know that the amino acid sequence, the only aspect of the ribonuclease structure left intact by this treatment, determines the structure of a fully folded protein.

Thus, if the surrounding chemical conditions are appropriate, a protein molecule automatically assumes its characteristic secondary and tertiary structure because of the attractive forces between various parts of the molecule. This phenomenon demonstrates the *principle of self-assembly* because the protein assumes its characteristic three-dimensional shape without any directing influence from outside the molecule.

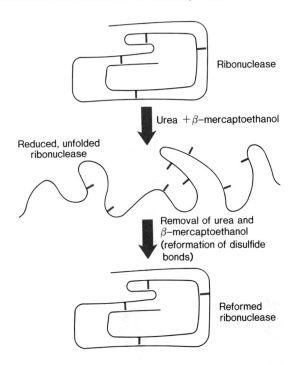

Figure 3.13 The reformation of the original folded structure from unfolded, reduced ribonuclease. When ribonuclease is treated with β-mercaptoethanol and urea, its disulfide (—S—S—) bonds (*color*) are broken and it is unfolded. If these reagents are carefully removed, the polypeptide chain will refold to form the native, catalytically active enzyme again. Thus, the sequence of amino acids in the unfolded chain determines the final structure of the refolded protein.

Many proteins are constructed of two or more polypeptide chains, each with its own primary, secondary, and tertiary structures. These individual chains, called **subunits** or **protomers,** are associated with one another in specific ways. The spatial relationships among them comprise the **quaternary structure,** the fourth level of protein structure. Proteins that are combinations of several individual polypeptides are called **oligomeric** ("few unit") **proteins.** Interestingly, like the peptide chains in Anfinsen's experiments, oligomeric proteins form spontaneously from their subunits, which are usually held together by noncovalent forces (although covalent bonds are also involved in some proteins). The formation of quaternary structure in proteins is another demonstration of the principle of self-assembly.

Hemoglobin, the oxygen-transporting substance in blood (see chapter 12), is an example of a protein made up of several (four in this case) subunits (see figure 3.15b and c).

Fibrous Proteins

On the bases of structural organization and solubility, we can divide proteins into two major groups: **fibrous proteins** and **globular proteins.** Fibrous proteins are mainly structural proteins, characterized by parallel polypeptide chains lined up along an axis to form fibers or sheets. Fibrous proteins are major components of the connective tissues of many living things. Globular proteins generally have more dynamic functions involving interactions with other molecules (for example, enzymes are globular proteins).

Collagen is a good example of a fibrous protein. It is the most abundant protein in vertebrate animals (animals with backbones, including humans), comprising as much as a quarter to a third of the total amount of protein in their bodies. Collagen is ideally suited for its connecting and supporting functions in the body because of its flexibility and great tensile strength (resistance to stretching).

Collagen fibers are found in all types of connective tissue, including ligaments, cartilage, bone, tendons, and the cornea of the eye. In these tissues, collagen is arranged in stable extracellular fibers to make the tissues resistant to tearing. In fact, it is possible to skin vertebrate animals primarily because of the large amount of collagen in the lower layer (dermis) of the skin; the collagen makes the skin sufficiently tough that it can be pulled away from underlying tissues without being torn.

How do stable extracellular collagen fibers develop? Once again, the construction of this protein depends upon the principle of self-assembly. However, the self-assembly of collagen fibers takes place in spaces outside the cells. In the assembly process, three long polypeptide chains coil together to form a triple-stranded, helical, ropelike aggregate (figure 3.14). These triple-stranded protein aggregates, called **tropocollagen,** are rod-shaped and about 280 nm (1 nanometer = 10^{-9} meter) long. The three polypeptide strands are held together by hydrogen bonding, by London dispersion forces, and by covalent cross-links.

Collagen is composed of parallel tropocollagen units stacked end to end. Each tropocollagen unit overlaps with the next about a quarter of the way down its length, and the tropocollagen subunits are covalently cross-linked to one another. Such an arrangement makes collagen fibers very strong and stable.

Globular Proteins

In contrast to fibrous proteins, the polypeptide chains in globular proteins are coiled into compact, spherical shapes. Globular proteins are much more flexible and fulfill more dynamic functions in organisms than do fibrous proteins such as collagen. Enzymes, antibodies, and transport proteins are all globular proteins.

The oxygen-transporting protein hemoglobin is a good example of a globular protein. Details of the three-dimensional structure of hemoglobin were determined through X-ray analysis by Max Perutz and his colleagues at Cambridge

Figure 3.14 Collagen structure. (a) Electron micrograph of skin collagen fibers that have been coated with metal. The repeat pattern (with a spacing of 70 nm) is created by the stacking of tropocollagen molecules in staggered, parallel arrays (magnification × 25,194). (b) The structure of collagen. Single polypeptide chains with often-repeated sequences (especially ones involving the amino acids glycine, proline, and hydroxyproline) are coiled to form a helix. Three of these helical polypeptide chains are then wound about one another to form a large, triple-stranded tropocollagen molecule. Tropocollagen molecules finally associate together in a staggered, parallel arrangement to produce a collagen fiber such as those in (a).

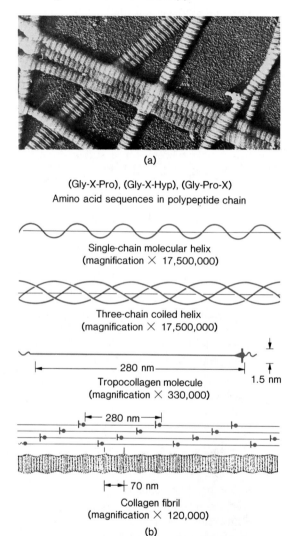

(a)

(Gly-X-Pro), (Gly-X-Hyp), (Gly-Pro-X)
Amino acid sequences in polypeptide chain

Single-chain molecular helix
(magnification × 17,500,000)

Three-chain coiled helix
(magnification × 17,500,000)

280 nm

1.5 nm

Tropocollagen molecule
(magnification × 330,000)

280 nm

70 nm

Collagen fibril
(magnification × 120,000)

(b)

University. Hemoglobin is an oligomeric protein with four subunits or polypeptide chains—two α- and two β-chains. The hemoglobin molecule is roughly spherical, with each α-chain in close contact with the two neighboring β-chains. Attached to each subunit, in a specific position, is a single **heme** group, containing an iron atom that is capable of binding with and separating from an oxygen molecule (figure 3.15).

Figure 3.15 Hemoglobin structure. (a) The heme group of hemoglobin has an iron atom complexed with its central nitrogens. The iron atom reversibly binds oxygen molecules in oxygen transport. (b) The structure of oxygenated hemoglobin. The heme groups are pictured as rectangles with spherical iron atoms at their centers. The binding of oxygen to one heme group changes the interactions of the subunits with one another and thus stimulates the binding of oxygen by the other heme groups. (c) A computer-generated model of the three-dimensional structure of a hemoglobin molecule. Heme groups are colored red, and each of the four polypeptide chains is shown in a different color. (d) Red blood cells of a person suffering from sickle-cell anemia (*top*) compared with normal red blood cells (*bottom*). One amino acid substitution in the hemoglobin β-chains changes the properties of sickle-cell hemoglobin and causes cell distortion that can result in severe circulatory problems.

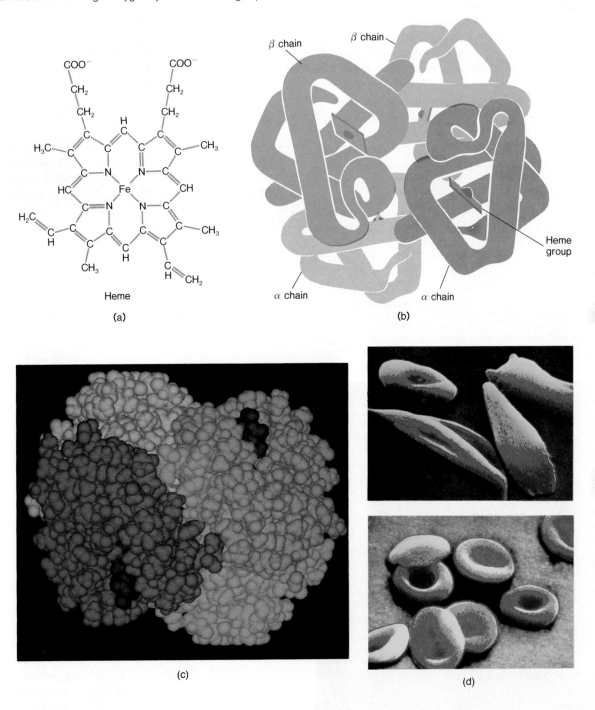

Heme

(a)

β chain

β chain

Heme group

α chain

α chain

(b)

(c)

(d)

Even though the amino acid sequences of hemoglobin polypeptide chains vary from species to species, the overall conformations of all hemoglobin molecules are very similar. However, any substitution of certain amino acids in the polypeptide chains of hemoglobin can have serious consequences.

A striking example of the effect of an amino acid substitution in hemoglobin is **sickle-cell anemia.** In this human disease, a single amino acid substitution in the β-chains (valine instead of the glutamic acid normally present) has disastrous effects. When oxygen molecules leave the hemoglobin of a person with sickle-cell anemia, the defective hemoglobin molecules change shape, altering the shape of entire red blood cells. The normally spherical cells elongate into "sickles" that cannot pass through small vessels easily (figure 3.15d). This difficulty can lead to a painful sickle-cell crisis, which occurs when circulation to various body areas is blocked, resulting in tissue damage. The dramatic consequence of the substitution of a single amino acid in the β-chains of hemoglobin clearly shows that higher levels of protein structure arise from the primary structures, the specific amino acid sequences, of polypeptides.

Nucleic Acids

Nucleic acids are macromolecules that carry genetic information. Structurally, nucleic acids are linear polymers of nucleotides (figure 3.16). Each nucleotide is composed of three substances: a nitrogen-containing organic base (either a **purine** or a **pyrimidine**), a five-carbon sugar (pentose), and phosphoric acid. There are two major types of nucleic acids, **deoxyribonucleic acid (DNA)** and **ribonucleic acid (RNA).** We will examine nucleic acids in more detail in chapters 23 and 24.

Lipids

Lipids, including fats, oils, phospholipids, and steroids, are structurally much more diverse than polysaccharides, proteins, or nucleic acids. In fact, the term lipid is simply a convenient name for organic compounds that are insoluble in water but soluble in hydrophobic (see p. 31) solvents. Some lipids are large molecules, but unlike polysaccharides, proteins, and nucleic acids, they are not polymers (chains of repeated smaller units).

Figure 3.16 Nucleic acids. (a) A computer-generated side view of a DNA molecule. Blue spheres are nitrogen, red are oxygen, green are carbon, and yellow are phosphorus; hydrogen atoms are not shown. (b) Nucleic acids are polymers of nucleotides that are composed of purine or pyrimidine bases, pentose sugars (ribose or 2-deoxyribose), and phosphate. The diagram shows components of RNA. In DNA, the OH in the colored box is replaced by H.

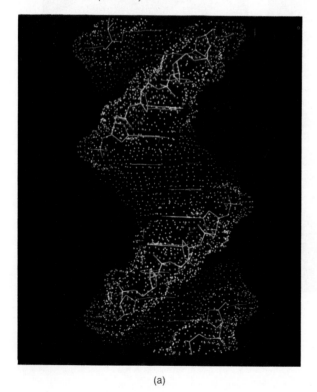

(a)

(1) Base: The purine and pyrimidine components of DNA or RNA

Adenine

(2) Nucleoside: The base plus the pentose sugar

Adenosine

(3) Nucleotide: The base plus the pentose sugar plus phosphoric acid

Adenosine–5′–phosphate

(b)

Lipids resemble polysaccharides in fulfilling both structural and storage-compound roles. Fats, which are condensation products of glycerol and long-chain fatty acids, are used as storage compounds (figure 3.17). Because of variations in the number of carbon atoms, in the number of double bonds (C=C bonds), and in the positions of these bonds in the fatty acids, there are many kinds of fat molecules.

Because the components of fatty acid chains are generally hydrophobic, fat molecules are insoluble in water. However, if one of the fatty acids is replaced by a hydrophilic group—that is, one that has an affinity for water—as is the case in phospholipids, the situation becomes more complex (figure 3.17b). As we will see shortly, the presence of both hydrophobic and hydrophilic tendencies in such compounds is important in the formation of biological membranes.

Biological membranes also contain lipids other than phospholipids. One of the most important is the **steroid** cholesterol (figure 3.18). Steroids contain four fused rings of carbon atoms in a characteristic arrangement. The basic steroid structure is also the foundation for many other biologically important molecules, such as the steroid hormones of vertebrate animals.

Figure 3.17 Lipids. (a) Glycerol and a fat. Biochemists call fats triacylglycerols or triglycerides because a fat molecule is formed, by condensation, from glycerol and three fatty acid molecules. (b) An example of a phospholipid, phosphatidylcholine (lecithin). Such phospholipids are important constituents of biological membranes.

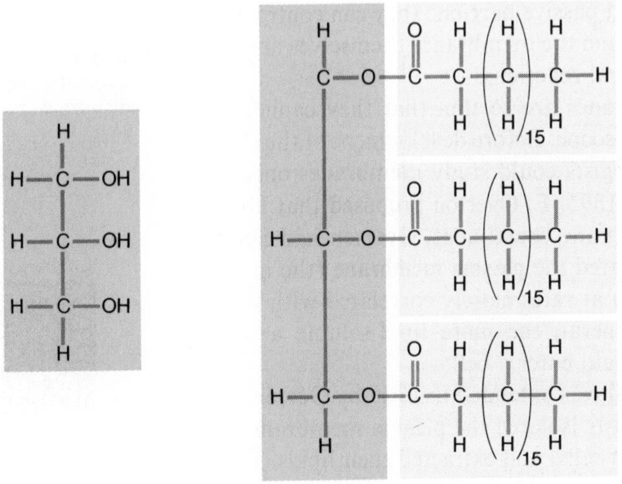

Glycerol

Tristearyl glycerol (a fat)

(a)

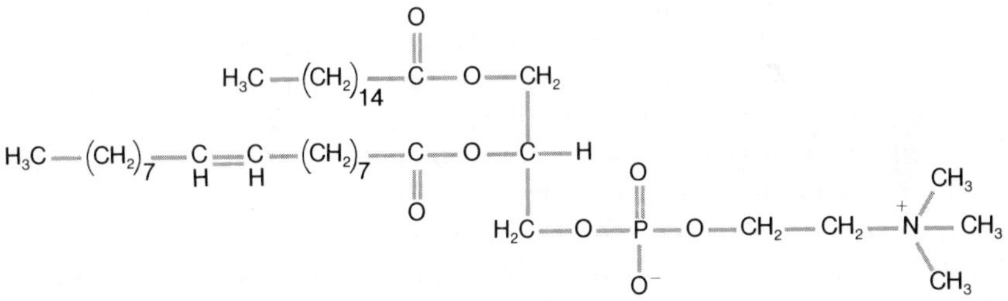

Phosphatidylcholine (lecithin)

(b)

Figure 3.18 Steroids. (a) Cholesterol. (b) Progesterone, a hormone. Progesterone and a number of other hormones are built around the basic cholesterol structure.

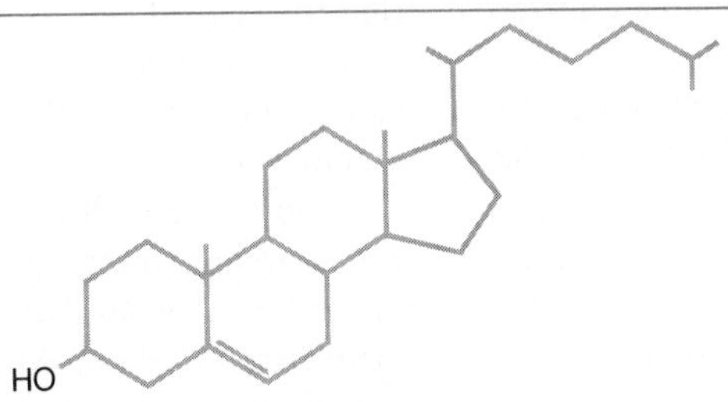

Cholesterol

(a)

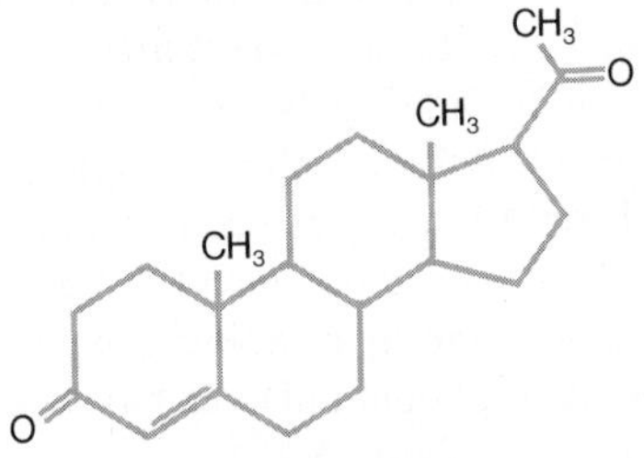

Progesterone

(b)

Membranes

Modern biology strongly emphasizes the importance of membranes in cell organization. The old-fashioned view of the cell as a sac of watery soup enclosed in a membrane has been replaced by the concept of an extremely complex array of membranes arranged throughout the interior of the cell as well as enclosing it. In the most general sense, a biological **membrane** is a thin, hydrophobic layer separating two areas in which materials are dissolved and suspended in water. Membranes separate a cell into a number of compartments, since water-soluble molecules in one area cannot readily pass through a hydrophobic membrane into another area. But membranes are not just passive barriers; they can control the passage of molecules, and the membranes themselves are the site of some biochemical reactions.

Biological membranes are so thin that they cannot be seen with a light microscope. Before development of the electron microscope, biologists could study membranes only by chemical analysis. In 1895, E. Overton proposed that lipids are an essential membrane constituent. Overton had noted that molecules penetrated the **plasma membrane** (the membrane enclosing a cell) at rates closely correlated with their lipid solubility. In general, the more lipid-soluble a substance, the faster it could enter a cell.

Overton's proposal was extended in 1925 by Gorter and Grendel. These scientists isolated the plasma membranes of erythrocytes (red blood cells) and extracted their lipids. They calculated that each erythrocyte membrane contained enough lipid to form a layer two molecules thick around the cell, and so they suggested that cell membranes are basically composed of a bimolecular lipid layer. Their model was somewhat oversimplified, however, since there was also evidence for the presence of proteins in membranes. In the 1930s, Davson and Danielli postulated that the lipid bilayer was covered on both surfaces with globular protein molecules, constituting a sort of protein-lipid "sandwich" (figure 3.19).

Since that time, direct chemical analysis of isolated membranes has shown that both proteins and lipids are indeed present in membranes. For example, the plasma membrane of an erythrocyte is 60 percent protein and 40 percent lipid by weight. Over half of the lipid in an erythrocyte membrane is composed of phospholipids, and another one-quarter is made up of the steroid cholesterol.

These membrane lipids have polar and nonpolar ends. The polar ends interact with water and are hydrophilic; the nonpolar, hydrophobic ends are insoluble in water and tend to associate with one another. In the Davson-Danielli model, the proteins and the hydrophilic ends of the lipids contact water. The hydrophobic tails are buried in the lipid bilayer, away from water (figure 3.19).

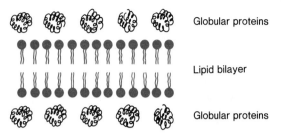

Figure 3.19 The Davson-Danielli membrane model. The hydrophilic lipid heads are indicated by circles and the hydrophobic tails by lines. The hydrophobic tails are buried inside the membrane, away from water.

Globular proteins

Lipid bilayer

Globular proteins

Figure 3.20 Electron micrograph of a portion of a plasma membrane showing the two outer electron-dense layers separated by a less dense central zone (magnification × 250,000).

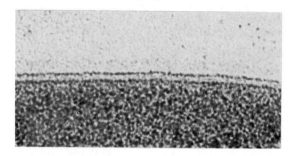

The Unit Membrane Model

Development of the electron microscope made it possible, at long last, to observe the plasma membrane directly. When membranes are treated with osmium tetroxide and viewed through an electron microscope, a membrane structure that looks somewhat like a sandwich, about 7.5 to 8.0 nm thick, can be seen. This membrane consists of two electron-dense layers separated by a less-dense central zone (figure 3.20).

J. D. Robertson used such observations to develop the **unit membrane model** in the late 1950s. This was basically an elaboration of the Davson-Danielli model. Robertson proposed that the two outer layers of a membrane consist of proteins in an extended (rather than globular) conformation, and that the central zone is a lipid bilayer with its molecules oriented with their hydrophilic ends outward. This three-layered structure was thought to be characteristic of all cell membranes.

Closer study, however, has uncovered structural differences among membranes. For example, the thickness of membranes varies considerably—they range in thickness from 5.0 to 10.0 nm. Moreover, the **freeze-etching technique** (see figure 3.21), employed in studies of membrane structure, makes it possible to sever a membrane down the center of its lipid bilayer, thereby splitting the membrane in half

Figure 3.21 (a) The freeze-etch procedure. (1) Cells or tissue are frozen in liquid nitrogen at −196° C. (2) The specimen is then cut with a very cold microtome knife, fracturing along lines of weakness, such as the center or face of membranes. (3) Next, the specimen is etched to expose more intracellular surfaces and details by subliming ice away at temperatures above −120° C. (4) A replica is made of the exposed structures by coating them with platinum and carbon. The replica can then be examined with the electron microscope. Particles within membranes are exposed and clearly visible in the replica. *A* and *B* correspond to the parts of the electron micrograph in (b). (b) Electron micrograph of a freeze-etched red blood cell plasma membrane showing the difference between the interior and the outer surface of the membrane. The interior (*surface A*) has numerous globular particles. In contrast, the membrane's outer face (*surface B*) is smoother (magnification × 27,170).

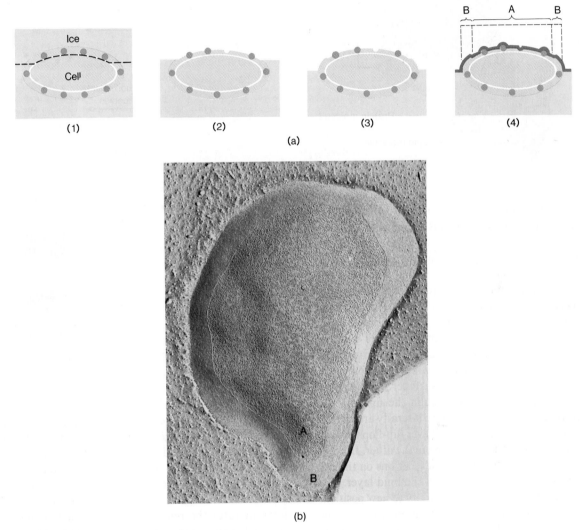

(a)

(b)

and exposing its interior. This procedure has demonstrated that many membranes—including the erythrocyte plasma membrane—have a more complex structure than the Davson-Danielli model predicted. Finally, the small, globular particles seen in these membranes (figure 3.21b) are now thought to be membrane proteins. This observation raises questions about the unit membrane theory because it indicates that some membrane proteins actually lie within the membrane lipid bilayer and do not simply coat the surface of the membrane, as was earlier supposed.

The Fluid Mosaic Model

These discoveries have led to alternate models of membrane structure. One of the most widely accepted is the **fluid mosaic model** of S. J. Singer and G. L. Nicholson.

All evidence still indicates that membranes have a bimolecular lipid layer. The lipids are arranged with their hydrophilic ends at the surface and their hydrophobic ends in the interior of the membrane, just as earlier models suggested. But membrane proteins definitely do not spread in sheets over the surfaces of the lipid layers—they are globular.

In a transmission electron microscope, an electron beam is focused on a very thin section or slice of the cell by means of electromagnets. After leaving the cell, the electron beam travels through more magnetic lenses, which magnify the image and project it on either a fluorescent screen (similar to a television screen) or photographic film.

In preparation for transmission electron microscopy, specimens are treated with heavy metals such as osmium or uranium, which attach to intracellular structures in the specimens. This heavy-metal treatment increases contrast among the cell components because electrons will not pass through the cell components where metal atoms are located. Thus, these components appear as dark (not fluorescent) objects on the screen.

The scanning electron microscope is used to study surfaces rather than thin sections of cells. SEM pictures, with their three-dimensional quality, reveal remarkable details of the surfaces of cells and other objects. Surfaces to be studied usually are coated with a very thin layer of metal, often gold. In the SEM, electron beams scan the surface of the specimen, and these striking electrons drive off electrons (called secondary electrons) from the surface atoms. The pattern of the secondary electrons landing on a detector is amplified, and the image is displayed on a cathode ray tube much like that in a television set.

Autoradiography

Autoradiography is another observation technique used to study the intracellular locations of cell constituents and the specific sites of particular reactions within cells. In autoradiography, cells are incubated with molecules that will be specifically incorporated into certain cell constituents. These molecules are radioactively labeled; that is, they contain one or several atoms that are radioactive isotopes. After incubation, the cells are washed and placed on glass slides. Each slide is coated with photographic emulsion similar to that on photographic film and left in the dark long enough to "expose" the emulsion to any radioactivity present. Wherever there is a radioactive site in a cell, the emulsion reacts as if light had struck it, and each radioactive disintegration produces a particle track of silver grains in the emulsion. In this way, biologists can determine the locations of specific molecules and reactions within the cell.

An example may help you to understand the usefulness of autoradiography. The pyrimidine thymine is found in DNA but not in RNA, so incorporation of thymine into cells indicates specifically that DNA synthesis is occurring. The nucleotide containing thymine is called thymidine. Cells can be incubated with thymidine containing the radioactive isotope tritium. These cells will then use the radioactive thymidine

Figure 4.7 Autoradiograph of chromosomes during cell division following exposure to a radioactive isotope. Cells were exposed to radioactively labeled thymidine and allowed to undergo cell division. They were then washed free of the labeled thymidine and allowed to begin a second division. After the chromosomes had condensed and lined up, this autoradiograph was prepared. The labeling pattern shows that DNA synthesis has occurred in the chromosomes because the radioactive thymidine was incorporated almost exclusively into the chromosomes.

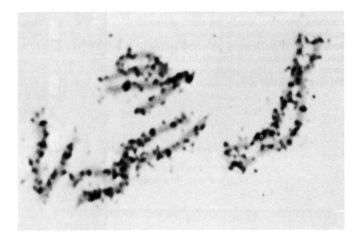

(*tritiated thymidine*) to synthesize DNA. After the treated cells are washed and coated with photographic emulsion, the great majority of the silver grains are found over the cell's chromosomes. This demonstrates clearly, then, that DNA is synthesized in the chromosomes within the nucleus rather than in the cytoplasm (figure 4.7).

Cell Fractionation

The cell research techniques we have described thus far—even such useful and informative techniques as electron microscopy and autoradiography—share one significant shortcoming: they all produce "stop-action" images of cell components. What is observed is a qualitative picture of the locations of cell constituents at the moment the cells were killed. Since cells and organelles are structurally and functionally dynamic, there is a need for techniques that allow dynamic, quantitative study of functions of organelles removed from cells. This need has led to the development of **cell fractionation** techniques. By these techniques, cells are broken open and their internal components are separated for study, under conditions that allow organelles to remain able to perform their specialized functions.

The first step in cell fractionation is to disrupt a cell and release its components without significantly altering their structure and activity. One disruption process is called **homogenization.** In this process, a glass sleeve or tube fitted with a plastic or glass pestle is used to break open cells (figure 4.8). After homogenization, cell components such as nuclei

Figure 4.8 Cell homogenization and fractionation by differential centrifugation. In general, the smaller the particle, the greater the centrifugal force required to drive it to the bottom of the tube.

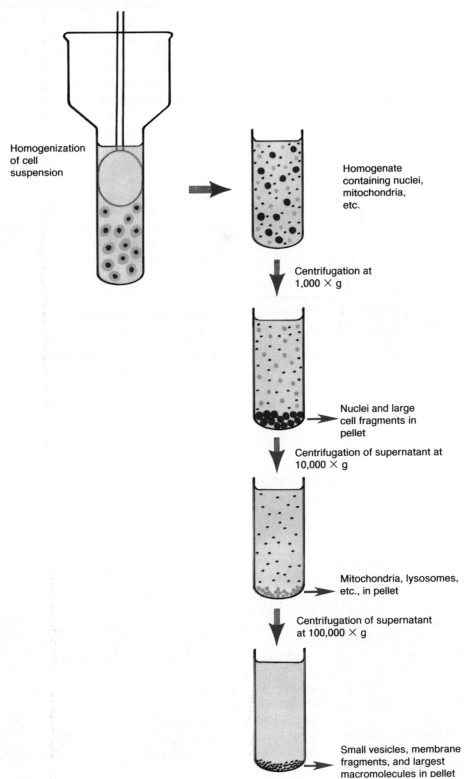

Homogenization of cell suspension

Homogenate containing nuclei, mitochondria, etc.

Centrifugation at 1,000 × g

Nuclei and large cell fragments in pellet

Centrifugation of supernatant at 10,000 × g

Mitochondria, lysosomes, etc., in pellet

Centrifugation of supernatant at 100,000 × g

Small vesicles, membrane fragments, and largest macromolecules in pellet

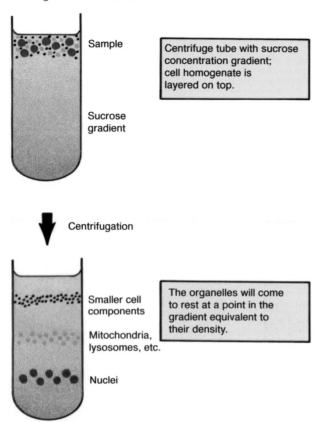

Figure 4.9 Cell fractionation by density gradient centrifugation. Sucrose in the tube increases linearly in concentration and therefore in density. The lowest concentration of sucrose is at the top of the tube and the highest is at the bottom.

Sample

Centrifuge tube with sucrose concentration gradient; cell homogenate is layered on top.

Sucrose gradient

Centrifugation

Smaller cell components

The organelles will come to rest at a point in the gradient equivalent to their density.

Mitochondria, lysosomes, etc.

Nuclei

and mitochondria can be isolated by **differential centrifugation.** In this separation technique, which makes use of centrifugal force, a container filled with a mixture of particles is spun in a centrifuge to drive the heavier particles to the bottom of the container. These heavier particles form a sediment, or **pellet,** while the lighter particles remain suspended in the surrounding solution, or **supernatant.** Nuclei, which are relatively large organelles, can readily be sedimented at forces of 1,000 times the force of gravity (g), while lighter particles, such as mitochondria, remain in the supernatant. Mitochondria can be collected by centrifuging the first supernatant again at 10,000 × g. If the resulting supernatant is centrifuged yet again, this time at forces around 100,000 × g, even the smallest organelles, membrane fragments, and the largest macromolecules are sedimented. The remaining supernatant, containing only dissolved molecules, probably represents the relatively featureless portion of the cytoplasm that fills the spaces between organized cell structures in electron microscope images of cells.

Another technique used to isolate cell constituents is **density gradient centrifugation.** In this technique, a homogenate is placed on top of a sucrose solution that progressively increases in concentration (and therefore in density) from the top to the bottom of the container (figure 4.9). When this sucrose gradient is centrifuged at high speeds, each particle in the homogenate will move down in the tube and come to rest at the point in the gradient where its density equals that of the sucrose solution. Thus, you would find the nuclei from a cell homogenate at a lower point in the gradient than the mitochondria, which are less dense.

Study of isolated organelles can provide information concerning their functions within intact living cells. Cell fractionation thus makes possible dynamic and quantitative functional studies of cell organelles—studies that complement the information obtained through techniques that give "stop-action" pictures of cells and organelles at single moments in time. All of these techniques, as well as others, have been used in developing our present knowledge of cells and their organelles.

Structural Components of Eukaryotic Cells

We can place all cells in one of two categories. Bacteria and cyanobacteria (blue-green algae) are **prokaryotic** (*pro* = before, *karyon* = nucleus) cells. Prokaryotic cells do not have distinct nuclei bound by membranes or membrane-enclosed organelles such as mitochondria. All other cells—the cells of plants, animals, protists, and fungi—are **eukaryotic** ("true nucleus") cells. Eukaryotic cells possess nuclei with membranes and are much more structurally complex because their cytoplasm also contains membranous networks and organelles that are enclosed in membranes.

The Plasma Membrane

The **plasma membrane** encloses the entire cell. As we will see, the plasma membrane also forms the cell's point of contact with its environment and determines what materials enter and leave the cell.

The plasma membrane resembles all cellular membrane structures (see chapter 3); it is an ordered but dynamic array of lipids and proteins. The proteins of the membrane rest in a lipid bilayer composed principally of phospholipids and steroids.

The plasma membrane is a versatile organelle that fulfills a number of roles. It is, of course, the cell boundary in a mechanical sense and therefore holds all the cell constituents together. It also functions as a selectively permeable barrier that allows certain atoms and molecules to enter or leave the cell, while preventing others from passing. Components of the plasma membrane also carry out active transport processes that move some substances into the cell and move others to the exterior (see chapter 5).

Plasma membranes are involved in many cell interactions in multicellular organisms. A plasma membrane can maintain an electrical potential and, in certain specialized cells such as nerve and muscle cells, can conduct impulses. In this way, plasma membranes are involved in integrating and coordinating the functions of whole multicellular organisms.

The Cytoplasmic Matrix

The organelles of a eukaryotic cell lie in a relatively homogeneous substance called the **cytoplasmic matrix,** or **cytosol** (sometimes also called the **hyaloplasm**). The cytoplasmic matrix is the "environment" of the organelles and is the site of a number of processes. It now seems that the cytoplasmic matrix is much more complex than we thought earlier. It is permeated by a network of fibrous elements that have complex connections to one another as well as to various cytoplasmic organelles (see p. 78).

Approximately 85 to 90 percent of the cytoplasmic matrix is water. A portion of this cellular water is *bound water,* held in hydration spheres around proteins and other macromolecules. Bound water is unable to act as a solvent for other intracellular molecules. The remainder of the matrix water is *free water* and can act as a solvent and participate in metabolic processes.

The supernatant remaining after a cell homogenate has been centrifuged at 100,000 × g presumably contains the constituents of the cytoplasmic matrix. Analysis of this 100,000 × g supernatant shows that it contains about a fourth of the cell's protein. Many of these protein molecules are enzymes that catalyze a number of metabolic reactions that take place in the matrix.

Some biologists think that the cytoplasmic matrix proteins are part of a special arrangement called a **colloid** or colloidal solution, which differs from ordinary solutions. Colloids exist in two different forms, each with different physical properties. In one form, molecules of the colloidal system associate and remain associated to form a semisolid or solid system. This form of colloidal system is called a **gel.** The other colloidal form, called a **sol,** is a fluid system in which any two particles interact with each other only momentarily. When you make a gelatin dessert, you are converting a sol to a gel. The hot, dissolved gelatin is a protein sol. When it is cooled in the refrigerator, the colloid is transformed into a gel.

There is considerable evidence that the cytoplasmic matrix, which is composed of proteins as well as other molecules, behaves as a colloid. For example, a number of cells

Figure 4.10 The proposed role of sol-gel transformation in amoeboid movement of a cell. The ectoplasm portion of the cytoplasm is a gel, and the endoplasm portion is a sol.

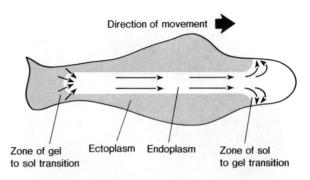

Direction of movement

Zone of gel to sol transition   Ectoplasm   Endoplasm   Zone of sol to gel transition

Figure 4.11 This nucleus has a clearly defined nucleolus (*nu*) and irregular patches of chromatin scattered throughout its matrix. The chromatin contains DNA and condenses to form chromosomes during cell division.

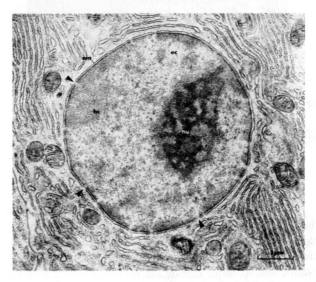

possess an outer gel layer (the **ectoplasm**) and an inner sol region (the **endoplasm**). The matrix may undergo reversible sol-gel transformations during amoeboid movement and other activities (figure 4.10).

The Nucleus and Nucleolus

The nucleus, the repository for most of the cell's genetic information, was discovered early in the study of cell structure because it is readily observed with the light microscope. Nuclei are membrane-enclosed spherical bodies about 5 to 7 μm in diameter. Relatively dense, fibrous patches of **chromatin** are the DNA-containing regions of the nucleus (figure 4.11). In nondividing cells, this chromatin exists in a dispersed condition, but during cell division, it condenses into **chromosomes.**

Figure 5.16 Examples of unicellular organisms in which cell = organism. In these cases all exchanges are made between cells and external environments. (a) The alga *Closterium,* an autotrophic organism that produces reduced carbon compounds photosynthetically. (b) An amoeba, a heterotrophic organism that ingests debris particles and small organisms, thereby obtaining reduced carbon compounds (magnification × 324).

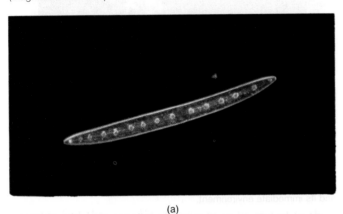

(a)

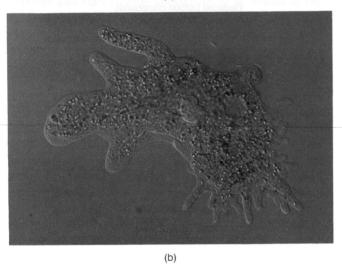

(b)

Closterium, an aquatic green alga, is an **autotrophic** ("self-feeding") organism; that is, it produces reduced carbon compounds through photosynthesis (figure 5.16a). The light energy needed to drive this process is absorbed directly from the environment. *Closterium* obtains mineral nutrients by ion uptake from the surrounding water, and gets rid of metabolic wastes by simple diffusion into the environment. Gas exchange is also made directly with the watery environment.

An amoeba, on the other hand, is a **heterotrophic** ("other-feeding") organism (figure 5.16b). It must take in reduced carbon compounds as well as all other nutrients from its environment. Amoebas feed by engulfing bits of debris or smaller organisms by phagocytosis. Furthermore, in an amoeba gas exchange occurs by simple diffusion, and metabolic wastes diffuse directly into the environment.

Whether autotrophic or heterotrophic, then, unicellular organisms must interact successfully with the external environment.

Loosely Integrated Multicellular Organisms

The interactions of some cells with their environments fall somewhere between the direct exposure of unicellular organisms to the external environment and the regulated internal environments maintained by more highly organized plants and animals. For example, some small plants known as liverworts are composed of fairly simple, flat plates of cells called **thalli** (singular: **thallus**). Certain exchanges can be made directly between individual thallus cells and the external environment, but there is still some degree of specialization among the cells. Photosynthetic cells near the upper surface produce the reduced carbon compounds that supply the energy requirements of all the cells in the thallus. Extensions, called **rhizoids,** on the bottom of the thallus reach into the soil and anchor the thallus in place. Rhizoids also absorb mineral nutrients from the soil, which are then made available to all cells (figure 5.17).

A sponge is an animal composed of a rather loose association of many cells. Some of a sponge's environmental exchanges—for example, gas exchange and waste removal—occur between individual cells and the external environment (the surrounding water). But certain other cells perform specialized functions. Beating flagella propel a stream of water through the sponge's body, and cells in the lining of the inner cavity of the sponge trap small particles from the water and enclose them in vesicles by phagocytosis. Some of these bits of food are then transferred to other cells in the sponge's body. In this way, the cells in the sponge's body that are specialized for food gathering obtain nutrients for all of the cells of the organism.

These two very different examples of loosely integrated organisms illustrate an intermediate degree of specialization in dealing with the environment. Some functions are carried out for the whole organism by certain cells, while other functions are handled as direct exchanges between individual cells and the external environment.

Complex Organisms with Regional Specialization

There is a distinct division of labor among the cells of more complex multicellular organisms. Many functions are localized in specific body areas. This *regional specialization* is obvious in the form of specialized exchange areas, which are localized boundaries across which exchanges of certain materials with the external environment take place. These localized boundaries help to control the internal environment in which the cells live.

Part 1 The Chemical and Cellular Basis of Life

Figure 5.17 Examples of loosely integrated organisms in which some functions are carried out for the entire organism by specialized cells, while other functions are left entirely to individual cells. (a) The liverwort *Marchantia*. This picture shows a number of thalli, some of which bear stalked reproductive structures. (b) Cross section of part of a *Marchantia* thallus. Chloroplasts (*darker green*) are present only in cells of the upper layers, especially in cells in the air spaces. These cells produce reduced organic compounds for all cells in the thallus. Rhizoids anchor the thallus and absorb mineral nutrients. (c) Simple sponges. (d) Section cut through a sponge showing the types of cells that make up the sponge's body. Note especially the flagellated cells lining the sponge's inner cavity.

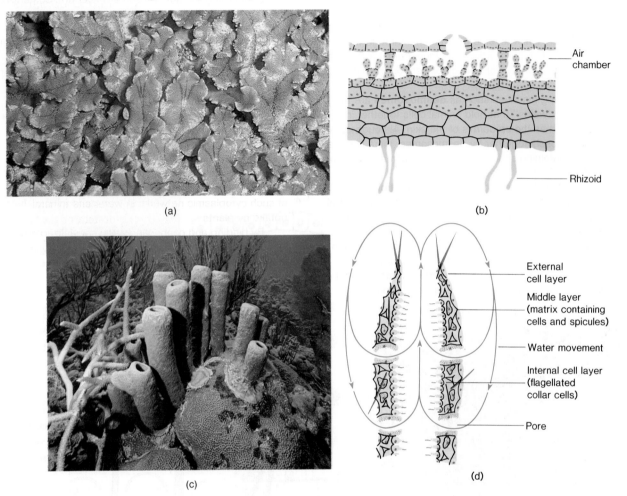

In plants, for example, roots take up water and minerals from the soil, while leaves are specialized for the light energy absorption and gas exchange necessary for photosynthesis. These two major exchange boundaries—leaves and roots—efficiently solve the basic problem of nutrient procurement in terrestrial plants. Regional specialization ensures that necessary exchanges with the external environment take place both above and below ground.

But how are cells in complex multicellular organisms organized to form such areas of regional specialization? What is the nature of the immediate cellular environment in such specialized regions? Early in the 1800s, biologists began to examine the way in which cells were put together in multicellular organisms. Theodor Schwann, one of the developers of the cell theory, discovered that organisms were more than mere aggregates of cells. He recognized that within multicellular organisms, groups of cells are organized as tissues. A **tissue** is a group of similarly specialized cells, along with intercellular material that binds them together, organized as a structural and functional unit. In order to better understand the regional specializations of complex organisms, let us briefly consider some important types of animal and plant tissues.

Animal Tissues

Animal tissues can be generally classified as epithelial, connective, muscular, or nervous tissue.

in **equilibrium** when the rate of the forward reaction (the formation of products) equals the rate of the reverse reaction (the formation of reactants) and the net concentrations of all four components change no further. This stable situation is represented by an **equilibrium constant,** symbolized as K_{eq}. We can calculate the equilibrium constant for the reaction by multiplying the product concentrations and the reactant concentrations and dividing the former by the latter:

$$K_{eq} = \frac{[C][D]}{[A][B]}$$

In a reaction that reaches equilibrium with a higher concentration of products than reactants, [C][D] will be larger than [A][B], and K_{eq} for the reaction will be greater than 1.0. Conversely, in a reaction that reaches equilibrium with a lower concentration of products than reactants, K_{eq} will be less than 1.0. The equilibrium constant thus represents in quantitative terms the final state of a reaction.

But what determines the final state of a reaction (and its equilibrium constant)? The equilibrium constant is directly related to the free-energy change that occurs during the reaction. In table 6.2 you can see that when a reaction has a negative free-energy change ($-\Delta G$), it is an exergonic reaction and will spontaneously proceed toward completion (its K_{eq} is greater than 1.0). However, if a reaction has a positive free-energy change ($+\Delta G$), it is an endergonic reaction and will not spontaneously move very far toward completion (its equilibrium constant is less than 1.0). Free energy must be supplied to the system to make an endergonic reaction proceed.

Reaction Rates

The concepts of thermodynamics can help us determine if a particular reaction will occur and what the concentrations of the individual components will be when equilibrium has been reached. Thermodynamic studies, however, do not indicate *how fast* a reaction will proceed. Even if we know a reaction is thermodynamically spontaneous, we cannot be certain that it will take place at a measurable rate. For example, the reaction in which hydrogen and oxygen combine to form water is highly exergonic, yet when these two gases are mixed together carefully, nothing happens. However, once the reaction begins (if, for instance, a match is lit), it proceeds so rapidly that an explosion results!

One way of understanding the rate at which chemical reactions take place is in terms of collisions between molecules. Reaction rates depend not only upon the frequency with which the reactants collide with one another, but also upon the energy with which they collide. Reactants must collide with a certain minimal amount of energy in order to react.

Table 6.2
The Relationship between the Equilibrium Constant (K'_{eq}) and Standard Free-Energy Changes ($G^{0\prime}$) of Reactions*

K'_{eq}	$G^{0\prime}$(kcal/mole)	
0.001	+4.09	
0.10	+1.36	Endergonic ($K'_{eq} < 1.0$)
1.00	0.00	
10.00	−1.36	
1000	−4.09	
1×10^5	−6.80	Exergonic ($K'_{eq} > 1.0$)
1×10^6	−8.18	

*The symbol $G^{0\prime}$ represents the standard free-energy change for a reaction. It is the amount of free-energy loss or gain per mole of reaction under certain specified conditions. Similarly, K'_{eq} symbolizes the equilibrium constant under the standard conditions used for such measurements.

This critical energy level is called the **activation energy.** In a reaction mixture, only a certain percentage of molecules are at this critical energy level because not all molecules at the same temperature have the same amount of energy. Increasing the temperature, however, increases the energy of individual molecules. Thus, a reaction normally proceeds at an increased rate if the temperature is raised.

Another way of understanding the rate at which chemical reactions occur is provided by the idea of an **activated complex** (or **transition state**). Figure 6.7 shows the progress of a single set of reactant molecules along the way toward becoming products. The activated complex (symbolized by $\pm$) represents an unstable intermediate condition in which the reacting substances are at the highest free-energy level reached during the reaction. We can liken the forming of an activated complex to having to push a boulder over a hump at the top of a hill. Once the boulder is over the hump, it will roll down the hill by itself.

The reaction rate is proportional to the concentration of the activated complex; the higher the concentration of the activated complex, the faster the reaction proceeds. If the activation energy is large, the concentration of the activated complex is small and the reaction is slow. This is exactly the situation that exists in our hydrogen and oxygen mixture. Hydrogen and oxygen do not react rapidly because very few of the molecules have sufficient energy to form activated complexes. However, a lighted match provides enough energy to trigger the reaction. The added heat energy raises the temperature and accelerates the movement of hydrogen and oxygen molecules so that more collisions occur. Many molecules then collide with enough energy to form activated complexes and a significant reaction begins. Because this reaction is highly exergonic, reacting molecules generate enough heat to raise the energy of still more molecules and the rate of the reaction increases explosively.

Figure 6.7 The progress of a single set of reactants on their way to becoming products. ‡ denotes the activated complex. Activation energy is required to achieve the activated complex. Then the reaction can proceed to product formation. The reaction diagrammed here is exergonic (negative ∆G) because it releases energy; that is, the products contain less energy than the reactants.

Figure 6.8 Catalysts provide a reaction pathway with a lower activation energy. Thus, the catalyst makes possible a large increase in the fraction of molecules having sufficient energy for the reaction to take place. As a result, the reaction moves toward equilibrium much more quickly.

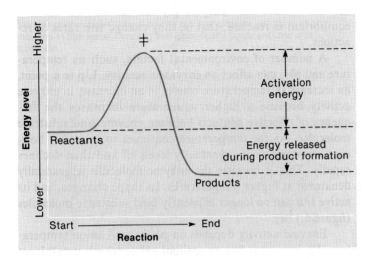

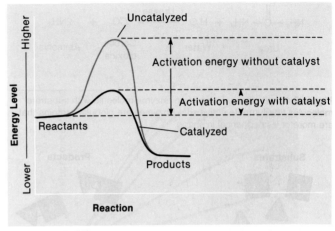

Reactions and Enzymes

Many important reactions in living things have relatively high activation energies. How then can they proceed at rapid rates as they do in many cases? Obviously, it is not possible to apply a great deal of heat energy to the reactants. Conditions inside living things are mild and stable compared to those used to promote reactions in the laboratory. Instead, in living organisms, reaction rates are increased by lowering activation energies through the action of catalysts.

Catalysts

Catalysts are substances that effectively lower activation energies. A catalyst affects the rate of a reaction without being changed by it; the catalyst is in its original form at the end of the reaction.

A catalyst normally affects a reaction rate by providing an *alternative reaction pathway,* a reaction mechanism different from that of the uncatalyzed reaction. In this alternative reaction pathway, the required activation energy is *lower* than it would be in the uncatalyzed reaction (figure 6.8). In a catalyzed reaction, then, activation energy is lowered, more activated complexes are formed, and the reaction proceeds toward equilibrium more quickly than it would without the catalyst. Returning to our boulder analogy, we can say that the hump at the top of the hill is lowered enough so that it takes less energy to roll the boulder over it. Reactions in living things are catalyzed by specific catalysts called **enzymes** that are produced by cells.

The Discovery of Enzymes

The study of enzymes—and to a great extent the study of the molecular nature of the cell—began in 1897 with the work of Eduard Buchner. Buchner was attempting to preserve an extract of yeast that he had prepared by grinding yeast cells with sand. He tried adding a large quantity of sucrose to the extract and observed with some surprise that carbon dioxide and alcohol were produced. In this way, Buchner accidentally discovered that fermentation (p. 166) can take place in an extract, isolated from intact living cells.

Prior to this time, most scientists, including even Louis Pasteur, had been convinced that processes such as fermentation took place only within intact living cells. Buchner's finding opened the door to the chemical analysis of individual metabolic processes isolated from other cellular activities. Such studies eventually led to the discovery of the enzymes that catalyze chemical reactions in living things.

Enzymology, the study of enzymes, came of age in August 1926, when James B. Sumner announced that he had successfully crystallized an enzyme. The enzyme he isolated and purified, urease, is a protein molecule. Sumner's enzyme preparation catalyzed the breakdown of urea to CO_2 and NH_3 at a rapid rate (figure 6.9). This first enzyme purification was the culmination of nine years' work under somewhat primitive conditions. Initially, Sumner did not even have an ice chest—he cooled his solutions by placing them on the window sill at night!

Sumner's work was very important because it provided a basic step in the developing awareness that most reactions in living things are catalyzed by enzymes. This has become one of the central concepts of modern biology.

Figure 7.18 The photosynthetic electron transport chain from water to NADP⁺. The positions of the components on the redox potential scale are only approximate. This standard way of representing these electron transports is called the "Z-scheme." Moving electrons to compounds with more negative redox potentials requires energy input. Light energy is used to move electrons "uphill" in these reactions. Cytochromes are proteins that contain an iron atom in a heme group. Cytochromes are important in chloroplasts and also in mitochondria, as we will see in the next chapter.

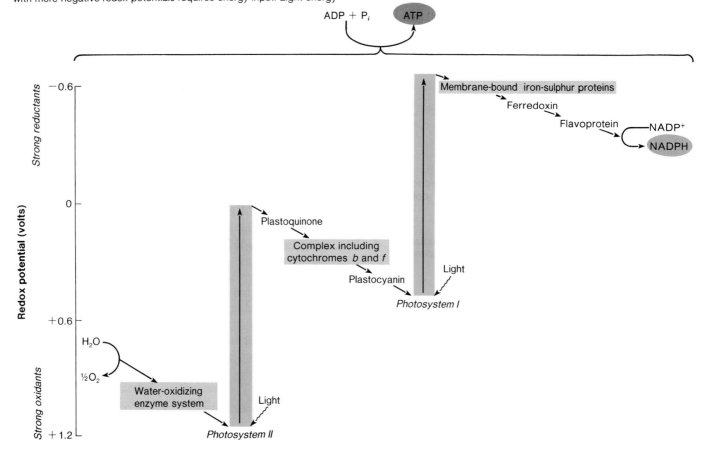

Electron Transport

As you saw in figure 7.18, water and NADP do not react directly with the excited reaction center chlorophylls of Photosystems II and I, respectively; neither does the reaction center chlorophyll of Photosystem II transfer its electrons directly to that of Photosystem I. Instead of single-step oxidation-reduction reactions, series of electron carriers transfer electrons through stepwise sequences of oxidation-reduction reactions, with electrons being passed from carrier to carrier. Electrons move in one direction (flow) through the series of carriers because each carrier is a stronger oxidizing agent (has a more positive redox potential [p. 125]) than the carrier immediately preceding it.

Despite intensive continuing research, there are still gaps in our understanding of the arrangement of the various electron transport components involved in photosynthesis; however, some important components are well known. **Plastoquinone** (figure 7.20) is a key part of the electron transport chain between Photosystem II and Photosystem I.

Plastoquinone is a relatively small molecule that can accept two electrons and two protons and be reduced to a quinol. Plastoquinone has a long hydrophobic side chain that makes it insoluble in water but soluble in the lipid phase of the membrane. Later we will see how the involvement of both protons and electrons in the reduction and oxidation of carriers such as plastoquinone is an important factor in the processes leading to the formation of ATP.

The primary electron acceptor of Photosystem I apparently is a protein-bound iron atom, and the electron from this acceptor is eventually transferred to **ferredoxin.** Ferredoxin is a soluble iron-sulfur protein found in the stroma matrix. The two iron atoms in ferredoxin are attached to the protein by bonding to sulfur atoms of the amino acid cysteine; the iron atoms are also connected to each other by sulfur atoms.

Figure 7.19 Action spectra of Photosystems I and II. Photosystem I functions more efficiently than Photosystem II at slightly longer wavelengths.

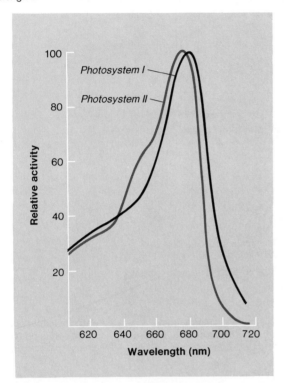

Figure 7.20 Plastoquinone and its reduced form (plastohydroquinone or plastoquinol). The long side chain is embedded in membrane lipids.

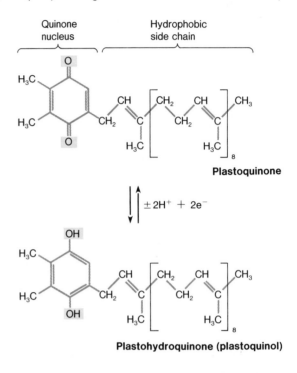

Ferredoxin is an important branch point for the electron transport chain. Although most reduced ferredoxin passes its electrons on to NADP via a specific **flavoprotein enzyme** (see figure 7.18), ferredoxin can also pass electrons to a cyclic pathway of electron flow around Photosystem I, which we will discuss later.

Our description thus far might lead you to think that most of the details of the light reactions are well understood, but this is far from true. For example, the splitting of water molecules in Photosystem II is a critically important chemical process because it supplies electrons to the photosystems *and* because it continually replenishes the supply of atmospheric oxygen that the majority of living things require for life. Furthermore, this reaction was the original source of the oxygen in the earth's atmosphere. However, little is known about the chemical nature of the components involved in splitting water molecules. Biologists researching this problem know that manganese is involved in some way, but many questions remain to be answered.

Photophosphorylation

As we have seen, photosynthetic light reactions supply not only the NADPH, but also the ATP that is used in CO_2 incorporation during the dark reactions. Illuminated chloroplasts produce ATP. This light-related ATP formation is called **photophosphorylation** and is coupled with the electron flow that occurs in the photosystems (see figure 7.18).

Coupling of electron flow and the phosphorylation of ADP to form ATP in chloroplasts (and in mitochondria) is a crucial part of the energy exchange process in living things. However, the nature of the coupling mechanism remains one of the central problems of modern biology. What is the actual link between the transfer of electrons from one carrier to another in the internal membranes of chloroplasts and the combination of ADP and inorganic phosphate to form ATP?

The hypothesis that seems to fit best with observed facts is one proposed by Peter Mitchell, who was awarded a Nobel Prize in 1978 for his work on phosphorylation. This hypothesis is called the **chemiosmotic hypothesis,** or simply the Mitchell hypothesis. The major concept of Mitchell's chemiosmotic scheme is that the intermediary between electron transport and the phosphorylation of ADP to form ATP is a *transmembrane electrochemical gradient of H^+ ions.* The

Figure 7.21 Models of the chemiosmotic hypothesis explaining ATP production during electron transport in chloroplast membranes. (a) The thylakoid membrane encloses the intrathylakoid space and separates it from the stroma matrix. Hydrogen ions are moved across the thylakoid membrane during light-coupled with ATP production. ATP production is catalyzed by a membrane-bound ATP-synthesizing system (ATP synthetase complex). (b) How reduction and oxidation of a carrier (Q) at opposite sides of a membrane could cause the transport of hydrogen ions across the membrane according to Mitchell's hypothesis. An electron carrier on one side of the membrane reduces Q. This requires two electrons and two protons that are taken up from the stroma matrix. The reduced carrier (QH$_2$) then diffuses across the membrane. On the other side, it donates its electrons to another electron carrier. Two hydrogen ions are released on that side. Finally, oxidized Q diffuses back across the membrane to complete the cycle.

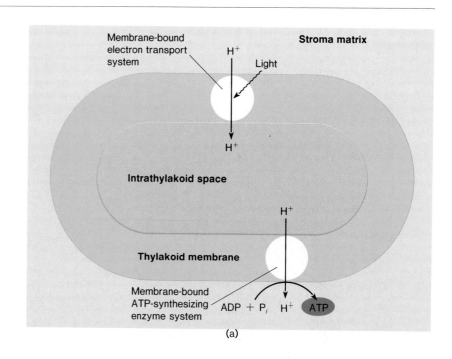

(a)

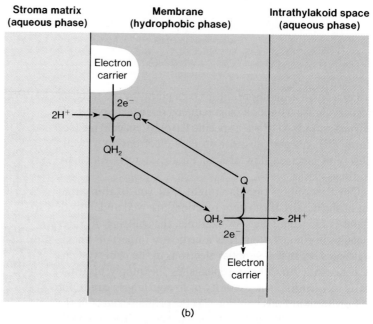

(b)

gradient exists in conjuction with a membrane-bound transport system. When such a gradient is established, the differences in proton concentration (the pH) between the two sides of the membrane, combined with the electrical forces across the membrane, tend to cause the positively charged H$^+$ ions to move in a particular direction. But how is such a transmembrane proton gradient established, and how is it associated with ATP production?

Mitchell proposes that as electrons are being passed from carrier to carrier through electron transport systems in thylakoid membranes, hydrogen ions (H$^+$) are carried along in some of the transfers. He suggests that the carriers are arranged in the membrane so that they pick up hydrogen ions from the stroma matrix and release them into the intrathylakoid space (figure 7.21). This results in a buildup of hydrogen ions inside the thylakoids.

Figure 7.22 Relationships between electron transport, proton gradients, and ATP formation in chloroplasts. (a) Data obtained by Raymond P. Cox demonstrating that illuminated chloroplasts take up protons, and this can change the pH of the medium in which they are suspended. In the dark, the protons leak out again, and the pH returns to the original value. (b) Jagendorf's "acid-bath" experiment, showing that an artificially generated pH gradient can cause ATP formation. This experiment was conducted in the dark.

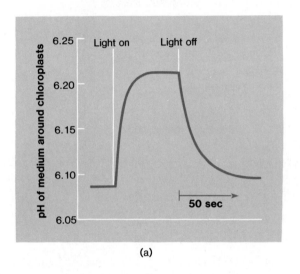

(a)

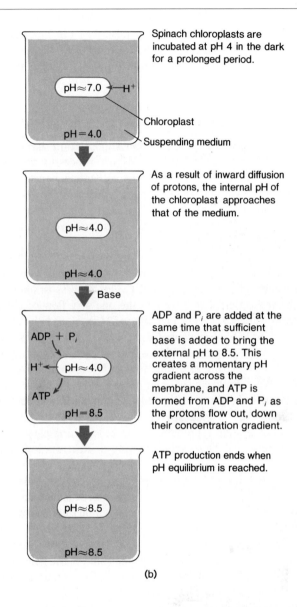

Spinach chloroplasts are incubated at pH 4 in the dark for a prolonged period.

As a result of inward diffusion of protons, the internal pH of the chloroplast approaches that of the medium.

ADP and P_i are added at the same time that sufficient base is added to bring the external pH to 8.5. This creates a momentary pH gradient across the membrane, and ATP is formed from ADP and P_i as the protons flow out, down their concentration gradient.

ATP production ends when pH equilibrium is reached.

(b)

Because H⁺ ions are more concentrated in the intrathylakoid spaces, they tend to diffuse out if a way is open to them. Furthermore, there is an electrical charge difference between the inside and the outside because so many positively charged hydrogen ions have been moved to the inside. Thus, there are forces that tend to drive H⁺ ions out through the thylakoid membranes. Mitchell hypothesizes that when H⁺ ions do flow out through the membrane, energy is released that results in the formation of ATP by a special enzyme system also located in the membrane.

There is no doubt that electron flow in chloroplasts can be linked to proton transport. Illuminated suspensions of thylakoid membranes take up protons into the intrathylakoid space. It is also known that an artificially generated pH gradient is capable of causing ATP synthesis in the dark. This was shown in a famous experiment by Andre Jagendorf, in which he allowed chloroplasts to come to equilibrium with a buffer at pH 4, so that the proton concentration inside the chloroplasts was high. The external pH was then increased, and ADP and phosphate were added to the external solution. ATP was produced until pH equilibrium was again reached (figure 7.22).

Not all biologists are convinced that the Mitchell hypothesis fully and accurately describes the mechanism of photophosphorylation, but the majority feel that further data will support rather than refute the central concepts of the hypothesis.

Noncyclic and Cyclic Photophosphorylation

We have described the photophosphorylation that occurs in association with electron transport processes involved in NADP reduction, but there is another mechanism of ATP production in chloroplasts that is separate from NADP reduction. These two types of photophosphorylation in chloroplasts are known respectively as **noncyclic** and **cyclic photophosphorylation.**

Noncyclic photophosphorylation refers to ATP production that occurs as electrons flow from water to NADP through Photosystems II and I. This process is noncyclic in the sense that it is linked to the one-way flow of electrons from water to NADP (see figure 7.18).

Cyclic photophosphorylation, however, involves a somewhat different process. Photosystem I, by itself, produces ATP without donating electrons for NADP reduction. Although light absorption causes chlorophyll excitation, there is no net production of oxidized or reduced compounds. Electrons return to their original source, the reaction center chlorophyll of Photosystem I, probably via ferredoxin and plastoquinone (figure 7.23). This cyclic pathway of electron flow involves a large enough free-energy change to form ATP. This process is called cyclic photophosphorylation not because phosphorylation is cyclic, but because electrons return to their original source (the P700 chlorophyll of Photosystem I). In itself, cyclic photophosphorylation *does not* result in water oxidation with the release of oxygen or NADPH production. It *does,* however, provide a mechanism for the generation of ATP that is separate from the linear electron-transport pathway through Photosystems II and I. Cyclic photophosphorylation makes an important contribution to the energy economy of leaf cells in many plants.

Carbon Dioxide Incorporation

The energy payoff of photosynthesis is the conversion of light energy to chemical energy in carbohydrate molecules. Recall that Daniel Arnon proved that chloroplasts supplied with carbon dioxide, ATP, and NADPH can produce carbohydrates, even in the dark. How then are ATP and NADPH used during **CO_2 fixation,** the incorporation of carbon dioxide into organic molecules?

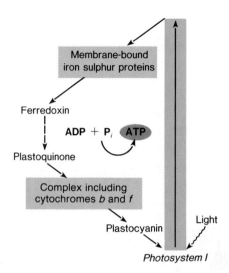

Figure 7.23 Cyclic photophosphorylation. Note that electrons return to the reaction center chlorophyll of Photosystem I, making this process a cycle.

The NADPH and ATP produced in the light reactions are used to reduce CO_2 during its incorporation into carbohydrate molecules in reactions that occur in the stroma matrix of chloroplasts. CO_2 fixation is a several-step process that is catalyzed by a series of soluble enzymes with a specific enzyme catalyzing each reaction.

When the radioactive isotope of carbon, [14]carbon (^{14}C), was made available for biological research during the 1940s, it became possible to analyze the reactions involved in CO_2 incorporation. Melvin Calvin and his colleagues in California carried out a series of experiments on CO_2 incorporation. They illuminated a suspension of cells of a unicellular green alga, *Chlorella,* to which they supplied [14]CO_2, carbon dioxide containing radioactive carbon. After varying lengths of time, the algae were killed by dropping the cells into boiling alcohol (methanol), which instantly stopped all chemical reactions in the cells (figure 7.24). Carbon-containing compounds were then extracted from the cells and separated by paper chromatography for analysis and identification.

Calvin and his coworkers were able to trace the reaction pathway through which CO_2 is incorporated by measuring the amount of radioactive carbon in various compounds after different periods of exposure to [14]CO_2. For example, if the

Figure 7.24 The apparatus used by Melvin Calvin for his classic experiments. The algae were contained in the flat flask, which was illuminated by the two lamps. $^{14}CO_2$ was added with a syringe, and the algae were killed after various periods by opening the stopcock and allowing them to fall into the beaker containing methanol.

were exposed to CO_2 for a very short time, only a few compounds contained radioactive carbon; after a slightly longer time, more compounds contained radioactive carbon; and so forth. Calvin reasoned that the compound containing ^{14}C after the shortest possible exposure would be the first stable compound formed during CO_2 incorporation, and that further steps in the pathway could be determined by using progressively longer exposure periods.

After a very short exposure to $^{14}CO_2$, the first radioactive compounds detected contained three carbon atoms. For this reason, the set of reactions involved in CO_2 fixation is often called the **C_3 pathway.** It also is known as the **Calvin cycle,** as well as the **Calvin-Benson cycle,** after Calvin and his colleague Andrew Benson. It would seem reasonable to think that CO_2 is bound to a two-carbon compound to make the three-carbon product, but Calvin found that this is not the case. The immediate precursor of the C_3 compound is a C_5 compound. Thus, the "carbon arithmetic" of the CO_2 fixation reaction is

$$C_5 \quad + \quad C_1 \longrightarrow 2C_3$$
$$\text{(Five-} \qquad \text{(CO}_2\text{)} \qquad \text{(Two three-carbon}$$
$$\text{carbon} \qquad\qquad\qquad \text{molecules)}$$
$$\text{molecule)}$$

A six-carbon intermediate molecule is formed during the incorporation process, but it exists for such a short time before splitting into two three-carbon molecules that it has not been possible to isolate it.

The five-carbon molecule to which CO_2 is attached is a sugar with two phosphates called **ribulose bisphosphate (RuBP).** The enzyme that catalyzes this reaction is **ribulose bisphosphate carboxylase,** and the two three-carbon molecules produced are molecules of **3-phosphoglycerate (PGA).**

This is one of the most important of all biochemical reactions. The great majority of carbon atoms in organic compounds in living things are derived from CO_2 through this reaction.

We have said that ATP and reduced NADP (NADPH) are required for CO_2 fixation. How are they used? PGA is not a final product of photosynthesis that can be used to make other compounds in living cells. It must be converted to another, slightly different substance called **glyceraldehyde-3-phosphate (GAP),** and this is where ATP and NADPH come in. One ATP is used and one NADPH must give up its two electrons each time PGA is converted to GAP. The 3-phosphoglycerate is phosphorylated using ATP to produce **1,3-diphosphoglycerate,** which is then reduced using NADPH to glyceraldehyde-3-phosphate (GAP).

H$_2$CO$\textcircled{P}$
|
HCOH   3-phosphoglycerate
|
C=O
|
O$^-$

— ATP
— ADP

H$_2$CO$\textcircled{P}$
|
HCOH   1,3-diphosphoglycerate
|
C=O
|
O$\textcircled{P}$

— NADPH
$\textcircled{P}$ ← → NADP$^+$

H$_2$CO$\textcircled{P}$
|
HCOH   Glyceraldehyde-3-phosphate (GAP)
|
C=O
|
H

Since combining each molecule of CO_2 with ribulose bisphosphate gives rise to two molecules of 3-phosphoglycerate, this sequence of reactions requires two ATP and two NADPH units for each molecule of CO_2 that is incorporated into carbohydrate.

RuBP + CO$_2$ ⟶ 2PGA

2PGA ⟶ 2 1,3-diphosphoglycerate
2ATP   2ADP   2$\textcircled{P}$

2 1,3-diphosphoglycerate ⟶ 2GAP
2NADPH   2NADP$^+$

GAP is the product of photosynthesis that can be used to make sugars, starch, fats, proteins, and all other compounds needed by the plant. Only part of the GAP produced by photosynthesis can be used for these things, however. Much of it must be used to maintain the process of CO_2 fixation.

For the CO_2 incorporation process to continue, an adequate supply of RuBP must be maintained, and the majority of the GAP molecules produced are used to regenerate the supply of RuBP molecules. This return to an original point means that the process is cyclic (hence the name Calvin *cycle*). ATP is also required for regeneration of the RuBP supply.

RuBP Regeneration

In the reactions we have just discussed, there must be one RuBP molecule to react with each CO_2 molecule. The products of the CO_2 incorporation reactions, three-carbon molecules, enter a series of Calvin cycle rearrangements to regenerate RuBP molecules (figure 7.25). In fact, a full five-sixths of the molecules produced in CO_2 incorporation must be used to keep the supply of RuBP constant. In simple terms, five three-carbon molecules are transformed to three five-carbon molecules. The five-carbon molecules produced by these rearrangements are **ribulose (mono) phosphate** molecules. A final step is needed to produce ribulose bisphosphate:

$$\text{Ribulose phosphate} + \text{ATP} \longrightarrow \text{RuBP} + \text{ADP}$$

Thus, the regeneration of each molecule of RuBP also requires one ATP.

ATP and NADPH Sums

Since the Calvin cycle requires two ATP and two NADPH for each molecule of CO_2 incorporated into carbohydrate, and since the production of ribulose bisphosphate requires one ATP, a total of three ATP and two NADPH are used for each CO_2 molecule incorporated into carbohydrate.

There is another useful way of looking at ATP and NADPH "arithmetic." A number of substances such as glucose are products of photosynthesis although they are not produced in the Calvin cycle itself. These various compounds can be produced, however, using glyceraldehyde-3-phosphate (GAP). But the withdrawal of GAP must be balanced with CO_2 incorporation if the pool of Calvin cycle compounds is to be kept constant so that the process continues to operate at a steady rate. How much does it cost in terms of ATP and NADPH to incorporate the three carbons of three CO_2 molecules and thus make one GAP molecule available for withdrawal?

If the incorporation of one CO_2 molecule requires three ATP and two NADPH, the incorporation of three CO_2 molecules will require three times as much energy. Therefore, nine ATP and six NADPH coming from the light reactions

Figure 7.25 The Calvin cycle. (a) Some of the GAP produced in photosynthesis is available for use by the plant, but some of it must be used to replenish the supply of RuBP and thus keep the CO_2 fixation process going. (b) Five out of every six product molecules must be converted back to the precursor molecule. By a series of rearrangements, five C_3 molecules are used to produce three C_5 molecules. The other one-sixth of the product molecules can either be used for synthesis of other compounds or to increase the amount of the precursor, and hence the rate of CO_2 fixation, during growth. (c) A skeleton outline of the Calvin cycle rearrangement reactions, showing how five C_3 molecules are used in reactions involving a number of intermediate compounds to produce three C_5 molecules. Each arrow intersection represents a specific, enzyme-catalyzed reaction. Each of the C_5 molecules produced is ribulose phosphate, and one ATP is needed to convert each of them to ribulose bisphosphate (RuBP).

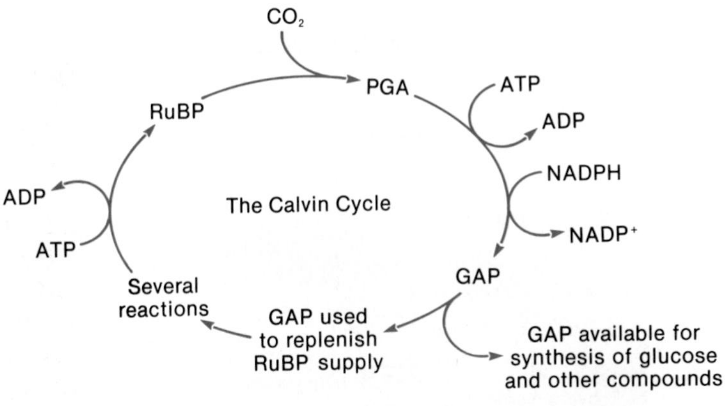

(a)

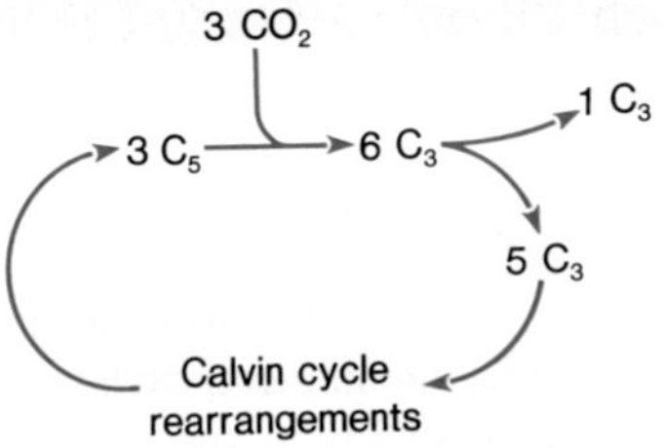

(b)

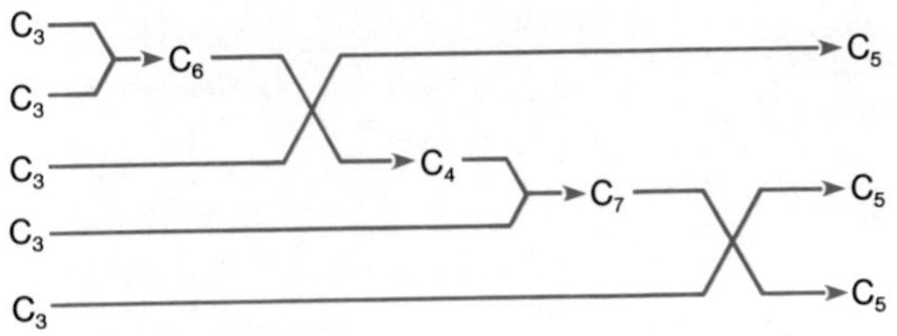

(c)

must be used in dark reactions to produce a net gain of one GAP. This GAP molecule can then be withdrawn from the Calvin cycle pool of compounds and used by the cell for other purposes.

Uses of GAP

As we mentioned earlier, at least five-sixths of the GAP produced by CO_2 fixation reactions must be used in Calvin cycle rearrangements to produce ribulose bisphosphate. What happens to the other one-sixth?

While a plant is growing, its capacity for photosynthesis must continually be increased, and this is possible if more than five-sixths of the GAP is converted back to RuBP. In such a case, the *rate* of CO_2 fixation increases because the size of the pool of Calvin cycle compounds is increased. When the requirements for increased photosynthetic capacity have been met, any surplus glyceraldehyde-3-phosphate can be used to form other compounds. Within the chloroplast, it may be converted to **glucose phosphate.** Glucose phosphate is then converted to starch, a glucose polymer that acts as a store of carbohydrate inside chloroplasts. Plants accumulate starch during the day, when light allows photosynthesis to occur, and break it down to release glucose during the night.

Fixed carbon seems to leave the chloroplast mainly in the form of **dihydroxyacetone phosphate (DHAP).** Thus, DHAP is a chloroplast's "export product" to the rest of the cell. DHAP is an isomer of GAP and can be formed from it. In the cytoplasm outside the chloroplast, DHAP can be used to make the six-carbon sugars glucose and fructose, which are then joined to form **sucrose.** Sucrose is the major form in which fixed carbon is transported from the leaf to other parts of the plant (figure 7.26).

Ordinary leaf mesophyll cells of C_4 plants lack the Calvin cycle enzymes; nevertheless, they fix carbon dioxide by joining it to a three-carbon acid, **phosphoenol pyruvate (PEP)**, to make the C_4 acid **oxaloacetate**. This reaction is catalyzed by the enzyme **phosphoenol pyruvate carboxylase (PEP carboxylase)**.

$$\text{Phosphoenol pyruvate } (C_3) + CO_2 \xrightarrow{\text{PEP carboxylase}} \text{Oxaloacetate } (C_4)$$

In many C_4 plants, the oxaloacetate is reduced to **malate**, another C_4 acid. Then malate is transported to the bundle sheath cells and broken down into CO_2 and **pyruvate** (C_3). To complete the cycle, the pyruvate is then returned to the mesophyll cells and converted back to PEP (figure 7.29) in a reaction that requires two molecules of ATP. The CO_2 liberated in the bundle sheath cells is incorporated into carbohydrates by the Calvin cycle enzymes, just as it is in C_3 plants.

The C_4 mechanism functions to increase the concentration of CO_2 available to the Calvin cycle enzymes in the bundle sheath cells, but it requires considerable energy expenditure to do so. ATP is expended in the transport processes that move organic molecules back and forth between mesophyll cells and bundle sheath cells as well as in the reactions that replenish the supply of phosphoenol pyruvate. What do C_4 plants gain by this seemingly "extra" work?

Part of the answer lies in the way in which the enzyme ribulose bisphosphate (RuBP) carboxylase functions. Carbon dioxide is a small molecule that does not contain charged groups or other distinctive features, and the active site of RuBP carboxylase does not completely distinguish it from O_2 (which has a similar shape). Under normal conditions (when the ratio of O_2 to CO_2 in the air is 21 percent:0.03 percent, or 700:1), although RuBP carboxylase preferentially binds CO_2, it can also act as an oxygenase, occasionally incorporating oxygen in place of CO_2:

$$\underset{(C_5)}{\text{RuBP}} + O_2 \rightarrow \underset{(C_3)}{\text{PGA}} + \underset{(C_2)}{\text{Phosphoglycollate}}$$

This interferes with photosynthetic efficiency because a molecule of RuBP is consumed and only one usable three-carbon molecule (PGA) is produced. Furthermore, when RuBP carboxylase incorporates O_2 instead of CO_2, the two-carbon molecule phosphoglycollate is produced in the reaction. Phosphoglycollate represents a problem for plant cells because it must be processed through complex, energy-requiring reactions (see box 7.2) in order to produce useful Calvin cycle compounds.

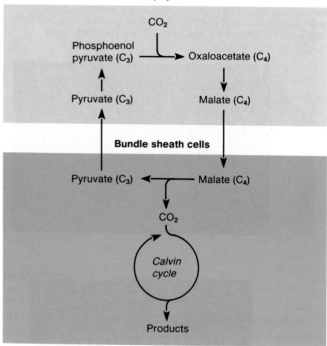

Figure 7.29 Summary diagram of C_4 metabolism in some C_4 plants. Other C_4 plants have a similar scheme in which the amino acids aspartate and alanine take the place of malate and pyruvate.

When stomata partially or completely close, much of the CO_2 available inside the leaf is used and the $O_2:CO_2$ ratio increases. Then the problem of oxygen incorporation by RuBP carboxylase becomes more severe. This means that stomatal closing during water stress situations can result in markedly decreased photosynthetic efficiency in C_3 plants.

C_4 plants essentially prevent oxygenase activity by RuBP carboxylase by doing the "extra" work to concentrate carbon dioxide in bundle sheath cells where their Calvin cycle enzymes are located. "Pumping" CO_2 into the bundle sheath cells increases the CO_2 concentration to such an extent that oxygen interferes very little with the efficiency of RuBP carboxylase and the Calvin cycle can operate at maximum efficiency.

Another factor is the activity of PEP carboxylase, which catalyzes the fixation of CO_2 in C_4 plants. PEP carboxylase has a very high affinity for CO_2 and does not show the kind of oxygenase activity seen with RuBP carboxylase. Thus, PEP carboxylase can continue to fix CO_2 effectively even when the CO_2 concentration is very low (and therefore the $O_2:CO_2$ ratio is very high). This means that C_4 plants can continue to obtain and utilize CO_2 in photosynthesis without opening their stomata to the extent required in C_3 plants.

Now that we have discussed some advantages of the C_4 mechanism of photosynthesis, we should consider a question about plant evolution: Why are the great majority of plants C_3 plants rather than C_4 plants?

The answer is not entirely clear, but it seems that the extra cellular work involved in the C_4 mechanism is so costly in terms of energy (ATP) that it may be advantageous only under certain circumstances. C_4 plants are found in environments with high light intensity. In such environments, there is abundant light energy available so that the extra ATP required by the C_4 mechanism can readily be produced by cyclic photophosphorylation. Areas with high light intensity also tend to have higher temperatures; plants growing in these areas have a great potential for water stress. Such conditions place a premium on efficient CO_2 fixation, which reduces a plant's water loss through open stomata. C_4 plants thrive in such environments; they are well equipped to absorb a great deal of the available light energy and convert it to chemical energy.

On the other hand, in cooler, moister environments, especially those with lower light intensities, C_3 plants flourish. Water loss associated with stomatal opening is less of a problem in such environments. Furthermore, C_3 plants can grow at lower light intensities than C_4 plants because the C_3 mechanism requires less light energy for ATP production.

Crassulacean Acid Metabolism

Many succulent plants growing in extremely arid environments, including some in the family Crassulaceae, are characterized by **crassulacean acid metabolism (CAM)** (figure 7.30). CAM is similar to C_4 metabolism in that CO_2 is first fixed into C_4 acids; but the two differ in that CAM plants do not have the characteristic *Kranz* leaf anatomy. The most important functional difference is that separation of initial CO_2 incorporation from Calvin cycle activity is temporal rather than spatial in CAM plants; in other words, the two mechanisms operate at different times in CAM plants rather than in different locations, as they do in C_4 plants. This temporal separation is advantageous for plants living in very arid climates such as deserts. When the danger of water loss decreases during the cool desert night, CAM plants open their stomata and fix large amounts of CO_2 into oxaloacetate. The oxaloacetate is then converted into malate or aspartate (another C_4 acid), both of which are stored in high concentrations in the cell vacuoles of the mesophyll cells. Plants using the CAM mechanism are succulent plants with thick, fleshy leaves containing large quantities of these mesophyll cells.

During the day, the C_4 acids in CAM plants are broken down, providing a source of CO_2 for fixation by the Calvin cycle enzymes (figure 7.30). To prevent water loss, the stomata are firmly closed during the day, and any daytime exchange of CO_2 and water is restricted to very slow movement through the cuticle.

Figure 7.30 (a) Summary of crassulacean acid metabolism. Note that the initial CO_2 incorporation occurs at night, while Calvin cycle activity occurs during the day. Malate accumulates during the night and is broken down in the daytime to make CO_2 available for Calvin cycle night to produce phosphoenol pyruvate (PEP). (b) A saguaro cactus from the desert Southwest. CAM photosynthesis occurs in the fleshy stems of cacti.

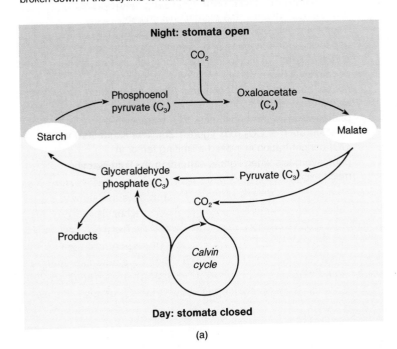

(a)

(b)

Box 7.2
Photorespiration and Photosynthetic Efficiency

Photorespiration is a process that occurs in photosynthesizing plant cells. It is the opposite of photosynthesis in that it results in the production of CO_2 and the consumption of O_2 in light. The discovery of this apparent cause of inefficiency in the photosynthetic mechanism aroused much interest. If ways of reducing or eliminating it could be found, the result might well be a dramatic improvement in agricultural productivity.

Photorespiration is intimately connected with photosynthetic mechanisms. It should not be confused with CO_2 production and O_2 consumption caused by the ordinary processes of cell respiration. The latter processes involve the mitochondria and also occur in the dark and in nongreen tissues, while photorespiration occurs only in photosynthesizing cells (box figure 7.2a).

Photorespiration is related to the functioning of the enzyme ribulose bisphosphate (RuBP) carboxylase. The enzyme also acts as an oxygenase, sometimes incorporating oxygen instead of CO_2. When this happens, the reaction does not produce two molecules of phosphoglycerate, but produces one molecule of phosphoglycerate and one of phosphoglycollate:

$$
\begin{array}{l}
H_2CO\,\text{\textcircled{P}} \\
| \\
C = O \\
| \\
HCOH \quad + \quad O_2 \\
| \\
HCOH \\
| \\
H_2CO\,\text{\textcircled{P}}
\end{array}
$$

Ribulose bisphosphate

$$
\begin{array}{ll}
H_2CO\,\text{\textcircled{P}} & O^- \\
| & | \\
C = O \quad + & C = O \\
| & | \\
O^- & HCOH \\
& | \\
& H_2CO\,\text{\textcircled{P}}
\end{array}
$$

Phosphoglycollate   Phosphoglycerate

Phosphoglycollate loses its phosphate group to become glycollate. Some algae excrete glycollate into their environments as a waste product, but the cells of higher plants must convert it back to intermediates in the Calvin cycle. This conversion takes place in a complex series of reactions involving mitochondria, chloroplasts, and other parts of the cell, especially the peroxisomes. At one stage during these reactions, two two-carbon molecules are converted into a three-carbon molecule, and one CO_2 molecule is released. This accounts for the CO_2 production that occurs during photorespiration.

Photorespiration processes lag a little behind those involved in photosynthesis. If a photosynthesizing leaf is suddenly put into the dark, a transient burst of CO_2 production takes place. Biologists measure this burst of CO_2 production to determine the amount of photorespiration occurring in various photosynthesizing leaves.

Photorespiration seems to be inevitable at the normal atmospheric O_2:CO_2 ratio. It is possible, though, to reduce photorespiration by altering this ratio in favor of CO_2. This is what happens in photosynthesizing C_4 plants because they concentrate CO_2 in their bundle sheath cells, the site of Calvin cycle reactions. In fact, photorespiration has not been detected in C_4 plants.

Knowledge of the effect of altered CO_2 ratios has been used commercially to a limited extent by enriching greenhouse atmospheres with CO_2.

Various factors limit the efficiency of photosynthesis under natural conditions, and the limiting factor varies with the conditions in the plant's environment. Photosynthesis may be limited by low light intensity, especially in the morning and evening, on cloudy days, and in shade. But increasing the light intensity beyond a certain level has no effect on photosynthetic rate at normal atmospheric CO_2 concentrations. Under such light-saturated conditions, however, the rate of photosynthesis can be increased by increasing the concentration of CO_2 (box figure 7.2b). This suggests that CO_2 concentration is indeed a limiting factor in photosynthesis when light is saturating the light-reaction mechanisms.

An Overview of Cellular Energy Conversions

It is best to begin a consideration of cellular energy conversions with the utilization of a carbohydrate, the sugar glucose. The first series of reactions in the oxidative breakdown of glucose is called the **Embden-Meyerhof (E-M) pathway** after two German biochemists who studied these processes in the 1920s and 1930s. (The term **glycolysis** ["glucose breakdown"] is sometimes used as a synonym for the Embden-Meyerhof pathway, but biochemists usually apply the term glycolysis to the E-M pathway plus a form of fermentation, which together lead from glucose to lactic acid.) The E-M pathway is a several-step process that occurs in the cell's cytoplasmic matrix. It can proceed in either the presence or absence of oxygen but, by itself, makes available only a small portion of the potential energy that could be extracted from glucose. The reactions of the E-M pathway are thought to be older in an evolutionary sense than the oxygen-requiring processes of respiration, because the E-M reactions could have occurred in the primitive environment of earth before the atmosphere contained free oxygen.

The formation of two three-carbon molecules of **pyruvic acid** is the final step of the E-M pathway (figure 8.4), so the process does not decrease the number of carbon atoms in organic molecules (glucose is a six-carbon molecule and pyruvic acid is a three-carbon molecule; $C_6 \rightarrow C_3 + C_3$). Different kinds of cells can further metabolize pyruvic acid in various ways.

Figure 8.3 ATP as the cellular energy currency. Oxidation of reduced organic compounds provides energy to combine ADP and phosphate to replenish the supply of ATP. Hydrolysis of ATP to ADP and phosphate provides energy to drive energy-requiring processes in cells.

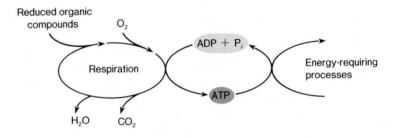

Figure 8.4 Preview of carbohydrate breakdown pathways using glucose as a starting point. For each molecule of glucose processed in the Embden–Meyerhof pathway, two molecules of pyruvic acid are produced. Alcoholic and lactic fermentation occur in certain cells under anaerobic conditions. The Krebs cycle is part of aerobic respiration because electrons from Krebs cycle oxidation reactions must go through the electron transport system to oxygen.

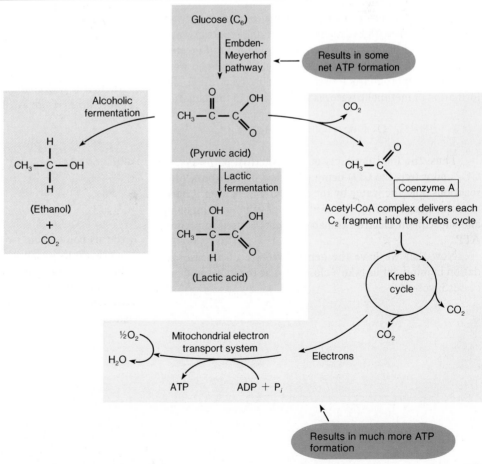

163

Under **anaerobic** conditions (conditions lacking oxygen) pyruvic acid is processed through one of several types of **fermentation** (figure 8.4). In **alcoholic fermentation,** which occurs, for example, in brewer's yeast, pyruvic acid is converted to ethanol (ethyl alcohol). **Lactic fermentation,** which occurs in some microorganisms and some animals cells, produces lactic acid. Neither kind of fermentation yields additional usable energy for the cells beyond that extracted in the Embden-Meyerhof pathway.

Under **aerobic** conditions (in the presence of oxygen), oxygen-requiring processes collectively known as **respiration** use much more of the potentially available chemical energy to form ATP. Pyruvic acid enters another series of reactions that occur inside the mitochondria called the **Krebs cycle,** named for Sir Hans Krebs, who clarified the cyclical nature of the reactions involved. In the Krebs cycle, organic molecules are further degraded by the removal of one-carbon fragments as CO_2. This process of CO_2 removal is called **decarboxylation.**

Oxidations of organic molecules in the Krebs cycle provide electrons that are transferred to electron carriers in the mitochondrial membranes. There, a series of oxidation-reduction reactions occurs as electrons are passed from carrier to carrier in an **electron transport system.** During these transfers, free-energy changes make possible the production of considerable amounts of ATP.

These electron carriers finally donate electrons to oxygen, which combines with hydrogen ions (protons) to form water. The reduction of oxygen, which serves as the *final electron acceptor,* with the subsequent formation of **metabolic water** is a fundamental characteristic of aerobic respiration. The formation of metabolic water is summarized in this equation:

$$O_2 + 4e^- + 4H^+ \longrightarrow 2H_2O$$

Thus, the final products of the oxidative pathways are CO_2 and water; the CO_2 being produced in decarboxylation reactions, and the water by the reduction of oxygen. Energy made available during the oxidation of organic compounds and subsequent electron transport events is used to produce ATP.

Now that we have the general picture of glucose oxidation in mind, let us take a closer look at the individual processes involved.

The Embden-Meyerhof Pathway

The progressive breakdown of glucose to produce pyruvic acid in the Embden-Meyerhof (E-M) pathway occurs in a series of nine reactions, each of which is catalyzed by specific enzymes. These reactions are shown in figure 8.5 where glucose molecules are represented in their normal ring form. In analyzing E-M pathway reactions, however, it is helpful to think of carbohydrate molecules as linear chains of carbon atoms, omitting the hydrogen and oxygen atoms of the molecules.

The process begins with preparatory reactions that "activate" the glucose molecule and make it possible for the succeeding reactions to take place. These preparatory reactions are phosphorylations (phosphate groups are attached to carbohydrate molecules), and use ATP; the ATP molecules donate their terminal phosphate groups in the reactions. Thus, the Embden-Meyerhof pathway reactions begin with an expenditure of ATP. But these priming reactions provide the necessary activation energy to make all subsequent reactions possible.

The hexokinase-catalyzed reaction of ATP and glucose, which results in the formation of **glucose 6-phosphate** (glucose with a phosphate group attached to the number 6 carbon) can be symbolized simply as:

C-C-C-C-C-C + ATP $\xrightarrow{\text{Hexokinase}}$ C-C-C-C-C-C- (P) + ADP
(Glucose) (Glucose 6-phosphate)

Two additional preparatory reactions follow the phosphorylation of glucose. A molecular reorganization yields **fructose 6-phosphate,** and then the transfer of another phosphate group from ATP produces **fructose 1,6-bisphosphate:**

C-C-C-C-C-C-(P) + ATP $\xrightarrow{\text{Phosphofructokinase}}$

(Fructose 6-phosphate)

(P)-C-C-C-C-C-C-(P) + ADP
(Fructose 1,6-bisphosphate)

To this point, then, two ATP molecules have been expended in preparatory reactions. So how does the E-M pathway produce an energy gain for the cell?

Subsequent reactions in the Embden-Meyerhof pathway are energy yielding, and thus the whole reaction series results in an ATP gain for the cell. Fructose 1,6-bisphosphate

Figure 8.5 The reactions of the Embden–Meyerhof pathway. The name of the enzyme that catalyzes each reaction is given by each arrow. Most enzyme names end in "-ase." The positions of the carbons in the molecules are represented symbolically as the "corners" of the ring. Note that acids such as pyruvic acid actually dissociate and give up a hydrogen ion (H$^+$) at cellular pH's. Thus, the pyruvate ion, which is produced by pyruvic acid dissociation, is pictured rather than the pyruvic acid molecule. (The "-ate" ending is used generally to indicate ionic forms of dissociated acids.) The oxidized form of NAD is symbolized NAD$^+$, while the reduced is NADH.

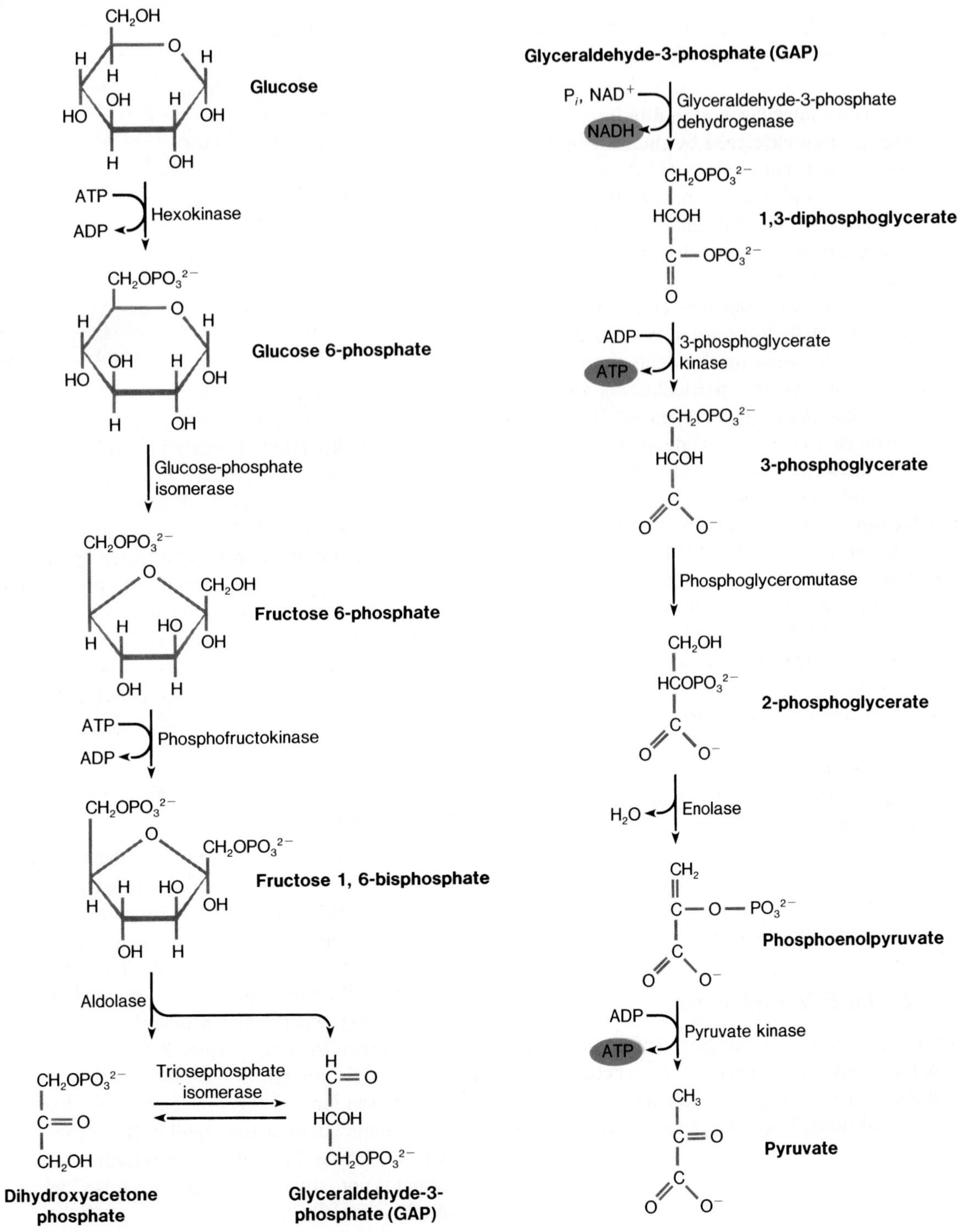

is split into two three-carbon molecules, **dihydroxyacetone phosphate** and **glyceraldehyde-3-phosphate (GAP):**

$$\text{\textcircled{P}-C-C-C-C-C-C-\textcircled{P}} \xrightarrow{\text{Aldolase}}$$
(Fructose 1,6-bisphosphate)

$$\text{\textcircled{P}-C-C-C} \quad + \quad \text{C-C-C-\textcircled{P}}$$
(Dihydroxyacetone phosphate and GAP)

These three-carbon molecules are readily interconvertible in a reversible reaction catalyzed by the enzyme **triose phosphate isomerase.** Since one product, GAP, reacts in the next step of the pathway and is used up, the dihydroxyacetone phosphate is converted to GAP. Thus, two GAP molecules must be accounted for in determining energy relationships of subsequent reactions.

Further processing of C_3 molecules begins with a dehydrogenation reaction. Dehydrogenation is an oxidation in which the transfer of electrons involves the simultaneous transfer of hydrogen ions. In this particular reaction, two electrons, with an accompanying hydrogen ion, are transferred to the electron carrier nicotinamide adenine dinucleotide (NAD). At the same time, a phosphate group is attached to the molecule, so the products of this reaction are **1,3-diphosphoglycerate** (a C_3 molecule with two phosphate groups) and a reduced molecule of NAD (NADH).

In the next reaction in the sequence, the free-energy change is large enough to allow a removed phosphate group to be transferred to ADP by phosphorylation, thus yielding a molecule of ATP. This is the first of two such transfers that are essential to energy retrieval in the E-M pathway. Because this phosphorylation is associated directly with a reaction involving a substrate in the pathway, it is called a **substrate-level phosphorylation.**

After two molecular reorganizations, there is a second phosphorylation, which occurs in the reaction catalyzed by pyruvate kinase. This reaction yields **pyruvate** (the ionized form of pyruvic acid that occurs in solution). Pyruvate is a key compound in cellular metabolism because it stands at the junction of several biochemical pathways.

Energy Yield of the E-M Reactions

How much energy (ATP) has the cell gained after all of these reactions? That depends upon whether the reactions have taken place under aerobic (oxygen present) or anaerobic (oxygen absent) conditions. Thus, we can give only a partial

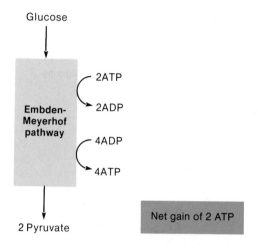

Figure 8.6 ATP arithmetic of the Embden–Meyerhof pathway. Two ATPs must be spent to start the process. Four ATPs are produced by substrate-level phosphorylations, giving a net gain for the cell of two ATPs.

answer at this point. It costs two ATPs to get the process started. Then two ATPs are produced for each of the two three-carbon molecules produced from each glucose molecule. In short, two ATPs are spent and four are produced, leaving a net gain of two ATPs (figure 8.6).

What happens to the pyruvate produced by the E-M pathway reactions? This question also has several possible answers, basically depending upon whether conditions are aerobic or anaerobic. (This is true as well for questions about the fate of the reduced NAD [NADH] produced in the E-M pathway.) Let us look first at fermentation, which occurs under anaerobic conditions.

Fermentation

Under anaerobic (oxygen absent) conditions, fermentation converts pyruvate into one of several products, depending upon the kind of cells and the enzymes they possess.

Some cells, such as yeast cells, convert pyruvate to ethanol (ethyl alcohol). Alcoholic fermentation is very important in the brewing and wine-making industries, as well as in baking, where carbon dioxide produced by yeast cells causes bread to "rise" (figure 8.7).

In lactic fermentation, other cells convert pyruvate to lactate (the ionized form of lactic acid that occurs in solution). Some bacteria that spoil milk do this. Animal muscle cells, including those in human muscles, also conduct lactic fermentation during vigorous exercise, when the oxygen supply is insufficient.

Figure 8.7 Bread dough rising in a commercial bakery. Carbon dioxide produced by yeast cells causes the dough to rise.

Figure 8.8 Reactions in alcoholic and lactic fermentation shown side by side. White boxes indicate positions where hydrogen ions attach when electrons are donated by reduced NAD (NADH). The pathway from glucose to lactate, which involves both the Embden–Meyerhof pathway and lactic fermentation reactions, is often called glycolysis (literally, "glucose breakdown").

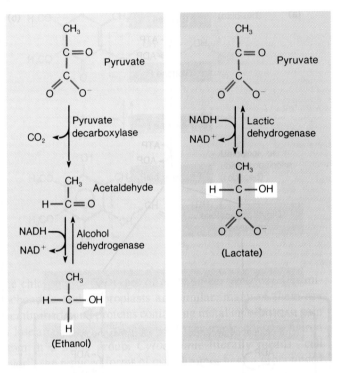

In both alcoholic and lactic fermentation (figure 8.8), there is a reaction in which reduced NAD (NADH) donates electrons. These electrons are accompanied by hydrogen ions (note the colored H's on ethanol and lactate molecules in figure 8.8). This is critical in order to keep the Embden-Meyerhof pathway reactions going in these cells—oxidized NAD (NAD⁺) must be available or the whole process will stop at the step where glyceraldehyde-3-phosphate is converted to 1,3-diphosphoglycerate. The total quantity of NAD available in cells is very small, so constant recycling is necessary (figure 8.9). However, this recycling does eliminate the possibility of using the energy potentially available in a reduced NAD molecule.

We can draw two conclusions about energy conversion in anaerobic metabolism by examining figure 8.9. First, only a portion of the free energy that could be made available if glucose were completely broken down to CO_2 and water is released in the conversions from glucose to lactate or ethanol (both of which are still reduced carbon compounds). Second, only about one-third of this released energy is captured in the form of ATP. For every molecule of glucose that undergoes these conversions, there is a net gain of only two molecules of ATP. Clearly, there is more potential energy in glucose molecules that might be retrieved by more efficient energy-conversion mechanisms. Aerobic respiration is just such a mechanism.

Aerobic Respiration

In aerobic respiration (which occurs inside mitochondria), pyruvate, the product of the Embden-Meyerhof pathway, is not used in fermentation and converted to lactate or ethanol, as it would be under anaerobic conditions. Instead, it is oxidized further under aerobic conditions with a great increase in the ATP yield. The primary factor making this possible is a mechanism that uses the energy potentially available in reduced coenzymes such as NADH to phosphorylate ADP molecules, thus producing ATP. Coenzymes reduced during the further oxidation of pyruvate donate electrons to electron transport systems, which in turn transfer them to oxygen, and ATP is formed in the process.

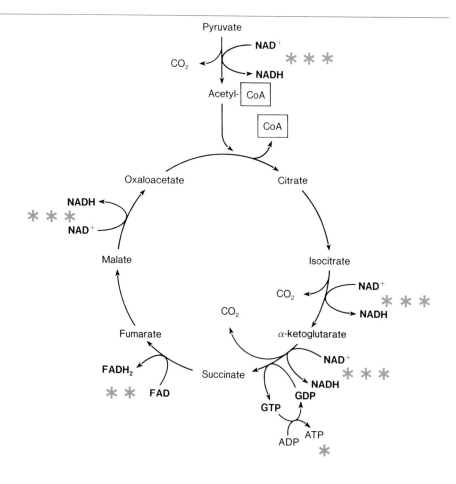

Figure 8.16 ATP arithmetic of the preparatory reactions and the Krebs cycle. Each red star represents one ATP formed. Each reduced NAD molecule permits 3 phosphorylations in the electron transport system. A reduced FAD permits 2 phosphorylations in the electron transport system. A substrate-level phosphorylation forms GTP, which can transfer phosphate to ADP to produce ATP. Thus, starting from pyruvate, 15 ATPs are formed.

Overall Energy Sums

Now we can summarize the energy return from the complete oxidation of a glucose molecule under aerobic conditions. We can also compare the efficiency of aerobic and anaerobic processes.

Under aerobic conditions, the NADH produced in the Embden-Meyerhof reactions can transfer electrons to the electron transport systems inside mitochondria, thereby salvaging more usable energy than is possible under anaerobic conditions, in which the electrons are accepted during the production of organic end products (lactate or ethanol). To do so, however, electrons must be transferred into the interiors of mitochondria, where they become available to the electron transport systems located in the internal mitochondrial membranes.

However, reduced NAD cannot enter a mitochondrion directly, so it cannot pass its electrons directly to an electron transport system in the inner mitochondrial membrane. Instead, the electrons are passed to a shuttle carrier that takes them into the mitochondrion and donates them to an electron transport system.

There are several such electron shuttles. A common one is the **glycerol phosphate shuttle,** which picks up electrons from reduced NAD, carries them into the mitochondria, and donates them not to NAD, but to ubiquinone (figure 8.17). In this case the electrons do not start at the "top" of the electron transport system, so the possibility for one phosphorylation is bypassed (see figure 8.12). If mitochondria use this particular mechanism, a reduced NAD in the cytoplasm ends up being worth only two ATP molecules.

Nevertheless, being able to get some ATP production from the NAD reduced during the Embden-Meyerhof reactions makes the E-M pathway more efficient in terms of energy production when it operates under aerobic conditions. Under such conditions, the E-M pathway yields a net gain of six ATP molecules per glucose molecule processed (figure 8.18): four direct (substrate-level) phosphorylations *plus* four ATPs formed because of the two NAD reductions (one NAD reduction for each three-carbon carbohydrate molecule), *minus* the two ATP molecules used initially for activation. If the ATP output from the E-M pathway is added to the ATP yield from preparing and feeding pyruvate products into the Krebs cycle, thirty-six molecules of ATP are formed for each molecule of glucose oxidized completely to CO_2 and water

Figure 8.17 A generalized diagram of an electron shuttle that oxidizes NAD at the outer mitochondrial surface and delivers electrons to the electron transport system in the inner mitochondrial membrane.

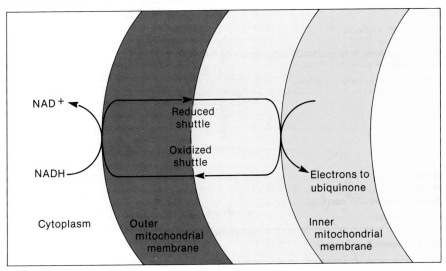

under aerobic conditions. Since anaerobic processes yield a net gain of only two molecules of ATP per molecule of glucose, aerobic respiration yields a net gain of eighteen times as much usable energy as did strictly anaerobic processes (table 8.1).

The Efficiency of Respiration

How efficient is aerobic respiration as a system for converting energy from one form to another? We know that thirty-six molecules of ATP are formed for every molecule of glucose metabolized in aerobic respiration. Of course, the energy relationships of individual reacting molecules are too small to be measured. Therefore, measurements in terms of *moles* of the substances involved are used, keeping in mind that the behavior of moles (gram molecular weights) is proportional to the behavior of individual molecules. The standard free energy of oxidation of a mole of glucose to CO_2 and water is -686 kcal (-2870 kJ). ATP production requires an energy input of 7.3 kcal (30.5 kJ) per mole. Thus, we can calculate an efficiency figure for respiration as follows:

$$\frac{36 \times 7.3}{686} \times 100 = 38\%$$

For several reasons, however, most biologists think this is a minimal estimate. First of all, the number of ATP molecules formed per molecule of glucose may actually be thirty-eight rather than thirty-six in some situations (when different mechanisms are employed to transfer electrons from

Figure 8.18 ATP production from the Embden–Meyerhof pathway when it is associated with aerobic respiration.

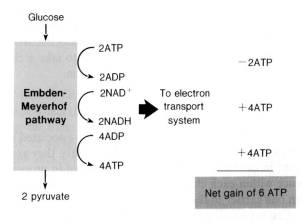

reduced NAD in the cytoplasm to the electron transport system inside the mitochondrion). Second, because of the effects of concentration and the efficiently coupled utilization of the ATP formed in respiration, ATP probably is worth more to living cells than 7.3 kcal per mole in terms of usable energy for cell work. Thus, respiration is probably more than 38 percent efficient.

Most of the remaining 60 percent or so of the available energy is released during the reactions as heat. We should not think of this heat as completely lost or wasted energy, however. Heat produced by these reactions affects the body temperature of organisms in which they occur.

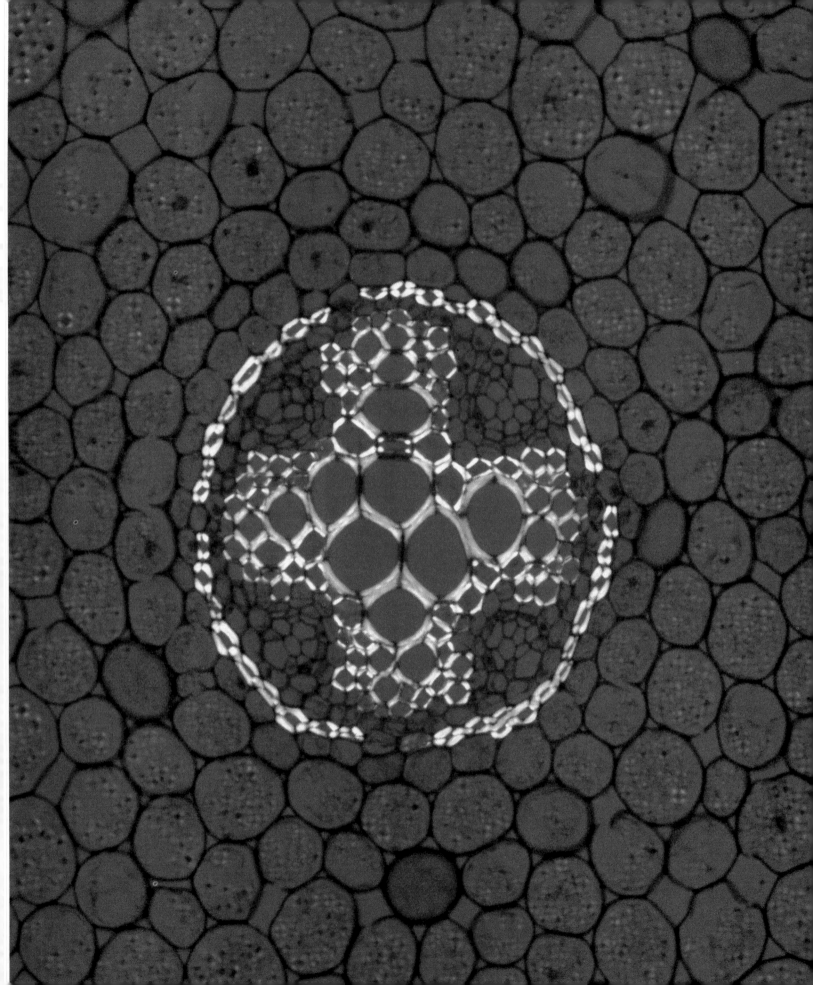

Roots, Water, and Mineral Nutrients

9

Chapter Outline

Chapter Concepts

1. A vascular plant's body is specialized to obtain and transport materials required for autotrophic existence.
2. Plant bodies consist mainly of molecules containing atoms of carbon and oxygen (from carbon dioxide) and hydrogen (from water).
3. Open stomata permit carbon dioxide diffusion into leaves, but at the same time, transpiration (evaporative water loss from leaves) occurs.
4. A stream of water moves continuously from a region of higher water potential to regions of lower water potential—from the soil into roots and up through the xylem of a vascular plant, replacing water lost by transpiration and carrying mineral ions from the roots to the rest of the plant.
5. A soil's water-holding capacity and amount of aeration help determine its suitability for plant growth.
6. Roots are specialized for absorption and transport. Soil water moves freely through spaces in the outer portions of a root, but must cross a plasma membrane and move through cytoplasm to reach the vascular elements in the root's center.
7. Under certain circumstances, mineral ions enter roots by diffusion, but most ions are absorbed by active transport that requires the expenditure of ATP.
8. Despite natural replacement processes, the application of fertilizer is necessary to restore depleted mineral nutrients in agricultural soils.

Facing page   A portion of a cross section of a buttercup root photographed with a microscope using plane-polarized light. Plants' roots are specialized for water and mineral uptake and are involved in the transport of materials upward to aboveground plant parts. Roots are also specialized for nutrient storage.

not readily pass through the membrane, there is a potential for water movement; there will be net osmotic movement of water from the region containing pure water to the region containing solute.

All parts of living things contain dissolved and suspended material. Thus, osmotic water movement in living things takes place between solutions that differ in the relative proportions of water molecules and solute particles they contain.

The concept of osmotic water movement is adequate to describe much of water movement in animals, but another factor, pressure, plays an important role in plants. Recall that each plant cell is enclosed by a rigid cell wall that resists expansion of the cell's plasma membrane. Plant cells normally are surrounded by solutions with higher proportions of water molecules than the cell contents, which contain more dissolved material. However, net inward water movement is limited by the restraining force of the cell wall which limits cell expansion. Thus, there is a pressure relationship between a plant cell that tends to expand due to osmotic water entry and the restraint of its cell wall (see the discussion of turgor pressure on p. 92). The concept of water potential is especially useful when considering water movement in plants because it includes both osmosis and other pressure considerations.

Pure water is defined as having a water potential of zero at a given pressure; at that same pressure, all aqueous solutions are defined as having negative water potentials. Water tends to move from a region of high (relatively less negative) water potential to a region of low (relatively more negative) water potential (figure 9.5).

We can now apply the terminology of water potential in describing the transpiration stream through a plant. Soil water is usually such a dilute solution of mineral ions that its water potential is high (only slightly negative). Under virtually all conditions (except when the air is saturated with water vapor) the water potential of air is much lower (much more negative) than that of soil water. Thus, water moves from the soil through a plant to the atmosphere surrounding the plant's leaves.

With this in mind, we can now look more closely at the movement of water through plants, beginning with soil water and water entry into roots.

Soil and Soil Water

Soil water usually forms a thin film on the surface of soil particles and fills some of the smallest spaces among them. Water is held here by **capillarity,** the tendency of water to enter and adhere to the walls of small spaces. Following a rain, all of the spaces among soil particles fill with water, but with time, most of this rainwater percolates (drains) down through the soil, once again leaving air in many of the spaces.

The water-holding capacity of a soil depends on the kinds of particles present. For example, sandy soil with its large particles (20 to 2,000 μm in diameter), has less total available particle surface area than does silty soil with its numerous smaller particles (2 to 20 μm in diameter) or clay (particles less than 2 μm in diameter). The larger particles of sandy soils are also more loosely packed and have fewer small, water-holding spaces (capillary spaces) between particles. Thus, a dried-out, sandy soil provides only a barren and hostile environment for plant growth because its water-holding capacity is low.

A soil's water-holding capacity is not the only factor affecting its suitability for plant growth; root cells must have oxygen for respiration. In clay, for example, the very tight packing of small particles can reduce air spaces and gas diffusion to the point that aeration is not adequate for root respiration. Oxygen supply is also a problem for plants growing in water or in soil continuously saturated with water (figure 9.6).

For optimal plant growth, soil should have a mixture of various-sized particles so that there are enough small capillary spaces to hold adequate quantities of water and enough larger air spaces to provide adequate oxygen for roots.

Roots and Root Systems

One of the most striking characteristics of roots is that they are not static; they grow continuously. Growing root tips constantly move through the soil and squeeze their way into new capillary spaces among soil particles where moisture may be more plentiful.

In all young plants root growth begins with the growth of a single **primary root.** In some plants, the primary root grows straight downward and remains the dominant root of the plant, with much smaller **secondary roots** growing out from it. This arrangement is called a **taproot system** (figure 9.7a). In other plants, a number of slender roots develop, and no single root dominates. These slender roots and their lateral branches make up a **fibrous root system** (figure 9.7b).

Figure 9.5 All water movement in plants can be described in terms of movement of water from areas with relatively less negative water potential to areas with relatively more negative water potential.

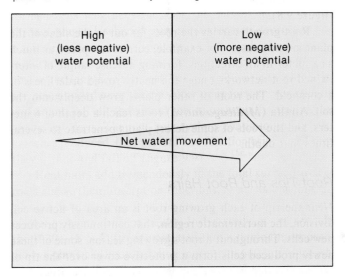

High (less negative) water potential	Low (more negative) water potential
Net water movement →	

Figure 9.6 The root system of the black mangrove obtains oxygen in waterlogged soil by means of special roots called pneumatophores that protrude above the surface. These roots have air channels through which oxygen diffuses to underground root tissues.

Figure 9.7 Root systems. (a) The taproot system of a dock plant. (b) The fibrous root system of a zinnia.

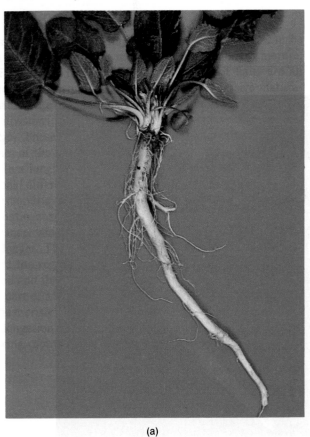

(a)

(b)

Figure 9.21 *Rhizobium* in root nodule cells. (a) Electron micrograph of bacteroids of *Rhizobium* inside vacuoles in a nodule cell of a soybean (magnification × 14,000). (b) Diagram showing the relationship between a *Rhizobium* bacteroid and the enclosing nodule cell. Nitrogenase actually is a complex of several proteins. NH_4^+ is incorporated into organic molecules before it is exported from the vacuole around the bacteroid. Nitrogen fixation is costly in terms of ATP; estimates for the number of ATP's used to fix one nitrogen atom range from 6 to more than 20.

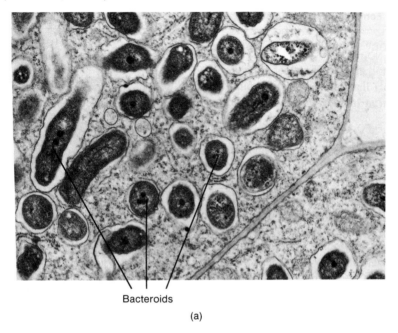

Bacteroids

(a)

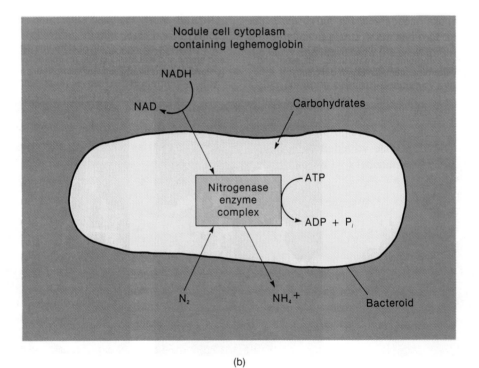

(b)

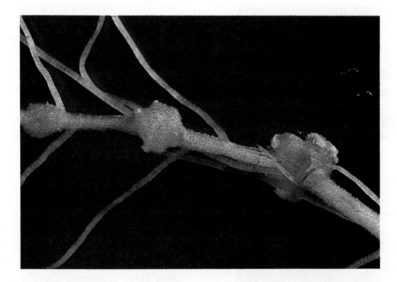

When a nutrient in short supply (sulfur, for example) cannot be mobilized and transported from one part of a plant to another, deficiency symptoms appear first in young, growing tissues, such as the newest leaves rather than in older tissues that may already contain adequate quantities of the nutrient.

Symbiosis and Mineral Nutrition

Nitrogen in a usable form is one of plants' most important nutritional requirements. Because nitrogen is a plentiful substance (about 78 percent of the air is nitrogen), this requirement would not seem to present any particular problem for plants. However, plants cannot use molecular nitrogen (N_2) directly; they require it in the form of nitrate (NO_3^-) or ammonium (NH_4^+) ions. Therefore, plants are dependent on the activity of soil microorganisms that incorporate atmospheric nitrogen by the process of **nitrogen fixation.** After nitrogen fixation has occurred, usable nitrogen compounds (NO_3^- and NH_4^+) are available for both the microorganisms and the plants.

Many of these nitrogen-fixing microorganisms are free-living in the soil, but several interesting symbiotic relationships (functional interdependencies) between plants and nitrogen-fixing microorganisms are important in the mineral nutrition of some vascular plants. For example, plant tissue and nitrogen-fixing bacteria are found together in swellings called **nodules** that develop on the roots of legumes, plants such as beans and peas. Bacteria of the genus *Rhizobium* enter root cells by a complex process in which the plant plays an active role. During the first ten days after infection, root cells are stimulated to divide and nodules develop. Membrane envelopes form and enclose groups of enlarged bacteria (bacteroids) inside vacuoles in root nodule cells (figure 9.21). These nodule cells produce *leghemoglobin,* an oxygen-binding substance similar to the hemoglobin found in animal blood (figure 9.22). Apparently, the binding of oxygen by leghemoglobin prevents oxygen from interfering with the nitrogen-fixing activity of the **nitrogenase** enzyme complex of the bacteria, which involves reduction reactions. These nitrogen-fixing bacteria convert molecular nitrogen (N_2) into forms required by the plant. At the same time, the plant supplies reduced organic compounds that meet the energy requirements of the microorganisms.

Another nutritionally important symbiotic relationship is the association of nonpathogenic fungi with roots in **mycorrhizae** (singular: **mycorrhiza,** "fungus-root"). The fungi envelop large parts of root systems, including virtually all absorbing surfaces. All materials absorbed into the roots first pass through these fungal sheaths. The fungus receives organic nutrients from the host plant. A mycorrhizal infection is beneficial to a plant in situations where nutrients are deficient or the plant faces strong competition from other organisms, because the fungus increases the surface area

Figure 9.23 A fly ensnared in a Venus fly trap. Venus fly traps and a few other plants, such as pitcher plants and sun dews, obtain nitrogen by trapping and digesting animals.

available for mineral and water uptake. The mycorrhizal relationship seems particularly important for certain trees, and some trees grow very poorly (or not at all) if the fungus is not present. Foresters deliberately infect some trees with mycorrhizal fungi when they introduce them into new habitats.

A few plants supplement their nitrogen supply in an entirely different way (figure 9.23). They trap and digest animals, thus obtaining a concentrated source of nitrogen from animal proteins.

Mineral Nutrition and Modern Agriculture

Mineral salts that are ionized in soil water are derived originally from the weathering of rocks; this physical and chemical breakdown of rocks accounts for the origin of most of the important mineral nutrients. However, one nutritionally important element—nitrogen—is not available in rocks, and nitrogen nutrition is of central importance to plant productivity.

The basic source of nitrogen is atmospheric nitrogen gas (N_2). However, as we said earlier, plants cannot absorb and use N_2; they are dependent upon the activity of microorganisms that incorporate nitrogen through the process of nitrogen fixation, which yields usable nitrogen compounds (NO_3^- and NH_4^+). Once nitrogen is incorporated into living material, however, it moves from organism to organism as part of a *nitrogen cycle* (see chapter 35). Nitrogen compounds in organisms are returned to the soil by decay processes following death.

Other elements also are recycled constantly through decay processes that return them to the soil after plants and animals die. Despite these natural replacement processes, intensive modern agriculture depletes soil mineral elements so drastically that they must be replaced through other means to maintain productivity of the land.

Nitrogen, phosphorus, and potassium are the three macronutrients that most frequently must be supplemented by fertilization of agricultural soils. Fertilizers are commonly labeled with a formula indicating the percentage of each of these three elements. The formula *20–10–10* indicates that the fertilizer contains 20 percent nitrogen, 10 percent phosphoric acid, and 10 percent soluble potassium (as K_2O). Often, a fourth element, calcium, is included in fertilizers, and occasionally other nutrients are added as well.

Fertilization, however, is not the simple solution it appears to be for replenishing these elements. Heavy nitrogen fertilization is absolutely essential for maintaining high levels of agricultural production, but the energy input required to produce and deliver nitrogen fertilizers is becoming a critical problem. Producing reduced nitrogen fertilizers consumes large amounts of fossil fuel, and transporting them from production sites to fields requires additional heavy energy input.

New technologies may bring new solutions in the future. For example, many biologists are now working with genetic engineering techniques in the effort to transfer the complex of bacterial genes responsible for nitrogenase system proteins into plant cells, but this is a difficult task that may take many years to accomplish. In the near future, however, the constraints of soil chemistry and plant mineral nutrition coupled with increasing energy costs and decreasing fuel reserves will continue to complicate the problem of meeting the food requirements of our growing world population.

Summary

Plants are autotrophic organisms; they can produce all the organic compounds they require using only inorganic substances. Vascular plants have regions specialized to obtain necessary materials from the environment. They obtain water and mineral ions from the soil through root systems. Leaves absorb light energy and are the sites of photosynthesis, the process by which carbon dioxide from the atmosphere is incorporated into organic compounds. Vascular tissues—xylem and phloem—connect roots and leaves and function in the transport of materials throughout a plant's body.

Carbon dioxide diffuses into leaves through open stomata, but stomatal openings also allow the evaporative loss of water from leaves, which is called transpiration. This lost water is replaced by the movement of large quantities of water up through the xylem from the roots. This transpiration stream also carries minerals absorbed by the roots up through the plant.

Water potential, a measure of the tendency for net water movement, is useful in describing the transpiration stream. Because the water potential of the atmosphere is lower (more negative) than that of soil water, water tends to move into roots, upward through a plant, and out into the atmosphere surrounding leaves.

Soil water is usually located in the small spaces among soil particles. A soil's water-holding capacity and amount of aeration affect its suitability to support growing plants.

An important part of root function is the continuous growth of root tips into new soil areas. Along a root's region of maturation, root hairs increase the absorbing area of a root; they function only briefly, but are constantly being replaced.

Roots are specialized for absorption and transport. The epidermis, with its numerous root hairs, exposes a relatively large surface area to soil water. The bulk of the cortex is composed of loosely packed cells that are surrounded by a network of intercellular air spaces. The innermost layer of the cortex, the endodermis, has waxy endodermal cell walls that fit together snugly and form a barrier, the Casparian strip, between the cortex and the stele. The stele lies in the center of the root and contains xylem and phloem.

Water moves readily through the apoplast of the cortex, but the only route from the cortex into the stele is through the cytoplasm of endodermal cells.

In some cases, mineral ions pass through the plasma membranes of root cells by simple diffusion or facilitated diffusion, but the bulk of ion uptake is accomplished by active transport, which requires ATP energy input and can work against concentration gradients.

Ions diffuse through the cytoplasmic network of the cortex and endodermal cells into the stele. Ions probably are actively transported into the apoplast of the stele where the transpiration stream carries them into and up through the xylem. Some ions, such as nitrate, are first incorporated into organic compounds before they are transported through the xylem.

Mineral nutrient requirements have been studied by chemical analysis of plants and by hydroponics experiments. The macronutrients have been known for some time and their deficiency symptoms are well recognized, but micronutrient requirements have been more difficult to determine. Plants deprived of various required nutrients often show characteristic deficiency symptoms.

Several kinds of symbiotic relationships contribute to nutrient uptake by roots. Nitrogen-fixing microorganisms live symbiotically in root nodules of some plants and provide a supply of usable nitrogen compounds to the plants. Fungus tissue enclosing roots in mycorrhizae increases the efficiency of ion absorption.

Nitrogen deficiency commonly limits plant growth and productivity. Intensive modern agriculture requires that the natural usable nitrogen supply be supplemented, but economic and energy problems are affecting our ability to provide nitrogen fertilizers.

Questions for Review

1. How much of the 74 kg of dry weight gained by van Helmont's growing willow plant was made up of carbon and oxygen?

2. Explain the positive aspects of transpirational water loss.

3. Why would millet be a more desirable food calorie producer for a semiarid area than potatoes?

4. What is water potential? Use this term in describing the transpiration stream through a plant.

5. Explain (a) the relationship between soil particle size and soil water-holding capacity and (b) the significance of soil particle size and aeration.

6. Explain the significance of the Casparian strip in the movement of water and mineral ions within the root.

7. Describe the structure and functional role of the symplast (cytoplasmic network) in roots.

8. Name the elements classified as macronutrients.

9. Explain the difficulties involved in determining whether or not particular elements are micronutrients.

10. What do the numbers 27–5–5 on a bag of lawn fertilizer mean?

11. Why do plants often suffer from nitrogen deficiency even though nitrogen is plentiful in the environment?

Transpiration Pull/Cohesion of Water Hypothesis

Root pressure alone, however, is simply not strong enough to move water very far upward in a tall plant. But if root pressure does not generate enough force to raise water to the tops of large plants, then how is it done?

Near the beginning of the eighteenth century, Stephen Hales, an English botanist, suggested that transpiration pulls water up through plants. Though this idea was debated for some time, it is now widely accepted. The overall driving force for this pull is the difference in water potential between the air around leaves and the soil water around roots. Dry air has a great water-holding capacity that strongly promotes evaporation. But how can evaporating water molecules that pull away from leaf cell surfaces exert a pull sufficient to draw water all the way up through the xylem from the soil below?

Water is raised through vessels and tracheids as a result of the pull applied to water at the top of the xylem and because of the very strong tendency of water molecules to "stick together." This attraction among molecules of the same kind is called **cohesion.** In liquid water, the molecules do not simply float randomly around one another; there is considerable attraction between the hydrogen atoms of one water molecule and the oxygen atom of an adjacent molecule. This hydrogen bonding (see p. 29) results in a network of mutually attracted water molecules.

The cohesive force among water molecules is so great that we can think of the column of water inside a xylem vessel as behaving almost like a chain of water molecules that can be pulled up through the xylem. In fact, it is easier to pull apart the molecules in fine wires made of some common metals than it is to disrupt a column of water in a small-diameter, airtight tube. Cohesive forces among water molecules are great enough to hold a column of water together even when it is being pulled with enough force to raise it to the top of a large tree. Furthermore, water also tends to adhere to the walls of xylem vessels. This **adhesion,** an attractive force between unlike molecules, is due to hydrogen bonding between water and cellulose in vessel walls. Adhesion helps to prevent air bubble formation when the column of fluid in the xylem is subjected to tension by pulling from above.

In 1894, Dixon and Joly coupled the idea of transpiration pull with knowledge about the cohesion of water and put forth the **transpiration pull/cohesion of water hypothesis,** which states that evaporation pulls away some of the molecules adhering to cellulose in the walls of leaf mesophyll cells. Water is then drawn from within those leaf cells to replenish the surface water films. This lowers the water potential inside the leaf cells and causes osmotic movement of water from the xylem into the leaf cells. Thus, the pull of transpiration is transmitted in a chain reaction as water is drawn through leaf cells from the xylem in the leaf veins. Because the xylem contains continuous tubular structures running from the leaf veins down through the stem to the roots, this pull (tension) is transmitted down through the entire plant, so that whole columns of water are literally pulled up from the top.

Given that the cohesion of water molecules could hold together a tall column of water under tension, researchers have tried to demonstrate that transpiration applies a strong enough pull to lift water columns to the height required in tall plants. Transpiration pull can be measured experimentally using either a living shoot or a physical apparatus (figure 10.14). To quantify transpiration pull, a pan filled with mercury is placed under a tube that holds a twig with leaves or a tube with a porous clay bulb. Mercury rises in the tube as the tube is evacuated by the withdrawal of water from the top, and the lifting force of transpiration pull can be expressed in terms of the height to which the mercury column is lifted. In the case of simple suction (figure 10.14b[1]), the mercury column rises to a height of only 76 cm (the height supported by one atmosphere of pressure). However, when a transpiration pull is applied, either in the form of a twig with leaves or a porous clay bulb (figure 10.14b[2 and 3]), the mercury column rises considerably higher. These data indicate that the pulling force of transpiration is more effective than suction in raising a column of fluid.

Even if there were a mechanism by which plants could apply suction at the top of the xylem, the column of fluid could rise no higher than the height supported by one atmosphere of pressure. Transpiration pull, however, provides an adequate pulling force to raise water to the treetops.

Unfortunately, it is extremely difficult to measure the actual pulling force inside xylem vessels of living plants; the process of measuring disrupts the system so much that the validity of the measured values must be doubted. Most biologists agree that adequate pulling force can be generated by transpiration and that cohesive forces among water molecules are adequate to hold a column of liquid together under tension *if* the columns of liquid in the xylem are continuous. Many xylem tubes, however, seem to have at least some space occupied by water-saturated air. However, at any given time, continuous water columns exist in an adequate proportion of the xylem vessels to supply the needs of the plant. Also, when an air bubble blocks the flow through a given xylem vessel, water can move through pits in the lateral walls of xylem elements and pass from that vessel to a nearby vessel, and then continue its upward flow.

Figure 10.14 Experiments on water movement through vascular plants. (a) The transpiration rate in an intact plant is measured by an air bubble's movement through the calibrated tube of a manometric device. The water supply in the dish can be replenished from the reservoir at the same time. (b) Measurement of the forces involved in transpiration pull: (1) Simple suction—Mercury is raised 76 cm up an evacuated tube by atmospheric pressure. (2) Transpiration pull—Leaf transpiration pulls water through the xylem of a living stem, and mercury rises considerably higher. (3) Transpiration pull—A physical model with water evaporating from a porous clay cup also exerts more pulling force than simple suction.

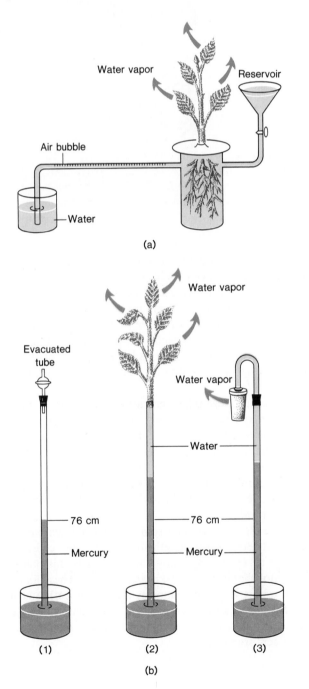

(a)

(b)

The number of fluid-filled xylem vessels varies with conditions and the time of year; for example, water content of stems is usually higher in spring and summer than in fall. In fact, loggers find that some kinds of logs float at certain times of the year but not at others when a high percentage of xylem vessels are filled with water.

Thus, a century after it was proposed, the transpiration pull/cohesion of water hypothesis still seems the best explanation for fluid movement through the xylem. It seems clear now that evaporative water loss from leaves is not just a necessary evil that plants must endure in order to obtain carbon dioxide for photosynthesis. Transpiration is vital to the life of plants because it provides the pulling force that draws xylem sap up from the roots to the upper parts of the plant. Yet intriguing questions about xylem transport remain, and further research is needed.

Phloem

During early spring growth, organic solutes, especially sucrose, are translocated from plant storage areas to young, actively growing areas of the plant that are not yet synthesizing adequate quantities of carbohydrates to meet their own needs. Later in the season, organic molecules produced in mature leaves are available for storage and are translocated to storage tissues, which develop in modified stems or roots, or to seeds, which also become very significant storage depots. How are these organic compounds moved from their **sources** (the production sites in the leaves) to **sinks** (the actively growing areas or storage tissues of seeds, roots, or stems)?

Phloem was identified as the translocating tissue long ago when the results of **girdling** (removal of a ring of bark) were analyzed. Girdling removes the phloem while leaving the xylem intact, and it eventually kills a tree. If a tree is girdled below the level of most of its leaves, the bark, with the underlying phloem, swells just above the cut, and sugar accumulates in the swollen tissue. This finding led biologists to the conclusion that the phloem is the carbohydrate translocating tissue.

Radioactive tracer studies have confirmed this. When ^{14}carbon-labeled CO_2 is supplied to mature leaves, radioactively labeled sugar produced in the leaves is soon found moving down the stem and into the roots, and this labeled sugar is found mainly in the phloem, not in the xylem.

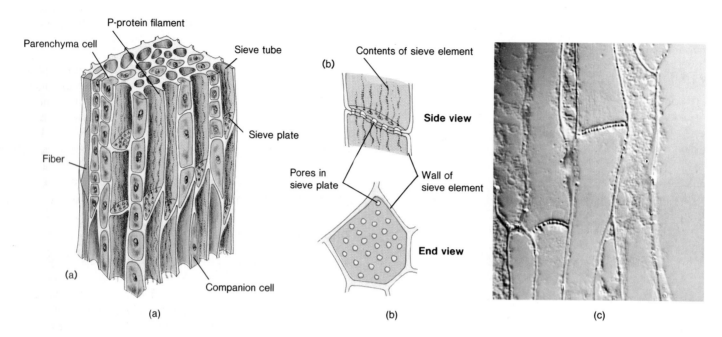

Other radioactive isotope studies have confirmed the phloem's role in transport of other organic molecules, such as amino acids, and in mineral ion transport. Mineral nutrients can be mobilized in one part of the plant and moved to another part. For example, some mineral nutrients are translocated out of leaves before the fall in the autumn. In some cases, hormones also are transported through the phloem from their production sites to target areas where they exert their regulatory influences.

Phloem Cells

Material moved through the phloem is translocated through **sieve tubes. Sieve tube elements,** the cells that make up sieve tubes, are highly specialized cells arranged in linear arrays running vertically through the length of the phloem. In flowering plants, sieve tube elements are lined up end-to-end with flat areas of contact where their end walls abut one another (figure 10.15).

There are prominent holes in the end walls of sieve tube elements, and adjacent cells contact each other directly through those holes. Because of its sievelike appearance, this contact area is called the **sieve plate.** (The name sieve tube also is derived from this structure.) Cytoplasmic connections run through the pores of each sieve plate, connecting the cytoplasm of the sieve elements adjacent to the plate. Thus, there is cytoplasmic continuity from cell to cell in the sieve tubes.

Individual sieve tube elements are structurally different from the majority of plant cells, because a mature sieve tube element has neither a vacuole nor a nucleus. Thus, functional phloem tissue contains end-to-end rows of nonnucleated, nonvacuolated, but still-living sieve tube elements. This special structural arrangement is very different from that of xylem vessels and tracheids, which consist of empty, nonliving cell walls.

Sieve elements contain some special filamentous structures that seem to run lengthwise through the cells. When the sieve tubes are physically damaged, these filaments disperse into what appear to be beaded chains. These filaments are known as **P-protein filaments** (P for phloem), and some researchers think that they are involved in phloem transport. Most other biologists, however, favor a different hypothesis for phloem transport.

Figure 10.16 Aphids. (a) An aphid feeding on a plant stem. A droplet of "honeydew" is being exuded from the animal. (b) This section of the stem shows that the aphid stylet has penetrated to a phloem sieve tube cell. Biologists have studied phloem function by cutting aphids away from their stylets. The stylets remain as small open channels into the phloem.

(a)

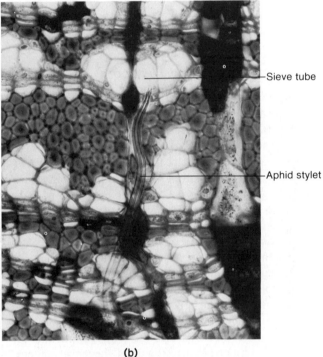

Sieve tube

Aphid stylet

(b)

Because sieve tube elements lose their nuclei during their differentiation, they are irreversibly specialized for translocation. They are no longer capable of carrying out nucleus-directed protein synthesis or other functions controlled by nuclear genetic information, but continuing genetic control of sieve tube activity may come from another source. Each sieve tube element has one or more **companion cells** adjacent to it. Unlike sieve tube elements, companion cells possess all of the normal cellular components; they have nuclei and vacuoles as well as other organelles normally found in plant cells. It seems likely that the companion cells exert some control over sieve tube functioning, and almost certainly, companion cells can transfer material into sieve tubes through pores that connect them.

In addition to sieve tubes and companion cells, phloem tissue, like xylem tissue, contains sclerenchyma fiber cells that serve a supporting function and thin-walled parenchyma cells.

Studying Phloem Function

Phloem function is difficult to study because the cells involved are delicate and easily damaged. The slightest disturbances of sieve tubes lead to several types of physiologically preprogrammed injury responses that shut down phloem translocation and conserve phloem sap. In addition to the formation of beaded chains of P-protein filaments, a slimy plug made of a substance called **callose** (a glucose polymer) also develops in each plate pore when the phloem is injured. The size of the callose plug formed in experiments varies dramatically, depending upon the way in which the phloem cells are killed. The callose plug apparently develops in injured phloem in nature just as it does in experiments.

Because injury responses in phloem occur quickly, researchers studying phloem function needed to find a way to study phloem without injuring it. One of the most clever methods of doing this was first applied by T. E. Mittler and J. S. Kennedy in the 1950s. Their method takes advantage of the feeding habits of **aphids,** small insects that insert long, pointed feeding devices, **stylets,** into the phloem and feed on phloem sap. The aphid stylet penetrates the tissue lying over the phloem and terminates in an individual phloem sieve tube (figure 10.16). An aphid stylet can enter a sieve tube without causing the injury response. Thus, an aphid can remain "plugged in" to a sieve tube and feed on phloem sap at leisure for hours. In experimental studies, aphids are allowed to begin feeding and then are anesthetized and cut away from their stylets. This leaves experimenters with small open channels through which phloem sap can be withdrawn from sieve tubes without disturbing phloem function and causing an injury response.

Characteristics of Phloem Transport

Any explanation of phloem transport must account for the movement of fairly large amounts of organic material over long distances (many meters in some trees) in relatively short periods of time.

Calculated rates of translocation indicate that the rates and amounts of organic material moved in plants cannot be explained by simple diffusion. For example, Dixon and Ball measured the total amount of carbohydrate in potato tubers and the total cross-sectional area of the phloem in the stem through which the stored carbohydrate had moved. They assumed that the phloem sap was a 2 percent sugar solution and then calculated a minimum flow rate required to deliver the amount of carbohydrate through the phloem to the tubers during the time the tubers were growing. They concluded that diffusion alone was several thousand times too slow to account for movement of the required amounts of carbohydrate. Actually, their estimate of the sugar content of the phloem sap was too low because phloem sap contains up to 30 percent sucrose in some plants, but even this correction does not alter the conclusion that diffusion is an inadequate explanation for phloem transport.

Several experimental techniques have been used to measure the rate of fluid movement in the phloem directly. If small sections of phloem are heated slightly, sensitive thermocouples (temperature-measuring devices) applied further down the stem can detect the arrival of the warmed sap. In other experiments, the rate of movement of [14]carbon-labeled sugar has been determined by analyzing sap withdrawn through aphid stylets at two levels of the stem. Both of these techniques for measuring rates of movement, as well as other studies, have yielded rates for phloem transport that are many times faster than simple diffusion. Materials appear to be moved through phloem at point to point rates of 60–100 cm/hr and possibly up to 300 cm/hr.

Over the years, various researchers have proposed that **cyclosis,** the streaming movement of cytoplasm, occurs in sieve elements and augments diffusion in phloem transport, but even the highest estimates of cyclosis movement rates are inadequate to explain the observed rates of transport through the phloem.

Another property of phloem transport must also be accounted for. The flow appears to be one-way in each sieve tube, but within a single bundle of phloem sieve tubes, some tubes may be translocating material in one direction while sieve tubes nearby are carrying material in the opposite direction. Also, an individual sieve tube can reverse its direction of transport.

A factor that clouds interpretation of direction studies is that there is some lateral movement of materials between adjacent phloem sieve tubes. There is even a little lateral movement between xylem and phloem, with materials moving through the cambium layer. Some of the confusing observations that suggest that some materials seem to move simultaneously in opposite directions through the same sieve tubes might be explained by this lateral transfer between various conducting elements. Despite these problems in interpretation, it is generally agreed that transport within each sieve tube involves one-way flow.

The Pressure-Flow Hypothesis

The most widely accepted hypothesis explaining phloem transport is called the **pressure-flow hypothesis,** first proposed by Münch in 1927. The hypothesis assumes a system bounded by selectively permeable membranes in which solute concentrations at two separated points in the system are significantly different (figure 10.17a is an experimental model of such a system). When this system is submerged in distilled water, water tends to enter osmotically through both selectively permeable membranes, but the tendency is much stronger at membrane (1) because the osmotic gradient across membrane (1) is so much greater. Thus, water moves into (1) much more rapidly than into (2), and pressure differences in the system soon result in a flow of water from (1) to (2). This flow occurs because pressure drives water out through the membrane of (2), even though water is being forced to move against a water potential gradient. As water flows through the system it sweeps along solute. This bulk flow of water tends to drive the model system toward equilibrium in a relatively short time.

Figure 10.17 Phloem transport. (a) Model system illustrating the pressure-flow hypothesis. (b) Proposed mechanism of flow of water and solute through the phloem. Pink arrows represent active transport of sucrose; blue arrows represent flow of water. At a source (for example, an actively photosynthesizing leaf), sucrose is actively transported into phloem sieve tubes. At a sink (for example, in a root), sucrose is actively transported out of phloem sieve tubes. So much water enters the source end of the phloem by osmosis that water is forced out the sink end. This results in flow of water through the phloem that sweeps sucrose along with it.

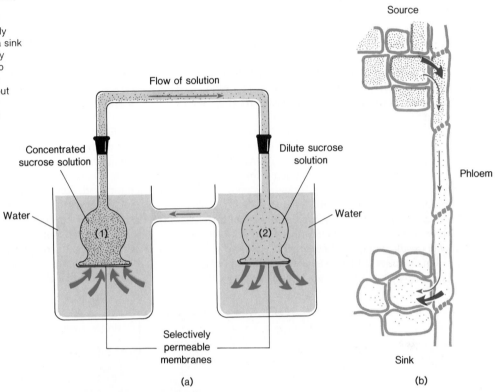

(a)

(b)

The same principles can be applied to the pressure-flow hypothesis of phloem transport. The hypothesis states that the activities of living cells in phloem can maintain the concentration differences between different parts of the system, so that instead of moving toward equilibrium as the model system illustrated in figure 10.17a does, the differential is maintained and bulk flow continues. During vigorous photosynthetic activity, for example, the leaf can serve as a "source" of sucrose that is loaded into the phloem sieve tubes by active transport across sieve tube element membranes (figure 10.17b). Then, at another point in the phloem, active transport processes remove sucrose from the sieve tubes and move it toward a storage "sink" in the starch-storing plastids of cells in storage areas such as roots.

Thus, you can think of the phloem sieve tubes as fluid-filled conduits, bounded by selectively permeable membranes, running through the plant from solute source to sink. A relatively more concentrated sucrose solution is maintained by adding sucrose at one end of the phloem transport system, and actively removing it at the other end. Thus, concentration differentials are maintained by energy-requiring active transport, and the bulk flow of phloem sap continues. The plant works to maintain a situation in which a spontaneous physical response sets up a flow through the system. Such a continuing flow would sweep along solutes at rates that biologists calculate would be adequate to agree with experimentally determined phloem flow rates.

Box 10.2
Proton Pumps That Power Transport

It has been known for some time that energy for active transport of solutes across cellular membranes is provided by the hydrolysis of adenosine triphosphate, the cellular energy currency. The breakdown of ATP to ADP (adenosine diphosphate) and phosphate is coupled with energy-requiring active transport such as that which occurs in phloem loading and unloading, in ion uptake in roots, and in many other places in plants. Now a fascinating new concept of how ATP is actually used to provide energy for active transport is emerging.

Recall that the energy potential of a hydrogen ion (proton) electrochemical gradient across membranes is thought to be employed for ATP production in chloroplasts and mitochondria (see pp. 145 and 176). There proton movement down an electrochemical gradient is coupled by an enzyme system with phosphorylation of ADP to make ATP.

It now appears that an almost exactly opposite set of processes is employed in cells to apply energy gained from ATP hydrolysis in the active transport of solutes. In a number of active transport processes that have been investigated, an ATP hydrolyzing enzyme system (ATPase) couples the breakdown of ATP with active proton pumping to produce an electrochemical gradient across a membrane. This electrochemical gradient then directly provides energy for active transport of solutes across the membrane.

For example, some actively transported substances are transported by membrane carrier molecules that function by simultaneously carrying one molecule of solute and one proton. This kind of transport, called cotransport, can continue as long as there is a proton gradient across the membrane; and it is ATP-driven proton pumping that maintains the gradient.

Other transport mechanisms also can be driven by a proton gradient, and it appears that this mechanism may be a widespread, if not universal means of applying the energy potential of ATP to active transport processes.

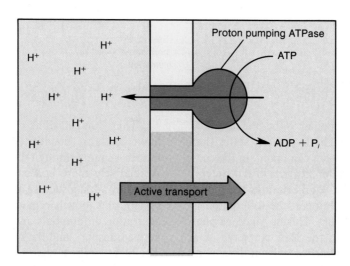

Box figure 10.2   Diagram showing the establishment of an electrochemical gradient across a membrane through proton pumping by ATPase. The proton gradient provides energy directly for active transport of solutes.

The pressure-flow hypothesis can also account for reversal of the flow through any given chain of sieve tube elements. All that is required is a reversal of the source-sink relationship; that is, sucrose is loaded into the opposite end of the phloem and actively removed from the end that previously was the source.

Many intriguing questions about the role of the companion cells in all of this remain unanswered. In the mature leaf, which is a source area, the ratio of the relative diameters of companion cells and sieve tubes is about 1:1; in midstem, on the other hand, the ratio is something like 1:100. The significance of this differential is not clear, but the nucleated companion cells, which supposedly regulate phloem function, are much more prominently developed in the area where the sieve tubes are being loaded with sucrose. It is very likely that companion cells help move sucrose from production sites in photosynthesizing leaf mesophyll cells to the sieve tubes, where it is loaded for transport.

If active, energy-requiring processes maintain the concentration differentials at the source and sink ends of phloem-conducting elements, and if these concentration differentials account for the maintenance of bulk flow through the phloem sieve tubes, then do the sieve tubes serve simply as living, but passive pipes through the plant? Maybe not, because if metabolic poisons are applied in a ring around a tree or if a ring is heated with steam, flow through the phloem stops. This implies that living, metabolically active phloem tissue is required for phloem transport to continue. However, the injury shutdown response of phloem could account for these effects of metabolic poisons as well. Clearly, there still is much to be learned about phloem transport and its control.

Summary

Water and mineral nutrients absorbed from the soil by roots are transported up through the xylem to aboveground parts of the plant. Carbohydrates and other organic molecules produced in leaves are transported to other plant parts through the phloem. Phloem also translocates inorganic ions, amino acids, and hormones from one part of the plant to another.

Stems are important transporting and supporting structures. Characteristically, annual and biennial plants have herbaceous stems, and perennial plants have woody stems.

Dicots and monocots are two major groups of flowering plants that have fundamental differences in stem structure, as well as other structural differences.

Vascular bundles separated by parenchyma are scattered throughout monocot stems. In herbaceous dicot stems, vascular bundles are arranged in a ring lying between the cortex and the large central pith. Dicot vascular bundles contain xylem and phloem separated by cambium.

Woody stems persist from year to year. Primary growth, initiated by buds, adds to stem length each growing season. Secondary growth, due to proliferation of cambium, increases the thickness of stems by adding new xylem and phloem tissue.

Each season's xylem growth accounts for one growth ring. Old phloem tissue is pushed outward and crushed against the bark that covers the stem surface. Only the phloem produced during the current growing season is functional transporting tissue.

Xylem contains two basic types of conducting tubes, the tracheids and the vessels. Both consist of empty cell walls when they are mature. Tracheids are elongate, spindle-shaped cells lined up with their ends overlapping. Water moves from one tracheid to the next through thin-walled pits. Xylem vessels are end-to-end rows of cell walls that are open at their ends to produce "cellulose pipes" running through plants.

The tendency of water to enter roots results in root pressure that can cause fluid to rise a short distance through the xylem.

The current explanation for the rise of water through the xylem is the transpiration pull/cohesion of water hypothesis. This hypothesis states that water evaporating from leaf cell surfaces pulls water out of leaf cells. The cohesion of water molecules causes this pulling force (tension) to be transmitted to the xylem where it pulls fluid upward from the roots.

Phloem translocates material from sources to sinks. The material moves through sieve tubes that consist of living but nonnucleated cells. Sieve tube elements have cytoplasmic connections through holes in the sieve plates located at the ends of cells. Nucleated companion cells apparently exert control over sieve tube functioning.

Phloem translocation is too rapid to result from diffusion and cyclosis (cytoplasmic streaming). Phloem transport seems to be best explained by the pressure-flow hypothesis, which proposes that solute is actively transported into phloem sieve tubes at a source and actively transported out of them at a sink. Water tends to enter sieve tubes osmotically at the source end with such pressure that it is forced out at the sink end of the tubes, and thus the flow is maintained. This flowing water carries solutes with it.

The liver also plays an important role in processing the body's extra amino acids. During times of adequate nutrient supply, there may be excesses of particular amino acids. Liver cells deaminate these amino acids to produce usable carbohydrates. Ammonia, a product of deamination, is quite toxic and must be incorporated into **urea** molecules by liver cells to prevent its buildup (chapter 14).

In addition to these and other metabolic functions, the liver collects and breaks down many toxic substances that enter the body, such as ethanol, many drugs, and other compounds.

The pancreas, like the liver, is a functionally complex organ that not only makes digestive contributions but also contains specialized cell populations, known as the **islets of Langerhans,** that produce the hormones **insulin** and **glucagon.** These pancreatic hormones are involved in the control of blood glucose level and several related aspects of metabolism. We will examine the pancreatic hormones in more detail in chapter 15.

Problems with the Digestive Tract

The gut tube, for all its twists and expansions, is really just a tunnel through the body. In the strict sense, the contents of the gut are outside the body and remain so unless they pass across cell membranes of gut-lining cells. Indeed, the contents of the gut must not come into direct contact with the blood or other interior tissues of the body, because the chyme contains bacteria and other potentially hazardous components. The nature of this risk is graphically illustrated by two rather common digestive ailments—appendicitis and ulcers.

The **vermiform appendix,** or just appendix for short (figure 11.25), is a short, blind sac off the beginning of the colon, just past its junction with the small intestine. The human appendix contains lymphoid tissue and might, therefore, be involved in resistance to infection, but its function is poorly understood. The appendix all too often becomes infected and develops an inflammation called **appendicitis.** Once severe appendicitis is diagnosed, standard medical procedure calls for surgical removal of the infected appendix. If untreated, the inflamed appendix can rupture, allowing gut contents to come into contact with the **peritoneum,** the tissue that covers all of the digestive organs and lines the body cavity. The material that escapes from a ruptured appendix causes **peritonitis,** a life-threatening infection.

Ulcers are breaches, of varying depth and position, in the wall of the gut. Most are located in the duodenum and are called **duodenal ulcers.** Somewhat more rarely, **gastric ul-**

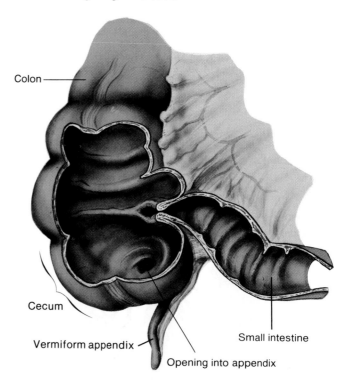

Figure 11.25 The vermiform appendix is a short blind sac off the cecum at the beginning of the colon.

Colon

Cecum

Vermiform appendix

Opening into appendix

Small intestine

cers may form in the stomach itself (figure 11.26). Ulcers are caused by the combined effect of acid and enzymes eroding the protective lining of the gut. However, what is not well understood is why in some individuals the normal mechanisms that protect the linings of the digestive tract function adequately for a lifetime, and in others they do not. It is widely believed that worry, stress, or frustration can contribute to ulcer development because gastric juice secretion is at least partly controlled by the nervous system. In fact, one of the more drastic treatments for serious ulcers is to cut the vagus nerve, the nerve that stimulates stomach secretion.

Most ulcers are rather mild, albeit painful, abrasions in the gut lining. Healing can be rapid because the cells lining the gut tube are quickly replaced, but the causative features (often stress) must be removed and some patients may switch to a bland diet to reduce acid secretion. The real dangers come when ulcers begin to bleed or when they perforate (produce a hole directly through the gut wall). Peritonitis can develop following perforation unless prompt medical care is received. Such serious cases require removal of the source of stomach acid. This can be accomplished by cutting the vagus nerve or by removing part of the stomach. An interesting alternative technique accomplishes the same thing without

Figure 11.26 An ulcer in the lining of the stomach.

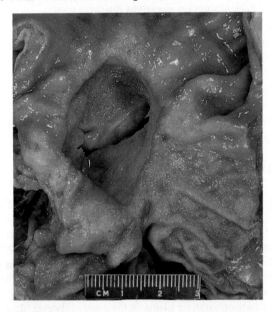

surgery. The patient swallows a balloon and then alcohol, at a temperature cold enough to freeze and kill stomach lining cells, is poured into the balloon. This destroys a large percentage of the acid-secreting cells and has the same effect as surgical treatments. By the time these cells are replaced, the ulcer has had a chance to heal.

Appendicitis and ulcers are not the only digestive ailments. Problems can also occur in the digestive glands. For example, the liver and its bile-storing structure, the gallbladder, are subject to several types of problems. One common condition is **gallstones,** which result from the precipitation of bile salts in the gallbladder or in the bile duct. Blockage of the bile duct may cause bile to back up and ultimately lead to jaundice. Surgical removal of the gallbladder is the usual solution.

Jaundice also may be a symptom of various types of liver failure, a much more serious condition given the overall importance of this organ. Inflammation of the liver is known as **hepatitis,** and it may result from a virus infection (viral hepatitis), poison (toxic hepatitis), or protein deficiency (deficiency hepatitis). The liver has considerable regenerative ability, but when liver cells are destroyed, they are sometimes replaced by scar tissue. This scarring is called **cirrhosis** of the liver. All of these liver problems weaken the liver's metabolic capacities and greatly reduce its ability to deal effectively with toxins entering the body. Thus, persons suffering from liver disease, whatever the original cause, are strongly urged not to drink alcohol because alcohol is nor-

mally metabolized almost entirely by liver cells. A damaged liver may well succumb to the onslaught of alcohol in the blood, with death as the likely result.

In addition to being susceptible to problems such as appendicitis, ulcers, gallstones, and liver disease, the digestive tract provides an ideal environment for the growth of many disease organisms, from viruses to tapeworms.

Despite all of these potential problems, most human digestive tracts function normally, and except for occasional minor upsets, seldom cause much concern. A normally functioning digestive tract is essential if the body is to receive an uninterrupted supply of vital nutrients.

Summary

Heterotrophic organisms must obtain organic compounds from their environment to meet their energy and raw material requirements. The foods that animals eat to provide these organic compounds are processed in digestive systems. Large organic molecules are broken down into absorbable forms in animal digestive tracts.

Carbohydrates are important nutrients mainly because they provide chemical energy sources. Much ingested carbohydrate consists of polysaccharides that must be hydrolyzed to absorbable sugars.

Animals must ingest proteins to obtain amino acids for their own protein synthesis. Animals' cells can synthesize some amino acids from other compounds, but other amino acids are essential dietary constituents.

Fats are highly reduced organic compounds that are important for energy and synthesis of structural components. Certain animals cannot synthesize some fatty acids, and these fatty acids must be included preformed in the diet.

Some minerals are required in large quantities, while only trace amounts of others are needed.

Vitamins are relatively simple organic compounds that must be present in the diet to maintain health. Vitamin deficiencies cause complex sets of serious problems.

All unicellular heterotrophs absorb nutrients directly from their environment. In multicellular heterotrophs, only the cells that line their digestive tracts are specialized for nutrient procurement.

Incomplete digestive tracts have only one opening, a mouth. Complete digestive tracts have both a mouth and an anus, and thus food can move through and be processed in assembly-line fashion in differently specialized regions.

All animals secrete some enzymes into digestive cavities where enzymes hydrolyze food molecules. Such digestion, outside cells and in a digestive cavity, is called extracellular digestion. Intracellular digestion is the processing of nutrients inside vesicles in individual gut lining cells.

Modifications of the basic tubular plan of a complete digestive tract reflect the need for food storage, the need for food maceration, the type of food ingested, and the relative need for large absorption surfaces.

In the human digestive tract, food is chewed and manipulated in the mouth. Then it is swallowed, passing through the esophagus to the stomach.

The stomach stores food and mixes it with mucus and gastric juice to produce chyme. Protein digestion begins in the stomach.

Chyme passes through the pyloric valve to the duodenum. There, bile, pancreatic juice, and intestinal gland secretions are added to the chyme and begin to act upon it. Enzymes hydrolyzing all of the organic nutrients act in the small intestine.

Most nutrient absorption takes place in the small intestine, but some water and mineral ions are absorbed in the colon. Feces contain water, bacteria, undigested material, fats, waste products, inorganic material, proteins, and dead intestinal lining cells.

Neural and hormonal regulation of digestive tract secretions ensures that digestive substances such as enzymes are released only when food is present.

The liver is involved in chemical processing of absorbed food and in maintenance of stable concentrations of nutrients such as glucose in the blood. The liver also processes nitrogenous wastes and breaks down toxins. The pancreas produces hormones that are involved in the control of carbohydrate metabolism in the body.

Problems that can develop in the digestive tract include infection by a wide range of organisms, as well as erosion of the tract's protective lining by its own secretions, which results in ulcers. These problems threaten the supply of vital nutrients and the maintenance of homeostasis.

Questions for Review

1. Name three kinds of polysaccharides included in animal diets that are polymers of glucose.
2. Why can a protein deficiency disease such as kwashiorkor develop even when a person is regularly eating considerable amounts of plant protein?
3. Define the terms scurvy and beriberi.
4. Why are overdoses of vitamin C much less likely to cause serious problems than overdoses of vitamin A?
5. Explain the advantages of a complete digestive tract over an incomplete digestive tract.
6. Explain the difference between extracellular and intracellular digestion and contrast the relative importance of each in planarian and human digestion.
7. Name five structural features that provide increased surface area for absorption in various digestive tracts.
8. Define the term zymogen.
9. What normal protective mechanisms guard against self-digestion of digestive glands and the linings of the digestive tract?
10. What are the steps in the digestion and absorption of starches, proteins, and fats in the human digestive tract?
11. What are bile pigments, and what is their relationship to jaundice?

Questions for Analysis and Discussion

1. Explain why ascorbic acid is classified as a vitamin (vitamin C) for humans, but is not essential in the diet of rats.
2. Which would you expect to be longer, the digestive tract of a two-kilogram cat or that of a two-kilogram rabbit? Why?
3. Even though certain amino acids are relatively rare in plant proteins (see p. 236), it is possible for a vegetarian to obtain adequate quantities of them through careful planning to include appropriate plants in their diet. However, excess amino acids are not stored in the body, but are deaminated and metabolized for energy. What additional requirement does this fact place on the planning of a strictly vegetarian diet?

Suggested Readings

Books

Eckert, R., and Randall, D. 1983. *Animal physiology.* 2d ed. San Francisco: W. H. Freeman.

Guyton, A. C. 1984. *Physiology of the human body.* 6th ed. Philadelphia: W. B. Saunders.

Schmidt-Nielsen, K. 1983. *Animal physiology.* 3d ed. New York: Cambridge University Press.

Articles

Baldwin, R. L. 1984. Digestion and metabolism of ruminants. *BioScience* 34:244.

Harpstead, D. D. August 1971. High-lysine corn. *Scientific American.*

Harris, M. January/February 1986. The 100,000-year hunt. *The Sciences.*

Hauser, J. T. 1984. Nematode-trapping fungi. *Carolina Tips* (Carolina Biological Supply Co., Burlington, N.C.) 47:37.

Jarvis, W. T. 1983. Food faddism, cultism, and quackery. *Annual Review of Nutrition* 3:35.

McMinn, R. M. H. 1977. The human gut. *Carolina Biology Readers* no. 56. Burlington, N.C.: Carolina Biological Supply Co.

Moog, F. November 1981. The lining of the small intestine. *Scientific American.*

Sernka, T. J. 1979. Claude Bernard and the nature of gastric acid. *Perspectives in Biology and Medicine* 22:523.

Sherlock. S. 1978. The human liver. *Carolina Biology Readers* no. 83. Burlington, N.C.: Carolina Biological Supply Co.

Thorn, R. G., and Barron, G. L. 1984. Carnivorous mushrooms. *Science* 224:76.

Gas Exchange and Transport

12

Chapter Outline

Chapter Concepts

1. Gas exchange is an urgent, continuing requirement for animal life.
2. Size, shape, internal transport capabilities, and metabolic requirements of an animal, along with the type of gas-carrying medium in which the animal lives, dictate the method of gas exchange that is used.
3. Small animals can exchange gases directly through body surfaces, but larger animals have specialized gas exchange mechanisms.
4. Aquatic animals generally exchange gases through gills, while terrestrial animals have tracheal systems or lungs.
5. Special gas transport mechanisms in the blood make possible internal transport of adequate quantities of oxygen and carbon dioxide in large animals such as humans.
6. Because lungs are warm, moist, blind sacs, they are particularly susceptible to infection and to degenerative diseases caused by inhaled irritants.

Diving beetles are very much at home in a freshwater pond (figure 12.1). Some of these active, streamlined insects are voracious carnivores that prey on snails, tadpoles, and even small fish, while others are scavengers that feed on bits of debris. In their search for food, diving beetles must spend considerable time underwater. Yet they are air-breathing insects. How do they manage to get oxygen while they are submerged?

Diving beetles have a marvelously efficient mechanism that allows them to "breathe" underwater. At the water's surface, a beetle traps a bubble of fresh air. This bubble is carried along and serves as the insect's source of oxygen underwater. Furthermore, as the insect consumes the oxygen in its bubble, the bubble's oxygen concentration falls, and oxygen diffuses from the water into the bubble. Some kinds of diving beetles can remain underwater for as long as 36 hours because their oxygen supply continues to be replenished in this way.

While oxygen is not essential for all forms of life on earth (it is actually toxic for some members of the kingdom Monera), many microorganisms and all plants and animals need oxygen to live. Because these organisms generally have very limited gas storage capacities, continuing **gas exchange** is a necessity of life for all of them.

Gas exchange supplies cells with oxygen that is continuously being used as an electron acceptor in metabolic oxidation-reduction reactions that produce ATP for a variety of vital energy-requiring processes. Another requirement for gas exchange results from the decarboxylation of carbon compounds, which yields carbon dioxide as a metabolic waste product. This CO_2 must be released to the environment as rapidly as it is produced because excessive CO_2 accumulation in the cellular environment inside the body is potentially harmful.

In this chapter, we will discuss the mechanisms by which animals accomplish this vital gas exchange.

Factors Affecting Methods of Gas Exchange

Several factors determine what mechanisms are adequate to allow an animal to obtain and deliver enough oxygen to the cells of its body, including:

1. The availability of oxygen in the organism's immediate environment. This depends largely upon whether the gas-carrying medium is air or water.
2. The organism's size and shape.
3. The presence or absence of an internal gas transport mechanism.
4. The organism's metabolic rate.

Oxygen Availability: Air versus Water

Air has three distinct advantages over water as an oxygen source. First, air contains roughly 21 percent oxygen (210 parts per 1,000) while water, at the most, contains only 6 to 8 parts oxygen per 1,000. Natural aquatic systems may contain considerably less oxygen than that under many circumstances (tables 12.1 and 12.2). Second, oxygen diffuses approximately 300,000 times faster in air than it does in water. Thus, an organism in air seldom faces the problem of depleting the oxygen supply immediately around its body, while aquatic organisms often experience oxygen depletion of the water immediately around them. Third, air is a much less dense medium than water, and much less energy is required to move air over gas exchange surfaces than to move water. Mammals use only 1 or 2 percent of their total energy output for breathing, but fish use up to 25 percent of theirs to propel the much more dense (and much less oxygen-rich) water over their gills.

However, air has one significant disadvantage as an oxygen-carrying medium. Moist gas exchange surfaces continually lose water by evaporation. Animals that exchange gases with air, then, must have some kind of protection against

Figure 12.1 Some diving beetles trap air bubbles under their wings. Others, such as this scavenger water beetle, also carry a silvery film of air retained by numerous hairs on the ventral side of their body.

Table 12.1
Composition of Dry Atmospheric Air at Sea Level

Gas	Percentage	Partial Pressure (mm Hg)*
Nitrogen	78.09	593.5
Oxygen	20.95	159.2
Carbon dioxide	0.03	0.2
Others (mostly argon)†	0.93	7.1
Total	100.00	760.0

*Partial pressure is that part of atmospheric pressure exerted by a gas. For example, oxygen's partial pressure is 0.2095 × 760 = 159.2. Actually, all atmospheric air contains water vapor in varying amounts, and the partial pressures of other gases are reduced as the partial pressure of water vapor increases.

†The less common noble gases—helium, neon, krypton, and xenon—together constitute only 0.002 percent of the atmosphere.

Table 12.2
Amount of Oxygen Dissolved in Fresh Water and Seawater in Equilibrium with Atmospheric Air

Temperature (°C)	Fresh Water (ml O_2/liter of water)	Seawater (ml O_2/liter of water)
0	10.29*	7.97
10	8.02	6.35
15	7.22	5.79
20	6.57	5.31
30	5.57	4.46

*These values are very low compared to the oxygen content of air, which is the same at all of these temperatures and is equal to 209.5 ml O_2 per liter of air.

this dessication. The majority of terrestrial animals have internal gas exchange areas and their body surfaces are protected by dry, relatively impermeable skins.

Body Size and Shape

The size and shape of an animal's body also influences the characteristics of its gas exchange mechanism. Once oxygen enters an animal's body, it diffuses very slowly through the watery medium in which it is dissolved. Thus, only animals with small bodies in which all cells are relatively near the gas exchange surfaces can depend upon diffusion as the sole means for gas movement to and from body cells.

Some animals have very flat bodies in which all body cells are relatively close to the body surface. The appropriately named flatworms, such as planarians, are excellent examples of this arrangement. Other animals, such as cnidarians like *Hydra,* are thin-walled and hollow, and their cells exchange gases both with water around them and with water in an internal cavity (figure 12.2).

Gas Transport Systems

In contrast to relatively small animals that depend solely on diffusion of gases between body surfaces and individual body cells, all large animals must have the means for getting oxygen to internal body cells and carrying away carbon dioxide. Many of these animals have **circulatory systems** with gas transport mechanisms that carry oxygen from specialized exchange areas to cells throughout the body.

The evolution of circulatory systems has generally paralleled the evolution of specialized gas exchange organs located in specific areas of the body. In most cases, these exchange organs, such as **gills** and **lungs**, are richly supplied

Figure 12.2 Body shape and adaptations that reduce oxygen diffusion distances from outer body surfaces to body cells. (a) The cells of planarians and other flatworms are spread out in their flattened bodies so that no cells are far from an outer body surface. (b) *Hydra* has a hollow internal cavity. Some cells exchange gases with water inside this cavity.

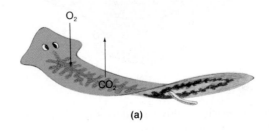

(a)

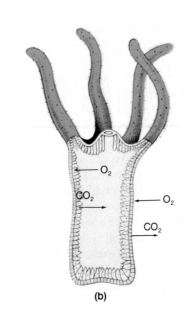

(b)

Figure 12.3 A circulatory system brings blood to a highly vascular gas exchange surface; then blood is carried to body tissues where cells exchange gases with blood.

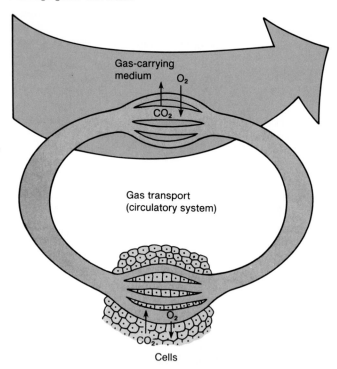

with blood vessels (highly vascularized). Thus, a great deal of blood can be brought near the surfaces where gas exchange with water or air takes place. The circulatory system then carries the blood into the immediate environment of the body cells that exchange gases with it (figure 12.3).

The effectiveness of circulatory gas transport is enhanced by the fact that most animals that depend on circulatory gas transport have specialized **respiratory (oxygen-carrying) pigments** in their blood. These carriers reversibly bind oxygen molecules and greatly increase the oxygen-carrying capacity of the blood. In some animals, for example, blood containing a respiratory pigment can carry up to 50 or even 100 times as much oxygen as the blood could carry in dissolved form alone (table 12.3).

Insects and some other small animals do not depend upon their circulatory systems for gas transport. Instead, their bodies are penetrated by a network of fine tubules called a **tracheal system**. The tracheal system carries air deep into all parts of the body, and cells exchange gases directly with the air inside tracheal tubules. Because few insects are larger than several centimeters long, tracheal systems are adequate for their gas exchange needs. Most larger animals, however, have either gills or lungs coupled with a circulatory system that transports gases efficiently.

Metabolic Rates

To a great extent, the demands an animal puts on its gas exchange mechanism depend upon its **metabolic rate**. The metabolic rate is the sum of all cellular processes occurring in the body. For organisms using aerobic respiration, the metabolic rate can be measured as oxygen consumed per unit time, and this measure gives a good indication of the demands that must be met by the gas exchange system. Animals that consume more oxygen require more efficient gas exchange mechanisms than organisms that consume less oxygen.

There is a wide range of variation in metabolic rates among animals of different species and different sizes. In addition, an individual animal displays a substantial range of metabolic rates at different times, depending upon environmental factors and its activity level. For example, humans consume twenty or more times as much oxygen during vigorous exercise as they do at rest (figure 12.4). Thus, gas exchange mechanisms must be efficient enough to meet the greatly increased demands that occur during periods of vigorous activity.

Metabolic rates usually are measured when animals are relaxed and resting, but not sleeping. Such resting metabolism (the energy spent on ordinary life maintenance functions) is called **standard metabolism**. Physiologists usually use standard metabolic rates when making comparisons among animals.

Most animals are **poikilothermic** (from the Greek *poikilos* meaning "changeable"). Their body temperatures rise and fall with environmental temperatures, and because metabolic processes are temperature dependent, the standard metabolic rates of poikilotherms change as environmental temperatures change.

For example, low environmental temperatures can so depress a poikilothermic animal's metabolism that the animal is practically incapable of moving. This cold paralysis is relieved only when the animal's body is warmed. Trout fishers who catch grasshoppers for their day's bait usually try to fill their bait boxes in the morning chill before the sun warms the mountain meadows around their favorite trout streams. Chasing sun-warmed grasshoppers later in the day can be hard, discouraging work. When a poikilothermic animal is warm, its increased metabolic rate puts increased demand on its gas exchange mechanism (figure 12.5).

Birds and mammals are **homeothermic** animals (from the Greek *homos* meaning "same"). Homeotherms maintain stable, relatively high, body core temperatures. How is this possible? First, homeothermic animals have higher standard metabolic rates than poikilothermic animals of similar size. Metabolic reactions occur at a more rapid rate, and thus more

Table 12.3
Effect of Respiratory Pigments on Oxygen-Carrying Capacity of Blood

Pigment	Color	Site	Animal	Oxygen (ml O_2/100 ml of blood)
Hemoglobin	Red (iron pigment)	Erythrocytes (red blood cells)	Mammals	15–30
			Birds	20–25
			Amphibians	3–10
			Fishes	4–20
		Blood plasma	Annelids (for example, earthworms)	1–10
			Molluscs (some clams)	1–6
Hemocyanin	Blue (copper pigment)	Blood plasma	Molluscs	
			Gastropods (snails)	1–3
			Cephalopods (for example, octopus)	3–5
			Crustaceans (for example, lobster)	1–4
No pigment	—	—	Insects and other terrestrial arthropods*	> 1

*Circulatory systems of insects and other terrestrial arthropods do not transport oxygen because their cells exchange gases with air inside tracheal systems.

Figure 12.4 The energy expenditure of a person walking and running on three different grades.

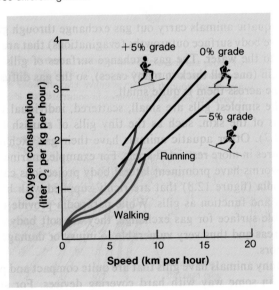

Figure 12.5 Rates of oxygen consumption of the Colorado potato beetle (*Leptinotarsa decemlineata*) at various temperatures between 7°C and 30°C. Before the experiment the animals had been maintained at 8°C. Note that oxygen consumption at 20°C is more than twice as great as it is at 10°C. At high temperatures, increased metabolic rates of poikilothermic animals place great demands on their gas exchange mechanisms.

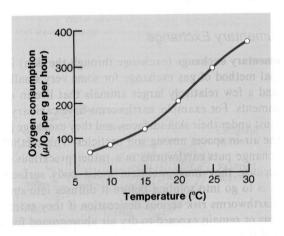

heat is released from body cells. This also means that homeothermic animals must take in more oxygen and release more CO_2 than poikilothermic animals. Second, insulation helps homeothermic animals to maintain a stable body core temperature. Feathers, hair, and layers of fat insulate the body and control heat loss to the environment.

In addition, homeothermic animals also have specialized responses, such as altering blood flow near the body surface, shivering, and sweating, that help them maintain stable body temperatures. All of these mechanisms require energy and also increase the need for efficient gas exchange and transport.

Figure 12.20 Gas exchange mechanisms in birds. (a) Birds have a number of large air sacs; some of them are posterior to the small pair of lungs, and others are anterior to the lungs. The main bronchus (air passage) that runs through each lung has connections to air sacs as well as to the lung. The schematic diagram below the bird shows these structures. Posterior and anterior sacs are sketched as single functional units to clarify their relationship to the lung and the bronchus. (b) The flow of air through a bird's respiratory system. Arrows indicate expansion and contraction of air sacs. The more darkly shaded portion in each diagram represents the volume of a single inhalation and distinguishes it from the remainder of the air in the system. Two full breathing cycles are needed to move the volume of gas taken in during a single inhalation through the entire system and out of the body. This arrangement is associated with one-way flow through the gas exchange surfaces in the lungs. (c) A section of a bird's (domestic fowl chick) lung showing the cylindrical tubes through which air flows. The diameter of each tube in this picture is slightly less than 0.5 mm (magnification × 37).

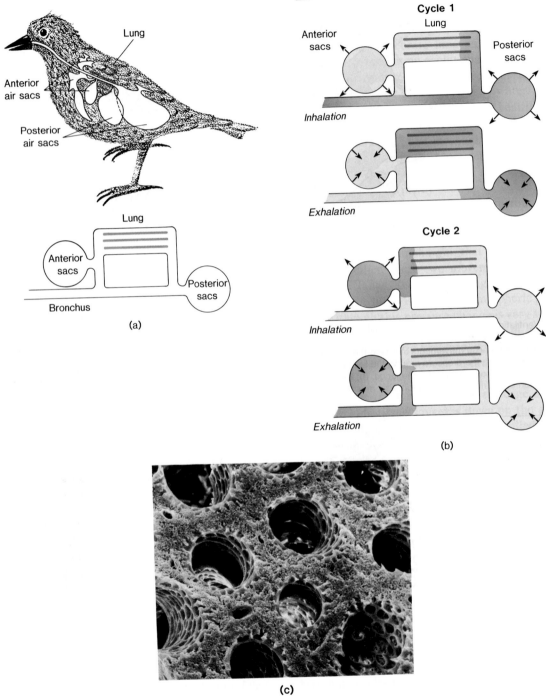

Part 2

Plants and Animals

Box 12.1
Fish That Can Drown

The African lungfish is a very curious creature. It has lungs and can breathe air. Being able to breathe air at the surface, it can survive in very stagnant water when it cannot obtain sufficient oxygen through its gills. Sometimes a lungfish can even wriggle across the ground to a new pond if its own pond is drying up.

You might conclude, then, that lungfish are hardy animals, able to withstand very bad conditions. However, biologists who went to Africa to study the physiology of lungfish encountered some peculiar problems that made them doubt this. Many lungfish died in the nets that were used to catch them, and those that were caught alive almost invariably died, no matter how gently they were handled, while they were being carried to the laboratory in buckets of water. This seemed strange because native fishermen who caught lungfish on lines and carried the fish slung over their backs were able to deliver them alive to the market. What were the biologists doing wrong?

Eventually the scientists discovered that the lungfish died in the buckets because they could not bend their bodies to reach the surface of the water. They are obligate air breathers; that is, they must breathe air to live. In the nets or buckets, lungfish were dying for lack of oxygen. In other words, African lungfish are fish that can drown!

Lungfish are also found on two other continents, South America and Australia. South American lungfish also are obligate air breathers. They must be able to surface and breathe air because they cannot survive using only their gills for gas exchange. Australian lungfish, on the other hand, can survive without breathing air. They seem to use air breathing only as an emergency means of supplementing gas exchange through their gills.

Another interesting aspect of the biology of lungfish is their response to periods of extreme drought. During such times, a lungfish goes to the bottom of its drying pond and secretes material that causes the mud around it to form a sort of cocoon; this cocoon hardens as the mud dries. The metabolism of the lungfish inside the cocoon slows to a fraction of its normal rate. In this inactive state, called **estivation**, the lungfish can survive long periods of hot, dry weather. When the rains come again and the pond refills, the lungfish emerges from its cocoon and resumes its normal activity.

Some biologists who study lungfish collect them in their cocoons and transport them to laboratories in other parts of the world. Secure inside their cocoons, lungfish can be kept on laboratory shelves until the biologists are ready to study them. Then it is a simple matter of putting the cocoon into water and waiting for the active lungfish to emerge. Lungfish have been stored in this way for years and still revived successfully. A great deal remains to be learned about the biology of these fascinating animals and their ability to go into this sort of "suspended animation."

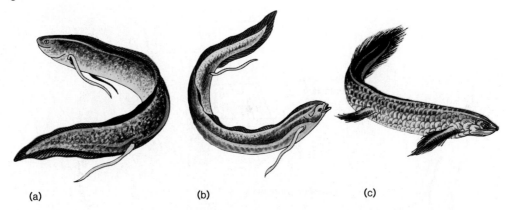

(a)   (b)   (c)

Box figure 12.1 Lungfish. (a) The African lungfish *Protopterus*. (b) The South American lungfish *Lepidosiren*. (c) The Australian lungfish *Neoceratodus*.

The Human Gas Exchange System

The human gas exchange system (**respiratory system**) is similar to the gas exchange systems of other mammals both in terms of basic structure (figure 12.21) and ventilation by a breathing mechanism that uses negative pressure inhalation.

Most people at rest generally breathe through their noses. During inhalation, air enters through **external nares (nostrils)** and passes through the **nasal cavities**. The linings of the nasal cavities near the nostrils have hairs, while the linings deeper in the cavities have cilia on their surfaces. Hair and cilia, as well as mucus on the lining surfaces, trap dust particles and other foreign material. Air is also warmed and moistened in the nasal passages. Blood vessels in the linings of the nasal cavities lose heat to the air, and water evaporates off the moist surfaces. By the time air enters the lungs, it has been cleaned and warmed, and is more than 99 percent saturated with water vapor. Conversely, air leaving the lungs cools as it passes out through the nasal passages, and as it cools, it loses water by condensation.

The nasal passages are separated from the mouth by the **hard** and **soft palates**. Air in the nasal passages is isolated from food in the mouth, so it is possible to chew and breathe at the same time. However, air passes from the nasal passages into the **pharynx**, where it crosses the path of food. Swallowing temporarily closes the way to the lungs while food is passing through the pharynx (see p. 251). It may seem inefficient for food and air to pass through the same space, and certainly there is danger of choking if food accidentally enters the passage leading toward the lungs, but this arrangement has an important advantage—it permits mouth breathing if the nostrils or nasal cavities become plugged. It also permits greater air intake during heavy exercise when greater gas exchange is required.

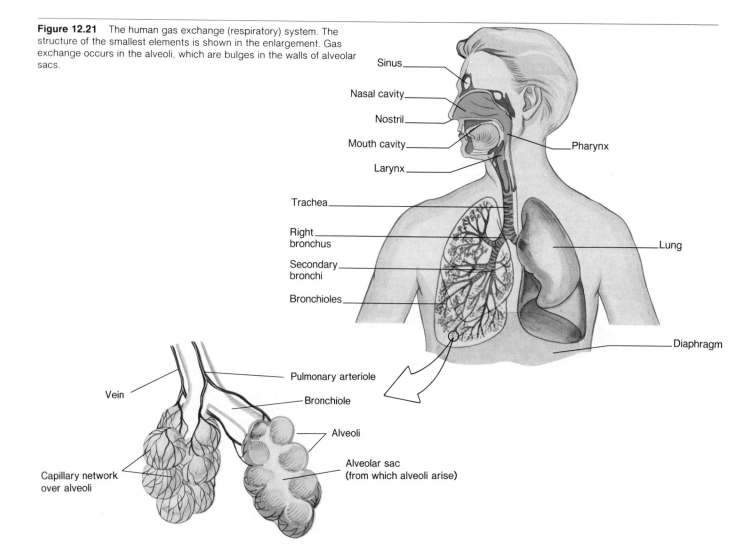

Figure 12.21   The human gas exchange (respiratory) system. The structure of the smallest elements is shown in the enlargement. Gas exchange occurs in the alveoli, which are bulges in the walls of alveolar sacs.

Air passes on from the pharynx through the **glottis,** an opening into the **larynx.** The larynx, or "voice box," lies at the top of the **trachea.** During swallowing, the larynx is raised and a flap, the **epiglottis,** covers the glottis, temporarily closing it. The larynx is a rigid, cartilage-supported structure. It contains a pair of elastic ridges, the **vocal cords,** that vibrate when currents of air pass between them and produce sounds. We produce different sound tones by varying the tension of the vocal cords.

Air passes on through the larynx into the trachea ("windpipe"). The trachea is a tube supported by a series of C-shaped cartilages in its wall that prevent it from collapsing. The tracheal lining is covered with cilia-bearing cells and mucus-secreting cells. The cilia beat in waves and move mucus and any trapped particles up toward the larynx (figure 12.22). From the larynx, mucus and foreign material can be cleared into the pharynx. This helps to protect the lungs from dirt particles. The tracheal lining cells also trap many potentially harmful microorganisms before they reach the lungs.

The lower end of the trachea divides into two **bronchi** (singular: **bronchus**) that carry air to the right and left lungs. Within the lungs, the bronchi branch and rebranch into a network of smaller passages called **bronchioles.** The bronchi and larger bronchioles have cartilage rings like those of the trachea, but the smaller bronchioles do not have such supports. The bronchi and bronchioles also have ciliated linings that function much like the lining of the trachea.

The smallest bronchioles terminate in **alveolar sacs** that have tiny bulges called **alveoli** (singular: **alveolus**) on their surfaces. Alveoli are chambers within which gas exchange between air and blood occurs. Although each alveolus is very small (human alveoli are about 0.25 mm in diameter), there are hundreds of millions of alveoli in a pair of human lungs, thus making the total surface area of the alveolar linings enormous. Alveolar surface areas in human individuals range from 50 to 100 m² with the average probably being about 70 to 80 m², an area about forty times as large as the skin surface area of a human adult.

Alveoli are well adapted for their role in gas exchange because they are highly vascular (see figures 12.21 and 12.23). More of their lining surface area is occupied by capillaries than by spaces between capillaries. The alveolar linings consist of a single layer of very flat cells, as do the capillary walls. Thus, the distance that gases must diffuse between air and blood is very small.

Human Breathing

Humans breathe by the same mechanisms as other mammals. The volume of the thoracic cavity is increased by muscular contractions that lower the diaphragm and raise the ribs (see figure 12.18). These movements create a negative pressure in the thoracic cavity around the lungs. Atmospheric pressure then forces air into the lungs. The walls of all of the air spaces in the lungs are elastic, and the inhaled air expands and stretches them. When rib and diaphragm muscles relax, air is exhaled as a result of increased pressure in the thoracic cavity and the recoil of the stretched elastic walls of the lungs.

The lungs have a very smooth, thin, membranous outer covering, and a similar smooth membrane lines the thoracic cavity. These lining membranes are called **pleura.** During breathing, the lungs move within the thoracic cavity. The

Figure 12.22 The tracheal lining. (a) Scanning electron micrograph of the tracheal lining showing cilia and mucus-secreting goblet cells (*GC*) (magnification × 567). (b) More detailed view of part of the tracheal lining, showing cilia (*Ci*) and goblet cells (*GC*). Goblet cells also have microvilli (*Mv*) on their surfaces (magnification × 2,075). (a) and (b) from Kessel, R. G. and Kardon, R. H.: *Tissues and Organs: A Text Atlas of Scanning Electron Microscopy.* © 1979 by W. H. Freeman and Company.

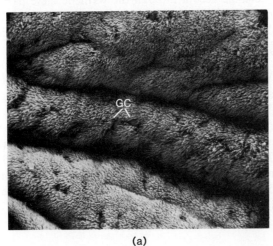

(a)

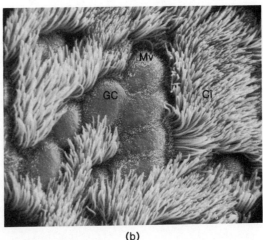

(b)

Figure 12.23 The exchange surfaces in alveoli. The enlargement shows the layers that lie between the air inside the alveolus and the blood. The epithelium lining the alveolus consists of a single layer of flattened cells. The endothelium (capillary wall) is very thin because it also consists of a single layer of flattened cells. Note the red blood cells (erythrocytes) inside the capillaries.

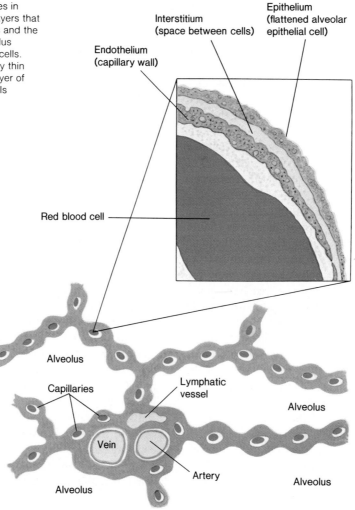

pleura secrete a small amount of lubricating fluid that allows the membranes to slide past one another very smoothly.

The total capacity of the lungs and respiratory passages of an adult human is about 6 liters of air, but a person has to inhale very forcefully to reach that capacity. If a forceful inhalation is followed by a forceful exhalation, 1¼ liters of air still remain in the lungs. It is not possible to force out this **residual volume** of air. Thus, humans have total breathing capacity of about 4¾ liters, but this **vital capacity** is reached only by maximum forceful inhalation and exhalation. A single, normal breath moves just a fraction of the vital capacity. The **tidal volume** moved by an ordinary breath is about 500 ml. However, only part of this air reaches the alveoli because the nasal passages, pharynx, trachea, bronchi, and bronchioles hold about 150 ml of air. Because very little gas exchange occurs through the walls of these passages, they represent dead space. Therefore, the **alveolar tidal volume** of 350 ml is the effective volume of gas movement in each breath.

A resting adult breathes about fourteen times per minute. With an alveolar tidal volume of 350 ml per breath, there is a functional gas movement into and out of the lungs of about 5 liters per minute. During vigorous exercise, the breathing rate may rise to 100 breaths per minute, and the volume of air moved during each breath may be near the vital capacity. This means that with fast, deep breathing, humans have the remarkable capacity to increase the volume of air moved from 5 liters to 475 liters of air per minute, nearly a hundredfold increase.

Table 12.4
Partial Pressures* of Oxygen and Carbon Dioxide (in mm Hg)

	Body Tissues	Blood Entering Lungs	Alveolar Air	Blood Leaving Lungs
Partial Pressure of Oxygen (Po$_2$)	40	40	100	100
Partial Pressure of Carbon Dioxide (Pco$_2$)	45	45	40	40

*These are approximate average values under normal conditions. The values given for blood are actually gas tensions. Gas tension is the amount of gas in a solution that is in equilibrium with an atmosphere having the stated partial pressure.

Dissolved gas as such exerts no measurable pressure, but in practice, it is easier to discuss all values using the term partial pressure.

The breathing rate and the volume of air inhaled and exhaled are controlled by a **respiratory center** in the brain. Changes in the breathing rate and in the volume of air being breathed are made automatically by the respiratory center in response to changes in the gas content of the blood, especially increases or decreases in the amount of carbon dioxide in the blood.

Gas Transport in Blood

In alveolar gas exchange, oxygen diffuses from the air inside the alveoli into the blood in capillaries, and carbon dioxide diffuses from the blood to the air. Thus, blood leaving the alveoli contains more oxygen and less CO$_2$ than blood arriving at the alveoli (table 12.4). Biologists describe the concentrations of gases in blood in terms of gas **tension** or **partial pressure** (partial pressure is explained in table 12.1). We will express partial pressures as millimeters of mercury (mm Hg) in our discussions of gas concentrations in blood.

Oxygen

Most of the oxygen entering the blood combines with hemoglobin in red blood cells to form **oxyhemoglobin**. Each molecule of hemoglobin contains four polypeptide chains, and each of these is folded around an iron-containing **heme** unit (see chapter 3 for details). Oxygen molecules bind to the iron atoms of the hemes; each molecule of hemoglobin can carry four molecules of oxygen. Since there are about 280 million hemoglobin molecules in each red blood cell (**erythrocyte**), each cell is capable of carrying more than one thousand million molecules of oxygen.

The oxygen-binding characteristics of hemoglobin are rather complex, and some of these complexities are functionally important because they affect the "loading" and "unloading" of oxygen carried by hemoglobin. Oxygen-binding characteristics of hemoglobin can be studied by examining oxygen **dissociation curves**, which show what percentage of the oxygen-binding sites of hemoglobin are

Figure 12.24 Oxygen dissociation curves for human hemoglobin measured at two different pH's. The lower pH encountered in actively metabolizing tissues causes oxyhemoglobin to dissociate more readily. This "Bohr effect" promotes "unloading" of oxygen in tissues.

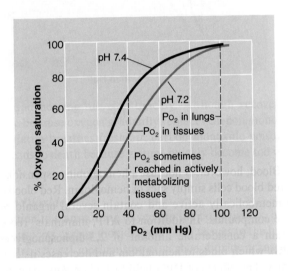

carrying oxygen at each of a series of oxygen partial pressures (**Po$_2$**) in the environment around the hemoglobin. Such dissociation curves are not linear, but S-shaped (figure 12.24).

At partial pressures such as those encountered in the lungs, hemoglobin quickly becomes practically saturated with oxygen. As blood moves through the circulatory system, it loses very little oxygen, even when it enters environments where oxygen partial pressures are around 60 mm Hg. However, when the blood flows through capillaries in body tissues where oxygen partial pressures are between 20 and 40 mm Hg, hemoglobin quickly gives up much of its oxygen. In fact, oxygen release is very sensitive to small Po$_2$ changes within that range, and is responsive to Po$_2$ decreases caused by increased oxygen consumption by cells, such as occurs during heavy exercise.

Figure 12.28 Summary diagram of circulation and gas exchange in lungs and tissues.

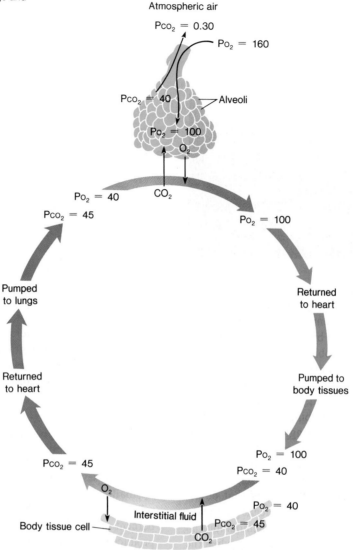

Atmospheric air

$Pco_2 = 0.30$

$Po_2 = 160$

$Pco_2 = 40$ — Alveoli

$Po_2 = 100$

O_2

$Po_2 = 40$ CO_2
$Pco_2 = 45$ $Po_2 = 100$

Pumped to lungs

Returned to heart

Returned to heart

Pumped to body tissues

$Pco_2 = 45$ $Po_2 = 100$
$Pco_2 = 40$

O_2 $Po_2 = 40$

Interstitial fluid $Pco_2 = 45$

Body tissue cell —

CO_2

inside red blood cells, this reaction is catalyzed by an enzyme, **carbonic anhydrase,** and it occurs much more rapidly.

At the pH levels found in blood, carbonic acid tends to be dissociated to hydrogen ions (H^+) and bicarbonate ions:

$$H_2CO_3 \longrightarrow H^+ + HCO_3^-$$

The following equation summarizes what happens when carbon dioxide enters the blood flowing through body tissues:

$$H_2O + CO_2 \rightleftharpoons H_2CO_3 \rightleftharpoons H^+ + HCO_3^-$$

The dissociation of carbonic acid slightly lowers the pH of the blood and has an important effect on delivery of oxygen to the tissues because of the Bohr effect mentioned earlier.

However, body cells cannot withstand significant changes in blood pH, and therefore, such blood pH fluctuations are prevented by buffers. A large part of the buffering capacity of the blood is provided by hemoglobin and other blood proteins.

In the blood, hemoglobin (Hb) is in the form of an almost completely ionized potassium salt:

$$KHb \rightleftharpoons K^+ + Hb^-$$

Carbonic acid reacts with potassium hemoglobin to produce potassium bicarbonate ($KHCO_3$), which is almost completely ionized as K^+ and HCO_3^-, and acid hemoglobin (HHb):

$$H^+ + HCO_3^- + K^+ + Hb^- \rightleftharpoons K^+ + HCO_3^- + HHb$$

This buffering system amounts to exchanging one acid for another; that is, carbonic acid for acid hemoglobin. How does such an exchange prevent large pH changes? Acid hemoglobin is a much weaker acid than carbonic acid, and therefore, within the normal pH range of blood, very little acid hemoglobin is dissociated. Thus, this buffer system, which involves formation of acid hemoglobin at the expense of carbonic acid, reduces the number of hydrogen ions in solution in the blood.

In body tissues, carbon dioxide diffuses into the blood, but in the lungs, the opposite occurs. Carbon dioxide pressure in alveolar air is lower than it is in the blood. Therefore, CO_2 diffuses out of the blood. This loss of CO_2 favors reversal of all of the chemical reactions that occur in the tissues when CO_2 is entering the blood and results in rapid release of CO_2 from the blood (figure 12.28).

Control of Breathing Rate

As we mentioned earlier, the breathing rate and the volume of air inhaled and exhaled are controlled by a respiratory center in the brain (more specifically, in the medulla—see p. 414). Several other automatic reflex centers in the brain also interact with the respiratory center. Although breathing rate and depth of breathing can be consciously controlled, breathing itself is automatically under the control of these reflex centers when attention turns to other things and during sleep or unconsciousness.

The respiratory center controls the action of muscles in the ribs and diaphragm that are responsible for breathing movements. Increases in breathing rate and volume are made automatically by the respiratory center in response to changes in the gas content of the blood.

Blood gas content is monitored by **chemoreceptors** called the **aortic bodies** and **carotid bodies,** specialized structures located in the walls of major arteries (figure 12.29). Various chemoreceptors detect changes in CO_2, H^+, and O_2 concentrations and send information to the respiratory center. The respiratory center itself is also sensitive and responsive to changes in the chemical content of the blood reaching it.

Figure 12.29 Location of the chemoreceptors that detect changes in CO_2, H^+, and O_2 concentrations in the blood. The aortic bodies are in the aorta, the major artery carrying blood from the heart to all body tissues. The carotid bodies are in the carotid arteries, which supply the brain and other parts of the head. Sensory nerves carry information to brain reflex centers that control breathing functions.

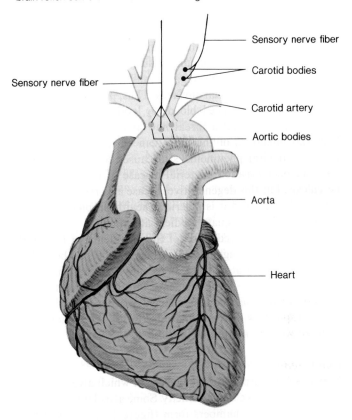

Sensory nerve fiber

Carotid bodies

Sensory nerve fiber

Carotid artery

Aortic bodies

Aorta

Heart

The main chemical stimulus that changes the breathing rate is variation in the partial pressure of CO_2 (P_{CO_2}) and H^+ ion concentration in blood. Even small increases in blood P_{CO_2} and H^+ concentration cause rapid increases in breathing. Oxygen receptors, however, are sensitive only to large changes in blood P_{O_2}; they become stimulated only when the P_{O_2} of arterial blood has fallen below 70 mm Hg. Since changes in blood P_{CO_2} and P_{O_2} normally are proportional to one another, breathing is generally regulated by CO_2 receptors because they are much more sensitive than the O_2 receptors.

Figure 13.18 Electron micrographs of a cross section of a capillary in heart muscle. (a) The wall of a capillary consists of a single layer of flattened endothelial cells. This picture shows parts of two endothelial cells: the nucleus and part of the cytoplasm of one cell, and a small part of the cytoplasm of another cell, in the lower right part of the capillary. The pinocytotic vesicles have transported fluid across the cell from the capillary and are releasing their contents through the endothelial cells' plasma membranes into the interstitial fluid. (b) Enlarged view of part of this capillary wall showing the place where two endothelial cells join (*arrow*). Water and solutes flow through pores in these junction regions. Note the number of pinocytotic vesicles.

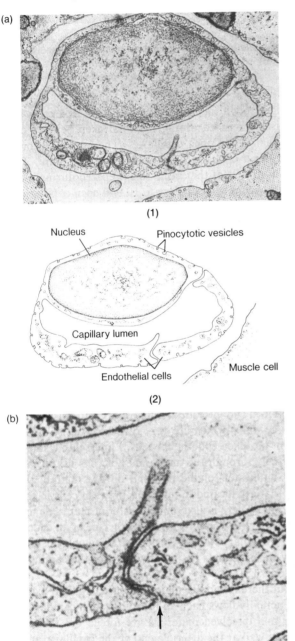

(a)

(1)

Nucleus

Pinocytotic vesicles

Capillary lumen

Endothelial cells

Muscle cell

(2)

(b)

Two forces cause this fluid movement through the capillary wall. The first, **hydrostatic pressure,** is the outward pressure exerted by a liquid against the walls of a container that holds it. In this case, the container is the capillary, and hydrostatic pressure is the blood pressure inside the capillary. The hydrostatic pressure at the arteriole end of capillaries is about 35 mm Hg. Pressure decreases as blood flows through the capillary so that near the venule end, the hydrostatic pressure has fallen to 15 mm Hg.

The second force involved in fluid movement through the capillary wall is **osmotic pressure.** The total concentration of ions and small molecules is very similar in blood and interstitial fluid, but the fluid portion of the blood, the blood **plasma,** has a relatively high concentration of protein molecules. These protein molecules do not pass freely through the capillary wall. The same kinds of protein molecules are found in interstitial fluid but in a much lower concentration. Because of this protein concentration difference, there is an osmotic pressure differential; the osmotic pressure of blood plasma is about 25 mm Hg higher than that of interstitial fluid.

Hydrostatic pressure tends to force fluid out through the pores in the capillary wall. This is a *filtering* process because water and small dissolved molecules are forced out while protein molecules and blood cells are kept in. Osmotic pressure, on the other hand, tends to cause water to move from the interstitial fluid to the blood because of the osmotic pressure differential. Thus, fluid movement through the capillary wall depends upon the relative strength of these two opposing forces. At the arteriole end of a capillary, osmotic pressure (the force tending to cause water to move in) is overcome by hydrostatic pressure (the force tending to drive water out). Therefore, water flows outward, carrying dissolved materials with it. Midway along the capillary, where hydrostatic pressure is lower, the two forces essentially cancel one another, and there is no net movement of water. At the venule end, osmotic pressure is greater than hydrostatic pressure, and water moves inward (figure 13.19). Not quite as much fluid returns to capillaries as is forced out, however. There is a small net loss of fluid from the blood as it passes through the capillaries. We will account for this fluid left in body tissues when we discuss the lymphatic system.

Fluid movement through capillary walls carries nutrients and other dissolved materials from the blood to the interstitial spaces and conveys metabolic wastes to the capillary. These essential fluid movements depend upon the maintenance of normal blood pressure and osmotic pressure relationships, and changes in either can alter fluid movement, with potentially serious consequences.

The Lymphatic System

We have noted that most, but not quite all, of the fluid forced out of capillaries returns to them. What happens to the fluid that remains in the interstitial spaces? Vertebrate animals have a special system of vessels, the **lymphatic system,** that drains fluid from interstitial spaces in the tissues and returns it to the blood, thus maintaining a balance between blood volume and interstitial fluid volume in the body.

In all parts of the body there are **lymph capillaries** that are closed at one end. Their walls are permeable to water and small molecules, and very permeable to protein molecules. Not only does the lymphatic system return excess interstitial fluid to the blood, it also transports proteins from the interstitial spaces to the blood. This mechanism is important because a few protein molecules leak through the capillary walls, and they must be returned to the blood if the normal and necessary osmotic differential between blood and interstitial fluid is to be maintained.

Once interstitial fluid enters the lymphatic system, it is called **lymph.** Lymph moves through lymph capillaries and **lymph vessels** as a result of pressure applied by muscle contractions near the vessels and a system of valves that prevent backward flow (figure 13.20). Some vertebrate animals have

Figure 13.19 Forces causing flow of water and solutes through a capillary wall. Near the arteriole end, hydrostatic pressure (*dark blue arrow*) is greater than the osmotic pressure difference (*yellow arrow*), so there is a net movement of water and solutes (*light blue arrow*) out of the capillary. Midway along the capillary, the forces cancel each other, and there is no net movement. Near the venule end, the osmotic pressure difference is greater than hydrostatic pressure, and water moves into the capillary.

At arteriole end

Hydrostatic pressure > Osmotic pressure

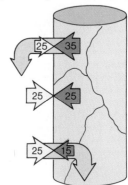

Osmotic pressure > Hydrostatic pressure

At venule end

Figure 13.20 Photomicrograph of a valve in a lymph vessel. Lymph is forced through lymph vessels as contracting muscles exert pressure on vessel walls. Valves maintain one-way flow through lymph vessels by preventing backward flow (magnification × 45).

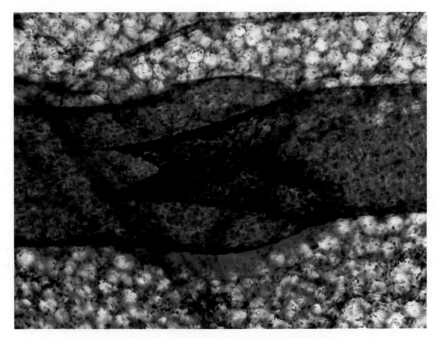

Figure 13.21 The lymphatic system. Lymph vessels flow into two main channels. The thoracic duct drains lymph vessels from all parts of the body except the upper right portion. Vessels from that area drain through the right lymphatic duct. These two large lymphatic ducts drain into the subclavian veins, thus returning fluid from the tissues to the circulatory system. Nodes filter debris and bacteria out of lymph. The inset is an enlargement of part of the intestinal lining showing the location of lacteals in the intestinal villi. Products of fat digestion pass through the lacteals to lymph vessels and eventually to the circulatory system.

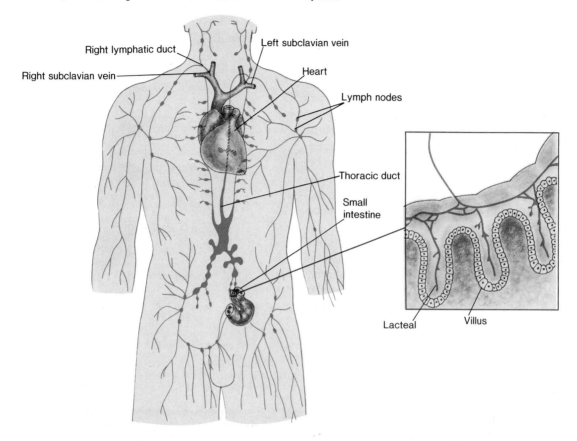

Right lymphatic duct

Right subclavian vein

Left subclavian vein

Heart

Lymph nodes

Thoracic duct

Small intestine

Lacteal

Villus

pulsating lymph hearts that move lymph along through their lymphatic systems, but human lymph movement is due to the pressure of muscles contracting adjacent to the vessels. Smaller lymph vessels unite to form larger vessels. Finally, two major **lymph ducts** empty into large veins near the heart (figure 13.21).

In addition to returning fluids from the interstitial spaces to the circulatory system, the lymphatic system has several other important functions. Lymph capillaries of the digestive tract, known as **lacteals,** are distributed in the intestinal villi where they are involved in transport of absorbed fats during digestion.

The lymphatic system also plays major roles in body defenses. **Lymph nodes** located along the lymph vessels contain a network of connective tissue fibers with **phagocytic cells** scattered among them. These phagocytic cells engulf dead cells, cell debris, and foreign objects such as bacteria as the lymph filters through the lymph nodes. Cells active in other defense responses (chapter 29) also reside in the lymph nodes. Because the lymph nodes are very active in body defense responses, they often become swollen and tender during infections. The so-called swollen "glands" that may accompany a sore throat are not glands, but active lymph nodes.

Anything that interferes with the flow of lymph through a particular lymph vessel or lymph node causes fluid accumulation in that part of the body. Not only is fluid drainage impaired when this occurs, but inadequate protein removal alters the osmotic balance between blood and interstitial fluid so that even more fluid accumulates in that tissue. Such fluid accumulation is called **edema,** and it can be visible externally as a swelling in that part of the body.

A very dramatic example of edema can be seen in the tropical disease **filariasis.** When infected mosquitoes bite humans, they transmit a parasitic worm into the bloodstream.

The larvae of these filarial worms invade the lymph vessels and interfere with (or completely block) the flow of lymph. Tissues in the area of the blocked lymph vessel can swell to enormous proportions. The condition is appropriately called **elephantiasis** (see figure 40.27).

Blood

Blood is a red liquid made up of formed elements (cells and cell fragments) and the extracellular fluid, the blood plasma. Plasma makes up about 55 percent of blood, while the cells and cell fragments constitute the remaining 45 percent. The formed elements of blood are **erythrocytes (red blood cells), leukocytes (white blood cells),** and **platelets**.

Red Blood Cells

Erythrocytes (red blood cells) are the most numerous of the formed elements. Usually there are four to six million red blood cells per cubic millimeter of human blood. Men generally have higher red cell counts than women. A healthy man has about 5.4 million erythrocytes per cubic millimeter, while a healthy woman has about 4.8 million per cubic millimeter, but these values vary with time and activity. For example, erythrocyte counts increase during exercise and following meals. At these times additional blood cells are mobilized from storage sites, especially from the **spleen,** an organ near the stomach that normally stores considerable quantities of blood cells. Longer term changes occur in people living at high altitudes; they consistently have higher red cell counts than people living at lower altitudes (see p. 280).

A human erythrocyte is a thin, biconcave disc with a very thin center and a thicker rim. A mature circulating erythrocyte does not have a nucleus because the nucleus degenerates during the erythrocyte's development (figure 13.22). Approximately one-third of an erythrocyte's volume is taken up by about 280 million hemoglobin molecules.

The hemoglobin inside erythrocytes reversibly binds with oxygen to form **oxyhemoglobin** (see p. 279). When most of its hemoglobin is in the form of oxyhemoglobin, blood takes on a bright red color. Such red, oxygenated blood is found in the pulmonary veins, which return blood from the lungs, and in the systemic arteries. Deoxygenated blood with relatively little oxyhemoglobin has a darker, more purple color. Characteristically, blood in systemic veins is deoxygenated because it has passed through capillary beds where it has given up oxygen to body tissues.

Having hemoglobin packed inside erythrocytes is adaptive; if hemoglobin were simply suspended in the plasma, blood would be more viscous (thick and syrupy), and blood pressure would have to be higher to force blood through the

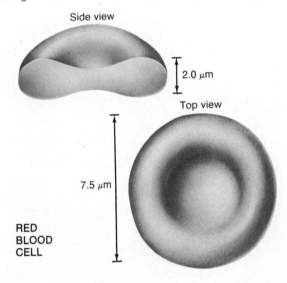

Figure 13.22 Erythrocytes. Human erythrocytes are flattened, biconcave discs. The side view shows how thin the center of an erythrocyte is. Erythrocyte shape apparently is an adaptation for more efficient gas diffusion.

Side view

2.0 μm

Top view

7.5 μm

RED BLOOD CELL

circulatory system. With hemoglobin located inside erythrocytes, human blood can contain enough hemoglobin to carry adequate quantities of oxygen while remaining a free-flowing fluid.

Erythrocytes normally are quite elastic; even though they are bent as they pass through small blood vessels, they spring back into shape. As erythrocytes age, however, they become more fragile, and such bending can damage or even rupture them. Aged, damaged erythrocytes are withdrawn from circulation as the blood passes through the liver or spleen. Large phagocytic cells called **macrophages,** which are especially common in the liver, phagocytize and destroy damaged or ruptured erythrocytes, and the hemoglobin from the erythrocytes is broken down by liver cells (figure 13.23). Iron, recovered from the heme portions of hemoglobin molecules, is exported from the liver and used again for hemoglobin synthesis in developing erythrocytes. The remainder of the heme parts of hemoglobin molecules are broken down by the liver cells and excreted as the greenish brown bile pigments, which are part of the bile that passes through the common bile duct to the intestine (see p. 251).

How old is an "aging" erythrocyte? The average life span of human erythrocytes is about 120 days. New erythrocytes must be produced as fast as old ones are destroyed, and they develop in the **bone marrow,** which is located in cavities inside bones. In humans, about two million new erythrocytes are produced each second, a rate equal to that of red cell destruction in a normal healthy person.

Figure 13.23 A macrophage (*M*) ingesting two erythrocytes (*Er*). Cytoplasmic extensions (*arrows*) of the macrophage are in the process of surrounding the erythrocytes. Macrophages continually remove aging and misshapen erythrocytes from circulation (magnification × 900).

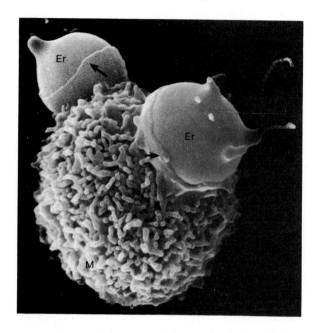

Figure 13.24 A diagrammatic representation of erythrocyte agglutination reactions in ABO blood typing. Plasma antibody types used in the test are indicated above the columns.

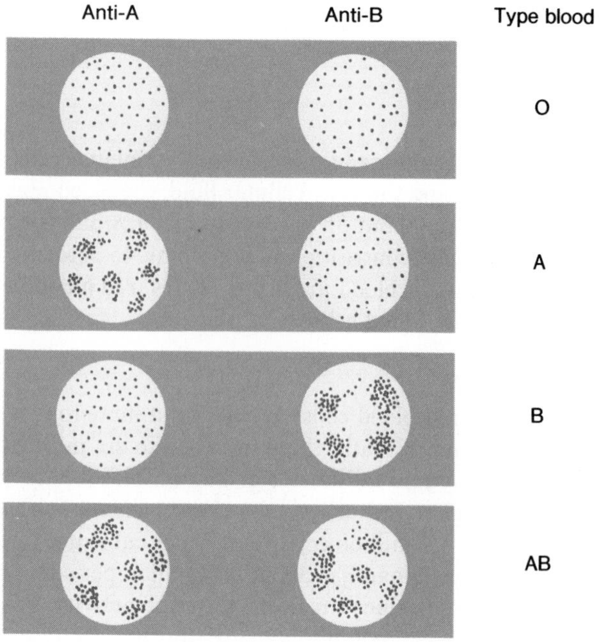

Because of the critical importance of erythrocytes for oxygen transport, their numbers must be precisely controlled if homeostasis is to be maintained. A major factor influencing the rate of erythrocyte production is a response by certain tissues to oxygen deficiency. An oxygen deficiency may develop if erythrocytes are not produced as rapidly as they are destroyed, thereby reducing the oxygen-transporting capacity of the blood. Kidney cells and, to a lesser extent, liver cells respond to oxygen deficiency by releasing a factor into the blood that causes a hormone, **erythropoietin,** to be released from the carrier molecule that normally binds it. Erythropoietin stimulates erythrocyte production in the bone marrow and thus increases the number of erythrocytes being added to the circulation.

Sometimes the mechanisms that normally balance erythrocyte production and destruction fail. For example, nutritional deficiencies or diseases cause slower than normal erythrocyte production. This results in a condition known as **anemia** in which there are fewer than the normal number of erythrocytes (or, in some cases, less hemoglobin in the erythrocytes). People suffering from anemia often are fatigued or chilled because their tissues lack adequate oxygen for cellular energy conversion or heat production.

Human Blood Types

The surfaces of human erythrocytes contain certain genetically determined sets of molecules, and people can be divided into groups or **blood types** depending upon which sets of molecules are present. Because these sets of molecules are genetically determined, a person's blood type is inherited. Two major classifications of human blood types are **ABO** grouping and the **Rh** system.

ABO grouping is based on differences in the type of glycoprotein molecule (see p. 58) present on erythrocyte surfaces. Type A people have A type glycoprotein molecules on their erythrocyte surfaces; type B people have B type glycoproteins; type AB people have both of these glycoproteins; type O people have neither of them.

The A and B glycoproteins function as **antigens;** that is, they combine specifically with **antibody** molecules. **Anti-A** antibody and **anti-B** antibody are protein molecules that occur in human plasma, and they combine specifically with the A and B antigens, respectively. We speak of antigens and the specific antibodies with which they combine as being **complementary.** When antigens combine with complementary antibodies, erythrocytes clump together. This happens because antibodies can combine with antigens on the surfaces of two different red cells, thus causing the cells to be bound

Table 13.2
Antigens and Antibodies in Human ABO Blood Groups

Blood Type	Antigen Present on Erythrocyte Membranes	Antibody in Plasma	Incidence of Type in the United States	
			Among Whites	Among Blacks
A	A	Anti-B	41%	27%
B	B	Anti-A	10%	20%
AB	A and B	Neither Anti-A nor Anti-B	4%	7%
O	Neither	Anti-A and Anti-B	45%	46%

to one another. Whole clumps of red cells can be stuck together in this way (figure 13.24). This cell clumping is called **agglutination.**

Normally humans do not produce antibodies to a particular antigen until they are exposed to it, but the ABO system is an exception. Humans routinely produce antibodies to those blood group antigens not present on their own blood cells without ever being exposed to them. Type A people have the anti-B antibody in their plasma; type B people have anti-A antibody; type AB people have neither of these antibodies; and type O people have both antibodies (table 13.2). Knowing this, you can see why it is vital to determine a person's blood type before giving him or her a blood transfusion. Donor and recipient must be compatible or a severe agglutination reaction can occur. For example, if type A blood is given to a type O person, the anti-A antibody in the recipient's blood combines with the A antigen on the surface of the donor cells and causes them to agglutinate. The clumps of cells formed can block blood vessels and cause serious circulatory problems.

Whenever possible, recipients should be given blood of the same ABO type as their own, but there are alternatives. AB people can receive blood of any of the ABO types because their plasma has neither of the antibodies. In an emergency, if the correct type is not available, type O blood can be given to a person having any ABO blood type because type O erythrocytes have neither of the antigens on their surfaces.*

The second major classification of human blood types is the Rh system. (*Rh* stands for **rhesus;** this factor was first identified in the blood of the rhesus monkey.) Rh classification is based upon the presence or absence of the **Rh antigen** on red blood cells. **Rh-positive (Rh⁺)** people have the Rh antigen while **Rh-negative (Rh⁻)** people do not. About 85 percent of all people in the United States are Rh-positive.

*Type O blood does, however, contain both anti-A and anti-B antibodies and therefore will cause a small amount of recipient blood cell clumping when given in a transfusion to persons with other blood types. Thus, giving type O blood to A, B, or AB persons is strictly limited to emergency situations in which blood of the proper type is not available.

In contrast to the situation with the ABO system, Rh-negative people do not normally have anti-Rh antibodies in their plasma. They must be exposed to the Rh antigen for the antibodies to develop. For example, anti-Rh antibodies develop in an Rh-negative person over a period of several months following an accidental transfusion of Rh-positive blood. Once this has happened, a second transfusion of Rh-positive blood can result in a severe agglutination reaction. Because such reactions occur only after two transfusions of blood of a wrong type, they are very rare.

A problem more likely to develop from Rh incompatibility is **erythroblastosis fetalis.** It can occur during a pregnancy in which an Rh-negative mother carries an Rh-positive fetus. This is a fairly common problem for Rh-negative mothers because the odds are high that their babies' fathers are Rh-positive, and the gene for development of the Rh antigen is dominant. Thus, these women's children are usually Rh-positive. Rarely does a problem develop during a first pregnancy of this sort because even if some fetal blood cells do reach the mother's bloodstream late in pregnancy or during delivery, antibody production is too slow to cause problems for the infant. However, a subsequent pregnancy can be a very different matter because a second exposure to Rh antigen results in much more rapid antibody production. Anti-Rh antibodies can then pass through the placenta and enter the fetal circulation, where they bind to fetal erythrocytes. This reaction causes the fetus to begin to destroy large quantities of its own erythrocytes, so that by the time of its birth, the infant is anemic and quite ill. It develops a jaundiced (yellowish) appearance because pigments produced during the destruction of its erythrocytes are circulating in its blood. These babies require a rapid series of transfusions to replace a large percentage of their blood.

A preventative for erythroblastosis fetalis is now available. An Rh-negative mother can be treated with a substance called RhoGAM after delivering her first Rh-positive child. RhoGAM binds Rh antigen so that it does not stimulate production of significant quantities of anti-Rh antibodies. This reduces the danger of erythroblastosis fetalis during the woman's second pregnancy.

Table 13.3
Leukocytes

Cell Type	Percentage of Leukocytes in Blood
Granular Leukocytes (polymorphonuclear cells)	
Neutrophils	55 to 65
Eosinophils	2 to 3
Basophils	about 0.5
*Agranular Leukocytes**	
Lymphocytes	25 to 33
Monocytes	4 to 7

*Agranular leukocytes are also called mononuclear cells because their nuclei are not subdivided into lobes, as are the nuclei of the granular leukocytes.

White Blood Cells

Leukocytes (white blood cells) contain nuclei and do not contain hemoglobin. They participate in body defense reactions and are capable of leaving the circulatory system by squeezing between the endothelial cells in capillary walls. In this way, leukocytes can act both inside blood vessels and in the spaces around tissue cells.

Leukocytes are far less numerous than erythrocytes. Normally, human blood contains 7,000 to 8,000 leukocytes per cubic millimeter. Thus, red cells outnumber white cells by about 700 to 1.

We can group leukocytes into two categories: granular and agranular (table 13.3). Granular leukocytes (granulocytes), so named because they contain distinctive cytoplasmic granules, are also called **polymorphonuclear leukocytes** because their nuclei have lobes and are highly variable in shape (*polymorph* = "many form" or "many shape"). The granular leukocytes, **neutrophils, eosinophils,** and **basophils,** are illustrated in figure 13.25.

Monocytes and **lymphocytes** are agranular leukocytes. These cells do not contain cytoplasmic granules and each has a single, relatively large nucleus. Monocytes regularly move into and out of the circulatory system through capillary walls. Many of them remain in body tissues where they enlarge to become important phagocytic cells, the **macrophages.** Lymphocytes, which also travel through the lymphatic system, are involved in very specific defensive responses to foreign antigens, such as those of infectious disease organisms. Some lymphocytes mount direct cellular attacks on foreign cells, while others differentiate to become **plasma cells,** the cells that manufacture specific antibodies against foreign antigens (see chapter 29).

Platelets

Platelets are small disc-shaped cell fragments that lack nuclei. They average 2 to 4 μm in diameter, and a cubic millimeter of blood contains between 200,000 and 400,000 platelets. Platelets are involved in the formation of blood clots and in this way function in preventing fluid loss from the circulatory system. Platelets are produced in the bone marrow, where they are pinched off as small fragments of large cells called **megakaryocytes.** Platelets must be produced in large quantities because their life span is only about one week.

Blood Plasma

Although blood plasma is 90 to 92 percent water, it is much more than just a liquid medium that suspends blood cells and platelets. Blood plasma contains a complex mixture of inorganic and organic molecules, including a number of inorganic ions, various organic nutrients, nitrogenous wastes, special products such as hormones, plasma proteins, and dissolved gases.

Ions and salts make up about 0.9 percent of the plasma. By far the most numerous ions are sodium (Na^+) and chloride (Cl^-) ions. Other important ions in the plasma are potassium (K^+), calcium (Ca^{2+}), magnesium (Mg^{2+}), bicarbonate (HCO_3^-), phosphate ($H_2PO_4^-$ and HPO_4^{2-}), and sulfate (SO_4^{2-}). These ions contribute to the normal osmotic pressure of the blood and are involved in pH regulation. Some of these ions are required in different amounts for normal cell functioning, and they must be delivered to all body tissues by way of the blood.

Some of the important organic nutrients in the blood are glucose, amino acids, and various lipids, including fats, phospholipids, and cholesterol. Nutrients are transported in the blood following absorption from the digestive tract. They also are transported by the blood from nutrient storage sites, such as the liver, to various body tissues.

Most of the nitrogenous waste in human blood is urea. When amino acids are deaminated, ammonia is produced. In the liver, this very toxic ammonia is quickly incorporated into urea, which is much less toxic (see p. 322). One very serious consequence of liver disease may be the loss of the liver's ability to accomplish this, and ammonia poisoning can result.

Plasma proteins make up 7 to 9 percent of the blood plasma. The three main types of plasma proteins are the **albumins,** the **globulins,** and **fibrinogen.** About 60 percent of the total plasma proteins are albumins, which are produced by the liver. The albumins are particularly important in maintaining the blood's osmotic pressure, which, as we saw earlier, helps to balance the fluid movement between capillaries and interstitial spaces. Some globulins function in transport because various molecules are transported while

Figure 13.25 A representation of human blood cells. Shown are erythrocytes (*Er*), three kinds of granular leukocytes (eosinophil—*Eo,* neutrophils—*N,* and a basophil—*B*), a monocyte (*M*), a lymphocyte (*L*), and platelets (*P*). The names of the granular leukocytes are based on the ways in which the cells stain when stained blood smears are prepared. Neutrophils are the most numerous of all leukocytes. Several types of leukocytes are phagocytic cells that engulf and destroy bacteria, viruses, and scraps of damaged tissue both inside the circulatory system and in spaces around tissue cells. We will examine the roles of lymphocytes in disease resistance and immunity in chapter 29.

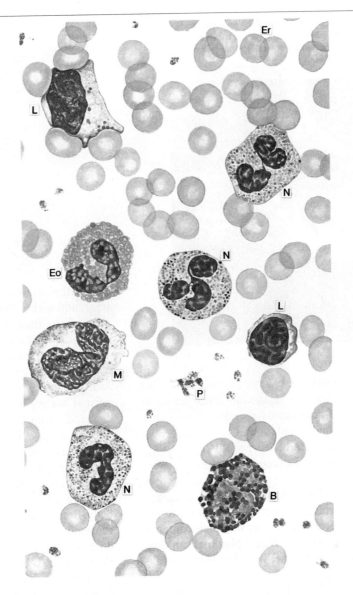

attached to them. Probably the most familiar of the globulins are the **gamma globulins.** Gamma globulins include specific antibody molecules, the **immunoglobulins.** Immunoglobulin molecules are produced by plasma cells in response to the presence of foreign antigens (see page 712). Fibrinogen, which is synthesized in the liver and is the largest of the plasma protein molecules, makes up only about 4 percent of the total plasma protein, but it plays a critical role in the blood-clotting process.

Blood Clotting
Blood must be a free-flowing liquid if it is to circulate easily through blood vessels under moderate blood pressure. However, this liquidity can cause serious problems. Any injury

that breaks blood vessel walls can quickly lead to serious blood loss. This life-threatening danger of excessive blood loss is countered by the complex **clotting (coagulation)** mechanism. Clots are plugs that form temporary barriers to blood loss until vessel walls have healed.

Blood clotting involves a chain of reactions, each reaction depending upon completion of the reaction preceding it in the sequence. More than thirty substances are known to be involved in blood clotting, and the effects and interactions of a number of these substances are very complex. Furthermore, a number of substances either promote or inhibit clotting. Promoting substances are called **procoagulants,** and inhibiting substances are called **anticoagulants.** The complexity of the blood clotting process seems to have evolved

Problems of the Circulatory System

The circulatory system is subject to a number of other diseases and problems, some of which affect the blood, while others affect blood vessels and the general circulation.

Infectious Mononucleosis

Mononucleosis is a virus-induced disease in which the total leukocyte count becomes abnormally elevated to about 15,000 per mm^3. Most of the extra leukocytes are oversized lymphocytes or abnormal monocytes. There is no specific treatment for mononucleosis, but most people recover in a few weeks if they rest and receive treatment for any additional problems that might have resulted from their reduced defenses against other invading organisms.

Leukemia

Leukemia is a form of cancer in which there is a continuing excessive production of abnormal white blood cells. Production of these extra white cells so dominates the function of the bone marrow that the normal production of erythrocytes and platelets is depressed. Thus, many leukemia victims become anemic, and they are also susceptible to hemorrhage because of platelet deficiency and the resultant faulty blood clotting. A common cause of death in leukemia cases is brain damage resulting from cerebral hemorrhaging. Infection is another frequent cause of death because there are insufficient normal leukocytes for body defense mechanisms.

Atherosclerosis

Atherosclerosis is an arterial disease that contributes to nearly one-half of all deaths in the United States. This common condition involves the formation of soft masses of fatty material in blood vessel linings (figure 13.29). These fatty masses, called **plaques,** contain large quantities of cholesterol. They often form in arteries, making the arterial lining much rougher than normal. This roughening tends to promote thrombus formation and leads to problems with embolisms. As a plaque develops, it decreases the diameter of the blood vessel and impedes blood flow. The formation of calcium deposits in the plaque and degenerative changes in the arterial wall lead to hardening of the artery. Hardened (sclerotic) arteries have lost their elasticity and are thus susceptible to rupture. The rupture of hardened cerebral arteries, which often results in brain damage, is called a **stroke.**

Plaque also can break loose and circulate until it blocks a small blood vessel or its rough surfaces cause formation of a clot that blocks a vessel. This is especially devastating if the blocked vessel is one of the coronary arteries of the heart. The portion of the heart muscle denied a blood supply is said to be **infarcted,** and the whole process is called **myocardial** (heart muscle) **infarction.** Often there are simultaneous disturbances of the impulse-conducting system so that rapid and uncoordinated heart beating occurs. These are symptoms of a heart attack. If the victim survives the initial crisis of a heart attack, there is a long recovery period during which dead muscle tissue in the infarcted area is replaced by fibrous tissue. Most people who recover from a myocardial infarction can return to normal activities, but they must be careful about engaging in activities that place heavy demands on their hearts. Their hearts' ability to meet extra circulatory requirements is reduced because part of the heart muscle has been permanently lost.

Hypertension

Blood vessel and heart problems all are related to or aggravated by another circulatory problem, high blood pressure (**hypertension**).

Hypertension is a very common disease affecting the heart and blood vessels. Hypertension is diagnosed when the diastolic blood pressure continually exceeds 95 mm Hg. It is estimated that as many as ten percent of American adults suffer from hypertension, and fewer than one-half of these people are aware of it. Hypertension increases the possibility of blood vessel rupture with subsequent damage to the tissue supplied by the vessel. High blood pressure is a major contributing cause of strokes.

About 85 percent of hypertension cases cannot be attributed to organic causes. The 15 percent that *can* be are due to hardening of the arteries, kidney disease, or hormonal imbalances. In these latter cases, hypertension is treated by treating the condition that causes it. However, most cases must be treated directly with drugs that reduce blood pressure.

Circulatory System Health

The circulatory system is prone to many types of problems. Nothing can be done about inherited circulatory system tendencies, but positive steps can be taken to improve and maintain circulatory system health.

Dietary moderation is a good general rule. Obesity increases circulatory demands and the heart's work load. Also, people whose diets contain large quantities of fat are more likely to develop atherosclerosis than people whose diets are low in fat. Reducing salt in the diet may be advisable, because high salt consumption may be correlated with various circulatory problems, especially hypertension.

Not smoking or stopping smoking also can preserve circulatory system health. Cigarette smoking is very strongly correlated with increased risk of heart and blood vessel disease.

Finally, sensible exercise programs can contribute substantially to maintenance of circulatory system health.

Figure 13.29 Atherosclerosis. (a) An X-ray picture taken after a substance that appears opaque on X-ray photographs was injected into the bloodstream. The arrow indicates an area where plaque formation has decreased the diameter of the femoral artery in the thigh. (b) A diagram showing plaque accumulation in an artery. The circular broken line indicates the approximate size of the normal lumen. Note that the lining in the plaque area is much rougher than normal endothelium. (c) Changes in arteries affected by atherosclerosis. (1) A normal artery. (2) Advanced atherosclerosis has led to hardening of this artery's wall. (3) This sclerotic artery is completely occluded by a clot.

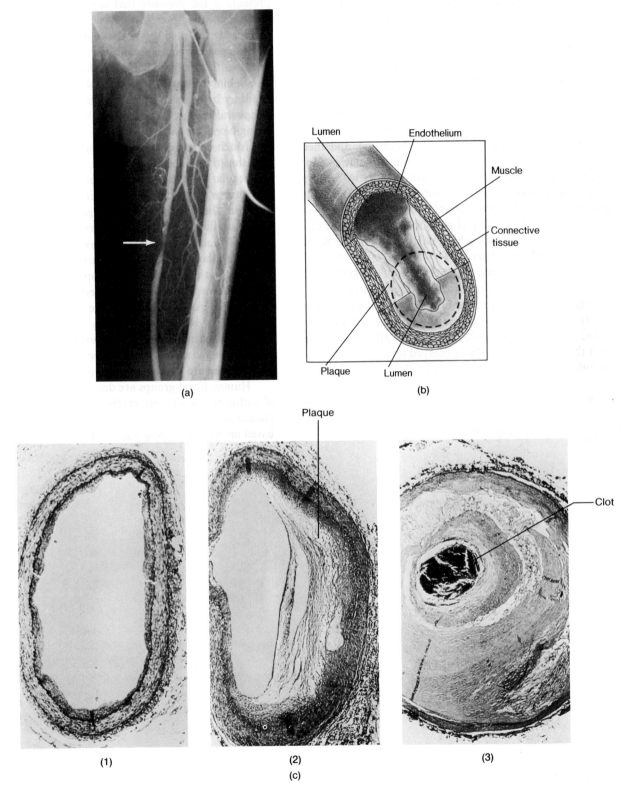

(a)

(b)

(c)

(1)   (2)   (3)

dioxide. The liver stores nutrients and processes metabolic waste products. Water, some ions, and some waste molecules are lost through the skin. The kidneys eliminate nongaseous body wastes, conserve body water and nutrient molecules, and are involved in maintenance of ion balance and in pH regulation in the body.

Although all of these functions are important, in this chapter we will focus on the contributions of the kidneys and other specialized animal excretory organs to the maintenance of body fluid homeostasis.

Excretion of Nitrogenous Wastes

Excretion is the process by which wastes, excess materials, and toxic substances are removed from the body's cells and extracellular fluids. Some of the most important substances that must be excreted from the body are the **nitrogenous** (nitrogen-containing) **waste** compounds produced by cellular protein metabolism. Amino acids, derived from proteins in food, can be used by cells for synthesizing new proteins or other nitrogen-containing molecules. Amino acids the body does not use for synthesis are either oxidized to generate energy or converted to fats or carbohydrates, which can be stored. Before this can happen, however, nitrogen-containing **amino groups** (-NH$_2$) must be removed from amino acids. This **deamination** (amino group removal) during amino acid breakdown is the source of the nitrogenous wastes (see p. 182). In humans, deamination reactions occur mainly in the liver. The following deamination reaction is a key step in nitrogenous waste production:

$$\text{Glutamate} + \text{NAD}^+ + \text{H}_2\text{O} \underset{}{\overset{\text{Glutamate dehydrogenase}}{\rightleftharpoons}} \alpha\text{-ketoglutarate} + \text{NH}_3 + \text{NADH}$$

Glutamate is an amino acid; glutamate dehydrogenase is the enzyme that catalyzes the deamination of glutamate; α-ketoglutarate is an organic acid that can be oxidized in the Krebs cycle; and **ammonia** (NH$_3$) is the nitrogenous waste molecule produced by this reaction.

Ammonia is a product of deamination; it is also the simplest nitrogenous waste compound. Because ammonia is quite toxic and very water soluble, it can be used as a nitrogenous excretory product only if there is a good deal of water available in which it can diffuse away. This limits direct ammonia excretion to organisms that live in water, such as protists, aquatic invertebrates, bony fish, and aquatic amphibians.

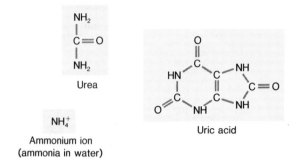

Figure 14.4 Chemical formulas of the three major nitrogenous wastes of animals. Note that at the pH's inside animals' bodies, ammonia gains a proton to become an ammonium ion.

Other organisms must excrete nitrogenous wastes in a different form. Many animals incorporate waste ammonia into organic molecules, mainly either **urea** or **uric acid,** which are then excreted (figure 14.4). Synthesis of these organic waste molecules, however, requires that these animals expend more energy on nitrogen excretion than ammonia-excreting organisms.

Urea is much less toxic than ammonia and it can be excreted in a moderately concentrated solution. Thus, body water is conserved. This is an important advantage for terrestrial animals with limited access to water. For this reason, a number of terrestrial organisms, such as mammals and adult amphibians, excrete urea as their main nitrogenous waste product.

Urea is synthesized in the livers of mammals and amphibians by a set of reactions known as the **urea cycle.** This cycle was discovered by Sir Hans Krebs, the same biochemist who played a major role in the discovery of the reactions of the Krebs cycle. As you can see in figure 14.5, two amino groups are incorporated into each urea molecule, but because several ATPs are used in the synthesis of each urea molecule, this is a much more metabolically "expensive" means of excreting excess nitrogen than is the use of ammonia. Presumably, for an organism whose supply of water is limited, the lower toxicity of urea and the need for water conservation make the use of urea as a nitrogenous waste molecule worth the "cost."

Uric acid (see figure 14.4) is the third major nitrogenous excretory product. Uric acid is not very toxic nor very soluble in water. This poor solubility is an advantage in terms of water conservation because uric acid can be concentrated more readily than can urea. In fact, uric acid leaves the bodies of animals that excrete it as a damp mass of semicrystalline material. Animals that excrete uric acid recover most of the water from this semisolid urine before it leaves their bodies,

Figure 14.5 The urea cycle is a series of reactions by which urea is formed. In vertebrate animals, urea is produced in the liver. The parts of the urea molecule being assembled in this reaction are enclosed in colored boxes. After fumaric acid has been formed from argininosuccinic acid, it can be converted back to oxaloacetate by the activity of Krebs cycle enzymes. The oxaloacetate is then ready to accept an amino group from an amino acid. The ammonia used in carbamoyl phosphate synthesis can arise from the deamination of glutamic acid, as well as from other sources. Note that this is an energy-requiring process because ATP must be used at several points.

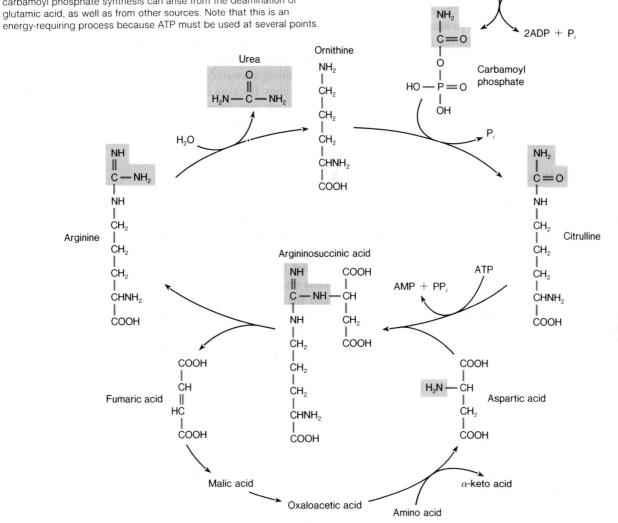

and thus, uric acid excretion is advantageous for animals living in very dry environments. Reptiles, birds, and insects excrete mainly uric acid. This form of excretion is adaptive even for birds that have access to plentiful water supplies, because liquid urine would be heavier for them to carry during flight.

One disadvantage of uric acid is that it is synthesized by a long, complex series of reactions that require expenditure of even more ATP than does urea synthesis. Here again, there is a trade-off between the advantage of water conservation and the disadvantage of energy expenditure for synthesis of an excretory waste molecule.

One other aspect of uric acid excretion adds to its adaptive value, however. Patterns of nitrogen excretion are related not only to the availability of water in the adult environment but also to the embryonic environments of animals. For example, reptile and bird embryos develop inside completely enclosed eggs. All nutrients and water required for metabolism and growth of the embryo must be inside the egg before embryonic development begins. Such **cleidoic** ("boxlike") **eggs** must also have a mechanism for nitrogenous waste storage because there is no way for nitrogenous waste to leave the confines of the shell until the embryo

The end of each nephron drains into a **collecting duct** that receives fluid from several nephrons in an area of the kidney, conveys fluid down through the medulla, and delivers urine to the pelvis. The loops of Henle and the collecting ducts give the pyramids of the medulla their striped appearance.

Each nephron is supplied by an **afferent arteriole** that branches into the capillary knot making up the glomerulus. The glomerular capillaries drain into an **efferent arteriole,** which subsequently branches into a second capillary network around the tubular parts of the nephron. These capillaries, called **peritubular** ("around the tubules") **capillaries,** pick up material reabsorbed from the tubules and deliver materials that are secreted into the tubules. The capillaries drain into venules, and these flow together into veins that carry blood to the single renal vein leaving the kidney.

Renal Function

The functioning of human nephrons in body fluid regulation involves three basic processes: filtration, reabsorption, and secretion. Fluid from the blood is filtered through the walls of the capillaries of the glomerulus into the space inside Bowman's capsule, a process called **glomerular filtration.** This fluid (**glomerular filtrate**) is identical to the blood in composition except that it does not contain blood cells or plasma proteins. As the fluid passes through the nephron, nutrients, essential ions, and water are reabsorbed and passed to capillaries. This **tubular reabsorption** recovers a number of vital substances. Certain other substances, delivered to the tubule area by the capillaries, are actively secreted by cells in the tubule walls. Through this **tubular secretion,** additional wastes are added to the fluid for eventual urinary excretion (figure 14.20). Let us look at each of these important processes more closely.

Figure 14.20 The principles of renal function. In the glomerulus, water and solutes are filtered into the kidney tubules while proteins and cells remain in the blood. As the glomerular filtrate passes through the proximal and distal convoluted tubules, solutes may be actively reabsorbed from the filtrate or secreted into it.

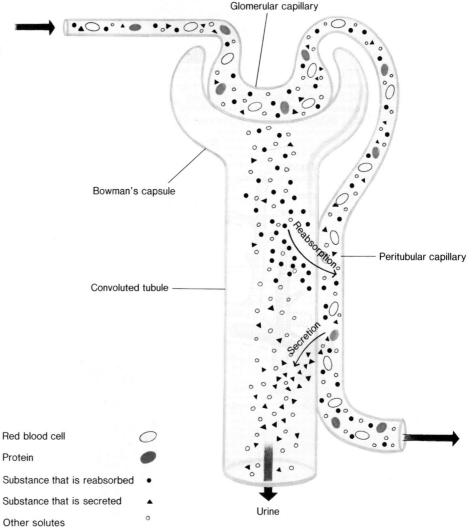

Glomerular capillary

Bowman's capsule

Reabsorption

Peritubular capillary

Convoluted tubule

Secretion

Red blood cell

Protein

Substance that is reabsorbed

Substance that is secreted

Other solutes

Urine

Glomerular Filtration

The glomerulus inside Bowman's capsule has a large surface area available for filtration because it consists of a great number of capillaries (figure 14.21). The walls of these capillaries are quite porous; they are about 100 times more permeable to water, small solute molecules, and ions than are the walls of most capillaries elsewhere in the body. Water and some solutes readily pass through the capillary walls and enter the lumen in the nephron in Bowman's capsule. However, blood cells and larger molecules, such as plasma proteins, cannot pass through the pores and thus are retained in the blood. This selective passage of materials through the glomerular capillary walls constitutes glomerular filtration.

The driving force for glomerular filtration is the blood pressure (hydrostatic pressure) inside glomerular capillaries. Blood pressure is very effectively applied to the process of filtration because the efferent arteriole is smaller than the afferent arteriole (see figure 14.21a), and it offers resistance to blood leaving the glomerulus. Blood pressure acts to push fluid out through glomerular capillary walls because it is stronger than the forces that oppose it (figure 14.22).

About 125 ml of fluid are filtered in the two human kidneys every minute. This means that about 180 liters of fluid leave a person's blood and enter the Bowman's capsules each day. We humans, of course, do not produce and void 180 liters of urine a day because our kidneys recover more than 99 percent of this glomerular filtrate and return it to the blood leaving the kidney.

Tubular Reabsorption and Secretion

Both solutes and water are reabsorbed from the nephron tubules into the surrounding capillaries. Most of this reabsorption takes place in the proximal convoluted tubule, where active transport mechanisms remove materials from the filtrate inside the tubule. Some substances recovered by active transport are nutrients (glucose and amino acids) and essential ions (Na^+, K^+, Ca^{2+}, Mg^{2+}, HCO_3^- and HPO_4^{2-}). Other substances pass through the tubule walls by diffusion.

Figure 14.21 The glomerulus. (a) A diagram showing the glomerulus as a knot of capillaries inside a Bowman's capsule. Filtration forces fluid out through the walls of the capillaries into the cavity of Bowman's capsule and on into the proximal tubule. The juxtaglomerular apparatus is sensitive to fluid pressure. It responds to a decrease in sodium or blood pressure. (b) A scanning electron micrograph of a section of kidney cortex showing a glomerulus inside a Bowman's capsule. The holes surrounding the glomerulus are cross sections of tubules.

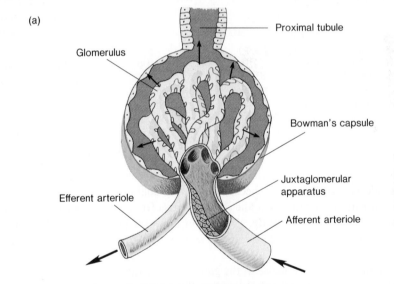

(a)

Glomerulus

Proximal tubule

Bowman's capsule

Juxtaglomerular apparatus

Efferent arteriole

Afferent arteriole

(b)

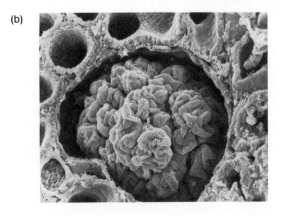

Figure 14.22 Diagrammatic representation of the forces involved in glomerular filtration. Each force favors fluid movement in the direction indicated by its arrow, and lengths of the arrows are proportional to the magnitude of the forces operating in the glomerulus. Glomerular capillary pressure (blood pressure) tends to force fluid from the capillary to Bowman's capsule. Blood pressure is opposed by a small fluid pressure in Bowman's capsule (P_B) and by an osmotic pressure factor (OP), which is due to blood protein molecules that cannot pass out of the capillaries into the Bowman's capsule space.

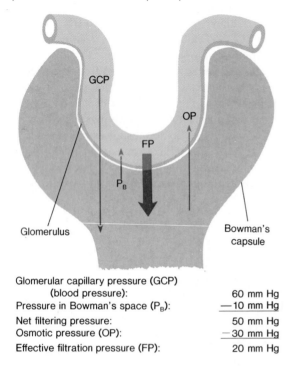

Glomerular capillary pressure (GCP) (blood pressure):	60 mm Hg
Pressure in Bowman's space (P_B):	—10 mm Hg
Net filtering pressure:	50 mm Hg
Osmotic pressure (OP):	—30 mm Hg
Effective filtration pressure (FP):	20 mm Hg

The reabsorption of materials by the proximal convoluted tubule is a vital function because the body could not afford to lose all of the nutrients and ions that pass from the blood into the glomerular filtrate. We can get a clearer picture of reabsorption by considering the case of one nutrient, glucose.

Normally, all of the glucose in the glomerular filtrate is reabsorbed in the proximal tubule, unless the filtrate contains more than 180–200 mg of glucose per 100 ml, which it does only if the blood glucose concentration is higher than that level. When the blood glucose level is higher, the active transport mechanism in the proximal convoluted tubule—even working at its maximum capacity—cannot recover all of the glucose from the filtrate, and some of it passes out in the urine. Therefore, there is a **renal threshold** for glucose; that is, there is a blood glucose concentration above which

glucose cannot be fully reabsorbed by the kidney. In a normal, healthy person, the renal threshold for glucose usually is exceeded for only short periods following ingestion of a great deal of sugar. However, blood glucose levels rise quite high in people with uncontrolled **diabetes mellitus** ("sugar diabetes"), and significant quantities of glucose regularly appear in their urine.

Because the kidney has a threshold for every substance it reabsorbs, the blood levels of a number of substances are automatically kept below specific limits. If the plasma concentration of a particular substance exceeds its renal threshold, the excess is excreted in the urine. This very important homeostatic mechanism acts to regulate the content of the blood.

When ions, glucose, and other solutes are reabsorbed from the tubule, the fluid inside the tubule is made more dilute than its surroundings. As a result, a large quantity of water leaves the tubule osmotically and is reabsorbed into the body.

Tubular secretion provides a means by which additional wastes can be added to the fluid passing through the tubules. Toxic substances, such as foreign acids and bases that have been absorbed in the gut, are eliminated by tubular secretion. The antibiotic penicillin and a number of other substances also are excreted in this way. But the process of tubular secretion is not limited to foreign substances. Many nitrogenous compounds produced in the body are expelled by active tubular secretion.

Production of Hyperosmotic Urine

All terrestrial animals, including humans, must conserve as much water as possible while at the same time excreting nitrogenous wastes and other materials. How is water conservation accomplished in the human kidneys?

Human kidneys produce concentrated urine and, if necessary, can eliminate body wastes in as little as 450 ml of urine a day. Most people with adequate access to drinking water, however, normally produce several times that amount of urine. A major portion of the water in the kidney tubules is reabsorbed from the kidney's collecting ducts. This reabsorption of water is not by active transport; there is no cellular water "pump" in the human kidney. Instead, water is recovered because a very salty environment is maintained in the medulla of the kidney where the collecting ducts are located. Because this environment is hypertonic to the fluid in the collecting ducts, water tends to move out of the collecting ducts by osmosis. Thus, this hypertonic environment maintained in the kidney medulla is the key to body water conservation.

Box 14.2

The Kangaroo Rat: A Mammal That Does Not Need to Drink

The kangaroo rat, which gets its name from its way of hopping on its hind legs, performs some remarkable feats of water balance maintenance in a very dry environment. This small rodent thrives in the desert regions of the southwestern United States, even in Death Valley where no drinking water is available. How does this animal manage to live in its harsh, dry environment?

The kangaroo rat lives in burrows that are deep enough to be much cooler and more humid than the desert ground surface in the daytime. A nocturnal animal, the kangaroo rat remains in its burrow during the warmer hours of the day and ventures out at night to search for the seeds that make up its diet. This avoidance of daytime heat and dryness minimizes evaporative water loss.

Some animals living in waterless environments survive by eating succulent leaves that have a high water content, but the kangaroo rat eats mainly dry seeds. The water content of the seeds is so low that they supply only 10 percent of the animal's daily water requirement. The remainder of the rat's water gain is provided by the metabolic water formed as a result of oxidation of food material. Oxidation of one gram of carbohydrate yields 0.6 g of water, and oxidation of a gram of fat yields 1.1 g of water. The seeds the kangaroo rat eats are high in fat and carbohydrate content and low in protein. The low protein content of the seeds is important, because when protein is used for energy more water is required for excretion of nitrogenous wastes.

Because the kangaroo rat's water supply is so meager, it must conserve water very efficiently. It reabsorbs so much water from its large intestine that its fecal material is almost completely dry. In addition, the kangaroo rat has exceptionally long loops of Henle in its kidneys. These enable the animal to produce urine more than three times as concentrated as human urine. Thus, a kangaroo rat loses very little water through its kidneys.

Biologists have demonstrated the efficiency of kangaroo rats' kidneys in experiments in which they have substituted soybeans, which are high in protein, for kangaroo rats' normal diet of high-fat, low-protein seeds. This substitution necessitated the use of more water for excreting nitrogenous waste. If the rats were supplied with seawater to drink, however, they managed nicely. We humans cannot maintain water balance while drinking

seawater because our kidneys do not produce urine that is sufficiently concentrated to avoid extra water loss as a result of salt intake. However, the more efficient water reabsorption mechanism of the kangaroo rats' kidneys enabled them to use seawater as a supplemental water source in the experiments.

Most of the total water loss from a kangaroo rat's body is by evaporation, mainly from its lungs. Even this loss is kept to a minimum because the animal's long nose allows exhaled air to be cooled. The exhaled air then carries out less precious body water than it would if it remained at body temperature.

Its minimal water needs and highly efficient methods of water conservation make the kangaroo rat remarkably well adapted to life in a dry environment.

Box figure 14.2 The kangaroo rat (*Dipodomys ordii*).

Figure 15.9 Relationship between the hypothalamus and the posterior lobe of the pituitary. Neurosecretory cells in the hypothalamus synthesize ADH and oxytocin. The hormones are transported down axons to the posterior lobe and released into the bloodstream there.

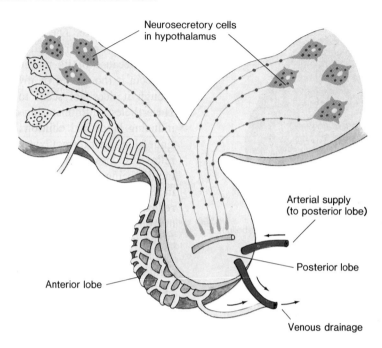

Neurosecretory cells in hypothalamus

Arterial supply (to posterior lobe)

Posterior lobe

Anterior lobe

Venous drainage

nally thought to be synthesized and released by the posterior lobe of the pituitary, but more recent studies have clearly demonstrated that they are products of neurosecretory cells in the hypothalamus, and that they travel down axons to the posterior lobe, where they are stored until their release (figure 15.9). Thus, ADH and oxytocin are actually products of brain cells, and the posterior lobe of the pituitary functions much like the insect corpus cardiacum in that both serve as release sites for hormones synthesized by neurosecretory cells of the brain.

The Anterior Lobe

The hormones of the anterior lobe have a wide range of effects on many processes in developing and adult mammals. However, the effects of **growth hormone** are perhaps best known because they are so obvious—and even dramatic, in some cases. If the secretion of growth hormone is insufficient during childhood and adolescence, an individual will become a pituitary midget (figure 15.10), while excess secretion during the same period produces a pituitary giant. Therapy for growth hormone deficiencies has been difficult because until recently, it has utilized growth hormone extracted from donated human pituitaries. Genetic engineering techniques (see chapter 25) are now producing a better supply of human growth hormone.

The oversecretion of growth hormone in an adult causes the toes, fingers, and face to resume growth. This results in a condition called **acromegaly,** which is characterized by distorted facial features, including an enlarged jaw and thick, heavy eyebrow ridges.

There are, however, complications in interpreting the effects of varying levels of growth hormone because one of its actions is stimulation of the liver to produce several polypeptide growth-promoting substances, including **somatomedin,** that appear to be mediators of growth hormone's effect. These substances are also known as **insulin-like growth factors;** they cause some of the effects of insulin (p. 364) in addition to many effects similar to those of growth hormone. Some very small people may actually produce adequate amounts of growth hormone, but apparently synthesize smaller quantities of these growth factors. This seems to be true, for example, of Central African pygmies.

Figure 15.10 Result of a childhood growth hormone deficiency. "General" Tom Thumb (Charles Stratton) grew to be only about 84 cm tall. Tom Thumb, like other pituitary midgets, had normal body proportions despite his small size. He is shown here with P. T. Barnum, the showman who exhibited him.

Growth hormone is also known as **somatotropic hormone (STH)**. The suffix *tropic* comes from a Greek word meaning "a turning"; these hormones give directions to their target cells.* The anterior lobe also produces **thyrotropic hormone,** commonly called **thyroid stimulating hormone (TSH),** and **adrenocorticotropic hormone (ACTH).** TSH stimulates the thyroid gland to produce its hormones, while ACTH stimulates the adrenal cortex to produce a group of steroid hormones.

Other tropic hormones from the anterior pituitary include the gonadotropic hormones, primarily **follicle stimulating hormone (FSH)** and **luteinizing hormone (LH),** which stimulate sex hormone production (estrogens and progesterone in females; testosterone in males) as well as egg or

*Many biologists instead use the suffix *trophic,* from a Greek word meaning to feed or cause to grow. You will probably encounter both endings when you read further about hormones in scientific literature.

sperm development. We will consider these relationships in chapter 27 in our discussion of human reproduction.

Another anterior lobe hormone, **prolactin,** is produced in the anterior lobes of both female and male mammals, but its function is clearly understood only in females. Prolactin (along with other hormones such as thyroxin and ACTH) acts on the mammary glands, causing them to produce milk after they have been prepared to do so by female sex hormones. The roles of prolactin and oxytocin differ in that the former is necessary for milk production, while the latter is necessary for milk ejection from the glands. More than thirty functions have been proposed for prolactin in male mammals, including the regulation of mating behavior, but these functions seem to vary from species to species.

The final anterior lobe hormone we will mention here is **β-lipotropin,** a polypeptide that functions in fat metabolism. β-lipotropin is especially interesting in that it is the precursor of several smaller hormone molecules (see p. 371).

The Intermediate Lobe

Not all vertebrates have a distinguishable, functional intermediate pituitary lobe. Adult humans, for example, lack this third lobe. In the vertebrates that do have it, however, the intermediate lobe is the primary source of a polypeptide hormone called **melanocyte stimulating hormone (MSH),** which is very similar in structure to ACTH. MSH causes the dispersal of pigment granules in the skin's pigment cells (melanocytes), thereby causing the skin to darken (figure 15.11). This response to MSH is especially important in amphibians that lighten and darken in response to changes in the color of their surroundings.

The Hypothalamus and the Pituitary

For years, certain facts have indicated that the brain might influence the pituitary. For example, it has long been known that stress can cause the pituitary to secrete more ACTH, thereby causing stimulation of the adrenal glands. But certainly the pituitary itself does not perceive and respond to stress. Similarly, in some species, the mere sight of a potential mate stimulates the release of pituitary gonadotropins, which in turn stimulates the production of sex hormones by the gonads. Again, how does this happen?

These responses are additional examples of neuroendocrine reflexes. There is a close relationship between the hypothalamus and the posterior pituitary; the two structures are connected by neurosecretory fibers. Years of careful anatomical studies, however, failed to show similar nervous

Figure 15.11 The effects of melanocyte stimulating hormone (MSH) in amphibians. (a) Frog (*Rana pipiens*) pigment cells stimulated by MSH. Pigment spreads inside the cells, making the skin appear darker (magnification × 228). (b) In the absence of stimulation, pigment contracts into small dots inside cells making the skin appear lighter (magnification × 228). (c) Control larva of the clawed frog *Xenopus laevis* with its pituitary intact. The pituitary is necessary for normal development and spreading of pigment cells. (d) *Xenopus* larva from which the pituitary gland has been removed. Note the absence of dark pigmentation.

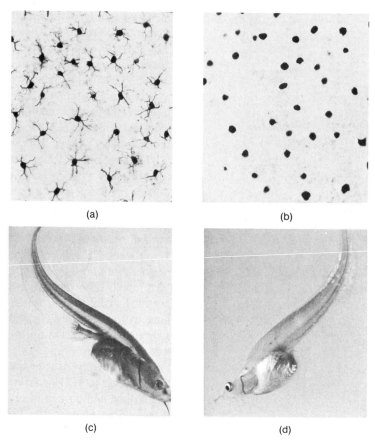

(a)

(b)

(c)

(d)

connections between the hypothalamus and the anterior pituitary. Then in 1936, researchers discovered blood vessels extending from the hypothalamus through the pituitary stalk to the anterior portion of the pituitary (figure 15.12). This network of capillaries and vessels has since been named the **hypothalamic-pituitary portal system.**

In 1945, G. W. Harris of Oxford University proposed that neurosecretory cells in the hypothalamus release substances into the blood vessels of the portal system that stimulate the anterior pituitary to secrete its hormones. The hypothetical substances were called **factors.** In 1955, scientists reported that extracts of the hypothalamus could, indeed, stimulate the anterior pituitary to secrete ACTH. This action was attributed to a small peptide called **corticotropin releasing factor (CRF).** In addition to CRF, a number of other hormones from the hypothalamus that are carried through the portal system and act upon the anterior pituitary have been discovered.

Two teams of researchers have led the effort to identify and synthesize the various factors from the hypothalamus that act upon the anterior pituitary, although many other investigators have contributed significantly to work in this area. The leaders of these teams, Roger Guillemin and Andrew Schally, shared the Nobel Prize for Physiology or Medicine in 1977 for the isolation and characterization of hypothalamic factors with Rosalyn Yalow, who developed analytical techniques that have dramatically changed hormone research and diagnosis.

Thyrotropin releasing hormone (TRH) was the first of the releasing factors from the hypothalamus to be isolated and purified. TRH is a small peptide composed of only three amino acids: histidine, proline, and glutamic acid. It causes the anterior pituitary to secrete thyroid stimulating hormone (TSH), which, in turn, acts upon the thyroid to stimulate the release of thyroxin. TRH was isolated in 1969 by Schally's research team using pig hypothalami and by Guillemin's team

Figure 15.12 The hypothalamic-pituitary portal system. Neurosecretory cells in the hypothalamus produce releasing and inhibiting factors and secrete them into a capillary bed in the hypothalamus. Vessels carry them from the hypothalamus to capillaries in the anterior lobe of the pituitary, where they control secretion of anterior lobe hormones.

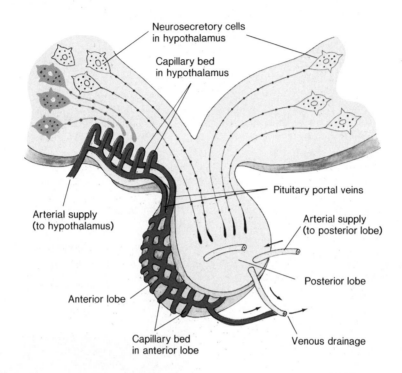

using sheep brains. As has been the case throughout the history of research on hormones, this was no easy task; Guillemin collected 500 tons of sheep brains to remove seven tons of hypothalami, and his team put in four years of intense work to produce a single milligram of TRH!

In 1971, both Guillemin's and Schally's research groups determined the structure of another releasing hormone—then called **luteinizing hormone-releasing hormone (LHRH)**. LHRH is composed of a sequence of ten amino acids. As its name implies, it stimulates the anterior pituitary to release luteinizing hormone, the gonadotropin necessary for ovulation (egg release from the ovary).

For some time it was thought that a second gonadotropin-releasing hormone, FSH-releasing hormone, would be found, but it now seems that the same releasing hormone affects the release of both gonadotropins under complex influences by other hormones. There is tremendous interest in this releasing hormone, now called **gonadotropin releasing hormone (GnRH)** in recognition of its dual role, because inhibiting its production or preventing it from acting can prevent ovulation in experimental animals. Possibly, this discovery could be applied in developing an entirely new method of birth control in the future.

Research on the regulation of growth hormone secretion has revealed a somewhat different pattern of control. In this case, there seem to be factors both stimulating and inhibiting the release of growth hormone. The existence of a releasing factor in the hypothalamus has been demonstrated, but this hypothalamic growth hormone-releasing factor has not yet been purified and isolated for study, although a peptide with 44 amino acids isolated from a tumor of the pancreas acts as a growth hormone-releasing factor and may be identical with the hypothalamic factor. However, a growth hormone-inhibiting hormone was found by Guillemin and colleagues in 1973 and by Schally and his group in 1976. This fourteen-amino-acid molecule, known as **somatostatin,** prevents the release of growth hormone from the pituitary.

Two other pituitary hormones, prolactin and melanocyte-stimulating hormone, also seem to be under this kind of dual control. Various studies have demonstrated the presence of a prolactin secretion-inhibiting factor (PIF) in the hypothalamus and have indicated that there may be a prolactin-releasing factor (PRF) as well. However, the chemical structures of the substances have not yet been determined.

Figure 15.13 Levels of control in mammalian hormonal systems. The hormones secreted by the various endocrine glands exert control, in turn, over the activity of the hypothalamus and the anterior pituitary. This relationship is explained further in figure 15.16.

Figure 15.14 The thyroid. (a) The structures of the two major thyroid hormones, thyroxin (T_4) and triiodothyronine (T_3). They differ by one iodine atom. (b) Thyroid tissue. The thyroid consists of follicles, hollow balls of follicle cells, which secrete thyroid hormones. The hormones are stored attached to glycoprotein molecules in the colloid that fills the inside of the follicles.

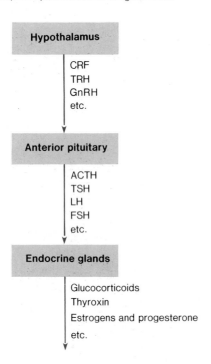

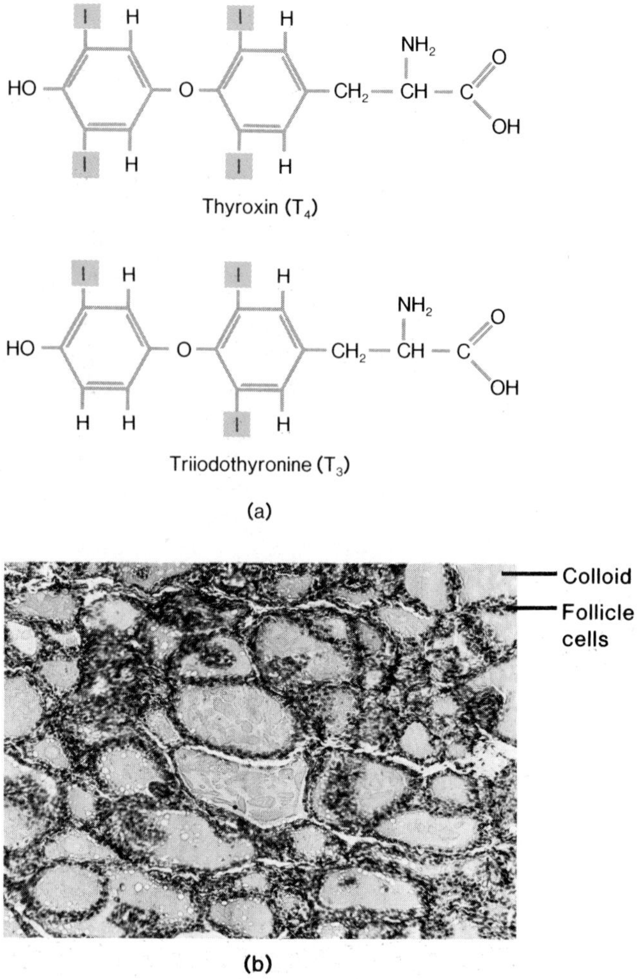

Hypothalamus

CRF
TRH
GnRH
etc.

Anterior pituitary

ACTH
TSH
LH
FSH
etc.

Endocrine glands

Glucocorticoids
Thyroxin
Estrogens and progesterone
etc.

Thyroxin (T_4)

Triiodothyronine (T_3)

(a)

Colloid

Follicle cells

(b)

Because prolactin controls milk production and perhaps other aspects of reproduction, isolation and synthesis of PIF and PRF (if there is one) could be useful in human and veterinary medicine.

Thus, the regulatory mechanisms involved in the hormonal control system operate at several points: in the hypothalamus, in the pituitary, and in the other endocrine glands (figure 15.13).

Hormones Controlling Metabolism

We can define **metabolism** as the sum total of all the biochemical processes that take place within an organism, and the **metabolic rate** is the rate at which these reactions proceed. In order for homeostasis to be maintained, all of these processes in various cells of the body must be regulated and balanced with each other. Several hormones are involved in control of metabolism.

Thyroid Hormones

Thyroid hormones, which are produced by the thyroid gland located at the base of the neck (see figure 15.6), are the principal hormones controlling metabolic rate. There are two major thyroid hormones, **thyroxin (T_4)** and **triiodothyronine (T_3).** Both molecules are synthesized from the amino acid tyrosine, and they differ only in that thyroxin contains four iodine atoms while triiodothyronine contains only three (figure 15.14). The two molecules have the same kind of effect on target cells, but T_3 (which is secreted in smaller quantities)

Figure 15.15 Metamorphosis in amphibians is a dramatic developmental process controlled by thyroid hormone. Two stages in the metamorphosis of the spring peeper frog (*Hyla crucifer*) are pictured here. Tadpoles gain legs, lose their tails, and undergo other changes, including digestive tract reorganization, all in preparation for life as adult frogs. If larval thyroids are inhibited, however, metamorphosis is prevented and tadpoles will grow without undergoing any of these changes. In the laboratory, thyroid-inhibited tadpoles will continue to grow until they are several times the normal size.

(a)

(b)

is somewhat more active than T_4. Both T_3 and T_4 are stored in the thyroid gland attached to large glycoprotein molecules until they are released into the bloodstream.

The thyroid hormones powerfully affect body cells. T_4 and T_3 increase the metabolism and oxygen consumption of cells throughout the body. They increase the rates of protein synthesis, including the synthesis of enzymes that catalyze metabolic reactions, and they stimulate carbohydrate metabolism by causing increased carbohydrate absorption and oxidation by cells.

Thyroid hormones are also instrumental in promoting normal growth and development, largely through the stimulation of protein synthesis in developing vertebrates (figure 15.15). In humans, thyroid deficiency during infancy and childhood causes a dwarfed condition known as **cretinism,** in which the victim never matures sexually and suffers severe mental retardation. Fortunately, cretinism can be prevented by administering hormones to children who have thyroid deficiencies.

As we saw earlier, TSH from the pituitary's anterior lobe controls the production of thyroid hormones. TSH promotes the uptake of iodine by the thyroid, controls the rate of synthesis of the thyroid hormones, and regulates their storage and release into the bloodstream. However, the rate of TSH production must be regulated so that only normal, adequate stimulation of the thyroid is achieved. TSH secretion is affected by the level of thyroid hormones in the blood. When thyroid hormone concentrations in the blood increase, TSH secretion temporarily slows down; when thyroid hormone concentrations in the blood decrease, TSH secretion temporarily increases. This relationship in which a substance (in this case, thyroid hormone) negatively affects the secretion of another substance that regulates it (TSH) is called a **negative feedback control mechanism.** Many such feedback mechanisms control the secretion of various hormones and they are vital for maintenance of homeostasis.

Figure 15.16 The negative feedback control mechanism that regulates secretion of the thyroid hormones. Increased thyroxin (T₄) and triiodothyronine (T₃) cause decreased TRH release from the hypothalamus and inhibit TSH release from the anterior lobe. When T₄ and T₃ in the blood decrease, inhibition of the hypothalamus is relieved. More TSH is released, and the thyroid is stimulated to release more T₄ and T₃. The negative feedback system normally balances the secretions of all elements and keeps the concentration of T₄ and T₃ in the blood within the proper range. Negative feedback control relationships also function in regulating the secretion of many other hormones.

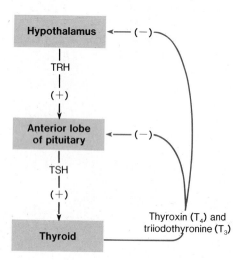

The feedback control relationship between thyroid and pituitary secretion is mediated by the hypothalamus (figure 15.16). Increased levels of T_4 and T_3 in the blood inhibit TRH production by the hypothalamus, while decreased T_4 and T_3 levels reduce inhibitions to TRH production and release, thereby causing increased TSH secretion.

The delicately balanced feedback control of thyroid secretion prevents excess thyroid secretion (**hyperthyroidism**), which causes weight loss, elevated body temperature, profuse perspiration, and nervous irritability. On the other hand, thyroid deficiency (**hypothyroidism**) leads to a lowered metabolic rate, obesity, physical sluggishness, mental dullness, and abnormal skin and hair conditions.

Clear evidence that the normal feedback mechanism is not functioning properly is seen when a condition called goiter develops. As you saw in figure 15.1, goiter is a noticeable expansion of the thyroid gland that is externally visible as a swelling at the base of the neck. Low thyroid secretion rates, caused by hormone synthesis problems or by a deficiency in dietary iodine, result in increased TSH secretion through the normal feedback relationship. The excess growth of the thyroid tissue, producing a goiter, occurs in response to continuing TSH stimulation.

Hypothyroidism and goiter are rare in humans in seacoast areas because dietary iodine is usually adequate when a considerable amount of seafood is included in the diet. Chronic thyroid deficiency problems can develop inland, however, where dietary iodine levels are low. Fortunately, this problem can be prevented by adding very small quantities of iodine to table salt used in low-iodine areas.

Pancreatic Hormones

Maintenance of a steady supply of energy to body cells is one of the most important aspects of homeostasis. This energy supply is assured through regulation of stable blood sugar (glucose) levels. The task of storing glucose and making it available to other body tissues falls primarily to the liver. In large part, glucose storage and blood sugar levels are regulated by two major hormones produced in the pancreas, **insulin** and **glucagon**.

The cells that produce insulin and glucagon are located in the **islets of Langerhans,** discrete patches of tissue scattered among the portions of the pancreas that secrete digestive enzymes (figure 15.17). The islets make up less than 2 percent of pancreatic tissue, and yet a human pancreas contains more than a million islets. The cells of the islets are glucose sensitive—they respond quickly to changes in blood glucose concentration. When glucose is abundant in the blood, insulin secretion is stimulated and insulin acts to lower blood sugar. When there is a decrease in blood glucose, glucagon secretion is stimulated and glucagon acts to raise blood sugar levels.

To maintain a steady supply of sugar to body tissues, the liver stores glucose in the form of glycogen, a glucose polymer. After a meal, when blood glucose concentrations rise rapidly as sugar is absorbed from the digestive tract, the liver conserves some of the glucose by producing glycogen. The reactions involved in glycogen production are collectively known as **glycogenesis** ("glycogen formation"). Later, glycogen can be hydrolyzed to glucose molecules, which are gradually released into the bloodstream. Glycogen hydrolysis is called **glycogenolysis.** At times when glycogen reserves are depleted, the liver can manufacture more glucose from other substances, most notably by using the carbon chains of certain amino acids, in a set of processes known as **gluconeogenesis** ("new glucose formation").

Glucagon is a polypeptide composed of twenty-nine amino acids. It promotes glycogenolysis and stimulates gluconeogenesis in the liver. Thus, glucagon helps ensure a steady supply of glucose in the liver, causes the conversion of glycogen to glucose, and raises blood sugar levels.

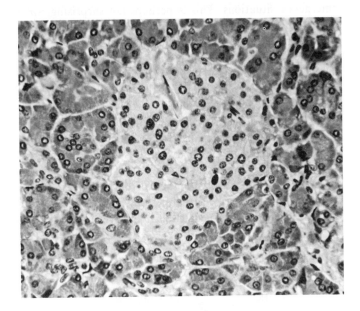

Insulin, a polypeptide composed of fifty-one amino acids, lowers blood sugar levels, mainly by promoting the uptake of glucose by body cells. In the presence of insulin, glucose enters cells about twenty-five times faster than it does without the hormone. Insulin also promotes glycogen synthesis in the liver and muscle cells and depresses glycogenolysis and gluconeogenesis in the liver. Moreover, it inhibits the breakdown of fats and proteins for use in cellular oxidation, thereby promoting the use of glucose in cells' energy-yielding reactions. Together these actions lower blood glucose levels.

A deficiency in insulin secretion results in the condition called **diabetes mellitus,** in which glycogen synthesis decreases and glucose absorption and utilization in body cells is inefficient. As a result, blood glucose levels rise so high that glucose is excreted in the urine. At the same time, more water is lost and the diabetic becomes dehydrated. Despite blood glucose excesses, body cells cannot meet their energy needs, and they therefore metabolize proteins and fats. These metabolic alterations quickly lead to a general weakness and susceptibility to disease.

Interest in insulin, and secondarily in glucagon, has been great ever since the discovery that diabetes mellitus is due primarily to insufficient insulin secretion by the pancreas. Conclusive proof of the role of the pancreas in blood sugar regulation was obtained in a landmark series of experiments done by Frederick Banting and Charles Best at the University of Toronto in the summer of 1921. The climax of their experiments came when they injected a pancreatic extract into a dog that was in a coma and near death following removal of its pancreas. Within an hour, the dog was awake and walking about the laboratory. A few months later, diabetic human patients were responding to experimental doses of pancreatic insulin extracts, and insulin therapy was soon to become a routine treatment for this life-threatening disease. In recognition of the importance of this discovery, Banting shared a Nobel Prize in 1923 with J. J. R. Macleod, who investigated insulin's action in humans.

Insulin can now be produced by genetic engineering techniques (see chapter 25), and as a result, research on diabetes and success in its treatment are progressing rapidly. However, the battle is far from won. There are still problems because prolonged insulin therapy has serious side effects for some diabetes mellitus patients.

In summary, glucagon and insulin are **antagonistic** hormones; they oppose and balance one another. Insulin promotes glycogen synthesis and stimulates body cells to increase their glucose uptake, thus lowering blood sugar levels. Glucagon, on the other hand, promotes glycogenolysis and gluconeogenesis in the liver, thus raising blood sugar levels. A normal, adequate supply of glucose for body cells and tissues depends to a large extent upon this antagonistic relationship between insulin and glucagon.

Adrenal Hormones

Some of the adrenal gland hormones also affect blood glucose levels and metabolism in general. The adrenal glands, one of which is located above each kidney, are really two glands in one. The two parts of an adrenal gland, the inner **medulla** and the outer **cortex** (figure 15.18a), are derived from different tissues in the embryo, and even though the two types of tissue grow together, they remain functionally distinct.

The adrenal medulla secretes the hormones **epinephrine** and **norepinephrine** (also called adrenalin and noradrenalin) (figure 15.18b). The effects of the two hormones are similar, though not identical. In general, epinephrine is secreted in response to stressful situations and prepares the organism physiologically to cope with such circumstances. The secretion of epinephrine causes an increase in blood sugar levels by stimulating liver glycogenolysis. Epinephrine also has many other effects not directly related to blood sugar (table 15.2).

Figure 15.20 Aldosterone, the principal member of the class of adrenal cortical hormones called mineralocorticoids, which are involved in ionic regulation in the body. Aldosterone promotes sodium reabsorption by the kidney. Compare the structure of aldosterone with that of cortisol, a glucocorticoid, which is shown in figure 15.18c.

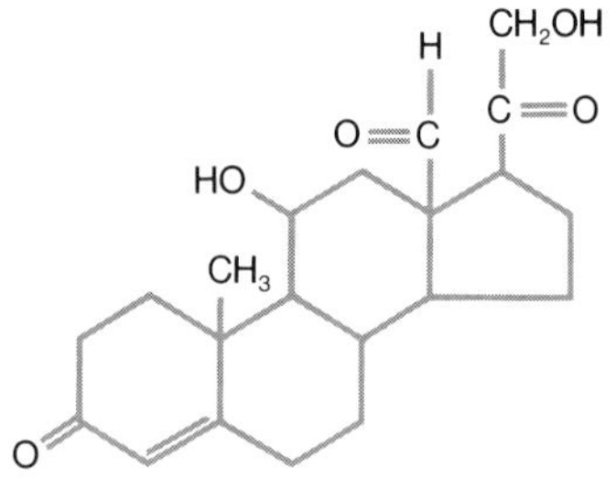

Aldosterone

Control of Calcium Balance

Calcium is especially important to the functions of nerve and muscle tissues. Death due to calcium imbalance can occur very quickly as the result of a massive nervous disruption or disturbance of heart muscle contraction. Consequently, blood calcium levels are closely regulated within a very narrow range. The **parathyroid glands** play a major role in this regulation.

Early endocrinologists assumed that the parathyroids were closely related to the thyroid in function because these small glands, usually four of them in humans, are located on the surface of the thyroid (see figure 15.6). Later they discovered that careful removal of the parathyroids from experimental animals causes very specific effects—including lowered blood calcium concentration and the disturbance of nerve and muscle function—which are different from the effects caused by thyroid removal.

The parathyroids produce a polypeptide hormone called **parathyroid hormone** that raises blood calcium levels and lowers blood phosphate levels. Parathyroid hormone acts on bone cells, causing them to remove calcium and phosphate ions from the matrix of bone and transfer them to the bloodstream. Since this raises both calcium and phosphate concentrations in the blood, the phosphate-lowering effect of parathyroid hormone depends on its action at another site, the kidneys, where it stimulates kidney tubule cells to excrete phosphate and retain calcium.

For many years parathyroid hormone was thought to be the only hormone that controls calcium balance in the body. Scientists thought that after parathyroid hormone had raised blood calcium levels sufficiently, the high blood calcium acted through a feedback mechanism to shut down parathyroid hormone secretion until calcium levels in the blood dropped again. This does happen, but in the early 1960s a second calcium-regulating hormone, **calcitonin** (also called **thyrocalcitonin**), was discovered. Calcitonin is produced by a special group of cells in the thyroid, the C cells. In other vertebrates, calcitonin is produced by separate glands, but in mammals, these specialized cells are incorporated into the thyroid gland during embryonic development.

Calcitonin is an antagonist to parathyroid hormone; that is, it has the opposite physiological effect. It lowers the blood calcium level by inhibiting bone cells from releasing calcium into the blood. Calcitonin secretion is stimulated by a rise in blood calcium, and it acts quickly to lower the blood calcium to a proper level.

Yet a third hormone is now known to affect calcium and phosphate levels in body fluids. This hormone, **1,25-dihydroxyvitamin D_3**, also known as 1,25-dihydroxycalciferol$_3$, is a derivative of a common form of vitamin D. Vitamin D_3 (calciferol) is either ingested in the diet or synthesized in the skin under the influence of ultraviolet radiation in sunlight. Vitamin D_3 is itself a **prohormone;** that is, it is a molecule that can be converted into an active hormone form. The first step in this conversion occurs when liver cells change the vitamin to 25-hydroxyvitamin D_3 (figure 15.21). This substance is then transported to the kidneys, where enzymes in the mitochondria of kidney cells change it to the active hormone, 1,25-dihydroxyvitamin D_3 (symbolized 1,25-$(OH)_2D_3$).

The major effect of this hormone is to stimulate the cells of the intestine to absorb more calcium, thus raising blood calcium levels. It is also thought to assist parathyroid hormone in the mobilization of calcium from bone. Finally, 1,25-$(OH)_2D_3$ has been shown to stimulate phosphate absorption in intestinal cells, thus increasing the blood's phosphate level as well. The discovery of the conversion of vitamin D_3 to a hormone has done much to clarify the role of vitamin D in mineral metabolism.

Figure 15.22 summarizes the regulation of blood calcium level by parathyroid hormone, calcitonin, and 1,25-dihydroxyvitamin D_3 .

Figure 15.21 Steps in the conversion of vitamin D_3 into a molecule that is active as a hormone. The carbon and hydrogen skeleton of the molecule does not change during these steps, so it is not shown in detail. Vitamin D_3, which is synthesized in the skin under the influence of ultraviolet radiation or ingested in the diet, is converted to 25-hydroxyvitamin D_3 in the liver and finally to 1,25-dihydroxyvitamin D_3, the active form of the hormone, in the kidney.

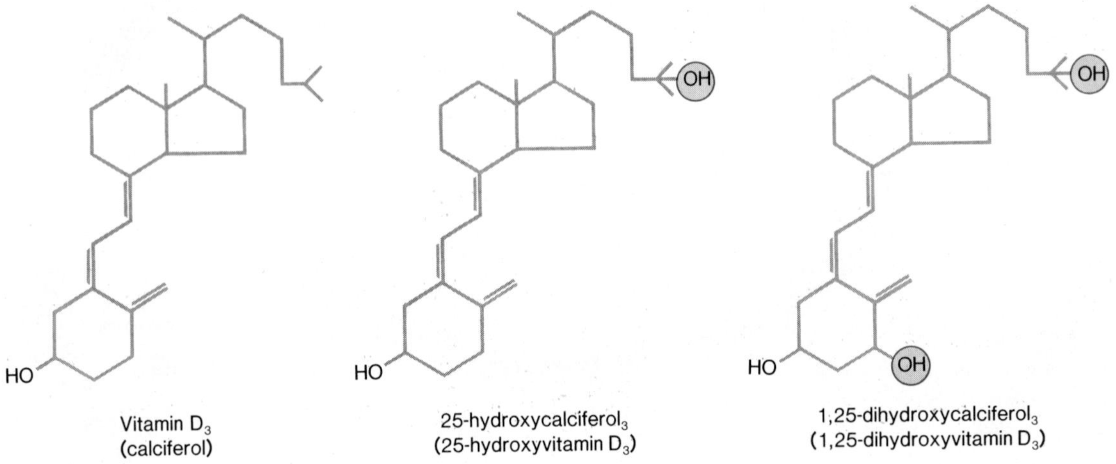

Vitamin D_3
(calciferol)

25-hydroxycalciferol$_3$
(25-hydroxyvitamin D_3)

1,25-dihydroxycalciferol$_3$
(1,25-dihydroxyvitamin D_3)

Figure 15.22 Feedback control diagram showing the regulation of blood calcium and phosphate levels by parathyroid hormone, calcitonin, and 1,25-dihydroxyvitamin D_3. Solid lines represent processes involved in increasing blood calcium concentration, while broken lines indicate factors that decrease blood calcium. Gray arrows represent negative feedback mechanisms.

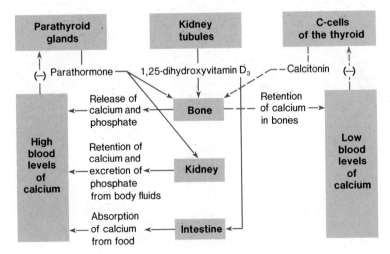

Figure 15.23 Conversion of proinsulin to insulin in the Golgi apparatus. The polypeptide chain of the proinsulin molecule (84 amino acids) is longer than the double polypeptide chain of the insulin molecule (51 amino acids). The active hormone is produced by removal of a peptide fragment that contains 33 amino acids.

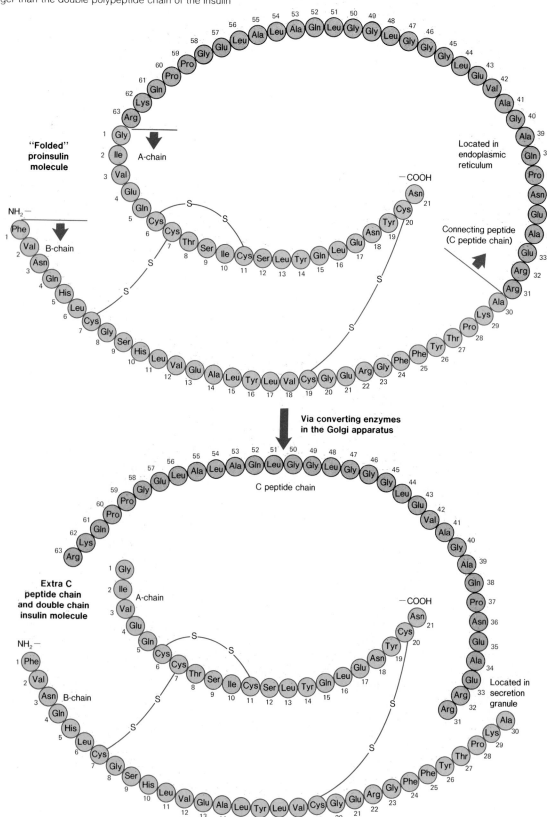

Hormone Synthesis

Some of the most interesting recent discoveries in endocrinology have concerned the synthesis of polypeptide hormones. Polypeptide hormone synthesis occurs in ribosome complexes on the rough endoplasmic reticulum of a hormone-producing cell. Coded information brought from DNA by messenger RNA directs the synthesis. Thus, the polypeptide produced could be called a gene product since it is made according to the specifications of the gene for that polypeptide. However, when biologists began to study the synthesis of these hormones using the sophisticated techniques of modern molecular biology, it soon became apparent that the gene product first produced in gland cells is *not* identical to the polypeptide hormone that the cells secrete. What are the differences between the originally synthesized polypeptides and the hormone molecules that are secreted, and how is one converted into the other?

Prohormones

A number of polypeptides have been isolated and described as **prohormones,** precursors in the synthesis of polypeptide hormones. These prohormones are larger molecules than the hormones, and they do not act on target cells in the same way. They are converted to the smaller, active hormones through the enzymatic cleavage of specific peptide bonds. Anything affecting the enzyme activity that converts a prohormone to an active hormone exerts influence over the hormone synthesis and secretion of gland cells.

One example of a prohormone is the single-chain polypeptide **proinsulin,** which is converted to insulin in the Golgi regions of pancreas cells. Insulin is then stored in granules to await secretion (figure 15.23).

All propolypeptide hormones appear to be larger molecules than the hormones produced from them. For example, proinsulin is larger than insulin and **proglucagon** is larger than glucagon. The **proparathyroid hormone** chain is several amino acids longer than the parathyroid hormone molecule. There are probably also *pro* forms for all of the anterior pituitary hormones, but they are especially difficult to extract and isolate because so many polypeptide hormones are produced in that gland.

Even larger molecules have been found in biochemical studies of hormone synthesis involving cell-free extracts and messenger RNAs for specific hormones. These larger molecules are called **preprohormones.** For example, **preproinsulin** has been isolated in studies using rat insulin messenger RNA, and it is basically the same as a proinsulin molecule with 23 additional amino acids attached. **Preproparathyroid hormone** has 115 amino acids, 25 more than the proparathyroid hormone molecule. The preprohormones are precursors of the prohormones, which are, in turn, precursors of hormones. Preprohormones may be the **initial gene products** in polypeptide hormone synthesis. Figure 15.24 outlines what is thought to be the synthesis process for an active hormone, although other intermediate steps may be possible.

One of the main problems in preprohormone research is that preprohormones have not been found in intact cells, only in cell-free biochemical systems. Preprohormones are probably extremely short-lived inside living cells; if they are converted rapidly to prohormones, they would be very difficult to extract and isolate from whole cells. Nonetheless, the concept of preprohormones as the initial gene product, or the first step in hormone synthesis, may prove vital in certain areas of genetic-engineering research. The goal of this research is to develop specialized microorganisms that can synthesize polypeptide hormones cheaply and in large quantities (see chapter 25).

Some discoveries about precursors of peptide hormones have raised many new questions concerning relationships among hormones and control of their production. A striking example is the polypeptide β-lipotropin (figure 15.25), which is a long chain of amino acids that includes the amino acid sequences that make up several endorphins and enkephalin (p. 373), as well as one form of MSH (p. 359). This large molecule seems to be synthesized and then "cut up" to yield the endorphins and MSH, but not enkephalin. Enkephalins are produced by cleavage of a different large polypeptide precursor.

However, the story is even more complex; β-lipotropin is synthesized as part of a still-larger polypeptide, proopiocortin (also called proopiomelanocortin), which also includes within it the sequence of amino acids making up the hormone ACTH, as well as another form of MSH. Because some of these sequences overlap, however, the active hormone finally produced is determined by how the giant peptide is "cut up." Questions abound in this area: How many kinds of cells synthesize these large precursor peptides? Does cleavage of the large precursor molecule yield only one active hormone molecule, or are several different hormones produced at once? In a particular situation, what determines how the precursor is "cut up?" What is the evolutionary significance of a common precursor molecule for several different hormones? These questions only scratch the surface, and you can see why research activity in this important and challenging area is intense.

Hormonal Regulation in Animals

The history of biology would probably read very differently if Charles Darwin had not dropped out of medical school. Darwin, who went on to make great contributions to evolutionary theory, originally set out to follow in the footsteps of his father and grandfather, both of them distinguished English physicians. However, not only did he find medical lectures dull, but after observing several surgeries he realized he could not bear the horrors of the operating room. At that time, the mid-1820s, surgery was still performed without anesthesia.

In Darwin's student days, surgery was limited to the treatment of superficial wounds, the setting of fractures, and amputations; surgery on internal organs was essentially impossible. A good surgeon was one who worked fast, so as not to prolong the agonizing pain.

Though the anesthetizing effects of some gases such as ether and chloroform had been known for some years, it was not until 1846 that an American dentist, W. T. G. Morton, demonstrated the use of general anesthesia in surgery (figure 16.1). Thus began a new era in medicine.

Anesthetics interrupt the functioning of one of the body's major regulating systems, the nervous system. Animals have two main types of regulation that coordinate the diverse, specialized activities of their various body parts: nervous (neural) coordination and chemical coordination, which involves chemical messengers called hormones. As we saw in chapter 15, hormones are transported in the blood, and minutes, hours, or even longer may be required for a chemical message to produce an effect. Communication through the nervous system, on the other hand, is extremely rapid, requiring only thousandths of a second. Both of these coordinating systems are vital for maintaining homeostasis.

Animal nervous systems are specialized to receive, transmit, process, and respond to information. An animal's nervous system gathers information about both the external and internal environments and transmits it to information processing areas in a central nervous system. We are all familiar with senses such as hearing, sight, touch, and so on—senses involving receptors that detect changes in the external

Figure 16.1 This painting by Robert Hinckley depicts W. T. G. Morton (holding a glass container in which is an ether-soaked sponge) demonstrating the use of general anesthesia in surgery at the Massachusetts General Hospital in 1846. The anesthetized patient was having a tumor removed from his neck.

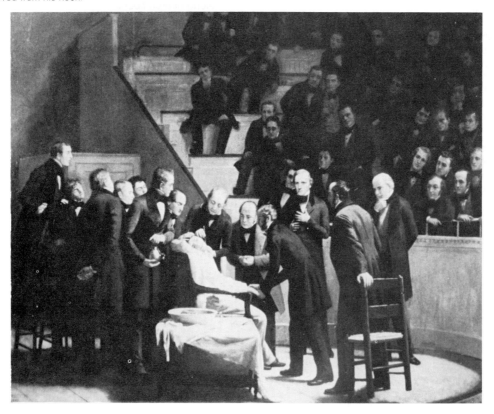

environment; but animals also have internal senses with receptors that monitor muscle tension, blood composition, and various other conditions in their bodies. The central nervous system collects, filters, and processes all of the information about environmental conditions that is constantly being gathered (figure 16.2). Information about trivial environmental changes is essentially ignored. Other information is stored in memory for later retrieval. Yet other information elicits specific, immediate responses that are effected through the activities of muscles and glands. The nervous system also generates some spontaneous activities that are not direct responses to environmental changes.

In this chapter, we will explore the structure and functions of nerve cells, emphasizing the cellular, molecular, and electrical bases for these functions, and the structural organization and function of nervous systems. Later, we will examine the activities of sensory receptors and the functions of central nervous systems (chapter 17), as well as some of the responses under nervous system control (chapter 18).

Neurons

The ability of nervous systems to receive, transmit, process, and respond to information is due to the activity of **neurons** (nerve cells), the functional units of any nervous system. Although neurons vary greatly in appearance and function, each neuron is capable of receiving some form of input from other cells or from the environment. It then responds to that input with an electrical change, and generates an output to one or several other cells.

Figure 16.3 shows an example of the type of neuron that directs the contractions of muscle cells that produce major body movements. Such a neuron consists of three basic parts. The **dendrites** are slender, branched extensions that constitute a major input region for the neuron; that is, dendrites receive messages from other cells. The **nerve cell body** contains the nucleus and the various organelles common to all cells. Finally, the **axon** is a process (fiber) extending from the cell body that carries messages (nerve impulses) on to the next cell.

Figure 16.2 Animal nervous systems are specialized to receive information via receptors, to transmit and process that information, and to respond appropriately. Effectors such as muscles and glands carry out responses directed by the nervous system.

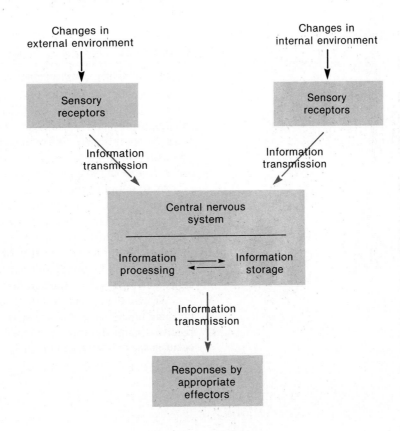

Chapter 16
Neurons and Nervous Systems
385

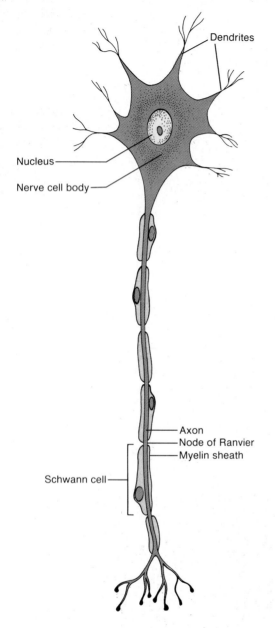

Figure 16.3 Structure of a vertebrate motor neuron. Note the branched dendrites and the long single axon that branches only near its tip. The myelin sheath is an insulating structure that is present around many neuron processes; it is produced by the Schwann cells (see p. 396). The nodes of Ranvier are areas between parts of the myelin sheath.

Dendrites

Nucleus

Nerve cell body

Axon

Node of Ranvier

Myelin sheath

Schwann cell

Figure 16.4 is a photomicrograph of neurons from a vertebrate spinal cord. The numerous processes seen in the figure are probably dendrites, since they show extensive treelike branching, but this is only one of many types of neurons, and dendritic branching is different in different types of neurons. Some large neurons in the brain, for example, have very intricately branched dendrites (figure 16.5a). With some neurons, it is difficult to distinguish between dendrites and axons because the cell body gives rise to only a single branched fiber (figure 16.5b).

Nervous Systems

In single-celled organisms (protists), the whole organism carries out sensory receiving and responding functions. In animals, however, specialized cells receive sensory stimuli, process information about the stimuli, and direct responses to them.

We must be cautious about drawing conclusions concerning evolution from comparisons of various contemporary organisms, because animals living concurrently should never be thought of as evolutionary ancestors and descendants. However, by comparing the nervous systems of various living animals, we can learn much about the fundamental functions of different nervous systems. These comparisons also help us visualize some possible steps in the evolution of complex nervous systems.

Invertebrate Nervous Systems

Cnidarians such as *Hydra* have a very simple nervous system of neurons interwoven into a diffuse network. This **nerve net** extends throughout all parts of the body (figure 16.6a). Beginning at any point, messages can be carried in all directions along this net. Since the nerve net has no central processing unit, however, its functions are severely limited and a *Hydra* shows only a very limited set of responses.

As you might expect, more complex animal bodies have more complex nervous systems. Such systems have clusters of nerve cells that make up a **central nervous system (CNS** for short). In humans, for example, the central nervous system includes the brain and spinal cord and is connected to the rest of the body by bundles of neuron fibers. These bundles, called **nerves,** make up the **peripheral nervous system (PNS).**

We can divide neurons into three broad categories based on their functions. Some are **sensory neurons,** which either receive stimuli directly or respond to specific changes that occur in specialized receptor cells. Others are **motor neurons,**

Figure 16.4 Photomicrograph of neurons from a vertebrate spinal cord. The two relatively large cells (in the upper portion of the photo) with numerous processes extending from them are motor neurons.

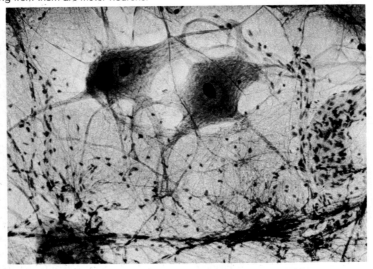

Figure 16.5 Examples of neuron diversity. (a) A brain cell (from the cortex of the cerebellum) with very highly branched dendrites. (b) A sensory neuron. The distinction between axons and dendrites is not as obvious in sensory neurons as it is in motor neurons (figure 16.3). Note that this sensory neuron has just one long fiber.

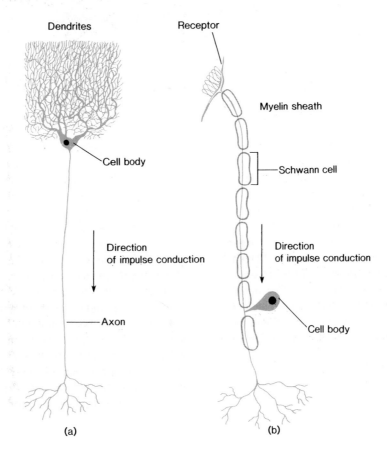

Figure 17.5 The limbic system, a system of functionally linked nuclei in the forebrain and brainstem that are involved in emotional and instinctive behavior as well as in memory storage.

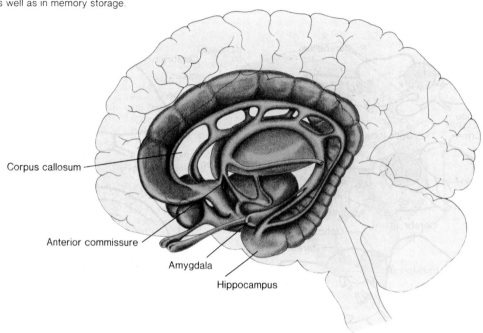

Corpus callosum

Anterior commissure

Amygdala

Hippocampus

The limbic system is a network of linked nuclei in various parts of the forebrain and midbrain (figure 17.5). It is believed that the limbic system contains the neuronal circuitry responsible for emotional and instinctive responses. Although many questions remain about the limbic system, some of its functions are at least partially understood. For example, direct electrical stimulation of the amygdala in experimental animals can provoke outbursts of extreme rage, while removal of the amygdala produces passive, docile behavior. Another structure, the hippocampus, appears to play an important role in memory processes. Major lesions in the human hippocampus have been known to severely impair the ability to store new memories. An individual with hippocampal damage may not even remember having read the beginning of a magazine article by the time he or she reaches its end.

Deep within the forebrain is the **thalamus,** another major relay center of the brain. It is an oval-shaped structure made up of a large number of nuclei that are responsible for collecting, processing, and relaying sensory inputs to appropriate areas of the cerebral cortex.

The Cerebrum

By far the largest and most striking part of a mammal's brain is the cerebrum. In humans, it accounts for more than half of the brain's total mass. The cerebrum is divided bilaterally into two **cerebral hemispheres.** The cerebrum's outer portion, the **cerebral cortex,** is a large mass of gray matter containing billions of nerve cell bodies. It is the cerebral cortex that is responsible for many of the most complex neurological capacities that characterize humans and other mammals.

During the evolution of mammalian brains, the amount of gray matter has increased, as has the number of ridges (called **gyri;** singular: **gyrus**) and grooves (called **sulci;** singular: **sulcus**) that characterize the cerebral cortex (figure 17.6). This surface folding is not random; within any given species, the locations of the gyri and sulci are always essentially the same. This regularity has made it possible to "map" the surface of the cortex. In humans and other mammals, it has been divided into four major areas: the **frontal, temporal, parietal,** and **occipital lobes.** As we shall see shortly, although different functions associated with each lobe have been mapped, much of the vast complexity of the cerebral cortex is yet to be unraveled.

Figure 17.6 (a) Photograph of a human brain showing the convoluted surface of the cerebral cortex. Each ridge is a gyrus; each groove is a sulcus. The surface area of the cortex is greatly increased by such folding. (b) Diagram showing the locations of the frontal, temporal, parietal, and occipital lobes.

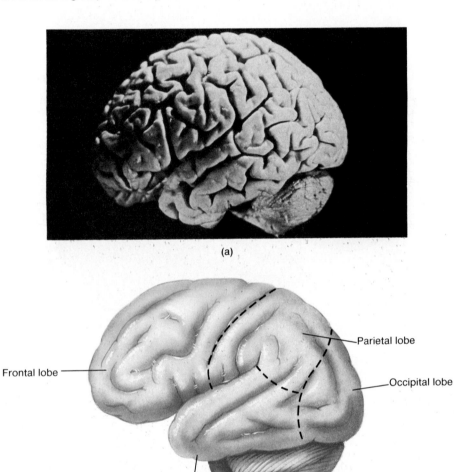

(a)

(b)

Underneath the gray matter of the cortical surface run the billions of myelinated fibers that make up the white matter. Some tracts carry signals from other parts of the central nervous system toward the cerebral cortex, and others carry signals away from it. Tracts also connect various regions within the cerebrum. For example, the right and left hemispheres of the cerebrum are connected by major transverse tracts called **commissures.**

Regional Specialization of the Cortex

Certain regions of the cortex are involved in receiving sensory information, while other regions are associated with motor functions. Still other areas are not specifically affiliated with either function. These latter regions are called association areas.

Figure 17.7 Diagram of the human cerebral cortex showing its four lobes and the primary projection areas of some senses—visual, auditory, and somatosensory—as well as the motor cortex of the frontal lobe (see figure 17.8). Note also Broca's area of the frontal lobe.

Wernicke's area lies posterior to Broca's area and includes parts of the temporal and parietal lobes.

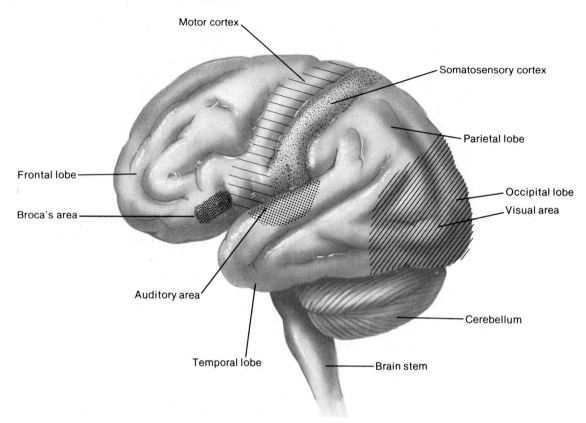

Motor cortex

Somatosensory cortex

Parietal lobe

Frontal lobe

Occipital lobe

Broca's area

Visual area

Auditory area

Cerebellum

Temporal lobe

Brain stem

Particular sensory-specific regions of the cerebral cortex process visual, tactile, and auditory information. These regions are known as **primary projection areas** and are named according to the senses with which they are affiliated (figure 17.7).

A number of details are known about one region of the parietal lobe, the **somatosensory cortex,** which handles information obtained through the sense of touch. Within the somatosensory cortex are subdivisions that receive sensory information from specific body regions. Each of these body regions is called a **receptive field.** The receptive fields associated with particular neurons in the somatosensory cortex can be determined by monitoring activity in single neurons or in groups of neurons when different parts of the body are stimulated. The somatosensory cortex can also be stimulated directly with electrical probes. When this is done, the subject perceives that a particular part of the body has been touched.

Both approaches furnish the same "map" of the body projected onto the cerebral cortex. These results are symbolically illustrated in figure 17.8. As you can see, those regions of the human body with large numbers of sensory receptors have the greatest representation in the projection area. The neural area serving the hand, for example, is as large as the area serving both torso and leg. In animals that rely less on forelimb or finger contacts and more, perhaps, on the tactile stimulation of whiskers, the forelimb would have a small projection and the nose area a larger one.

Specific sensory information also goes to the **association areas** of the cerebral cortex. Each of the four lobes of the cerebral cortex contains an association area. The association areas in the brains of humans and other primates (monkeys, apes, etc.) are much larger than in the brains of other vertebrate animals. In the primate brain, for example, about 80 percent of the cortex is made up of association areas, and only about 20 percent by projection areas. Although they

Figure 17.8 Somatosensory and motor regions of the cerebral cortex. The somatosensory projection area is located in the parietal lobe. The motor cortex is part of the frontal lobe (see figure 17.7). Only half of each cortical region is shown here, the left somatosensory area and the right motor cortex. Note the large proportion of both devoted to the face and hands and the relatively small areas devoted to the trunk and limbs.

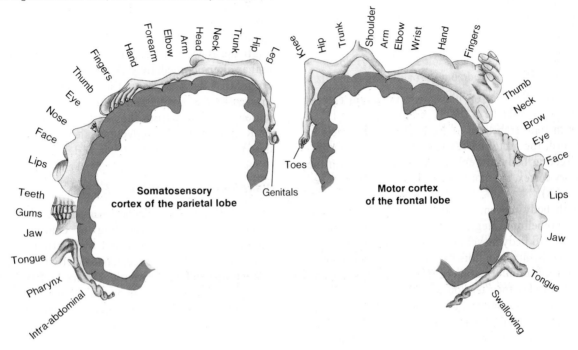

make up such a large and important part of the cerebral cortex, very little is known about the association areas. As their name suggests, association areas take specific information from the projection areas and *associate* the information into higher, more complex levels of consciousness. The association areas are linked to the projection areas by massive fiber tracts. They also are connected with one another in each hemisphere and are linked with areas on the other side of the brain by commissures.

The association areas are correlated with major neurological functions. One method of determining the functions of different association areas is to study neurological deficiencies caused by damage to specific areas. In humans, it has been possible to observe people who have suffered strokes or gunshot wounds that damaged small, localized areas of the brain.

For example, the parietal association area plays a role in the subjective "pictures" that we have of our spatial environment and body image, because visual, tactile, and positional information all converge in this area of the brain. Lesions of the parietal association area can cause a disorder

known as visual agnosia, a peculiar loss of the ability to recognize objects by sight. If people suffering from agnosia can handle the object in question, however, they can identify it using their sense of touch.

Researchers have found that the functions of the association areas of the cerebral cortex are unequally distributed between right and left hemispheres. In right-handed people, the finer, more precise movements of the right side of the body are controlled by the left hemisphere. The complex association functions related to language are even more strongly focused in the left cortex. Even tiny lesions in some areas can have disastrous effects. For instance, damage to **Broca's area** (see figure 17.7) can destroy the ability to speak but leave unharmed a person's understanding of the language and ability to write. Lesions in other areas, such as **Wernicke's area,** may destroy language comprehension. Persons who suffer injury in Wernicke's area usually have difficulty understanding the speech of others, and although they can speak words, they cannot put them together in meaningful sentences.

Figure 18.14 The structure of skeletal muscle. Each muscle (1) contains many muscle fibers (2 and 3). Muscle fibers contain myofibrils (3 and 4), each of which consists of many sarcomeres (5). The banding of sarcomeres arises from the arrangement of thick and thin filaments inside them (6). The cross section sketches (7) show the very precise geometric arrangement of the thin and thick filaments in small areas of the I and A bands. Each sarcomere contains hundreds of thick and thin filaments.

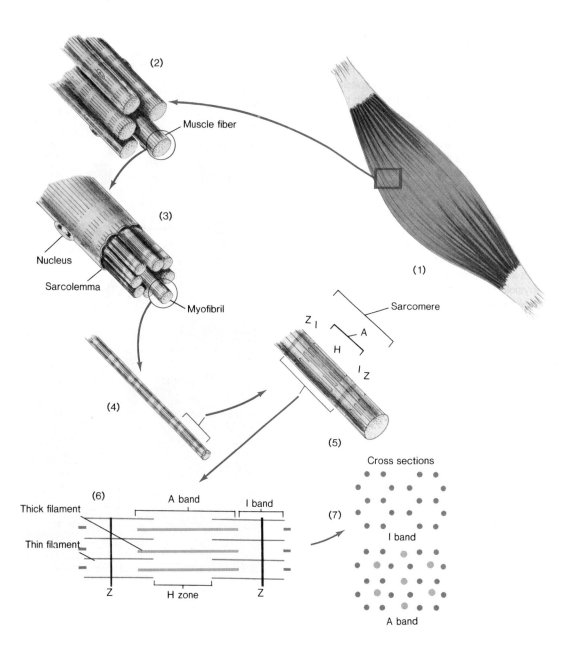

Figure 18.15 An electron micrograph of several myofibrils in a skeletal muscle fiber of a rabbit, showing the various regions represented diagrammatically in figure 18.14. Note the A band (*A*), I band (*I*), H zone (*H*), and Z line (*Z*) (magnification × 760).

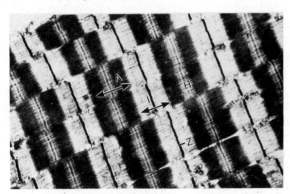

Figure 18.16 Sketches showing how filaments in a sarcomere slide during muscle contraction.

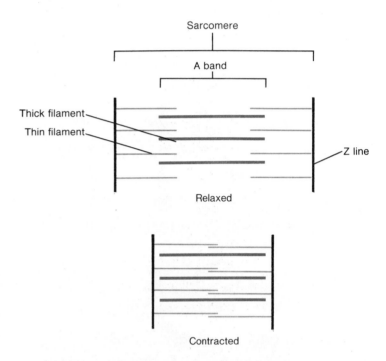

Each skeletal muscle fiber is striated (see figure 18.6); that is, it has a pattern of alternating light and dark bands. The striation of whole fibers arises from the alternating light and dark bands of the many smaller, tubular **myofibrils** contained in each muscle fiber (figure 18.14). The myofibrils have dark bands, called **A bands,** that alternate with lighter **I bands.** Each A band has a region in its center that is lighter than the rest of the band. This lighter, central region is called the **H zone** (also known as the M band). Finally, the middle of each I band has a distinctive, thin, dark line called the **Z line** (figure 18.15). The banding patterns of myofibrils reflect their functional organization, and the portion of a myofibril running from one Z line to the next is a single contractile unit called a **sarcomere.**

Electron microscopy and biochemical analysis of myofibrils have shown that the banding patterns of myofibrils result from the arrangement of two types of filaments located inside the myofibrils. Relatively thicker filaments made up of myosin run through the A band. Relatively thinner filaments, which contain actin (and several other proteins), run through the I band and also overlap with the thick myosin filaments in part of the A band. The portions of the A band in which the thick and thin filaments overlap are darker than the H zone, which is the central portion of the A band that contains only the thick myosin filaments (see figure 18.14). The Z line is a structure to which the smaller actin-containing filaments are anchored.

A. F. Huxley and H. E. Huxley proposed a model for the molecular events in muscle contraction that is now called the **sliding filament theory.** This theory states that in response to a stimulus, the thick myosin filaments and the thin actin-containing filaments slide by each other and increase the amount by which they overlap. This sliding draws the two Z

lines of each sarcomere closer together. The H zone in each sarcomere essentially disappears, and the I bands all along the sarcomere become narrower while the widths of the A bands remain constant (figure 18.16). Shortening of all of the sarcomeres within each myofibril shortens the entire myofibril, and simultaneous shortening of all of the myofibrils in a muscle fiber shortens the whole fiber. Of course, it is the shortening of many muscle fibers that causes the whole muscle to contract and exert a pulling force.

How do these thick and thin filaments slide and bring about muscle contraction? Filament sliding involves temporary connections that form between myosin in the thick filaments and actin in the thin filaments. The connections are flexible, temporary cross-bridges that are established when the globular "heads" of myosin molecules attach to binding sites on actin molecules in the thin filaments (figure 18.17). Once a cross-bridge forms, it bends, thus exerting a pulling force on the thin filament that slides it by the thick filament.

Although the details are not fully clear, it seems that the original binding between actin and myosin requires the presence of an ATP molecule. Hydrolysis of ATP occurs during cross-bridge formation and provides energy for the bending process. Then, the linkage between actin and myosin is released. The myosin "head" then returns to its prebending position and is ready to form a new cross-bridge with an actin molecule further along the thin filament.

Figure 18.17 The proposed mechanism of filament sliding during contraction. (a) Thick filaments contain myosin molecules, which have globular double "heads" that form temporary connections with actin molecules. Thin filaments consist of two chains of spherical actin molecules intertwined in a double helix. (b) Myosin "heads" bind with actin molecules and then flex, thus exerting a sliding force that moves the thin actin-containing filament. ATP is required for binding and flexing of the myosin "heads." After flexing, the myosin heads release from actin and return to their original shape. They can then bind with other actin molecules further along the filament.

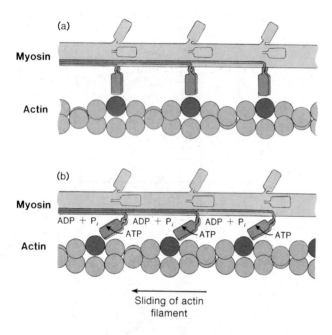

During each contraction, this cross-bridging cycle is repeated over and over by each of a huge number of myosin molecules. At any given instant there are numerous links between myosin molecules and actin molecules, and filament sliding is the composite effect produced by the bending of many of these cross-bridges.

After the contraction of a muscle fiber is complete, the fiber relaxes. During relaxation, the filaments slide back to their original positions relative to one another, and all of the changes in myofibril and muscle fiber length are reversed.

Control of Muscle Contraction

As we mentioned earlier, motor units consist of a motor neuron and the several muscle fibers innervated by the branched end of the motor neuron's axon. Nerve impulses (action potentials) conducted by the motor neuron's axon cause muscle fibers to contract. But what occurs between the passage of an impulse down a motor axon and the filament sliding, which is the essence of muscle contraction?

The end of a motor axon forms a synapselike junction with an area of the sarcolemma, the muscle fiber's plasma membrane. This specialized portion of the sarcolemma is the **motor end plate.** Such junctions are called **neuromuscular junctions** (figure 18.18). As in a synapse between nerve cells, a space separates the synaptic knob of the motor axon from the motor end plate.

When an impulse passes down to the tip of a motor axon, acetylcholine is released from vesicles at the surface of the synaptic knob. Acetylcholine then crosses the gap between the axon and the motor end plate of the sarcolemma and binds with acetylcholine receptors there. This, by the way, is the site where curare, the poison we mentioned at the beginning of this chapter, acts. Curare binds to acetylcholine receptors, making them unavailable for acetylcholine binding.

The sarcolemma, like the nerve cell membrane, is normally polarized; there is a resting potential across the membrane. Acetylcholine binding changes the permeability of the sarcolemma by opening chemically gated ion channels through it and initiates an action potential, which sweeps along the sarcolemma. So far, these events closely parallel those we saw in synapses between nerve cells. But how does an action potential traversing the sarcolemma cause filament sliding in all parts of all myofibrils of the muscle fiber at precisely the same time?

Transverse Tubules and Cisternae

In the late 1940s, L. V. Heilbrunn and his colleagues injected a number of substances into muscle fibers and found that, of all the substances tested, only calcium salts caused fiber contraction. This discovery led to a proposal that permeability changes that occur in the sarcolemma during the action potential allow calcium ions (Ca^{2+}) to enter the cell and that these calcium ions diffuse inward and set off contraction of the contractile proteins.

As more research was carried out in the years that followed, this hypothesis was challenged. One major problem was that very little calcium enters a muscle fiber as a result

Figure 18.18 Neuromuscular junctions. (a) Scanning electron micrograph showing the branched end of an axon forming neuromuscular junctions with several muscle fibers. (b) A neuromuscular junction from which the axon has been stripped away to reveal details of the underlying portion of the muscle cell's plasma membrane.

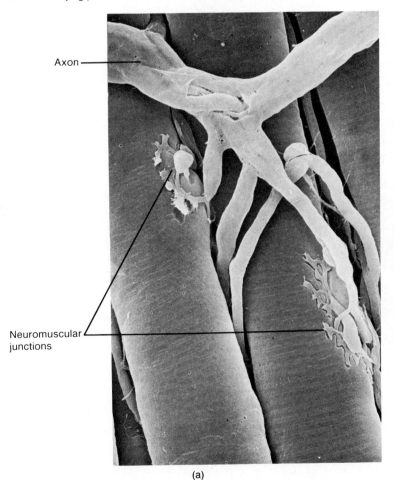

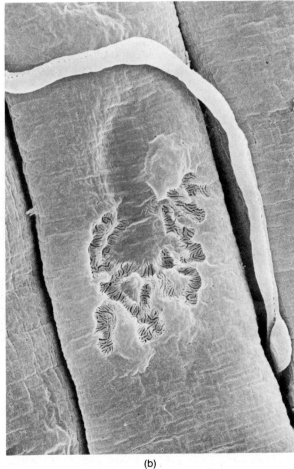

Axon

Neuromuscular junctions

(a)

(b)

of action potentials in the sarcolemma. Another objection was that calcium ions moving inward by diffusion would reach outer myofibrils before they reached the ones deep in the center of the cell, and yet, all of the myofibrils of a muscle fiber contract at the same time. It seemed, therefore, that if calcium ions actually do initiate contraction events in the myofibrils, some mechanism must permit simultaneous delivery of calcium ions throughout the interior of the muscle fiber.

Such a mechanism was found to exist. It is based on the functioning of networks of membranous structures located in the spaces among the myofibrils of a muscle fiber. One membranous network is the **sarcoplasmic reticulum,** the muscle fiber's endoplasmic reticulum. Expanded sacs of sarcoplasmic reticulum, called **terminal cisternae** (singular: **cis-**

terna) lie near the Z lines of the myofibrils (figure 18.19a). These cisternae contain a large quantity of calcium ions.

Another network of membranous structures is a system of hollow tubules whose walls are continuous with the sarcolemma. These tubules, which open to the outside of the cell, are called **transverse tubules (T tubules).** They penetrate all parts of the cell and their tips come into contact with the cisternae of the sarcoplasmic reticulum.

Figure 18.19 Control of filament sliding in muscle fibers. (a) A drawing showing the relationships of myofibrils, sarcoplasmic reticulum, and transverse tubules (T tubules) in a muscle fiber. (b) Simplified diagram of the control of actin and myosin filament sliding by calcium release and recovery. (1) Transverse tubules, which are continuous with the sarcolemma, branch extensively and penetrate all parts of the muscle fiber. (2) When action potentials pass down the transverse tubules, calcium is released from terminal cisternae, and filament sliding occurs. (3) Relaxation occurs when the terminal cisternae recover calcium ions. Cisternae membranes no longer permit calcium ions to flow outward, and cisternae quickly recover calcium by active transport inward across their membranes.

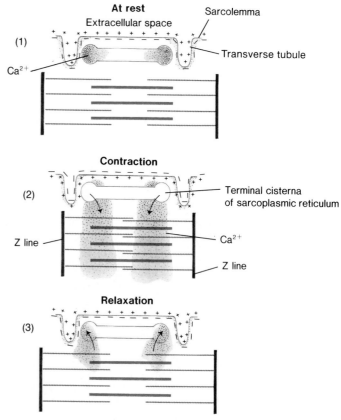

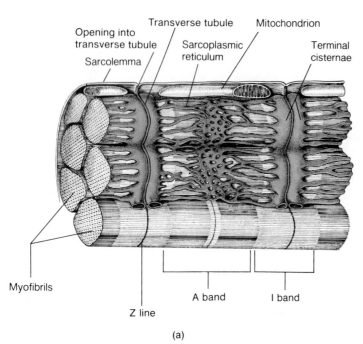

(a)

(b)

Each action potential passing along the sarcolemma results in depolarization of the membranes of all of the transverse tubules. When an electrical change in the membrane arrives at the tip of a transverse tubule near a terminal cisterna, it somehow triggers an abrupt change in the membrane of the cisterna, which then releases calcium ions (figure 18.19b). Because transverse tubules penetrate all parts of the muscle fiber, calcium release is essentially simultaneous in all parts of the fiber, and the calcium ions cause simultaneous contraction of all of the myofibrils.

After the calcium release, the membranes of cisternae again become relatively impermeable to calcium ions, an active transport system pumps calcium ions back into the cisternae, and contraction stops. This inward movement of calcium ions into the cisternae prepares the cell to respond to subsequent stimulations by the motor nerve.

Tropomyosin, Troponin, and Calcium

The mechanism by which calcium ions, which are released from terminal cisternae, initiate filament sliding is now quite well understood. Two regulatory proteins, **troponin** and **tropomyosin,** which are present with actin in the thin filaments, are involved in regulation by calcium. Each thin filament is essentially a double helix (two chains of linked actin

Figure 18.20 Proposed relationships of tropomyosin, troponin, and calcium ions in the control of filament sliding. (a) Rodlike tropomyosin molecules lie along the actin chains. Globular troponin molecules are closely associated with tropomyosin molecules. (b) Details of the proposed arrangement of tropomyosin, troponin, and actin. Tropomyosin rods lie over the sites (*cross-hatched areas*) on actin molecules that combine with myosin cross-bridges. Troponin molecules have calcium binding sites. (c) When calcium combines with troponin, its conformation is changed so that it moves tropomyosin and exposes sites on actin molecules that combine with myosin "heads." This permits the interactions between myosin and actin that cause filament sliding.

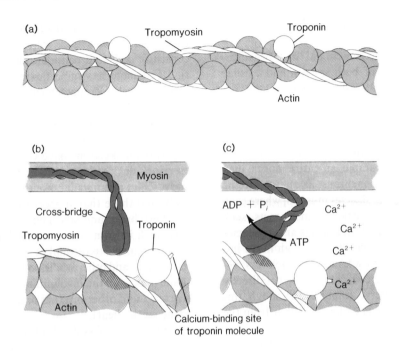

molecules wound around each other) (see figure 18.17). Actin molecules have combining sites for myosin "heads." Tropomyosin molecules, along with troponin molecules, function to block the formation of linkages between myosin "heads" and actin molecules (figure 18.20).

Rodlike tropomyosin molecules lie nestled in the grooves between the chains of the double helix of the thin filament. At rest, in the absence of stimulation, tropomyosin occupies a key position, relative to the actin molecules, that precludes binding by myosin "heads."

Troponin molecules are globular protein molecules that are closely associated with tropomyosin, and their conformation affects the positioning of tropomyosin relative to the myosin-binding sites on actin molecules. When the calcium ion concentration increases as a result of stimulation, troponin molecules bind calcium and, as a result, undergo conformation changes that shift the position of tropomyosin. This movement of tropomyosin presumably uncovers sites on the actin molecules and permits the linkage between actin and myosin that is essential for filament sliding and muscle contraction.

Contraction of Other Kinds of Muscles

Cardiac muscle and smooth muscle have functional characteristics that distinguish them from skeletal muscle. Cardiac muscle requires at least ten times as long as skeletal muscle to contract and relax. Smooth muscle works even more slowly, requiring as much as a hundred times longer than skeletal muscle to contract and relax.

Tonus is also different in smooth muscle because it is highly variable. Smooth muscle may remain slightly or strongly contracted for much longer periods than skeletal muscle.

Cardiac muscle differs sharply from skeletal muscle in that it has a long **refractory period,** the period following a stimulus during which it is unable to respond to another stimulus. This means that cardiac muscle contractions cannot run together as those of skeletal muscles can, and thus a heart cannot become locked in a single sustained contraction. The most interesting feature of vertebrate cardiac muscle, however, is its inherent rhythmicity. Individual heartbeats are

Imagine a walk through a meadow in spring or early summer. Flowers are everywhere. In fact, the ground is virtually a carpet of brilliant flowers (figure 19.1). Returning to the same meadow only a week or two later, however, you find that although everything is green and growing, there is hardly a flower in sight. The meadow plants' flowering period is over and they have moved on to other phases of their seasonal activity.

Every phase of a terrestrial vascular plant's growth, function, and reproduction involves interactions with a changing environment. Because plants remain in a single place throughout their lives, they cannot avoid adverse changes in their environment by moving to more favorable locations. They are directly exposed to all of the seasonal changes in light, temperature, and moisture availability, as well as to the possibility of being eaten by animals. They must adjust to seasonal changes in their environment and occasionally, to the loss of parts of their bodies.

Plants develop in an organized way; flowering and seed production are completed during a specific part of their growing season, and then they prepare for winter dormancy. How is all of this coordinated? How do seeds germinate at an appropriate time? What factors control the shape, size, and organization of the plant body? How do whole fields of a given species flower during the same brief period? What factors cause plants to change their activities so that their vulnerability to harsh winter conditions is minimized?

Throughout the year, changes in light conditions are the most reliable indicators of changing seasons. Seasonal day length and light intensity changes are the same every year, while temperature and moisture conditions can vary considerably. This reliability of seasonal light changes has probably been a major factor in the evolution of plant responses to environmental change. Thus, light is the dominant external factor affecting seasonal and shorter-term plant responses.

We will consider three general categories of plant responses to light: phototropism, photomorphogenesis, and photoperiodism.

Phototropism ("light turning") involves a growth curvature of plants that is specifically determined by the *direction* from which light strikes them. Houseplants that "turn toward the light" provide good illustrations of phototropism.

Photomorphogenesis is a general name for plant responses initiated by light stimuli that are not specifically directional or periodic. Photomorphogenic responses affect seed germination, stem elongation, leaf unrolling, chloroplast development, and many other structural and functional features in the life of plants.

Figure 19.1 Events in the lives of plants are precisely regulated in a continuing interaction with the environment. This picture of a mountain meadow illustrates the simultaneous flowering of many individual plants that occurs during only a brief period in the growing season.

Photoperiodism refers to the responses of plants to day length changes; that is, changes in the twenty-four-hour cycle of alternating light and darkness. Photoperiodic responses usually occur over a relatively longer period of time and involve qualitative changes in the plant; for example, the change from the nonflowering to the flowering condition.

Plant Hormones

The orderly and properly timed responses of plants to light, as well as to other environmental factors, involve the action of plant hormones. A plant hormone is like an animal hormone in that it is produced in one part of the organism and causes a response in other parts of the organism, even though it is present in only a very low concentration. However, there are some important differences between hormonal regulation in plants and animals.

Animal hormones are produced by cells specialized for hormone production, and in many cases, hormone-producing cells are clustered in specific organs, the endocrine glands. Animal hormones circulate throughout the body, but they act on only one or a few kinds of cells, their target cells. While plant hormones also may be produced in localized parts of plant bodies, the cells that produce them are not specialized for hormone production alone. Hormone-producing cells of plants often are, for example, actively dividing cells in a meristematic region of the plant. Also, responses to plant hormones are not restricted to a few kinds of target cells; practically all plant cells respond in some way to each of the plant hormones, although some cells respond differently or more extensively than others.

Vascular plants have only about a half-dozen major hormones, but this small number does not limit the number of types of responses they cause. In many cases, different hormone combinations cause different responses. Thus, plant regulation depends heavily on changing balances of a few hormones, instead of specific actions by a larger number of individual hormones, as is the case in animal regulation.

The modes of cellular action of plant and animal hormones seem to be significantly different. Animal hormones bind with specific receptor molecules of target cells. Then the hormone-receptor complex causes a specific response of the target cell, either by acting directly or by stimulating production of a second messenger, such as cyclic AMP, which causes the specific response of the target cell. But there is some question whether such specific hormone receptor molecules are involved in plant cells' responses to hormones. Some molecules in plant cell membranes do bind plant hormones, but it is not clear whether this binding is directly involved in normal hormone actions. Also, there is very little cyclic AMP in plant cells and no evidence, as yet, that cyclic AMP functions in plant cells as it does in some animal target cells, or that there are other second messengers in plant cells. Rather, plant hormones appear to produce reactions in plant cells directly.

Finally, and most significantly, almost all hormonal effects in plants modify growth patterns. Plant hormones are regulators of the lifelong growth of plants and are not as closely associated with short-term, reversible physiological adjustments as many animal hormones are.*

Five major types of plant hormones have been identified: auxins, ethylene (a gas), cytokinins, gibberellins, and abscisic acid.

*Because substances commonly called plant hormones function mainly in growth regulation, and because they differ markedly from animal hormones in their modes of action, some biologists suggest that they be called plant **growth regulators** rather than plant hormones.

Auxins

Auxins were the first of the plant hormones to be identified. Their discovery resulted from their role in various *tropisms* (turning responses to environmental stimuli), especially phototropism. Auxins are now known to be involved in a great variety of plant cell responses.

Phototropism

Scientific analysis of phototropism began with the experiments of Charles Darwin and his son Francis around 1880. The Darwins used grass seedlings to investigate the bending of plants in response to unidirectional light. They found that seedlings failed to bend toward the light after their tips were cut off or covered with black caps. A seedling would bend toward the light, however, if its tip was exposed while the rest of the seedling was buried in fine black sand. The Darwins concluded that curvature of grass seedlings in response to light depended on some "influence" transmitted from the seedling tip to the rest of the seedling (figure 19.2).

Subsequent work on phototropism focused on the role of the seedling tip. Oat (*Avena*) seedlings were used in most of these experiments. An oat seedling has a sheath, called the **coleoptile,** that encloses the first leaves, and the tip of this coleoptile is the key to the phototropic response of the young seedling (figure 19.2a).

P. Boysen-Jensen, working in Denmark about thirty years after the Darwins' investigation of phototropism, conducted experiments in which he removed the tips of oat coleoptiles. When he did this, the coleoptile stopped growing. Next he placed a piece of gelatin on the cut surface and set the cut tip on top of the gelatin. A short time later the coleoptile resumed growing, and in this condition it showed the normal phototropic response to light from one side. Boysen-Jensen's experiments demonstrated that the influence of the coleoptile tip did not depend on the tip's being in its normal place on the plant because the influence could pass through a gelatin block (figure 19.2b).

A few years later in 1918, Arpad Paál in Hungary cut off coleoptile tips and placed them to one side of the cut surfaces. When these plants were allowed to grow in the dark, Paál observed a curvature very similar to the normal response of intact seedlings to unidirectional light (figure 19.2b). Thus, asymmetrical placement of the tip imitated the effect of light shining on the plant from one side. Paál reasoned that the tip of a normal coleoptile produces a growth-promoting substance that travels downward and that light must make it travel asymmetrically, thus causing greater growth on the shaded side of the seedling.

Figure 19.2 The coleoptile and the discovery of auxins. (a) A photograph of two oat (*Avena*) seedlings, one with its husk removed (*left*), showing the coleoptile (the structure growing up above the seed) and the roots. The first leaves are rolled up inside the coleoptile. (b) A diagrammatic summary of some of the important experiments that demonstrated the existence of auxins. The Went and Paál experiments were both conducted in the dark.

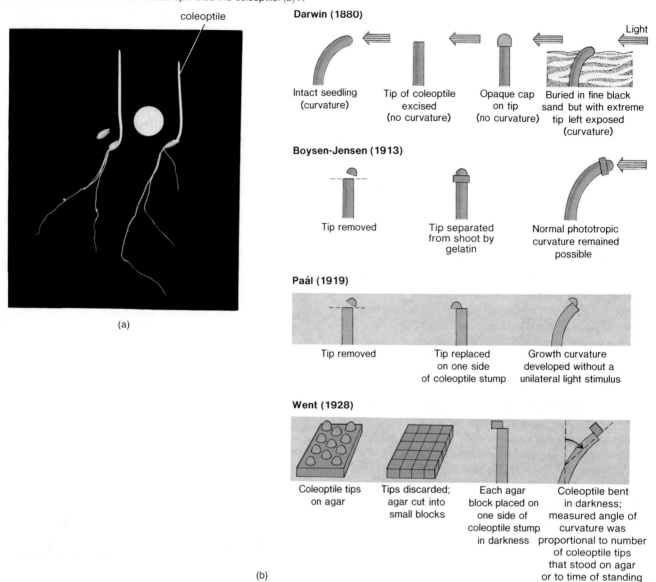

coleoptile

(a)

Darwin (1880)

Intact seedling (curvature)

Tip of coleoptile excised (no curvature)

Opaque cap on tip (no curvature)

Buried in fine black sand but with extreme tip left exposed (curvature)

Light

Boysen-Jensen (1913)

Tip removed

Tip separated from shoot by gelatin

Normal phototropic curvature remained possible

Paál (1919)

Tip removed

Tip replaced on one side of coleoptile stump

Growth curvature developed without a unilateral light stimulus

Went (1928)

Coleoptile tips on agar

Tips discarded; agar cut into small blocks

Each agar block placed on one side of coleoptile stump in darkness

Coleoptile bent in darkness; measured angle of curvature was proportional to number of coleoptile tips that stood on agar or to time of standing

(b)

Then in Holland in 1926, Frits Went demonstrated that Paál's proposal was right: a substance produced in the coleoptile tip causes growth responses in the rest of the coleoptile. Went cut off coleoptile tips, placed them on an agar block and left them there for varying periods of time. He then cut up the agar and put pieces of it on "decapitated" coleoptiles that were kept in the dark. A piece of this agar placed squarely on top of a cut coleoptile caused the seedling to grow straight upward. If Went placed the agar to one side of the top of the coleoptile, however, the seedling curved just as it did in Paál's experiment with the tip itself. This seedling curvature response in the dark, and in the complete absence of the coleoptile tip, proved conclusively that the effect of the coleoptile depends not on the physical presence of the tip, but on a diffusible chemical substance that can accumulate in the agar (figure 19.2b).

Figure 19.3 The oat (*Avena*) coleoptile curvature bioassay for auxin (a), which is derived from Frits Went's original study. The amount of auxin diffusing into an agar block can be estimated with reference to a standard curve (b) prepared from responses of coleoptiles to known concentrations of auxin. The tests are conducted in the dark.

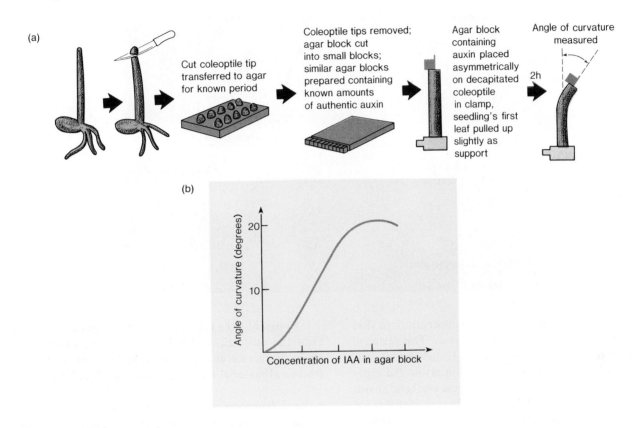

Went also found that the angle of seedling curvature in the dark depended on the number of tips that had been on the agar and the length of time they were left there. In other words, the degree of seedling curvature is proportional to the amount of growth substance in the agar. Thus, the degree of curvature can be used as a quantitative test for determining the amount of growth substance present in a test sample. Such tests employing specific biological responses are known as **bioassays,** and Went's coleoptile curvature bioassay (figure 19.3), as well as various other bioassays, have been instrumental in the search for the plant hormones.

The growth hormone produced by oat coleoptile tips was given the name **auxin,** from the Greek word *auxein* meaning "to grow." Went's bioassay became a key tool in the search for the chemical nature of auxin.

The Nature and Cellular Actions of Auxin

Early researchers thought they were working with a single substance in their experiments on growing coleoptiles. However, bioassays for auxin activity showed that auxin could be isolated not only from plants, but also from a number of other seemingly unlikely sources, including various molds and yeasts, and even animals. In fact, the first auxin to be identified chemically was not isolated from plants, but from human urine! This auxin is **indoleacetic acid (IAA),** a molecule whose structure is very similar to that of the amino acid tryptophan (figure 19.4).

Evidence now indicates that IAA is the principal naturally occurring auxin. Other compounds isolated from plants also act as auxins, as do a number of laboratory-synthesized compounds. What all of these compounds have in common is that they cause plant cells to elongate. Thus, cell elongation is the major effect of an auxin.

Some synthetic auxins are economically important. For example, **naphthaleneacetic acid** (figure 19.4) is widely used in horticulture for plant propagation because it promotes root development on stem cuttings. It also is used in orchards to prevent early fruit drop. A commonly used **herbicide** (plant killer), **2,4-dichlorophenoxyacetic acid,** or **2,4-D,** is also a synthetic auxin. It causes abnormal growth and metabolic

Figure 19.4 Indoleacetic acid (IAA) is the principal naturally occurring auxin. The formula of the amino acid tryptophan is given to show its close similarity with IAA. Naphthaleneacetic acid and 2,4-dichlorophenoxyacetic acid (2,4-D) are synthetic auxins.

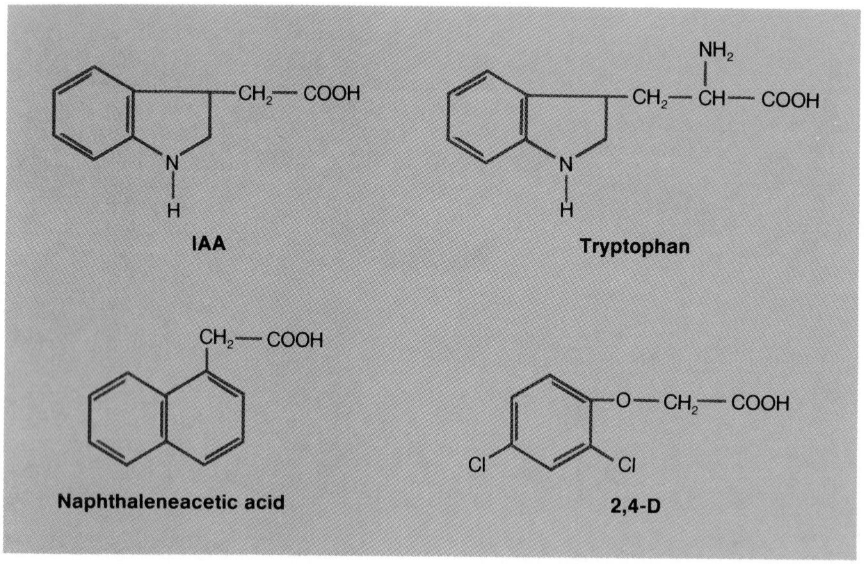

responses in dicots (broadleaf plants) at concentrations that do not harm monocots, such as lawn grasses. This selective killing of "broadleaf weeds" makes 2,4-D and similar synthetic auxins practical for weed control in lawns and especially in fields of cereal grain. Although these synthetic auxins have been used as weed killers for many years, it still is not known why a given concentration of 2,4-D kills broadleaf plants while not harming grasses.

Cellular responses to auxin seem to occur in two phases. The first phase is a relatively rapid response that takes place within ten to fifteen minutes after cells are exposed to appropriate quantities of auxin. A responding cell actively pumps hydrogen ions (H$^+$) out through its plasma membrane into the surrounding cell wall via an energy-requiring process (using ATP). The cell wall contains cellulose fibers that are chemically cross-linked to form a firm, boxlike structure just outside the plasma membrane. The hydrogen ions that are pumped out lower the pH in the cell wall, and the decreased pH activates enzymes that break down the cross-bridges linking the cellulose fibers. This weakens the wall and allows the normal turgor pressure of the plant cell to push the wall outward, thus expanding the cell.

A second, slower phase of the response to auxin begins after thirty to forty-five minutes. In this phase, auxin stimulates synthesis of certain enzymes involved in cellulose production.

Apparently the initial cell expansion phase stretches and thins the cell wall, and the second phase, which includes auxin-stimulated enzyme synthesis, leads to production of structural material that thickens the enlarged cell walls.

Lateral Transport and Auxin-Induced Responses
As you have seen, elongation on the side of a plant opposite a unidirectional light source causes the curvature response in phototropism. Many years ago, it was proposed that cells elongate more on the shaded side because the auxin concentration is greater there (figure 19.5). How does the auxin concentration differential that causes the greater elongation on the shaded side come about?

In the 1920s, N. J. Cholodny and Frits Went proposed that a growth substance is laterally transported in response to a stimulus such as asymmetrical lighting. Other researchers, however, proposed that light simply inhibited auxin synthesis on the lighted side of the plant. Furthermore, it was later discovered that bright light promotes the natural breakdown of auxin. However, light destruction of auxin requires very bright light, much brighter than that required by some phototropic responses.

In the 1950s, W. R. Briggs and his colleagues at Stanford provided additional support for the Cholodny-Went hypothesis of lateral transport when they showed that auxin produced in the coleoptile tip is indeed transported to the shaded side of the coleoptile. Their key experiment compared corn coleoptile tips that were prepared in two slightly

Figure 19.5 Greater cell elongation on the shaded side, a response to greater auxin concentration, causes coleoptile curvature in phototropism.

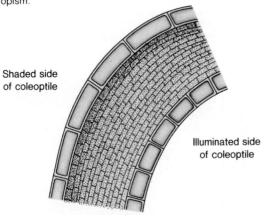

Shaded side
of coleoptile

Illuminated side
of coleoptile

Figure 19.6 Evidence for lateral transport of auxin in corn (*Zea mays*) coleoptile tips in response to unilateral illumination. (a) Lateral transport of auxin is possible when a thin glass barrier only partially divides a coleoptile tip. (b) When a barrier completely divides the coleoptile tip, auxin cannot be transported laterally. The numbers represent the degree of curvature of coleoptiles in a Went-type bioassay for auxin content in the agar blocks that were under the halves of the tip in each case. The experiment demonstrated that the asymmetrical distribution of auxin produced by unidirectional illumination is due to lateral transport of auxin, not to light-induced destruction of auxin on the lighted side.

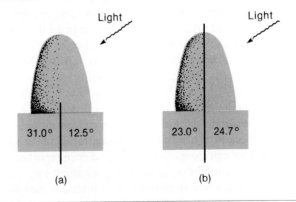

Light | Light

31.0° | 12.5° | 23.0° | 24.7°

(a) | (b)

Figure 19.7 Geotropism. When a seed germinates in the soil, the stem and roots grow out of the seed (a) and respond to gravity, with the stem growing upward and the roots growing downward (b). Stem curvature results from differential auxin concentration. Root curvature involves several factors.

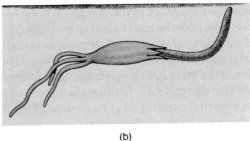

(a)

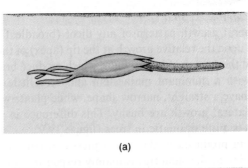

(b)

different ways (figure 19.6). Coleoptiles were first set on agar blocks. Then in one group, each coleoptile and its accompanying agar block were completely divided by a thin glass barrier. In another group, the barriers were placed to split the agar block but only part of the coleoptile. Next, unidirectional light was applied for three hours, and the pieces of agar block below the coleoptiles were removed for testing in a Went-type bioassay. The piece of agar below the shaded side of the partially split coleoptile showed three times as much auxin activity as the piece below the lighted side. In the completely split coleoptile, however, the auxin concentration in the two halves of the agar block was essentially equal; lateral transport had been blocked.

These and other experiments indicate that auxin produced in the coleoptile tip is indeed transported to the shaded side. This in turn causes differential cell elongation and the resultant seedling curvature. Despite this evidence that auxin is transported laterally, little is known about the transport mechanism.

Geotropism
Light is not the only factor that affects the growth direction of seedlings. Growing seedlings also bend in response to gravity. **Geotropism** (response to gravity) is very important in early seedling growth because it orients seedlings properly and uniformly in the soil. In geotropism, stems curve upward and roots curve downward. Thus, stems are **negatively geotropic** because they curve away from the earth's gravity center, and roots are **positively geotropic** because they curve toward it (figure 19.7).

Figure 19.11 Auxin released from seeds influences fruit growth. (a) A normal strawberry fruit. (b) A fruit of the same age from which all seeds had been removed. (c) Magnified view of a fruit from which all seeds except one had been removed. (d) A strawberry fruit that developed after all seeds were removed and replaced with a lanolin paste containing a synthetic auxin.

(a)   (b)

(c)   (d)

Figure 19.12 The structure of ethylene.

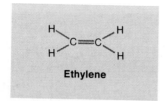

Ethylene

Ethylene

Ethylene is the only currently known plant hormone that is a gas at normal environmental temperatures and pressures (figure 19.12). It has been known for many years that plants produce ethylene, and as early as 1935 there were suggestions that it might function as a hormone. However, the very small quantities of ethylene involved in plant responses were difficult to detect and measure until some years later.

The effects of ethylene and auxin are interrelated; when auxin exceeds a certain concentration in plant tissues, the tissues begin to produce and release ethylene. This makes interpreting responses to auxin more complex, because in some processes ethylene is an antagonist of auxin; that is, ethylene causes responses opposite to those caused by auxin.

The best-known response to ethylene is fruit "ripening." Of course, as K. V. Thimann, one of the leaders in plant hormone research, has pointed out, "ripeness" is more a subjective human judgment than a specific physiological state of fruits. Basically, we say fruits are ripe when we regard them as suitable for eating. Ethylene promotes hydrolysis of starch, with a resulting increase in sugar concentration. Ethylene also causes dissolving of the middle lamellae (see p. 80) that help to hold cell walls together. This loosens cells and softens tissue during fruit ripening.

Ethylene production is "contagious"; that is, when a fruit releases ethylene, it stimulates other fruits to begin to produce it as well. Thus, the old saying that "one bad apple spoils the barrel" is quite accurate. To counteract this contagious ripening, apple producers store their harvest in sealed compartments with a high concentration of atmospheric carbon dioxide (CO_2 is an antagonist of ethylene). On the other hand, ethylene can be used to stimulate ripening in fruits that are picked green and shipped long distances.

Cytokinins

In the early 1940s, biologists attempted to grow plant tissues in culture vessels using media containing auxin and all known plant nutrients. Something seemed to be missing, however. Plant cells in these cultures would enlarge, often to spectacularly large sizes, but cell divisions were rare. Thus, a search began for substances that would promote cell division in cultured plant cells. Eventually, by trial and error, it was discovered that when coconut milk, which is a liquid endosperm (nutrient storage tissue) was added to a culture medium, cell division in cultured cells was greatly stimulated.

Other preparations, such as crude yeast extracts, were also found to provide the necessary stimulus for cell division. Coconut milk and yeast extracts are complex mixtures of many substances, however, and biologists needed to determine what factors contained in these complex mixtures promote cell division. Folke Skoog and his colleagues at the University of Wisconsin sought an answer by testing the effectiveness of nucleic acids extracted from various sources.

They found that DNA preparations from yeast promote cell division in cultured plant cells. As they proceeded to test other nucleic acid preparations, however, they obtained some curious results. They got their best results with an old nucleic acid solution that had been stored in the laboratory for some time. Old preparations, in fact, were very much more effective cell-division promoters than freshly prepared ones. Furthermore, they found that nucleic acid solutions sterilized by intense heating in an autoclave (a device that sterilizes with steam under high pressure) were much more effective than ones sterilized more "gently" by filtration.

These results suggested that the division-promoting factors are similar to the breakdown products produced when nucleic acids are stored for long periods or heated to high temperatures. Skoog and his colleagues found that a number of compounds that were chemically similar to components of nucleic acids were indeed effective cell-division promoters. They called these compounds **cytokinins** (from *cytokinesis,* meaning cell division) and named the active one that they isolated **kinetin** (figure 19.13). However, although kinetin effectively promotes cell division in tissue cultures, it is not a naturally occurring compound. Years later, in 1964, D. S. Letham and his associates in New Zealand announced that they had isolated a natural cytokinin from *Zea mays* (corn), which they appropriately named **zeatin.** Finally, in 1967, it was discovered that coconut milk contains zeatin and another structurally similar cytokinin called **zeatin riboside.** Thus, the mystery of coconut milk's cell division promoting effect on cultured plant cells was solved at last, some twenty-five years after the search for it began.

In cultured plant tissue, there is a complex interaction between auxin effects and cytokinin effects on cells (figure 19.14). When auxin and cytokinin concentrations are balanced, the tissue grows as an undifferentiated mass called a **callus;** when the auxin to cytokinin ratio is increased, roots develop; when the auxin to cytokinin ratio is decreased, shoots and leaves develop. Similarly, in intact plants, regulation of various growth processes involves changing ratios of auxin and cytokinin concentrations.

As with all plant hormones, cytokinins are known to affect many different processes in plants. For example, cytokinins promote development of lateral buds. If kinetin is applied directly to lateral buds, they are released from apical dominance and begin to grow as lateral branches. Cytokinins are involved in breaking the dormancy of embryos in germinating seeds, and they promote flowering and fruit development in some plants. Cytokinins also retard senescence (aging) of leaves and other plant parts.

Figure 19.13 Cytokinins. Kinetin was isolated as a hormonally active fraction in breakdown products of nucleic acids. All cytokinins are structurally related to the purine adenine. Zeatin is a natural cytokinin isolated from corn (*Zea mays*) seeds. Zeatin riboside occurs, along with zeatin, in coconut milk. Note that zeatin riboside has the same basic structure as zeatin, but with a sugar (ribose) attached.

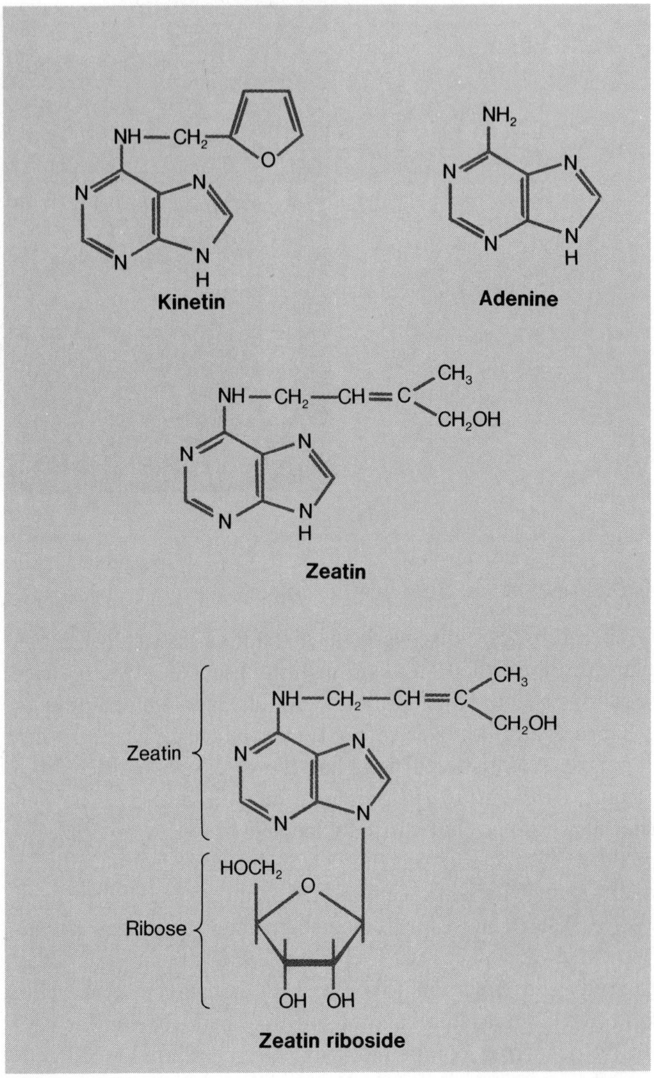

Figure 19.14 Interactions between auxin (IAA) and cytokinin (kinetin) in cultures of tobacco tissue. When IAA and kinetin concentrations are balanced, cultures grow as undifferentiated calluses (*middle rows*). Higher IAA-to-kinetin ratios result in root growth (*upper right*). Lower IAA-to-kinetin ratios result in shoot growth (*lower left*).

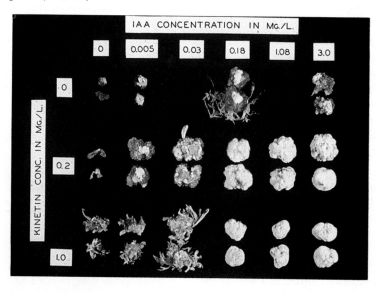

Gibberellins

Gibberellins were discovered as a result of research on a rice plant disease called "foolish seedling disease." Affected rice seedlings become unusually tall, spindly, and weak, and usually break and fall over before they produce ripe rice grains. Japanese scientists learned that the disease is caused by a fungus, *Gibberella fujikuroi*. In 1926, E. Kurosawa showed that a substance that caused excessive growth in rice plants could be extracted from the fungus, or even from a culture medium in which the fungus had been grown. Because of World War II, however, biologists in other countries did not become aware of this work until the 1950s. Then they began to study the effects of this extracted substance, **gibberellic acid (GA),** on vascular plants and to search for similar naturally occurring compounds that they assumed were produced in vascular plants. Eventually, chemically similar compounds that caused such growth responses were found in vascular plants, as well as in algae and in other fungi. Collectively, these compounds are known as **gibberellins,** and they are named according to their chemical structure and the order in which they were discovered. For example, the gibberellin most widely used in experimental work is a gibberellic acid that is abbreviated GA_3 (figure 19.15).

One of the most striking effects of gibberellic acid (GA) is the stem elongation that it causes in genetic dwarf varieties of certain plants. When GA is applied to such dwarf plants, they grow to normal size (figure 19.16). This response is quite specific because the dwarf plants do not elongate in response to auxins. It has been hypothesized that plants grow to their normal height because of the action of internally produced (endogenous) gibberellins and, therefore, that the dwarf mutants are short because they are unable to produce normal quantities of gibberellins. Reasonable as this hypothesis may be, it must be tested by further research before it can be accepted or rejected.

Another interesting effect of gibberellic acid is the response it causes in plants growing as **rosettes.** A rosette is a compact growth form where leaf attachments (nodes) are very close together on the stem. Cabbage, for example, is a biennial plant that grows in a rosette form (the familiar cabbage "head") during its first growing season. During its second season, however, cabbage grows tall, flowers, and produces seeds. In order to grow tall during the second growing season, a cabbage plant normally must experience chilling during the winter between its first and second growing seasons. Such chilling is the normal stimulus for stem elongation. Applying GA to a first-year plant, however, can have the same effect as chilling. It causes a rapid elongation of the stem known as **bolting** (figure 19.17).

Figure 19.15 The structure of gibberellic acid (GA₃). Subscript numbers indicate the order in which the compounds were discovered.

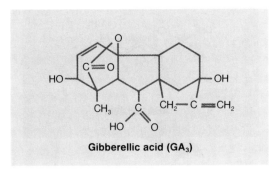

Gibberellic acid (GA₃)

Figure 19.16 Effect of application of gibberellic acid (GA) on growth of normal corn (*Zea mays*) plants and a genetic dwarf variety. The normal plant on the left received no treatment. The normal plant second from the left received GA, but GA has little effect on the height of normal plants. The dwarf plant third from the left received no treatment. The plant on the right is a dwarf plant that received GA treatment. GA was added during the growth of the plant at 3 to 5 day intervals with the total amount applied being about 500μg. Note that it has grown to the same height as normal plants.

Figure 19.17 Bolting caused by gibberellic acid. The cabbage plants on the left were untreated. Treatment with gibberellic acid caused bolting and flowering in the plants on the right.

Box 19.1

Florigen: The Long Search for a Flowering Hormone

For more than one hundred years, biologists have searched for a "flowering hormone," a substance that causes the growth changes leading to flower development and subsequent reproductive events. Two discoveries—the photoperiodic induction of flowering and the phytochrome system—have provided additional motivation for the search, but the flowering hormone has not, as yet, been isolated.

What evidence is there for a flowering hormone? Evidence has been accumulating since 1936 when M. H. Chailakhyan and his colleagues in the Soviet Union experimented with photoperiodic induction of flowering and found that a flowering stimulus appeared to be transmitted from one part of a plant to another. They removed the leaves from the upper part of chrysanthemums, which are short-day plants, and placed a light barrier between the upper and lower parts. Then they exposed the leafless upper part and the intact lower part to different photoperiods. When the upper (leafless) part was given long days and the lower part (with leaves) was given short days, the plant flowered. But the opposite treatment, in which the upper part received the short-day treatment, did not result in flowering (box figure 19.1A). Chailakhyan concluded that the leaves receive the photoperiodic stimulus, and he proposed that leaves

produce a chemical flowering substance that is transported through the plant. He suggested that the substance be called "florigen" (roughly meaning "flower maker").

Subsequently, K. C. Hamner and James Bonner extended and refined the concept of florigen in their experiments on cockleburs. They found that exposure of any part of a cocklebur, even a single leaf, to light-dark cycles with appropriately short days and long nights induces the plant to flower. Thus, the flowering stimulus can be transmitted from a single leaf to the whole plant. Another of Hamner and Bonner's experiments showed that the flowering stimulus can be transmitted from one plant to another. They induced a plant to flower by exposure to short days. Then they returned it to long-day photoperiods and grafted it to another plant that had been kept in a long-day environment. This second plant also flowered even though it had not experienced inducing photoperiods itself (box figure 19.1B). Furthermore, grafting a photo-induced short-day plant to a *long-day plant* in a noninducing environment causes the long-day plant to flower. This indicates that the flowering substance is the same in both cases, even though the plants respond to different photoperiodic stimuli.

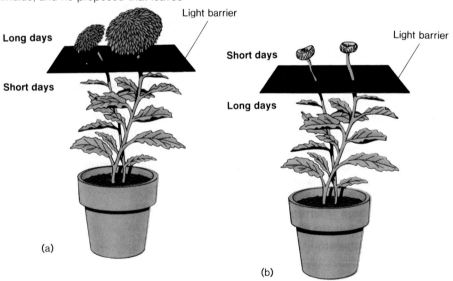

Box figure 19.1A Chailakhyan removed leaves from the upper parts of a plant and placed a light barrier between the upper and lower parts. (a) When leaves were exposed to the inducing (short-day) photoperiod, the plant flowered. (b) Exposing the leafless upper part to short days had no effect.

Other experiments have shown that interruption of the phloem by girdling (removing a ring of stem tissue that includes phloem) prevents movement of the stimulus from one part of the plant to another. Since the phloem is the normal route of long-distance transport of organic molecules inside a plant, this is further evidence that a chemical flowering substance exists. There have also been reports that extracts of leaves from photo-induced plants applied to noninduced plants cause the noninduced plants to flower, but these results are not always repeatable.

However, not all biologists are convinced that there is a flowering hormone; they think that the flowering stimulus caused by photoperiodic induction might simply be a change in the ratios of other hormones—auxins, cytokinins, or gibberellins. Ethylene also might be involved, because it promotes flowering in some plants, such as pineapples. However, this may be related to auxin effects, since plants begin to produce ethylene when auxin concentrations are high. Thus, research on hormonal control of flowering and the search for the elusive flowering hormone continue.

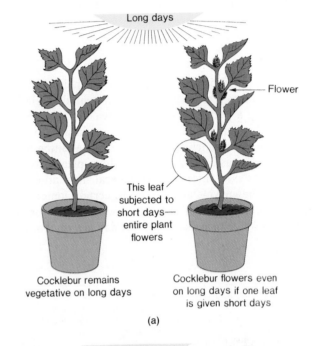

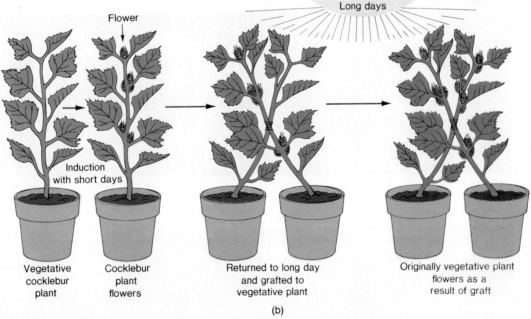

Box figure 19.1B Hamner and Bonner's experiments on flowering in cockleburs. (a) Short-day treatment of a single leaf results in flowering, even though the rest of the plant receives long days. (b) A cocklebur plant induced to flower by short days was returned to long days and grafted to a vegetative plant that had been kept in a long-day photoperiod. This second plant flowered in response to the graft. In fact, in some cases, grafting even a single leaf from an induced plant to a noninduced plant causes the plant to flower.

Biological Rhythms

The nature of the internal clocks of living things is one of the most intriguing and puzzling questions in modern biology. Obvious daily cyclic fluctuations (**rhythms**) in many physiological processes are expressions of timekeeping by these biological clocks.

Daily Rhythms

Plants in nature show obvious daily cycles. For example, beans and many other plants spread their leaves horizontally in the daytime, exposing a greater leaf surface area for light absorption. At night their leaves fold up or down in what has been called a sleep movement (figure 19.27). In addition to such externally obvious daily cycles or rhythms, plants have many subtle, rhythmic, internal physiological fluctuations. For example, some enzyme activity levels, certain ion concentrations in internal fluids, and sensitivities to many drugs and other chemicals change throughout the day in a rhythmic

fashion. All plants and all animals, including humans, have such daily rhythms. Of course, it is not surprising that life processes vary in a rhythmic manner with the time of the twenty-four-hour **solar day,** because all organisms have evolved in a decidedly rhythmic world where light and darkness cycle with great regularity. Other physical factors, such as temperature and barometric pressure, also fluctuate, but not with nearly the same regularity as the daily light cycle.

What *is* surprising is that the solar-day rhythms of many living things continue even when the organisms are deprived of obvious external information about the time of day. The cycles continue when organisms are maintained in laboratories in constant light, constant temperature, and even constant barometric pressure. For example, bean plant sleep movements continue for days when a bean plant is left undisturbed under constant conditions (figure 19.28). Biologists have concluded that this persistence under constant conditions must indicate that organisms possess internal clocks that time rhythmic processes even in the absence of obvious external cues about the time of day.

Figure 19.27 A photographic record of rhythmic leaf movements in a cocklebur plant (*top of each frame*) and a bean plant (*bottom of each frame*). The plants were photographed at hourly intervals from noon to noon. Note that the bean leaves drop more sharply and later in the evening than do the cocklebur leaves.

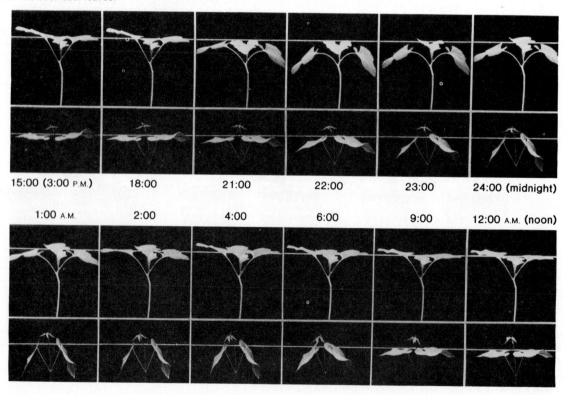

15:00 (3:00 P.M.) 18:00 21:00 22:00 23:00 24:00 (midnight)

1:00 A.M. 2:00 4:00 6:00 9:00 12:00 A.M. (noon)

Another surprising result in this research is that, in many cases, temperature change has very little effect on the rates at which rhythmic processes go through their daily cycles. Biologists have demonstrated that the **period length,** a measure of the rate at which a biological rhythm "runs," is quite temperature independent in most cases (figure 19.29). This result was unexpected because rates of many biological processes are strongly affected by temperature increases. Relative temperature independence is a striking and commonly observed property of biological rhythms.

Another feature of most rhythms in organisms kept under constant conditions in the laboratory, especially in constant light or constant darkness, is that period lengths usually are slightly longer or slightly shorter than exactly twenty-four hours. We call these **circadian** (literally, "about a day") **rhythms** (figure 19.30). Sometimes, biologists use this name much more generally and speak of all clock-timed rhythmic phenomena as circadian rhythms or circadian clocks. Technically, however, the term circadian describes only rhythms that have period lengths slightly different from twenty-four hours when measured under constant conditions. Circadian periods are seen only under laboratory conditions, where organisms are deprived of information about normal light/dark cycles. In nature, the daily light/dark cycle keeps period lengths of daily rhythms exactly twenty-four hours long. This external control by environmental factors that keeps rhythms precise is called **entrainment,** and the environmental factors that entrain rhythms are called **zeitgebers** (from the German *Zeitgeber,* meaning "time-giver"). Thus, rhythms become circadian in the laboratory only when deprived of normal entrainment by environmental zeitgebers.

Figure 19.29 Biological rhythms and temperature. (a) The period of a rhythm is the time from a point in one cycle to the same point in the next cycle. Amplitude is the difference between the high point and the low point within one cycle. (b) Temperature relationships. Increased temperature would be "expected" to change both period length and amplitude of rhythms. Although amplitude changes are as predicted, period length changes little, if at all, in response to increased temperature.

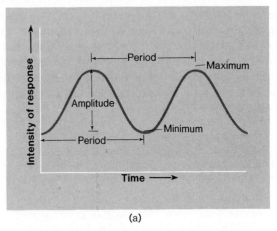

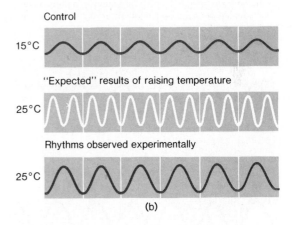

(b)

Figure 19.28 The technique originally used for measuring bean sleep movements under constant laboratory conditions. The leaf is attached to a lever to which is attached a pen that makes a tracing on a paper fastened to a rotating drum. The results of several days of recording are shown.

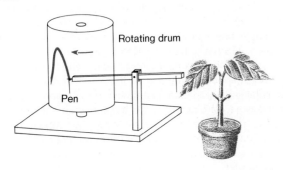

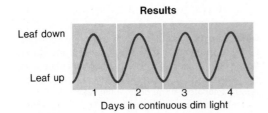

Rhythms and Clocks

Most or maybe all of the biological rhythms, such as plant sleep movements, that have been described and studied are overt, external expressions of timekeeping by an underlying timer (**biological clock**). You might think of rhythms as representing the hands of the clock—they are resettable just as the hands of an ordinary clock can be reset to any time setting. In all plants and animals there seems to be an underlying cellular clock, and the resettable rhythms (the hands) are linked to it. For example, jet lag following intercontinental travel disappears after a few days. Environmental zeitgebers cause resetting of rhythms.

Interpreting the results of biological rhythms experiments is very difficult. Experimental results might relate to a property of only the clock's hands (observable, overt rhythms), the clock's basic timing mechanism, the linkage between the two, or some combination of all three of these things. One certainty, though, is that there are some challenging unanswered questions about both the underlying cellular clock and its relationship to the overt rhythms that play such a large part in the lives of all plants and animals. Possibly the most fundamental of these questions involves the nature of the cellular clock (or clocks) and its location (or locations) in the organism.

Figure 19.30 A comparison between an exact twenty-four-hour daily rhythm and a circadian rhythm. (a) An exact twenty-four-hour solar day rhythm. This is the characteristic pattern of all daily rhythms under natural entrainment. (b) A circadian rhythm with a period length of less than twenty-four hours. This slightly exaggerated model shows how peaks of rhythmic activity occur earlier on successive solar days. Such circadian rhythms are observed under laboratory constant conditions when organisms are deprived of normal entrainment, such as the daily light-dark cycle.

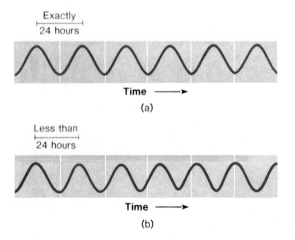

Biological Clocks: Internal or External Timing?

There are two opposing, but not entirely dissimilar, viewpoints on the nature of the biological clock. One hypothesis proposes that the clock is an entirely *internal*, biochemical oscillator mechanism operating at the cellular level. This proposed cellular clock works like the pendulum that measures time in a pendulum clock. Of course, pendulum clock hands (rhythms) can be reset to any time on the face of the clock. The hypothesis also proposes that the oscillator mech-

---

Box 19.2
Plants That Move

A growth response such as geotropic bending of a young seedling's stem permanently changes the shape of a plant. However, a bean plant's sleep movements are clearly reversible. Many other plants also exhibit sleep movements, as well as other kinds of reversible movements.

The "sensitive plant," *Mimosa*, moves quickly and dramatically in response to being touched or shaken. Within seconds after being disturbed, leaflets fold up and leaves droop. Such movements involve turgor pressure changes in cells in specialized regions. Control of this movement is not well understood, however, since a stimulus applied to one small area of a *Mimosa* sometimes causes leaf drooping in a large part of the plant.

Other reversible movements play very different roles in the lives of some plants. The Venus flytrap, for example, has highly modified leaves that fold up and trap insects which are then digested to provide nutrients for the plant. Other plants are heliotropic ("sun turning"). They continuously turn toward the sun, thereby maintaining a particular orientation to it. An entire field of sunflowers, for example, track the sun so that their leaves remain perpendicular to the sun's rays throughout the day. Such movements also involve turgor pressure changes in certain cells, but as yet their control is not understood.

Box figure 19.2 Movement in plants. (a) Photographs of the sensitive plant, *Mimosa pudica*, before (1) and after (2) the plant has been touched. Note how the leaflets fold up in response to being disturbed. (b) A specialized cluster of cells called the pulvinus, located at the base of the leaf stalk (1) is responsible for leaf drooping. In the resting condition, cells in the pulvinus are turgid and the leaf stalk stands erect (2). Following disturbance of the plant, however, cells in the lower part of the pulvinus lose

anism may be regular, cyclical changes in some cellular process, such as a complex enzyme reaction series, differences in membrane permeability, or possibly even repeated transcription (see chapter 24) of a segment of a DNA molecule.

The other hypothesis also suggests that the cellular clock is internal, but that its basic timing information comes from the *external* environment. This proposed cellular clock is like the motor of an electric clock. Again, the hands (rhythms) of an electric clock are resettable, but basic time measure-

ment is by an electric motor driven by alternating current supplied from the outside. Scientists who support this hypothesis propose that regular daily fluctuations of physical environmental forces, such as magnetic or electrostatic field intensities, cosmic radiation, or some combination of these or other physical forces, provide basic timing information, despite all efforts by investigators to maintain constant laboratory conditions (figure 19.31).

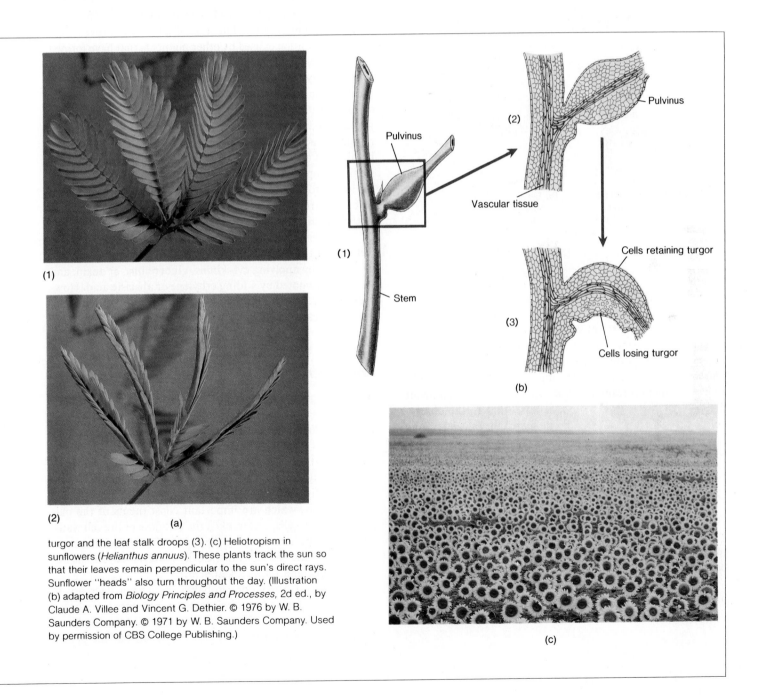

turgor and the leaf stalk droops (3). (c) Heliotropism in sunflowers (*Helianthus annuus*). These plants track the sun so that their leaves remain perpendicular to the sun's direct rays. Sunflower "heads" also turn throughout the day. (Illustration (b) adapted from *Biology Principles and Processes*, 2d ed., by Claude A. Villee and Vincent G. Dethier. © 1976 by W. B. Saunders Company. © 1971 by W. B. Saunders Company. Used by permission of CBS College Publishing.)

Figure 19.31 Two models of the internal cellular clock. The "hands" can be reset (entrained by environmental zeitgebers) in either case, but what is the basic clock mechanism? (a) Is it an autonomous, internal biochemical oscillator that measures time like a pendulum? This hypothesis was developed by C. S. Pittendrigh, J. Aschoff, and others. (b) Or is it like an electric clock, in which time measurement depends on alternating current from the outside? This hypothesis was developed by F. A. Brown, Jr., and others.

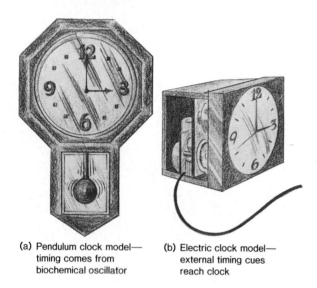

(a) Pendulum clock model— timing comes from biochemical oscillator

(b) Electric clock model— external timing cues reach clock

Either of these hypotheses can be used to explain almost every result of experiments on biological rhythms and almost every inferred property of biological clocks. It is difficult to conceive of experiments that would clearly distinguish between the two hypotheses regarding the nature of basic clock timing. All such studies are handicapped because all available experimental results provide us with information about the behavior of only the clock "hands," that is, the observable, overt rhythmic processes. The search for the underlying biological clock (or clocks) of living things is a challenging research problem.

Season-Ending Processes

Perennial plants (plants that are active during several to many growing seasons) cannot survive freezing winter temperatures in their active growing, summertime condition. What preparations do perennial plants make for winter dormancy?

In late summer or early fall, nutrients are transported underground to roots for storage, and tough scales or other protective devices form around the buds that will initiate the next season's growth. To most people, however, changing leaf color and the eventual loss of leaves are the most obvious and familiar signs of approaching winter in plants. Leaves contain relatively soft tissues that would be very difficult to protect from winter damage, and preparation for winter dormancy includes **senescence** (aging) and **abscission** (detachment) of leaves.

Leaf Senescence

During senescence, leaves change dramatically. They lose protein and chlorophyll. This loss of chlorophyll reveals the autumn coloration caused by other pigments that have formed in the leaves. Early in senescence, leaf oxygen consumption rises sharply and reaches a peak level two or three times as great as the presenescence level. Oxygen consumption falls off later, but this initial burst indicates that senescence is not just a slow, steady decline in activity, but a set of energy-requiring, active metabolic changes. Enzymes catalyze the breakdown of some constituents of leaf cells, and certain essential components, such as mineral ions, are exported from the leaves to other parts of the plant. This conserves valuable plant resources that otherwise would be lost with the falling of the leaves.

Hormonal control of leaf senescence is complex and not well understood, as yet. Experimentally, leaf senescence can be delayed by applying cytokinins, gibberellins, or auxin, and it can be promoted by adding ethylene or abscisic acid. However, more research is needed to clarify the roles of these hormones in causing or preventing leaf senescence in intact plants.

Abscission

While senescence changes most of the cells of a leaf, abscission is due to changes in a very narrow band of cells located, depending upon the type of plant, either at the base of the petiole (leaf stalk) or at the point where the blade of the leaf joins the petiole. This band of cells forms an **abscission zone,** the breaking point where the leaf eventually detaches from the stem (figure 19.32). During abscission, enzymes break down pectins, which are important constituents of the material between cells, and weaken the cellulose of the cell walls. These enzyme actions cause the layers of the abscission zone to separate, and the leaf falls. Corky material covers the abscission zone and forms a **leaf scar.**

Control of abscission probably involves a decrease in auxin production in leaves, since auxin prevents leaf abscission in most plants. This can be demonstrated by simple experiments in which the leaf blade is cut off, but the petiole

Figure 19.32 The abscission zone. (a) Abscission zones may be located at the base of a petiole, at the junction of a petiole and a leaf, or even at several different points. (b) Internal structure of an abscission zone (*arrow*) at the junction of a petiole and the stem of a silver fir.

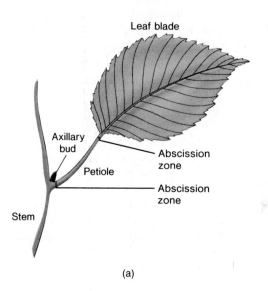

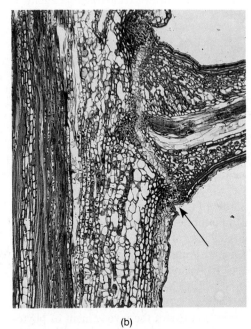

(a)

(b)

is left in place. Because young and mature leaf blades produce auxin, removal of the blade deprives the petiole of its major auxin source. Without further treatment, the petiole goes through normal abscission and falls. If auxin is applied to the cut end, however, the petiole remains in place. Therefore, it was originally thought that the control of abscission was fairly easily explained as the result of decreased auxin production in aging and senescing leaves. But even the role of auxin in leaf abscission now seems more complex. Cells of the abscission zone may respond to differences in auxin concentrations on the stem side and the leaf-blade side of the abscission zone rather than simply responding to decreased leaf auxin production. Furthermore, ethylene is known to be a powerful abscission promoter, and senescing tissues produce ethylene.

Control of abscission, like so many other processes in plants, probably involves complex interactions and changing balances of several plant hormones. Thus, the whole life of a vascular plant, from seed germination to aging and senescence, involves continuing interactions between the plant and its environment, and plant hormones mediate the plant's responses at every step along the way.

Summary

A terrestrial vascular plant remains in one place and interacts with its changing environment throughout its life. These interactions are mediated by plant hormones.

Light is the dominant environmental factor affecting growth responses and seasonal changes in plants. Plants display phototropic responses to directional light, photomorphogenetic responses to changing light quantity and quality, and photoperiodic responses to changing light/dark cycles.

The cells that produce plant hormones are not specialized for hormone production alone, nor are responses restricted to a few kinds of target cells; practically all plant cells respond in some way to each plant hormone.

Auxin was discovered as a result of experiments on phototropic responses of grass seedlings. The principal naturally occurring auxin is indoleacetic acid, but other substances, including some synthesized in the laboratory, act as auxins. Auxin's principal effect is to cause cell elongation.

Phototropic responses result from the lateral transport of auxin to the darker side of a plant, where greater cell elongation occurs. Stem geotropism is similar in that auxin stimulates greater elongation on the lower side. Auxin is also responsible for apical dominance because auxin produced in the shoot apex inhibits the growth of lateral buds.

Ethylene, a gas, is an important fruit-ripening agent. It promotes hydrolysis of starch to sugar and causes fruit to soften.

Cytokinins, such as kinetin, are cell-division promoters that act along with auxins in general regulation of growth responses. Zeatin and zeatin riboside are naturally occurring cytokinins.

Gibberellins, which were first extracted from a fungus, stimulate stem elongation in dwarf plants and probably play a role in stem elongation in normal plants. Gibberellic acid, a gibberellin, causes bolting in rosette plants. Gibberellic acid is important in seed germination because it stimulates synthesis of α-amylase, which hydrolyzes stored starch and thus makes sugar available to the growing seedling.

Abscisic acid, an inhibitor, is involved in a plant's preparation for and maintenance of dormancy. Abscisic acid acts by inhibiting RNA and protein synthesis. It also promotes abscission of various fruits and seeds, and is involved in stomatal closure during wilting in some plants.

Many light responses are promoted by red light and opposed by far-red light because of the involvement of phytochrome, a pigment that is reversibly converted from one form to another upon the absorption of red or far-red light.

Photoperiodism is particularly important in timing reproductive activity (flowering) and in the onset of dormancy. A short-day plant flowers when nights are longer than its particular critical length, while a long-day plant flowers when nights are shorter than its particular critical length.

Photoperiodic responses involve time measurement that depends on an internal clock mechanism, which also times rhythmic daily changes in various physiological functions. However, the exact nature of the underlying cellular clock mechanism remains obscure.

Perennial plants prepare for winter dormancy in several ways. Nutrients are stored underground, and protective scales form around delicate buds. The most obvious sign of preparation for winter, however, is leaf senescence and abscission. These processes also are under hormonal control, as are all of the environmental responses in the lives of vascular plants.

Questions for Review

1. Explain the adaptive value of plants' responding to photoperiodic changes rather than seasonal temperature changes to time events during their life cycles.
2. Define phototropism and geotropism.
3. Explain the Went bioassay for auxin and discuss the general usefulness of bioassays in hormone research.
4. What is the role of auxin in apical dominance?

5. Describe an etiolated plant and explain how etiolation occurs.
6. How does "one bad apple spoil the barrel?"
7. Describe the role of ABA in a short-term, homeostatic physiological response.
8. What is P_{fr}?
9. Distinguish between short-day and long-day photoperiodic responses.
10. What are circadian rhythms?

Questions for Analysis and Discussion

1. Gibberellins are not required for the growth of the fungus that causes "foolish seedling disease" in rice. What adaptive value do you suppose gibberellin production has for the fungus?
2. Propose a relationship between the phytochrome system and florigen production (the flowering stimulus) in cockleburs.
3. Why do you think that some short-day plants near the northern edge of their range are able to produce mature seeds and fruits during some growing seasons but not during others?

Suggested Readings

Books

Kendrick, R. E., and Frankland, B. 1983. *Phytochrome and plant growth.* Studies in Biology no. 68. London: Edward Arnold.

Noggle, G. R., and Fritz, G. J. 1983. *Introductory plant physiology.* 2d ed. Englewood Cliffs, N.J.: Prentice-Hall.

Thimann, K. V. 1977. *Hormone action in the whole life of plants.* Amherst, Mass.: University of Massachusetts Press.

Wareing, P. F., and Phillips, I. D. J. 1981. *Growth and differentiation in plants.* 3d ed. New York: Pergamon.

Wilkins, M. B., ed. 1984. *Advanced plant physiology.* Marshfield, Mass.: Pitman.

Articles

Albersheim, P., and Darvill, A. G. September 1985. Oligosaccharins. *Scientific American.*

Cleland, C. F. 1978. The flowering enigma. *BioScience* 28:265.

Hendricks, S. B. 1980. Phytochrome and plant growth. *Carolina Biology Readers* no. 109. Burlington, N.C.: Carolina Biological Supply Co.

Hillman, W. S. 1979. Photoperiodism in plants and animals. *Carolina Biology Readers* no. 107. Burlington, N.C.: Carolina Biological Supply Co.

Palmer, J. D. 1984. Biological rhythms and living clocks. *Carolina Biology Readers* no. 92. Burlington, N.C.: Carolina Biological Supply Co.

Rubery, P. H. 1981. Auxin receptors. *Annual Review of Plant Physiology* 32:569.

Satter, R. L., and Galston, A. W. 1981. Mechanisms of control of leaf movements. *Annual Review of Plant Physiology* 32:83.

Sisler, E. C., and Yang, S. F. 1984. Ethylene, the gaseous plant hormone. *BioScience* 34:234.

Smith, H. 1984. Plants that track the sun. *Nature* 308:774.

Weyers, J. 17 May 1984. Do plants really have hormones? *New Scientist.*

Genetics and Development
The Continuity of Life

Part 4

 The ability to reproduce is a fundamental property of living things. All multicellular organisms age and eventually die. But for a species, reproduction provides a potential means for increasing numbers and results in continuing replacement of aging individuals with young, vigorous individuals. In addition, sexual reproduction produces individuals with new genetic combinations. This contribution to genetic variability is fundamental to the evolutionary process.

Reproductive processes involve specific activities of individual cells. In unicellular organisms, one individual divides mitotically to produce two individuals. Multicellular organisms produce specialized individual reproductive cells such as eggs and sperm. Via these specialized cells, characteristics are transmitted from one generation to the next by specific genetic mechanisms. Each individual offspring has a new set of genes that is somewhat different from the set of each parent.

Expression of the genetic complement of a new individual occurs through developmental processes. Development continues throughout life in the form of growth, maturation, and aging. Developmental processes are also involved in continuing replacement of body cells, in healing of injuries, and in specific responses to infection and disease. Developmental processes gone awry as abnormal growth, however, can be life-threatening.

In chapters 20 through 29 we first will examine the cell division mechanisms involved in reproduction and development and the means by which genetic information is transmitted and expressed. Then we will consider some ways in which our growing knowledge of genetic mechanisms is being applied in such areas of biotechnology as genetic engineering. Finally we will look at some details of reproductive patterns in various organisms, and the nature of developmental processes that occur throughout life.

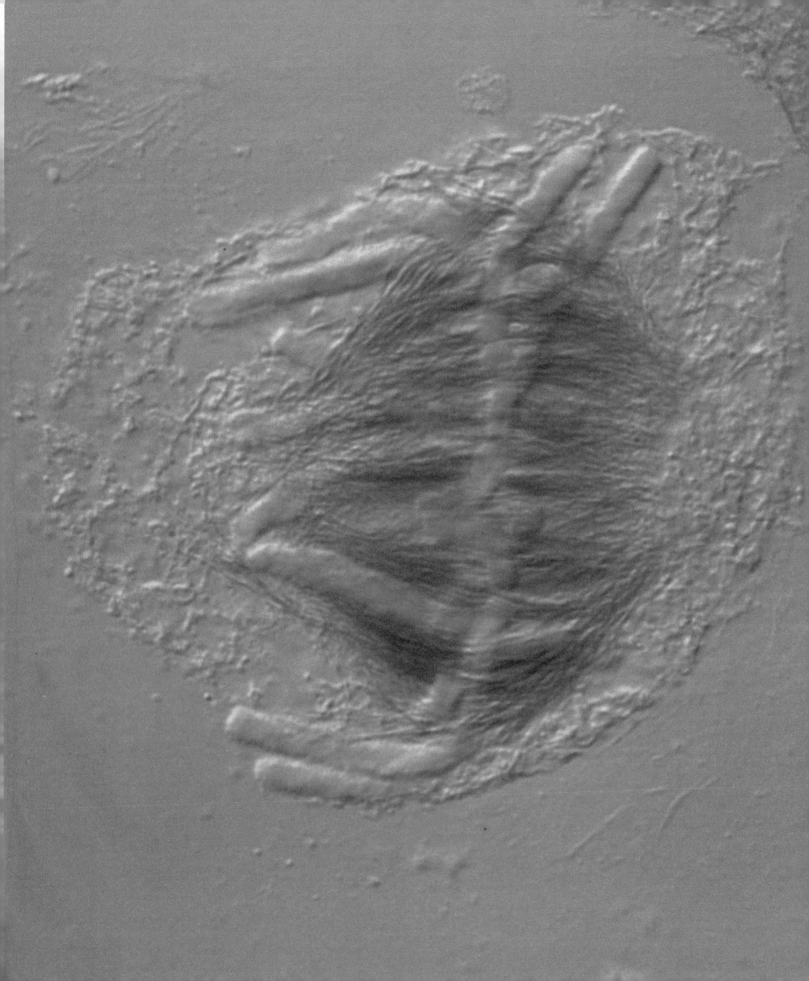

20

Cell Division

Chapter Concepts

1. The cell cycle involves growth, duplication of genetic information, preparation for division, and division.
2. Mitosis is the process whereby the duplicated genetic information of a cell is precisely divided to produce two genetically identical nuclei.
3. The cytoplasm of a dividing cell is separated by cytokinesis, a process that differs in plant and animal cells.
4. A sexually reproducing organism must reduce its chromosome number by half at some point in the life cycle. This prevents a doubling of the normal chromosome number each time fertilization occurs.
5. Meiosis is a type of cell division by which this reduction in chromosome number occurs.

Facing page   Cell division of a blood lily cell, photographed using the Nomarski differential interference contrast technique. This technique makes it possible to observe the structure and movement of the rodlike chromosomes during cell division. Cell division is fundamental to the continuity of life on earth because every living cell arises through division of a preexisting cell.

The idea that organisms are composed of cells became quite widely accepted in the centuries following Robert Hooke's first description in 1665 of his microscopic observation of small spaces in cork that he named "cells." But it was not until the 1830s that M. J. Schleiden and Theodor Schwann advanced the concept that all plants and animals are composed of cells and cell products.

Theories on how new cells originate were hotly debated in the decades following Schleiden and Schwann's work. Schleiden and Schwann, among others, thought that the nucleolus becomes a new nucleus, which in turn develops into a new cell. Other scientists thought that new cells either bud off from preexisting cells or form by an unknown process in the space between cells. Still others insisted that new cells form when a cell divides into two equal halves.

Observations over the years eventually confirmed that cells arise by a process of division, and this idea was well enough established in the 1850s for Rudolph Virchow to complete the formulation of the Cell Theory begun by Schleiden and Schwann. Schleiden and Schwann had said that all organisms are composed of cells and cell products. To this statement, Virchow added that all cells are produced by the division of previously existing cells. In this chapter we will explore how these cell divisions take place.

The genetic information that determines the characteristics of cells and controls their many functions is contained within chromosomes in cell nuclei. During cell division, duplicate copies of this genetic material are distributed to the new cells with great precision, thus ensuring the continuity of genetic information from cell to cell. The process by which the genetic material is distributed is called **mitosis.** Mitosis results in two new nuclei, each genetically identical to the original nucleus because each receives a set of chromosomes identical to that of the original cell. The cytoplasm is divided, more or less equally, by a separate process known as **cytokinesis.**

In any organism, growth due to an increase in the size of the component cells is limited, and thus extensive growth depends on mitotic cell division. However, cell division does not end when an organism's growth slows or stops at maturity, because cells produced by mitosis also serve as replacements for cells that die or are lost by the organism. In humans, for example, the outer layers of the skin are constantly being sloughed off and replaced by the division of cells lying beneath them. Continuing cell division that results in cell replacement is vital for the maintenance of homeostasis.

In some types of organisms, specialized cells produced by mitotic cell division may develop directly into new individuals that are genetically identical to the parent. This type of reproduction is called **asexual reproduction** (see chapter 26).

The other principal method by which organisms reproduce is **sexual reproduction.** Sexual reproduction requires a genetic contribution from two different cells and involves a type of cell division called **meiosis.** Let us begin our discussion with mitosis; then we will consider meiosis and the key role it plays in sexual reproduction.

The Cell Cycle

In the 1870s, the techniques of microscopy had been refined to the point where a number of scientists were able to provide detailed and accurate descriptions of the movement of chromosomes during mitosis. For a long time, then, the emphasis in studies of cell division was on the process of mitosis itself. Because there was little visible activity during the remainder of the cell's life span, the processes taking place between mitotic events remained largely unknown. In fact, the period during which mitotic activity was not occurring was viewed as a resting stage, an "interphase."

It was not until much later, however, with the development of increasingly sophisticated biochemical techniques in the 1950s, that biologists were able to discover and describe many of the processes that occur in cells between mitotic divisions. They learned that the period between divisions is indeed a very important and active time for the cell. The entire cycle—from the time a new cell is formed until it divides—is called the **cell cycle,** and mitosis can be fully understood only in the context of this cell cycle.

After it had been confirmed that deoxyribonucleic acid (DNA) is the primary genetic material (see chapters 3 and 23), it was clear that the genetic information contained in DNA must be copied (replicated) before each cell division. Biologists set out to learn exactly when DNA replication takes place. New techniques were developed to study the dynamics of DNA function and duplication throughout the cell cycle. Two major methods of studying DNA are Feulgen staining, which indicates the presence and relative amounts of DNA, and autoradiography (see p. 66), a process by which DNA can be radioactively labeled and then visualized on photographic film. A combination of these two techniques revealed that the DNA of a cell is replicated during a specific period of the interphase between mitotic divisions, frequently hours before mitosis occurs. The discovery of the timing of DNA replication led to a description of four distinct, functional phases of the cell cycle (figure 20.1).

Figure 20.1 The cell cycle. Mitosis (*M*) and the synthesis of the genetic material (*S*) are separated by two gap phases (*G₁* and *G₂*).

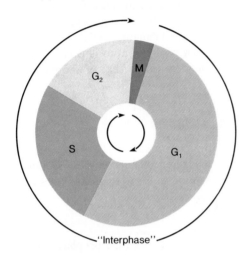

"Interphase"

The phase in which the DNA is synthesized is designated the **S** (for synthesis) **phase**. The period of time after a cell is produced by mitosis and before it enters the S phase is called the **gap 1 (G_1) phase**. (The G_1 phase is a "gap" only in the sense that it does not include major cell reproductive events involving DNA.) During this G_1 phase, growth and synthesis of many compounds other than DNA occur. After the genetic material is duplicated in the S phase, another gap occurs, the **G_2 phase**. Events in the G_2 phase are not fully understood, but G_2 apparently involves the synthesis of various enzymes and structural proteins necessary for mitosis. The fourth phase, called the **M phase** (for mitosis), is the nuclear division process itself.

The length of time a particular cell remains in each stage of the cell cycle varies depending upon the type of cell involved. In human cells cultured at 37° C, the G_1 stage lasts for eight hours, the S phase lasts for six hours, the G_2 phase takes about 4½ hours, and mitosis is completed in just one hour. In the cells of vascular plants, the entire cycle lasts from ten to thirty hours, with the M and S phases comparatively longer than in animals. Many protozoa, fungi, and most embryonic cells have much shorter cell cycles with greatly reduced G_1 phases. The length of the G_1 phase is apparently correlated with the amount of growth and synthesis necessary for the cell. Cells that are reproducing very rapidly do not have an extended G_1 period. Furthermore, variations in temperature may increase or decrease the duration of each phase.

The transition from the G_1 phase to the S phase is a critical point in the cell cycle. A cell that does not enter the S phase does not divide; in fact, a cell can remain in G_1 indefinitely. However, once a cell enters the S phase, it is committed to go on to mitosis and complete the cycle in most cases. This "point of no return," which occurs late in G_1, is called the **restriction point** and it is a key to the entire cell cycle. The precise factors that determine whether or not a cell enters the S phase remain to be discovered, however.

The majority of cells in adult organisms remain in the G_1 phase, permanently blocked from entering the S phase. However, as we noted earlier, some cells in mature adult organisms continue to divide periodically, or even routinely, to replace cells that are lost or cease to function. Such dividing cells make the critical transition from G_1 to the S phase.

A better understanding of what occurs at the restriction point, and thus what determines whether cells enter the S phase and go on to mitosis or remain in G_1, might offer clues for the treatment and control of abnormal cell division. Cancer cells, for example, grow without restraint or control. They are probably derived from cells that are normally blocked from entering the S phase; that is, cells that normally do not divide. Some factor, external or internal, transforms these cells so that they continue to enter the S phase in each cell cycle and thus are committed to uncontrolled cell division (see chapter 29). Much of the current basic research on cancer is aimed at determining why cancer cells are not under the constraints that prevent normal cells from dividing in an uncontrolled fashion.

Cell Division in Prokaryotes

Cell division in prokaryotic cells (cells that lack nuclear membranes and membrane-bound organelles) and eukaryotic cells is similar in that in both cases, duplicated genetic material is apportioned to each new cell. In prokaryotes, however, mitosis does not occur. The process of cell division is simplified by the fact that the genetic information is not contained within a nucleus. In bacteria, for example, each copy of the replicated circular chromosome becomes anchored to an attachment point in the cell membrane (figure 20.2). The cell membrane elongates until the cell is approximately twice its original length. Then the membrane appears to pinch the cell into two segments, each of which is a new cell that becomes enclosed by a cell wall. Each new cell contains an identical copy of the original bacterial chromosome. In chapter 23 we will explore the details of genetic replication in prokaryotic cells.

Figure 20.2 Cell division in a prokaryotic cell. (a) A bacterial cell has a single chromosome which is a circular DNA molecule (with associated RNA and proteins) anchored to the plasma membrane at an attachment point. (b) Replication of the chromosome begins. (See chapter 23 for details of chromosome replication in prokaryotic cells.) (c) At the completion of replication, each chromosome has its own attachment point. (d) Additional plasma membrane is produced, the cell elongates, and the attachment points are moved apart. (e) Two new cells at the completion of cell division.

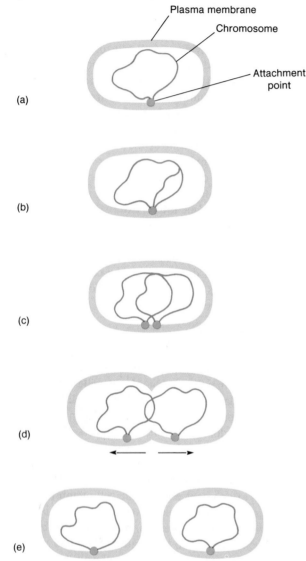

Plasma membrane

Chromosome

Attachment point

(a)

(b)

(c)

(d)

(e)

Mitosis

Mitosis is the division of the nucleus in eukaryotic cells that results in equal distribution of duplicated genetic material to the two new cells (called **daughter cells**). This division is very precise, and the end result of mitosis is two nuclei with genetic information identical to that of the parent cell's nucleus.

The Stages of Mitosis

Mitosis is a continuous process that is arbitrarily divided into four stages for convenience of description. In figure 20.3 you can see, in diagrammatic form, the four stages of mitosis as it occurs in animal cells.

Prophase

The first stage of mitosis is called **prophase.** A typical cell entering prophase from G_2 has an intact nuclear membrane and may have visible nucleoli, but the chromosomes themselves are not visible with a light microscope (figure 20.3a; see also figure 20.6a). During the early stages of prophase, the chromosomes begin to condense. This apparent **condensation** is actually a coiling process that results in a shortening and thickening of each chromosome. The coiling involves not only DNA molecules, but also chromosomal proteins. As we will see in chapter 22, these protein molecules, particularly **histones,** serve as "spools" around which a thread of DNA is wound.

Chromosome condensation seems to be a necessary part of the mitotic process because chromosomes in their normally extended condition might become quite tangled during mitotic separation. Chromosomes cannot remain permanently condensed, however, because genetic expression (the cell's utilization of the coded genetic information in the DNA) is complicated when the chromosomes are condensed. Thus, chromosomes coil at the beginning and uncoil at the end of each cell division.

In the cells of most organisms, the nuclear membrane begins to break down during prophase. Later, this allows the chromosomes to be apportioned between the two new nuclei that are formed. Also nucleoli in the dividing cell disappear during the early part of prophase.

The coiled chromosomes of prophase are easily visible under a light microscope after they have been stained (figure 20.3b; see also figure 20.6). Each chromosome consists of two subunits that are attached to each other. Each chromosome

Figure 20.3 Stages of mitosis beginning with a cell just prior to entering the M phase of the cell cycle. Note the beginning of cytokinesis in late telophase.

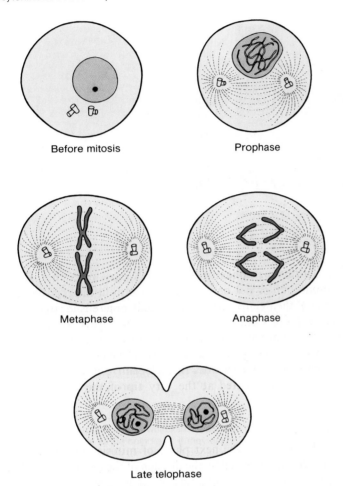

Before mitosis

Prophase

Metaphase

Anaphase

Late telophase

Figure 20.4 Electron micrograph of a replicated human chromosome showing the two chromatids joined at the centromere (magnification × 33,824).

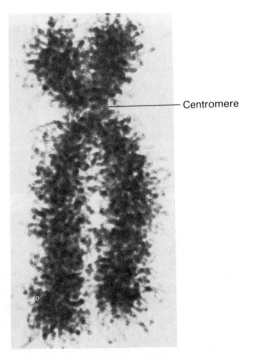

Centromere

subunit, called a **chromatid** (figure 20.4), contains one double-stranded, helical DNA molecule (as well as various proteins). These two DNA molecules (one in each chromatid) are the products of DNA replication during the S phase of the cell cycle. The point at which the two chromatids are joined is the **centromere,** a distinctly constricted area of the chromosome.

Also during prophase the **spindle apparatus** begins to form. The spindle apparatus consists of a set of microtubules (see p. 77) that are associated with chromosome movement later in mitosis. In many kinds of cells, another obvious occurrence at this time is the movement of the **centrioles,** which were duplicated earlier during the S phase of the cell cycle.

During prophase, one pair of centrioles moves to one side of the nucleus and the other pair moves to the opposite side. As they move, the centrioles are accompanied by a halo of microtubules called the **aster,** part of which remains around the centrioles in their new locations on opposite sides of the nucleus. The spindle apparatus develops between these pairs of centrioles (figure 20.5). Thus, the centrioles lie near the **poles** (ends) of the spindle apparatus. Despite the proximity of all these structures to one another, there does not seem to be any actual physical connection between the centrioles and the microtubules of either the aster or the spindle apparatus.

Because of their prominence and their location in some cells, centrioles were originally thought to be responsible for the organization of microtubules into the spindle apparatus. However, there is experimental evidence that centrioles do not play a vital role in cell division—removing centrioles from cells that normally have them does not prevent those cells from forming a spindle apparatus and dividing.

Figure 21.19 Results of crosses between two breeds of white chickens—white leghorns and white wyandottes. The *I* gene masks the expression of the *C* gene in the F₁ generation, but colored chickens are produced in the F₂ generation.

P₁　　　White leghorn × White wyandotte
　　　　　　IICC　　　　　　　　　*iicc*

F₁　　　　　　　　　White
　　　　　　　　　　IiCc

F₂　　　　*IC*　　*Ic*　　*iC*　　*ic*

	IC	*Ic*	*iC*	*ic*
IC	*IICC*	*IICc*	*IiCC*	*IiCc*
Ic	*IICc*	*IIcc*	*IiCc*	*IIcc*
iC	*IiCC*	*IiCc*	Colored *iiCC*	Colored *iiCc*
ic	*IiCc*	*Iicc*	Colored *iiCc*	*iicc*

What happens when members of these two white breeds are crossed with each other? The results of such a cross are shown in figure 21.19. All members of the F₁ generation are white because of the presence of the *I* gene, which masks the expression of genes for color. But in the F₂ generation, some colored chickens indeed are produced.

Pleiotropy

We have now seen some cases in which several genes affect the phenotypic expression of a single characteristic. There is an entirely different kind of situation in which a single gene affects several phenotypic characteristics. This is called **pleiotropy.** In humans, an example of pleiotropy is the gene that, in the homozygous recessive condition, results in the disease phenylketonuria (PKU). Untreated individuals with this disease have unusual amounts of the amino acid phenylalanine in their blood, and also lower mental ability, somewhat larger heads, and lighter hair color. All of these characteristics are determined by a single gene.

Other examples of pleiotropy include the following: blue-eyed white cats are generally deaf, and humans with Eddowe's syndrome have brittle bones and a blue coloration of the normally white sclerotic coat of the eye. An extreme example is a gene appropriately called polymorph (polymorph means "many form" or "many shape") in *Drosophila,* the fruit fly. This single gene affects eye color, body proportions, wing size, wing vein arrangement, body hairs, size and arrangement of bristles, shape of testes and ovaries, viability, rate of growth, and fertility.

The fact that pleiotropy is very common reflects the way in which genes act through effects on biochemical processes. An alteration in a single biochemical pathway, resulting from the expression of a single gene, may affect many developing characteristics.

Quantitative Genetics

The alternate traits that we have discussed so far, such as tall vs. short plants, green vs. yellow seeds, A vs. B blood types, and so on, are clearly different from each other. This kind of variation is called **discontinuous variation,** because the variants fall into discrete, nonoverlapping sets. With regard to other traits, however, organisms may exhibit **continuous variation,** in that individuals vary from each other only slightly. Differences among these individual phenotypes are *quantitative;* that is, they are differences in degree or amount that require precise measurement to detect—as opposed to *qualitative* differences, which require only simple observation for sorting the phenotypes into discrete groups. The study of the inheritance patterns of these traits is called **quantitative genetics.**

Many sets of alleles may be involved in the quantitative inheritance of certain characteristics, and the environment may play a large role in determining the phenotype. Because large numbers of genes may be involved, quantitative inheritance is sometimes called **polygenic** or **multiple factor inheritance.** Phenotypes of quantitatively inherited characteristics often are distributed in bell-shaped (normal) curves. Human height, for example, is a quantitatively inherited trait determined by many genes, and the distribution of human heights follows a normal curve (figure 21.20).

Another example of this kind of distribution was found in a study of the inheritance of ear length in corn (figure 21.21), which began with crosses between plants that produce short ears and plants that produce long ears. The bell-shaped distribution of offspring in the F₁ generation was retained in the F₂ generation. This contrasts with the pattern seen when characteristics are determined by single genes. Because of the law of segregation, we would expect that if ear length were determined by a single gene, then a number of the original parental phenotypes would appear in the F₂ generation. This clearly did not happen.

Figure 21.20 Height distribution of 175 men recruited for the Army around 1900. Human height is a quantitatively inherited trait determined by many genes.

Number of individuals	1	0	0	1	5	7	7	22	25	26	27	17	11	17	4	4	1
Height in inches	58	59	60	61	62	63	64	65	66	67	68	69	70	71	72	73	74

Figure 21.21 Results of an experiment showing the inheritance of ear length in corn. These results illustrate polygenic inheritance. The relative heights of the bars indicate the proportion of ears that are each length.

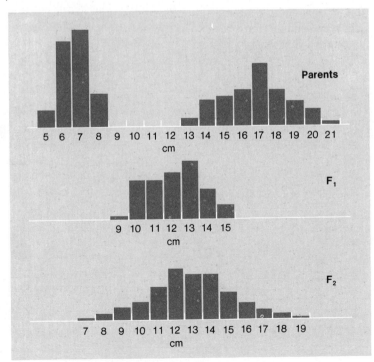

However, polygenic inheritance does not represent a conflict with the principles of Mendelian genetics. We can demonstrate this using a very simple, hypothetical example of polygenic inheritance. Consider a hypothetical organism in which the genotype $A^1 A^1 B^1 B^1 C^1 C^1$ (three different genes) produces a height (phenotype) of 20 centimeters. The addition of an allele denoted by a "2" results in a height increase of 1 centimeter. For example, an individual with the genotype $A^1 A^2 B^1 B^2 C^1 C^2$ would be 23 cm tall, while an individual with the genotype $A^2 A^2 B^2 B^2 C^2 C^2$ would be the maximum height—26 cm tall. If a cross were made between an $A^1 A^1 B^1 B^1 C^1 C^1$ individual and an $A^2 A^2 B^2 B^2 C^2 C^2$ individual, and members of the F_1 generation were crossed with one another, the height distribution of organisms in the F_2 generation would be as follows: $\frac{1}{64}$, 26 cm; $\frac{6}{64}$, 25 cm; $\frac{15}{64}$, 24 cm; $\frac{20}{64}$, 23 cm; $\frac{15}{64}$, 22 cm; $\frac{6}{64}$, 21 cm; $\frac{1}{64}$, 20 cm. Only two out of sixty-four offspring would be the parental types. This is due to the fact that three genes are involved. When still more genes are involved, even a smaller proportion of the offspring reflect the original parental types.

Many of the geneticists of Mendel's time were unknowingly studying polygenic inheritance, and because quantitative inheritance patterns resemble a blending pattern, these scientists were unable to recognize the laws of segregation and independent assortment or to appreciate the significance of Mendel's work.

Many of the economically important characteristics of agricultural plants and animals are inherited in a quantitative fashion. Some examples are milk production in dairy cattle and yield in corn. For this reason, those people involved in breeding agricultural plants and animals must have a thorough understanding of quantitative genetics.

Human Genetics and Genetic Counseling

Over the years, considerable progress has been made in the study of inheritance of many human characteristics. A list of Mendelian traits in humans compiled in 1978 includes 736 known dominants, 753 suspected dominants, 521 recessives, 596 suspected recessives, and 107 traits on the X chromosome. (These are called X-linked genes and will be discussed in the next chapter.) Most of these identified traits are related to various diseases, since as you might imagine, there is much more medical incentive to study inherited diseases than the inheritance of normal human characteristics.

Many couples ask such questions as, "Will one of these genetic disorders that we (or our relative, or our previous children), display be likely to occur in our future children?" Answers to such questions should be sought from a qualified genetic counselor. A genetic counselor (sometimes a physician) is trained in genetic principles and is familiar with the inheritance patterns of human diseases and abnormalities.

The pattern of inheritance of a trait in a particular family is commonly analyzed by constructing a **pedigree.** Figure 21.22 shows a sample pedigree along with definitions of some of the symbols used. If the information is complete enough, a genetic counselor may be able to reconstruct the genetic history of a particular trait. After studying the pedigree, the counselor can then inform the parents of the risk (probability) of having an offspring with the particular genetic defect. Once the risk has been determined, most genetic counselors encourage the prospective parents to decide for themselves whether or not to have children.

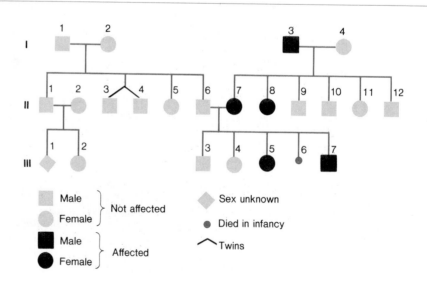

Figure 21.22 A typical human pedigree with the standard notation. For purposes of reference, Roman numerals to the left identify generations, and each individual within a generation has his or her own number. For example, II–7 indicates the seventh person in the second row.

Modern Genetic Counseling

Many pedigrees are very complex, and risks are often difficult to estimate. Sometimes, the risk must be estimated from past statistics rather than from knowledge about the inheritance pattern of a particular trait. But some factors make the job of a genetic counselor easier. With some traits, heterozygous individuals can be detected by medical tests. For example, individuals who are heterozygous for Tay-Sachs disease or sickle-cell anemia can be detected by tests. When such testing is available, risk to offspring of certain marriages can be determined much more precisely.

Marriage between Relatives

The risk of genetic defects in offspring is important not only when there is a history of genetic disease, but also when close relatives are considering marriage. The offspring of two individuals who are genetically similar are less likely to have a number of genes that are heterozygous and more likely to have a number of genes that are homozygous than are average members of the population. This greater homozygosity may have serious consequences because most individuals possess at least a few recessive alleles that, if homozygous, would be harmful, perhaps even lethal, to the individual. Because the expression of these alleles normally is masked in a heterozygous condition, their presence causes no harm to heterozygous carriers. Thus, the interbreeding of closely related individuals, called **inbreeding,** causes problems by increasing the frequency of homozygosity.

The increase in homozygosity in offspring is proportional to how closely related the parents are. Mating between parents and their children and between sisters and brothers is called incest. There is a taboo against incest in most societies, and most states and countries have laws forbidding marriage of such closely related people. A more common type of marriage is between first cousins. The frequency of first-cousin marriages varies from society to society. For instance, in the United States, almost half the states have laws forbidding first-cousin marriages, while other states permit them. However, in Japan, marriages between first cousins are encouraged and account for up to ten percent of the marriages in some areas.

Table 21.3 shows some of the effects of increased homozygosity in first-cousin marriages. The presence of harmful recessives is referred to as a type of **genetic load.** As you can see, the genetic loads of close relatives are more likely to include the same harmful recessive genes than are those of unrelated, or at least more distantly related, individuals. Clearly, there is a sound genetic basis for the ancient taboos against incest.

Heredity and Environment

One of the most basic, and also most often forgotten, principles of genetics is that the genotype and the environment interact to produce the phenotype (figure 21.23). It is frequently quite difficult to determine how much of a phenotype is determined by heredity ("nature") and how much by environment ("nurture"). An example of the complex interaction of heredity and environment is seen in the determination of human height. Many genes affect height,

Table 21.3
Mortality in Offspring of First-Cousin Marriages Compared with That in Offspring of Unrelated Parents

Trait	Unrelated	First Cousins
Stillbirths and neonatal deaths	0.044	0.111
Infant and juvenile deaths	0.089	0.156
Juvenile deaths	0.160	0.229
Postnatal deaths	0.024	0.081
Miscarriages	0.129	0.145

Adapted from *An Introduction to Human Genetics, Third Edition,* by H. Eldon Sutton. Copyright 1980 by Saunders College/HRW. Source: N. E. Morton, *Progr. Med. Genet. 1*:261, 1961. Reprinted by permission.

Figure 21.23 Example of the interaction of the environment and genetic expression in development. The water buttercup, *Ranunculus aquatilis,* is a plant that grows in shallow pools with some leaves submerged and some above the water's surface. Those leaves that develop below the water's surface are finely divided. They offer less resistance to water currents and favorable surface-to-volume ratios for absorption of CO_2 and minerals from the water. Leaves above the surface are coarsely lobed and more buoyant so that they float.

Table 21.4
Examples of Heritability of Traits in Various Organisms (estimated by various methods)

Trait	Heritability
Amount of white spotting in Frisian cattle	0.95
Slaughter weight in cattle	0.85
Plant height in corn	0.70
Root length in radishes	0.65
Egg weight in poultry	0.60
Thickness of back fat in pigs	0.55
Fleece weight in sheep	0.40
Ovarian response to gonadotropic hormone in rats	0.35
Milk production in cattle	0.30
Yield in corn	0.25
Egg production in poultry	0.20
Egg production in *Drosophilia*	0.20
Ear length in corn	0.17
Litter size in mice	0.15
Conception rate in cattle	0.05

From Francisco Ayala and John Kiger, *Modern Genetics*. (Menlo Park, Calif.: Benjamin-Cummings, 1980).

including those that control growth hormone, digestive enzymes, rate of calcium deposition in the bones, and many others. In addition, environmental factors, such as the presence or absence of adequate nutrition during critical growing periods, may strongly influence an individual's height. A combination of all of these factors ultimately determines how tall a person grows to be.

Geneticists attempt to assess the relative contributions of genetic factors by using a concept known as **heritability.** Heritability is the proportion of the phenotypic variation in a population that is due to genetic factors. If all of the variation is due to genetic causes, the heritability is 1.0 (100 percent). If all of the observed phenotypic variation in a population is due to the environment, the heritability of that variation is 0.0. Thus, all traits demonstrate heritability values between 0 and 1.0 (table 21.4).

Heritability is a concept that applies to populations, not to individuals. For example, the heritability of egg weight in the domestic chicken is 0.6. This does not mean that 60 percent of the weight is determined by the genes and 40 percent by the environment. It means that 60 percent of the total variation in egg weights among a group of chickens is due to genetic differences and the rest is due to environmental causes.

In economically important plants and animals, the heritabilities of different characteristics are estimated from careful breeding studies. In humans, of course, this cannot be done, but the heritabilities of some abnormalities have been estimated from the rate at which they occur among relatives of a person with the condition. Some of these abnormalities and their estimated heritabilities are certain kinds of epilepsy (0.4), clubfoot (0.8), and harelip (0.7).

There are complicating factors in determining heritability. For example, we have examined the roles of modifier genes and, especially, epistatic genes, and have seen that genes may be present without being expressed (recall the *C* genes in white leghorn chickens, p. 519). Thus, geneticists must be concerned about **penetrance,** the percentage of individuals carrying a given gene that actually express it. Furthermore, when a gene is expressed, the degree to which it is expressed may be determined by genetic factors (other genes being present or absent), as well as environmental influences. **Expressivity** denotes the degree to which genes are expressed and thus affect the phenotype. Mendel was fortunate in that the seven sets of genes that he studied in garden peas showed 100 percent penetrance and clear-cut expressivity. This is certainly not the case with many genes in many organisms, including humans.

In human genetic studies, twins occasionally provide an opportunity to study the relative roles of heredity and environment. There are two kinds of twins: **monozygotic** (identical) and **dizygotic** (fraternal). Dizygotic twins are produced from two different zygotes and are therefore no more similar genetically than brothers and sisters of different ages. Monozygotic twins are produced from the same zygote and, as a result, are genetically identical. In most cases, twins are raised in the same environment so that there is similarity in both the genetic and environmental components. Occasionally, however, monozygotic twins are reared apart from each other. In those situations, the individuals have identical genetic components, but different environmental components. A comparison of the phenotypes of such twins could provide some estimate of the relative genetic contributions to certain traits and therefore their heritability. However, cases of identical twins raised separately, in different environments, are rare, and even in such cases it is sometimes hard to determine how different the environmental influences have been for the two individuals. Thus, firm conclusions are difficult to draw.

The heritability of human traits, such as intelligence, is very difficult to measure. Furthermore, the study of heritability in humans has often led to controversy. For example, there was a study during the nineteenth century that compared the brain sizes of various human races. Brain size was determined by measuring the cranial capacity of skulls. The scientists who conducted this study concluded that Caucasians had larger skulls and brains and were therefore intellectually superior to other races, such as American Indians and black Africans. In the 1970s, however, Stephen Jay Gould reexamined the data on which these conclusions were based and found no appreciable difference in the cranial capacities of the various races. Apparently, the earlier study had been either accidentally or deliberately influenced by the prejudices of the investigators.

A scientific approach to analyzing the heritability of human traits, especially one such as intelligence, requires objectivity and freedom from prejudice, of one kind or another, that is sometimes difficult to achieve.

Summary

The modern science of genetics was founded by Gregor Mendel in the mid-1800s. From a series of experiments with garden peas, Mendel derived two basic laws of genetics that are still valid today.

Mendel's first law, the law of segregation, states that organisms contain two discrete factors (genes) for each characteristic and that these genes segregate so that a gamete has only one of the pair of alleles (forms of the gene) for each trait. Mendel's second law, the law of independent assortment, states that genes affecting different traits segregate independently of each other. These two laws can be used as a basis for predicting the proportions of different types of offspring from a specified cross.

The types of genes present in an organism constitute that organism's genotype. Interaction of the genotype with the environment results in the outward appearance (phenotype) of an individual.

Sometimes, one allele of a gene masks the expression of another allele. Because of this effect, which is called dominance, organisms with different genotypes may have the same phenotype. In other cases, both alleles of a pair are clearly expressed in the phenotype, resulting in incomplete dominance or codominance. Modifier genes and epistatic genes sometimes affect the phenotypic expression of entirely different genes that are not their alleles. Pleiotropy is where a single gene has many phenotypic effects.

A number of genes, each having a small effect, can combine to produce a particular characteristic. The study of this kind of inheritance is called quantitative genetics.

The application of genetic principles to human populations has made possible genetic counseling services. Couples may, in many cases, obtain estimates of the risk of their offspring having certain genetic diseases.

Both genotype and environment contribute to an organism's phenotype. The relative contributions of each, however, are difficult to resolve.

Questions for Review

1. What are alleles?
2. Briefly explain Mendel's law of segregation and his law of independent assortment.
3. Define the terms homozygous and heterozygous.
4. Explain what we mean by the terms genotype and phenotype.
5. How many different alleles can be present for a single gene in a diploid individual? Is there any limit to the number of alleles that might be present in a population of individuals?
6. How many types of gametes may be formed in an organism with the genotype $A^1 A^2 B^1 B^2 C^1 C^2$? (A, B, and C are genes located on separate chromosomes.)
7. Assume that right-handedness is dominant to left-handedness in humans. Explain how two right-handed people might have a left-handed child. If a right-handed person who had one left-handed parent married a left-handed person, what predictions could you make about their offspring?
8. Assume that in humans, normal pigmentation is due to a dominant gene (A) and albinism (lack of pigmentation) to its recessive allele (a). A normally pigmented woman marries an albino man and their first child is an albino. What are the genotypes of these three individuals? If they should have several more children, what are the possible genotypes and phenotypes of their children, and what are the probabilities that each of these genotypes and phenotypes will occur?
9. In cattle, polled (hornless) is dominant to horned. If a breeder of purebred cattle, all of which are polled, suspects that one of her prize bulls is heterozygous and carries the horned allele, how might she determine if her suspicion is correct?
10. Some genes are lethal when present in the homozygous condition. In chickens, when a gene known as creeper is present in the homozygous condition in a developing embryo, the spinal cord and vertebral column develop abnormally and the embryo dies inside the eggshell. When the creeper gene is present in the heterozygous condition, the chick hatches, but it has skeletal abnormalities and walks with a peculiar creeping, stumbling gait. This set of characteristics give the gene its name. Interpret the results of the following crosses:

Normal × Normal	96 normal
Normal × Creeper	51 normal, 48 creeper
Creeper × Creeper	54 creeper, 26 normal

11. In crosses between two crested ducks, only about three-quarters of the eggs hatch. The embryos in the remaining quarter of the eggs develop nearly to hatching and then die. Of the ducks that do hatch, about two-thirds are crested and one-third have no crest. What results would you expect from a cross between a crested and a noncrested duck?

Gregor Mendel thought that his "factors" (we now call them genes) were independent, discrete units within the cell. His law of independent assortment stated that genes segregate independently of each other during the formation of gametes. However, this is not true of all genes. Since the time of Mendel, much progress has been made in elucidating the relationships between genes and chromosomes. We now know that genes are organized in linear arrays as part of chromosomes and that each gene occupies a specific site on a chromosome. In this chapter we will examine chromosomes and the relationships between genes and chromosomes.

Chromosomes

At the time Gregor Mendel published his results in 1866, it was known that all living things are made up of cells. It was also known that plant and animal cell nuclei contain rodlike structures called chromosomes. However, Mendel did not know enough about the behavior of chromosomes in dividing cells to realize the significance of chromosomes for his theories of inheritance.

The Chromosome Theory of Heredity

When Mendel's work was rediscovered in 1900, a great deal more was known about the behavior of chromosomes in dividing cells. In 1902 two investigators, Walter S. Sutton of the United States and Theodor Boveri of Germany, suggested independently that the genetic material is contained in the chromosomes. This idea has developed into the **chromosome theory of heredity.**

Sutton and Boveri had several reasons for proposing that Mendel's independent factors, or genes, were located on chromosomes. They knew that within the nucleus, chromosomes exist in pairs. This coincides with Mendel's discovery that organisms have two copies of each gene. According to the chromosome theory of heredity, one copy of each gene is located on each one of a pair of chromosomes. Thus, an individual with a heterozygous genotype (for example, *Aa*) has one allele (*A*) on one chromosome of a given pair and another allele (*a*) on the other. It had also been noted that *gametes have only one of each pair of chromosomes.* The separation of the members of homologous chromosome pairs during meiosis (see chapter 20) explains how the alleles of a gene segregate, and only later, at the time of fertilization, recombine in the formation of zygotes. However, before we consider the arrangement of genes on chromosomes, we need to learn more about the structure and function of chromosomes.

Chromosome Structure

Prokaryotic and eukaryotic cells differ in the manner in which their genetic material is packaged within the cell and organized into chromosomes. Recall that one of the major distinctions between prokaryotic and eukaryotic cells is that the area of a prokaryotic cell that contains the genetic material is not enclosed by a nuclear envelope (membrane) as it is in a eukaryotic cell. Furthermore, the genetic material of prokaryotic cells is organized as a single, circular chromosome, while eukaryotic cells ordinarily have several or more essentially linear chromosomes, with the chromosome number varying from species to species. Let us take a closer look at the distinctions between chromosome organization in prokaryotic and eukaryotic cells.

Prokaryotic Chromosomes

The chromosome of the intestinal bacterium *Escherichia coli* has been studied extensively, and we can use it as an example of a chromosome from a prokaryotic cell. The *E. coli* chromosome was once thought to be a naked, circular DNA molecule. However, this conclusion was based on the study of DNA molecules extracted from lysed (broken open) bacterial cells, and we now know that the techniques used at that time to extract the chromosome stripped off RNA and protein molecules that are normally associated with the DNA of the chromosome. Furthermore, those early extraction techniques also disrupted the three-dimensional structure of the DNA molecule itself. Nevertheless, biologists did discover that the circular DNA molecule extracted from an *E. coli* cell was about 1,360 μm in circumference.

However, the discovery of the size of this DNA molecule raised some interesting problems because the dimensions of an *E. coli* cell are only about 1 μm by 2 μm. Furthermore, electron micrographs of *E. coli* cells revealed that chromosomal material is localized in a rather small area in the cell. This area, the **nucleoid** (see p. 921), occupies only about one-tenth of the volume of the cell. Clearly, the bacterial chromosome must be packed in a very compact form to fit into such a small space. How is this packing accomplished?

Researchers have developed techniques for disrupting bacterial cells that are less damaging to chromosome structure. By using these new techniques, they have been able to add considerably to our knowledge of chromosome organization in prokaryotic cells. In an *E. coli* chromosome, for example, the DNA is folded into loops that are linked together by RNA molecules (figure 22.1). Within each of these loops, the DNA double helix is twisted into a series of coils, an arrangement in which the molecule is said to be **supercoiled.**

Figure 22.1 Packed organization of the *E. coli* chromosome. (a) The double helix of the DNA molecule is represented as a single line. The chromosome is folded into 40 loops which are linked by RNA molecules. Only six loops are shown here. Within each loop, the DNA is supercoiled. (b) Electron micrograph of an *E. coli* chromosome released from a cell. The chromosome shows supercoiling (magnification × 30,000).

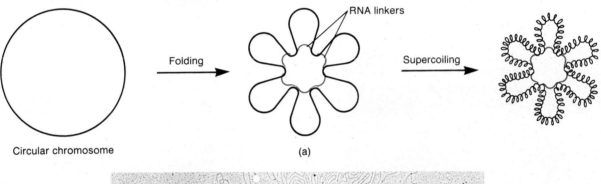

(a)

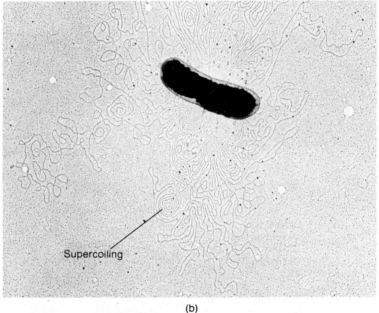

(b)

Apparently, the loops and the supercoiled arrangement at least partially account for the dense packing of the chromosome in the nucleoid region of a bacterial cell. It is not yet clear, however, whether or not the proteins associated with bacterial chromosomes are also involved in stabilizing the normal organization of the chromosome. It may be that these proteins are mainly involved in genetic regulation and expression rather than in the chromosome's structural organization.

Eukaryotic Chromosomes

In contrast to prokaryotic chromosomes, the organization of DNA in eukaryotic chromosomes is more complicated. DNA in eukaryotic chromosomes is associated with several kinds of proteins in very orderly arrangements. When they are condensed during cell division, the chromosomes in most eukaryotic cells are rod-shaped and stain quite strongly with certain dyes. Biologists call the stainable material that makes up these chromosomes **chromatin.** Chromatin consists of about 60 percent protein, 35 percent DNA, and 5 percent RNA. The chromatin that makes up a chromosome is fibrous and the chromatin fiber consists of a long molecule of DNA associated with basic protein molecules called **histones,** quantities of acidic and neutral nonhistone proteins, and several kinds of enzymes that are active in DNA and RNA synthesis.

The DNA and proteins in chromatin are organized in a very specific way. When chromatin is isolated, spread out, and examined with the electron microscope, it appears to consist of a series of "beads" that are about 10 nm in diameter and arranged in a chain (figure 22.2a). Each of these "beads" is a basic structural unit of a chromosome called a

Figure 22.2 Nucleosomes. (a) This photograph shows uncoiled chromatin material from chicken red blood cells. The "beads" in the photograph are the nucleosomes and the "string" is the DNA between nucleosomes. (b) The arrangement of histones and DNA in nucleosomes. Each nucleosome has an aggregate of eight histone molecules making up a core around which DNA is wrapped in a helical coil. The eight histones are two each of histones H2A, H2B, H3, and H4. Histone H1 is associated with spacer DNA between nucleosomes. (c) Proposed model of nucleosome packaging in a compact chromosome. A chromatin fiber (nucleosomes and intervening DNA) is arranged in a cylindrical coil, the solenoid, which is about 30 nm in diameter. The solenoid also is coiled, probably into a tight spiral about 200 nm in diameter.

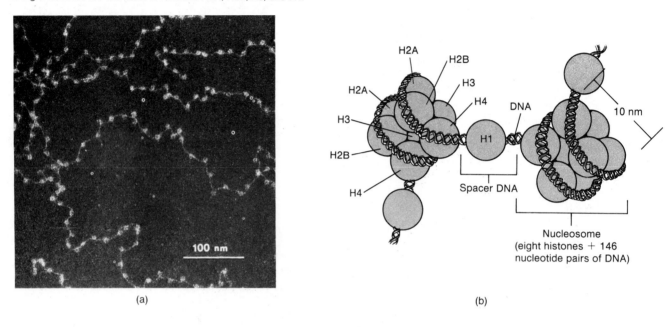

(a)

(b)

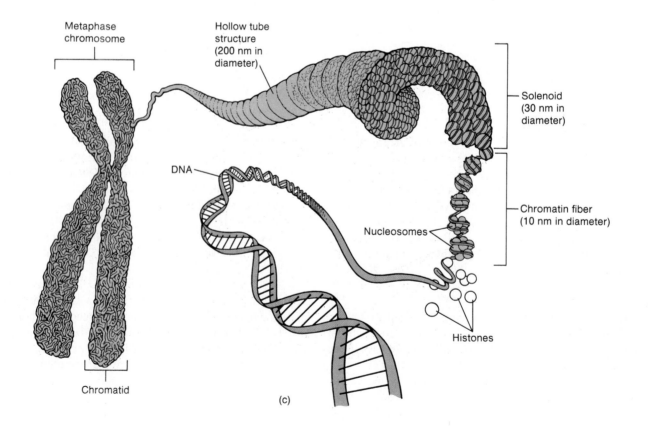

(c)

nucleosome. Each nucleosome is an aggregate of eight histone molecules, two each of four kinds of histones that are identified as H2A, H2B, H3, and H4. These eight histone molecules form a protein core around which a specific length of DNA—146 nucleotide pairs—is wrapped in a helical coil (figure 22.2b).

The nucleosome "beads" in chromatin appear to be held together by segments of DNA, called **spacer (linker) DNA,** that stretch from one nucleosome to the next. A fifth kind of histone molecule, histone H1, is associated with spacer DNA. This H1 histone apparently is involved in stabilizing another level of organization that makes the chromatin even more compact. The entire chromatin fiber is arranged in a cylindrical coil, called a **solenoid,** that is about 30 nm in diameter. The solenoid also is coiled, but here the picture is not quite so clear. One model suggests that the solenoid is in a tight spiral that forms a hollow tube about 200 nm in diameter. It may be that this spiral makes up the basic structure of chromosomes as they appear during metaphase (see figure 22.2c), the stage at which chromatin is most compact.

It is not yet clear how these various levels of spiral packing change as chromosome structure changes during the cell cycle. Likewise, the roles of various nonhistone proteins in the organization of eukaryotic chromosomes are still under investigation.

Chromosome Number

While the basic structural units in chromosomes, the nucleosomes, seem to be remarkably similar in all eukaryotic cells studied so far, there are major differences among the chromosome complements of various organisms. One of the most obvious differences is the number of chromosomes present.

Chromosome Numbers in Different Species

The number of chromosomes found in different organisms varies a great deal. The numbers range from 4 chromosomes in one plant, *Haplopappus gracilis,* and in several insects and other invertebrate animals, to nearly 500 chromosomes in some ferns (table 22.1). The number of chromosomes present is usually, but not always, consistent within a species and is not correlated with either the size of the organisms in a species or with the complexity of their organization.

Polyploidy

Up to this point, we have discussed mainly organisms that are diploid; that is, organisms that have two sets of chromosomes (two of each kind of chromosome). When an organism has more than two sets of chromosomes, the condition is called **polyploidy.** Polyploid organisms are named according to the number of sets of chromosomes they have:

Table 22.1
Chromosome Numbers of Some Common Plants and Animals

Species	Common Name	Chromosome Number
Triticum aestivum	Wheat	42
Zea mays	Corn	20
Lactuca sativa	Lettuce	18
Lycopersicon esculentum	Tomato	12
Gossypium barbadense	New World cotton	52
Phaseolus vulgaris	Kidney beans	22
Glycine max	Soybeans	40
Pyrus communis	Apple	34
Ipomea batatus	Sweet potato	90
Ophioglossum vulgatum	Fern	480
Apis mellifera	Honeybee	16
Bombyx mori	Silkworm	56
Cyprinus carpio	Carp	100
Rana pipiens	Leopard frog	26
Alligator mississippiensis	Alligator	32
Anas platyrhynchos	Mallard duck	78
Bos taurus	Cattle	60
Equus caballus	Horse	64
Felis felis	Cat	38
Canis familiaris	Dog	78

triploids (3N) have three sets, tetraploids (4N) have four sets, pentaploids (5N) have five sets, and so on. Polyploids with uneven numbers of chromosome sets—triploids for example—are usually sterile, because it is impossible for an uneven number of chromosomes to be distributed evenly during meiosis.

Polyploidy is quite common in plants but relatively unusual in animals. This is because the polyploid condition usually causes relatively little physiological or developmental disturbance in plants, and does not interfere with plant reproductive processes as long as the number of chromosome sets is even (for example, 4N, 6N, 8N, etc.). Many ornamental plants are bred as polyploids because they tend to have larger leaves and flowers. It is also possible to chemically treat seeds, seedlings, or shoot tips to induce polyploidy in growing plants (figure 22.3). In some cases, triploids are specifically developed because, since these plants are sterile and do not put any energy into seed production, they may produce larger flowers or highly desirable, seedless fruit. Furthermore, because plants can often be propagated asexually from "cuttings" taken from a parent plant with certain desirable characteristics, even plants with an uneven number of chromosome sets, which usually precludes sexual reproduction, can be used commercially.

Figure 22.3 Effects of polyploidy. The flowers of a diploid snapdragon (*left*) and a tetraploid snapdragon (*right*). In this case polyploidy was induced by chemical treatment.

In animals, where in many cases sex determination is based on a single pair of chromosomes (the sex chromosomes, which we will encounter later in this chapter), polyploidy disrupts normal reproductive processes.

Aneuploidy

While polyploidy has to do with the number of sets of chromosomes, **aneuploidy** is a general name for conditions in which the chromosome number of an individual varies from the normal number for the species by a small number of chromosomes. Usually, aneuploidy involves an excess or a deficiency of a single chromosome in otherwise normally diploid cells. For instance, if an individual has an extra copy of one chromosome, the condition is called **trisomy** and is designated as 2N + 1. Recall that 2N symbolizes the normal diploid number. Thus in humans, for example, who have a normal chromosome number of 46, 2N + 1 would equal 46 + 1, or 47 chromosomes. If an individual is missing one chromosome, the condition is called **monosomy** and is designated as 2N − 1.

As we noted in chapter 20, aneuploidy usually occurs due to nondisjunction of homologous chromosomes during cell division. If nondisjunction occurs during meiosis, it can have drastic consequences for the development of a zygote produced as a result of fertilization involving a gamete with a chromosome excess or deficiency.

A well-known example of aneuploidy is the human condition known as **Down's syndrome.** Many people with Down's syndrome have heart trouble and almost all of them are mentally retarded. This condition, sometimes referred to as "mongolism," results from the presence of an extra copy of the chromosome designated number 21 (figure 22.4). The extra copy of the chromosome results from nondisjunction during sperm or egg formation; either the duplicated pair of chromosomes fails to separate during the first division of meiosis or the chromatids of the duplicated chromosome fail to separate during the second division. In humans, the chances of producing an aneuploid child increase as the parents, especially the mother, grow older. This is an important consideration when a woman over age 35 considers having a child.

Down's syndrome and other chromosomal disorders now can be detected during pregnancy by using a technique called **amniocentesis.** A needle is inserted through the mother's abdomen into the uterus, and a sample of the amniotic fluid surrounding the developing fetus is withdrawn. Cells from the amniotic fluid are then cultured and examined for the presence of the extra number 21 chromosome, or other problems. Detection of a chromosomal abnormality does not solve the problem, however. Rather, it places the parents in the situation of having to make some terribly difficult decisions.

Human Chromosomes

Although we are now familiar with pictures of sets of human chromosomes such as that shown in figure 22.4, virtually all knowledge of human chromosomes has been obtained since the 1950s. Even the exact number of chromosomes in humans was not determined with precision until 1956. This seems extraordinary in light of the fact that chromosomes in other organisms had been observed and described for over a hundred years. However, the chromosomes of humans are quite difficult to prepare in a manner that permits accurate chromosome counts, and routine techniques for studying human chromosomes have been available for only a relatively short time.

Preparations of human chromosomes are usually made from certain types of white blood cells (lymphocytes). Figure 22.5 illustrates the procedure. When dividing cells are observed, metaphase chromosomes are photographed. The individual chromosomes are cut out of the photograph and lined up with one another in matching homologous pairs. This ordered arrangement of all the chromosomes in the nucleus is called a **karyotype.**

Figure 22.4 Down's syndrome (trisomy 21). (a) Chromosomes of an individual with Down's syndrome. Note the extra copy of chromosome no. 21. (b) A young man (*right*) with Down's syndrome. Affected individuals are shorter than average and have characteristic facial features, such as close-set eyes with narrow, slanting eyelids, a relatively small nose, and a large, furrowed tongue that may protrude from the mouth. In most cases they suffer from heart malformations and mental retardation.

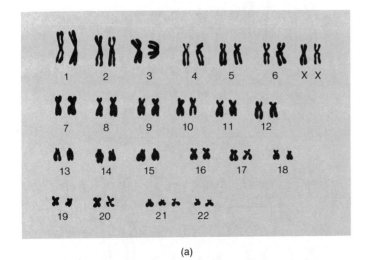

(a)

(b)

Figure 22.5 The procedure for preparing karyotypes of cultured human white blood cells. Because chromosomes are best observed in metaphase, colchicine, a substance that stops mitosis at metaphase, is added to the culture after some cell growth has occurred. This provides the person doing the work with many metaphase cells.

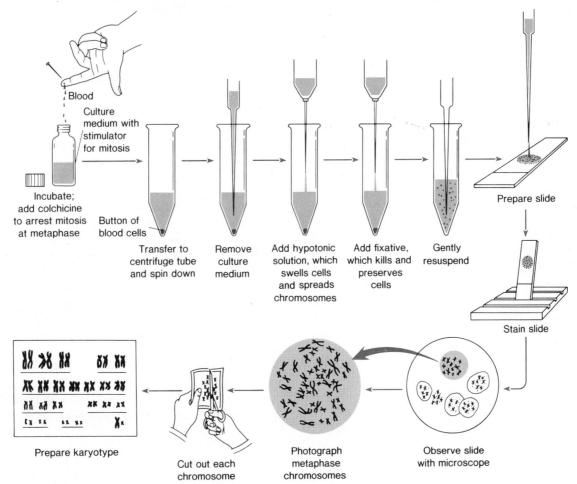

Figure 22.6 Photograph of the chromosomes of a human male in a cell at metaphase of mitosis.

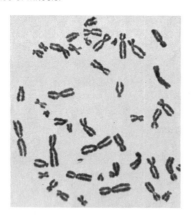

Figure 22.7 A human male karyotype. A human female karyotype would differ only in that it would have two X chromosomes rather than an X and a Y chromosome.

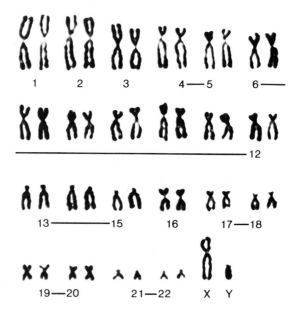

Figure 22.6 is a photograph of the metaphase chromosomes of a human male. Figure 22.7 is a karyotype made from the chromosomes shown in such a photograph. Such karyotypes show the characteristic diploid chromosome number of human body cells, which is forty-six (twenty-three pairs). The last pair of chromosomes in the karyotype, labeled X and Y, are not identical. These are the **sex chromosomes.** The human male has the differing X and Y chromosomes in this pair, while the human female has two X chromosomes.

The study of human chromosomes was greatly advanced in the 1970s with the introduction of new methods for staining chromosomes. These methods produce distinct patterns of dark and light stained bands. Figure 22.8 shows one such type of banding, referred to as G banding. The bands are called **G bands** because the Giemsa stain is used to prepare the chromosomes for examination. The visualization of this type of banding pattern is extremely important because it permits reliable identification of individual chromosomes. Many of the chromosomes are otherwise very similar and can be sorted only into general groups on the basis of similar appearance when routine karyotyping procedures are used without Giemsa staining (as they are in figure 22.7). Such staining also allows for easier identification of structural abnormalities in the chromosomes and helps to detect changes in the number of chromosomes. The mapping of the locations of particular genes on specific chromosomes also is aided by the presence of G bands.

Sex Chromosomes

In the karyotype in figure 22.7, you can see that the human male has an X chromosome and a Y chromosome, and twenty-two other pairs of chromosomes. The karyotype in figure 22.8 shows that the genetic makeup of the human female differs from the human male only in that the female has a pair of X chromosomes and no Y chromosome. As we noted earlier, the X and Y chromosomes are the sex chromosomes. The other chromosomes, those that are common to both males and females, are called **autosomes.**

Sex Determination

The pair of sex chromosomes segregates at meiosis so that normally each gamete receives just one member of the pair. In humans, each egg has an X chromosome, but a sperm cell can have either an X chromosome or a Y chromosome (figure 22.9). Therefore, the sex of a new individual is determined at the time of fertilization by the type of sex chromosome that is present in the sperm. If the sperm cell has a Y chromosome, the zygote (XY) will become a male. If the sperm cell has an X chromosome, the zygote (XX) will become a female.

Many animals have the same sex determination system as humans—an organism with two X chromosomes is a female, and an organism with an X and a Y is a male. However, other sex determination patterns also exist. In some

Figure 22.8 A karyotype of a normal human female, showing the different G-band patterns that permit reliable sorting of individual chromosomes. Note that the sex chromosome pair in females consists of two identical X chromosomes.

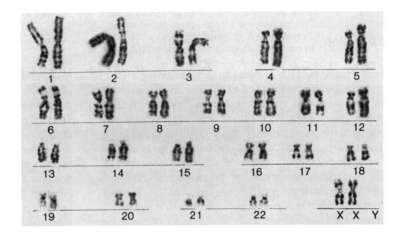

Figure 22.9 Sex chromosomes and sex determination in humans. The number of male zygotes produced should equal the number of female zygotes.

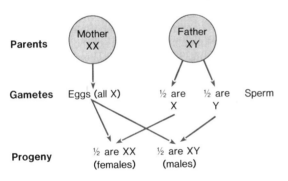

Figure 22.10 T. H. Morgan's experiments with *Drosophila* eye color.

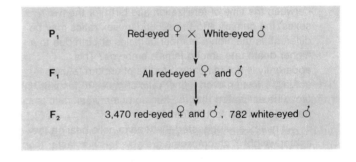

animals, Y chromosomes may or may not be present, depending upon the species. In grasshoppers, for example, there is no Y chromosome, so there are XX females and XO (a lone X chromosome with no other sex chromosome) males. In birds, moths, and butterflies, the sex chromosome pattern differs in another way; males have two like sex chromosomes, while it is the females that have the differing pair of sex chromosomes. In ants, wasps, and bees, there is yet another pattern. Some eggs develop parthenogenetically (without fertilization) into haploid males, while fertilized eggs develop into diploid females. In some animals that reproduce parthenogenetically, there is a mechanism for restoring the diploid number in developing offspring, but that is not the case in ants, wasps, and bees, where all cells in males are haploid (N), and all cells in females are diploid (2N). The development of different types of females, such as queens and workers in bees, is due to differences in nutrition, not to differences in chromosome complement.

Sex Linkage

Genes occupy specific sites on chromosomes. The specific part of a chromosome that is the site of a particular gene is known as its **locus** (plural: **loci**). During this century, **chromosome mapping** studies (the determination of relative positions—loci—of individual genes on specific chromosomes) have produced extensive maps of gene loci on chromosomes for many organisms. All of this work to determine the positions of gene loci on chromosomes began with a discovery that involved sex chromosomes.

Thomas Hunt Morgan and his colleagues at Columbia University were the first biologists to associate a specific gene with a specific chromosome when in 1910 they published the results of one of their many genetic experiments on the fruit fly (*Drosophila melanogaster*). In that experiment, they crossed female flies that had red eyes with males that had white eyes (figure 22.10). In the F_1 generation, all the flies

Figure 22.16 Another sex-influenced trait is the relative lengths of the index finger and fourth finger of the hand. Both individuals in the photograph are heterozygous for the trait. The hand on the right is that of a female, and the hand on the left is that of a male.

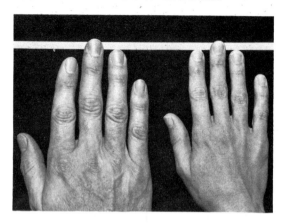

Figure 22.17 Chromosomal mutations. Arrows show points of chromosome breakage, and displaced chromosome fragments are colored. In deletions, the broken chromosome fragment does not reattach and is lost. In duplication, a broken segment from one chromosome attaches to its homologous chromosome. An inversion involves breakage and reattachment to the same chromosome but in a reversed position. A translocation is a transfer of a chromosome fragment to a nonhomologous chromosome. Radiation damage to cells increases the frequency of these problems because radiation causes greatly increased chromosome breakage.

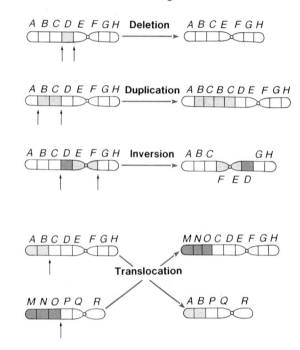

Chromosomal Abnormalities

Changes in chromosome structure can have important genetic consequences, and such changes often occur as a result of chromosome breakage. Chromosomes may be broken by radiation, various chemicals, or even viruses. Sometimes fragments of broken chromosomes are simply lost. Loss of a chromosome segment is called a **deletion.** Often, however, chromosome fragments rejoin in a different arrangement.

Changes in the genetic material of a cell that can be passed on to daughter cells are called **mutations.** Deletions and chromosomal reorganizations are called **chromosomal mutations** or **macromutations** to distinguish them from changes in individual genes (gene mutations), which we will discuss in chapter 24. Figure 22.17 illustrates deletions and various other types of chromosomal mutations.

A **duplication** results in the presence of the same chromosome segment more than once in the same chromosome. An example of the phenotypic expression of a duplication is the Bar duplication in the fruit fly *Drosophila.* This duplication reduces the number of facets in the insect's eyes and results in the development of small, bar-shaped eyes instead of normal round eyes.

An **inversion** occurs when a segment of a chromosome is turned around 180°. Inversions can cause reproductive problems for organisms that are heterozygous for an inversion because lethal, abnormal chromosome arrangements often are produced by meiotic crossing over.

Translocations result from the interchange of blocks of genes between two nonhomologous chromosomes. Translocation heterozygotes usually have reduced fertility. Because of the independent assortment of different pairs of nonhomologous chromosomes, some gametes may be produced that lack a particular chromosome segment.

Chromosomal mutations also cause difficulties in the normal genetic functioning of chromosomes, resulting in serious problems for the individual carrying them. For example, deletions usually are lethal if both homologous chromosomes are deficient. Furthermore, while a deletion in only one of the chromosomes of a homologous pair may not be lethal, it can drastically affect an individual's development and normal body functions. An example in humans is the **cri-du-chat syndrome** (cat's cry syndrome), which results from a deletion in the short arm of chromosome number 5. Infants that are heterozygous for this deletion have a high-pitched, mewing cry and usually exhibit severe growth abnormalities and mental retardation as well.

Finally, chromosomal mutations are very frequently observed in cancer cells. It appears that chromosome breakage and the subsequent reconnecting of chromosome segments in new (abnormal) arrangements may play a key role in the conversion of normal body cells into abnormally growing tumor cells (see p. 721).

Chromosome Number Problems in Humans

We have already discussed Down's syndrome, which is an example of the effects of aneuploidy in humans. Several other conditions result from aneuploidy in the autosomes. For example, **Patau's syndrome** is due to an extra copy of chromosome number 13. This syndrome is characterized by harelip and cleft palate, as well as other serious physical defects. Most infants with Patau's syndrome die during the first three months of life. **Edward's syndrome** results when an infant has an extra copy of chromosome number 18. This disease is associated with many organ system malformations, and affected infants live an average of only six months.

There also are a number of abnormalities that result from aneuploidy in the sex chromosomes. An individual with **Turner's syndrome** has only one X and no Y or second X chromosome. People with this condition are sterile females who lack ovaries and exhibit little development of secondary sex characteristics. Mental deficiency is seldom associated with Turner's syndrome.

Metafemales have three or more X chromosomes. These females have only limited fertility and are mentally retarded in most cases.

An individual with **Klinefelter's syndrome** has a Y chromosome with two or more X chromosomes. Individuals with this chromosomal condition are sterile males with some tendency toward femaleness.

Studies of human chromosomal abnormalities teach us some important things about the mechanism of sex determination in humans. During the early years of genetic research, before human chromosome studies began, sex determination had been studied thoroughly in *Drosophila*. Results of those studies indicated that in *Drosophila,* sex is essentially determined by the number of X chromosomes present, whether or not a Y chromosome is present. Presence of only one X chromosome produces a male. Thus, either an XY or an XO (a lone X chromosome with no other sex chromosome) genotype produces a male. A *Drosophila* individual with an XXY genotype becomes an essentially normal female. These results indicated that the Y chromosome is a basically neutral factor in *Drosophila* sex determination.

When chromosomal studies began on human sex determination, it was assumed that the human chromosomal mechanism of sex determination might be similar to the *Drosophila* model since both humans and fruit flies have the X-Y sex chromosome system. *But this is clearly not the case.* Turner's syndrome subjects have an XO genotype and an at least partly female phenotype. Individuals with Klinefelter's syndrome have an XXY genotype and a basically male phenotype. Clearly the Y chromosome is not a neutral factor in human sex determination, as it seems to be in *Drosophila*. Rather, it is a male-producing factor. Among the few genes borne by the Y chromosome are genes whose expression seems to be vital for the development of the male phenotype (see box 22.2).

Linkage and Chromosome Mapping

During meiosis, members of homologous chromosome pairs assort randomly into gametes. The gametes then combine to form the zygotes of the next generation. Mendel's law of independent assortment states that each gene assorts independently of other genes. Mendel assumed that genes behaved as independent particles in cells. Genes, however, are not independent particles, but are attached parts of linear arrays on chromosomes. It is actually the chromosomes that segregate independently during meiosis. Thus, if two gene loci are close together on the same chromosome, they tend to assort together during meiosis, and they do not obey Mendel's law of independent assortment. This tendency of certain genes to assort together is called **linkage.** Beginning with the work of T. H. Morgan and his colleagues, biologists have discovered that genes occur in distinct sets called **linkage groups,** each of which corresponds to one of the pairs of homologous chromosomes in that organism's cells. For example, in *Drosophila* there are four linkage groups, corresponding to *Drosophila's* four pairs of chromosomes.

Linkage and Crossing Over

You can see an example of linkage in figure 22.18 on page 543. If a test cross is made between an individual that is heterozygous for each of two genes and another individual that is homozygous recessive for both of those genes, each of the four possible genotypes is expected in equal ratios in the offspring. However, the example in figure 22.18 shows a strong deviation from this expected ratio.

We can explain this deviation from Mendelian expectations by assuming that the two gene loci involved here are on the same chromosome and thus cannot assort independently of one another; that is, these genes are linked.

Box 22.2

Active Y Chromosomes and Inactive X Chromosomes
The Functions of Sex Chromosomes in Body Cells

Although the Y chromosome in mammals carries few genes, it has a powerful effect on sex determination. One intriguing gene borne on Y chromosomes is identified as $H–Y^+$. This gene directs the synthesis of a specific kind of protein molecule, **H–Y antigen,** which is present in the membranes of virtually all cells in male mammals but not in those of females. The H–Y antigen may play a role in determining whether the gonads of a developing embryo become testes rather than ovaries. Once that determination is made, hormone differences cause other body parts to develop in a male direction (see chapter 28 for details of the process). In addition to a possible role in that first commitment to maleness in the developing embryo, the presence of the H–Y antigen may make male cells different from female cells for a lifetime, in ways that are as yet undetermined.

Another question about sex chromosomes that has interested biologists for some time is how males manage with only one "dose" of the genes present on the X chromosomes (sex-linked genes), while females apparently have a double "dose" of such sex-linked genes. That question turned out to have a rather unexpected answer. Apparently, cells in females also have only one functional X chromosome. How does this come about?

Years ago, M. L. Barr observed a consistent difference between nondividing cells from female and male mammals, including humans. Females have a small, darkly staining mass of condensed chromatin present in their nuclei. Males have no comparable spot of chromatin in their nuclei. This darkly staining spot in female nuclei is now called the **Barr body** or **sex chromatin body.**

In 1961 Mary Lyon, a British geneticist, proposed that the Barr body represents a condensed, inactive X chromosome whose genes are not expressed in cells of females, and that only genes on the other X chromosome are expressed. She further proposed that X chromosome inactivation occurs sometime during development and that once inactivated in a cell, a given X chromosome remains inactive in that cell and in all of its descendants.

The **Lyon hypothesis,** as it is called, has been tested and found to be valid. Early in development, X chromosomes are indeed inactivated in cells of female embryos. However, the inactivation does not consistently involve a single one of the two X chromosomes; that is, one X chromosome is inactivated in some cells, while the other is inactivated in others. As Mary Lyon proposed, once inactivation of a particular X chromosome occurs in an embryonic cell, that X chromosome forms a Barr body in each and every cell descended from the original cell. Because many sex-linked genes are heterozygous in any given individual, this X chromosome inactivation has interesting genetic consequences for body cells. One allele of a gene will be expressed in some cells, while another allele will be expressed in other cells. The female body, therefore, is a mosaic with "patches" of genetically different cells. It has been demonstrated, for example, that blood cells from heterozygous females belong to two separate subpopulations that produce either one or the other of two different forms of certain enzymes, depending on which allele of the sex-linked genes is being expressed. Cells from skin and other tissues have also been shown to be divided into two subpopulations with regard to expression of sex-linked genes. Thus, it now seems quite certain that female mammals, including human females, are genetic mosaics with reference to the expression of sex-linked genes.

Having a mixture of cells in which different X chromosomes are inactivated allows heterozygous females to escape the effects of certain harmful, or even lethal, genes. Although harmful alleles may be the only ones expressed in some body cells, normal alleles are expressed in other body cells and thus "cancel out" the effects of the harmful alleles.

The connection between inactivated X chromosomes and Barr bodies also has implications for genetic testing. Human cells have one less Barr body than the number of X chromosomes they contain. Thus, normal males have no Barr bodies. Normal females have one. Metafemales may have two or more Barr bodies because they have three or more X chromosomes. This correlation between

Figure 23.3 The Hershey–Chase experiment. Results of two experiments are combined in summary form in this diagram. Hershey and Chase labeled the protein coats of one batch of bacteriophage with ³⁵S. (Sulfur is found in protein but not in nucleic acids.) The phages were allowed to attach to bacteria for a time, and then were separated from the bacteria by being agitated in a blender. Radioactivity measurements indicated that little radioactive material entered bacterial cells; the ³⁵S-labeled protein remained outside. But the infection proceeded, and new virus particles were formed and released. Hershey and Chase labeled the DNA in another batch of phages with ³²P. (Phosphorus is an important constituent of nucleic acids but not of proteins.) Again, after a period of attachment, phages were separated from bacterial cells in a blender. This time the bulk of the radioactivity was found inside the host cells, indicating that it is DNA that enters host cells and infects them. These important experiments helped to prove that the primary genetic material is DNA, not protein.

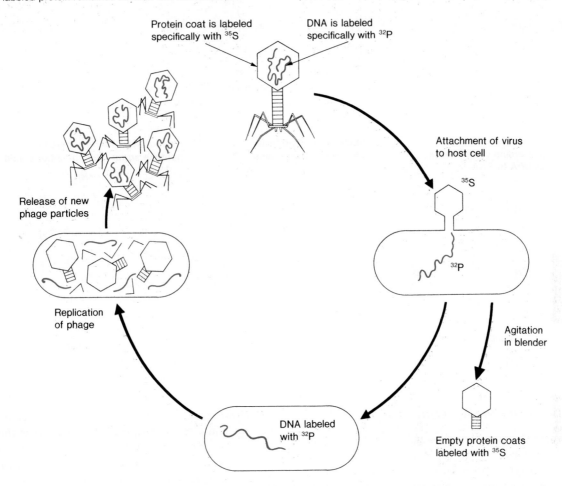

They found that it is DNA that enters and infects the cells, and that the viral protein coats remain on the outside of the bacterial cells. Once the infection has begun, the empty protein coat can be removed from a cell without affecting the progress of the infection (figure 23.3). Hershey and Chase's experimental results clearly indicated that DNA is the substance inserted by the T2 bacteriophage to take over the genetic mechanisms of the cell. Thus, it must carry the genetic information of the virus.

DNA: The Genetic Material

The third major line of evidence that confirmed DNA as the genetic material came when the structure of the DNA molecule was finally revealed through the work of Watson, Crick, Franklin, Chargaff, and Wilkins. The structure that they discovered demonstrated that DNA has a set of properties that make it ideal for the coding and transmission of genetic information, However, the search for the structure of DNA was long and difficult. Let us briefly retrace some of the major steps in that search, which culminated in a turning point in the history of biology.

Figure 23.4 Components of nucleic acid molecules. (a) Two purines and three pyrimidines found in DNA and RNA. Thymine is found in DNA, but not RNA. Uracil is a component only of RNA. (b) Nucleic acids are polymers of nucleotides that are composed of purine or pyrimidine bases, pentose sugars (ribose or 2-deoxyribose), and phosphate. The diagram shows components of DNA. In RNA, the H in the colored box is replaced by OH.

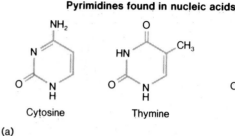

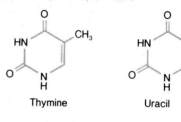

The Search for Nucleic Acid Structure

There are two major types of nucleic acids, **deoxyribonucleic acid (DNA)** and **ribonucleic acid (RNA).** Structurally, nucleic acids are linear polymers of **nucleotides.** Each nucleotide is composed of three substances: a nitrogen-containing organic base (either a **purine** or a **pyrimidine**), a five-carbon sugar, and phosphoric acid (figure 23.4).

There are several differences between DNA and RNA. DNA contains the sugar deoxyribose and the bases adenine, guanine, cytosine, and thymine. The composition of RNA differs from that of DNA in two major ways: the five-carbon sugar found in RNA is ribose rather than deoxyribose, and RNA contains uracil rather than thymine. As figure 23.4 indicates, adenine and guanine are purines, while cytosine, thymine, and uracil are pyrimidines. Nucleosides (each consisting of a sugar and a base) are linked together by phosphates to form nucleotide chains in both DNA and RNA (figure 23.5).

However, at the time the fundamental components of nucleic acids were determined, most scientists thought that DNA was a small molecule, with its four bases occurring in equal proportions and arranged in a fixed, unchangeable sequence. This may partly explain why biologists were confident for years that nucleic acids were too simple in structure to convey complex genetic information, and why they concluded that genetic information must be coded in proteins rather than in nucleic acids.

The 1940s witnessed the development of significant new analytical techniques in chemistry that led to great advances in the research on nucleic acid structure. One of the most important of these new developments was the invention of paper chromatography. By the late 1940s, the chemist Erwin Chargaff had begun using paper chromatography to analyze the base composition of DNA from a number of species. He soon discovered that the earlier conclusions about DNA structure were in error. Chargaff found that the base composition of DNA varies from species to species, just as one would expect in a substance carrying genetic material. Furthermore, he demonstrated that there was a consistency in

Table 23.1
DNA Base Composition

Organism	Molar Ratio of Bases				A + G
	Adenine	Guanine	Cytosine	Thymine	C + T
Escherichia coli (bacterium)	24.6	25.5	25.6	24.3	1.00
Saccharomyces cerevisiae (yeast)	31.3	18.7	17.1	32.9	1.00
Carrot	26.7	23.1	23.2	26.9	0.99
Rana pipiens (frog)	26.3	23.5	23.8	26.4	0.99
Human (liver)	30.3	19.5	19.9	30.3	0.99

DNA composition, regardless of the source: the adenine/thymine and guanine/cytosine molar ratios are 1.0 (table 23.1). In other words, the adenine content equals the thymine content ([A] = [T]), and the guanine content equals the cytosine content ([G] = [C]). These findings came to be known as *Chargaff's rules* and were the key to much later research.

Another turning point in the research on DNA structure came in 1951 when Rosalind Franklin arrived at King's College, London, and joined Maurice Wilkins in efforts to prepare highly oriented DNA fibers for study using **X-ray crystallography.** In this technique an X-ray beam is passed through a crystal of the substance being studied. Part of the X-ray beam is scattered (diffracted) as it passes through the crystal. The way in which it scatters depends upon the molecule's structure. A photographic plate on the other side of the crystal records a pattern of spots representing the intensity of the emergent X rays (figure 23.6a). This pattern reveals information about the locations of the various atoms in the crystal, which in turn, can be used to determine the three-dimensional shapes of molecules. Franklin and Wilkin's work culminated during the winter of 1952–53 with the production of Franklin's X-ray diffraction photograph of DNA (figure 23.6b).

Also in 1951, an American biologist, James Watson, arrived at Cambridge University and began to work with Francis Crick on a model of DNA structure. After Watson and Crick had made a series of unsuccessful attempts to unravel the structure of DNA, Franklin's data finally provided them with the necessary clues.

The cross-pattern of X-ray reflections in Franklin's photograph told Watson and Crick that DNA is helical. The black areas at the top and bottom of the photograph indicated that the purines and pyrimidines are regularly stacked next to each other, at a distance of 0.34 nm. Franklin had already concluded that the phosphate groups lay to the outside of the helical structure. It also became clear from her X-ray data and her determination of the density of DNA that the helix contained two strands, not three or more (as some chemists had proposed).

Figure 23.5 A short stretch of DNA. Each nucleoside in the polymer is linked to neighboring nucleosides by phosphate groups. These phosphates connect the 3′ carbon of one sugar with the 5′ carbon of the adjacent nucleoside sugar. For purposes of identification, carbon atoms in each sugar molecule are numbered 1′, 2′, etc., to distinguish them from the numbered carbons (not shown) in the adjacent base. The 3′ and 5′ carbons of the sugars are used in describing the direction a polynucleotide strand runs in a molecule. The structure of an RNA strand is similar except that ribose replaces the deoxyribose.

DNA polynucleotide strand structure

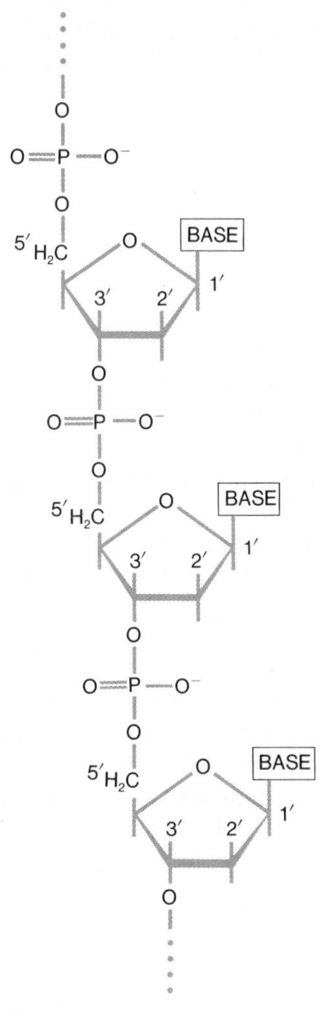

Figure 23.6 X-ray crystallography of the DNA molecule. (a) Diagram of the technique. (b) X-ray diffraction photograph of DNA taken by Rosalind Franklin. The crossing pattern of dark spots in the center of the picture indicates that DNA is helical. The dark regions at the top and bottom of the photograph show that the purine and pyrimidine bases are stacked on top of one another and are 0.34 nm apart. This photograph provided many clues for Watson and Crick in their search for the structure of DNA. (a) From *Biochemistry 2/e* by Lubert Stryer, W. H. Freeman and Company. © 1981. Reprinted by permission.

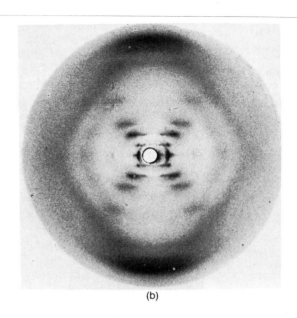

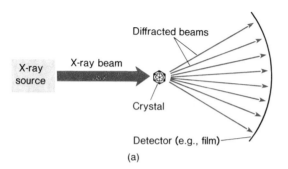

(a)

(b)

Watson and Crick gathered information on DNA from diverse sources and set about constructing a model of DNA structure. They combined Chargaff's discoveries on base composition and information from Franklin's X-ray photographs (figure 23.6) with various predictions of how genetic material might logically be expected to behave. Chargaff's rule that [A] = [T] and [G] = [C] would hold true if there were two strands of nucleotides in a DNA molecule and if each purine (adenine or guanine) was hydrogen-bonded with its corresponding pyrimidine (thymine or cytosine). Watson and Crick found that only the A-T and G-C combinations would hydrogen-bond properly and produce a smooth structure with a constant diameter. Finally, the model had to be compatible with the data obtained from Franklin's X-ray photographs. After putting together and analyzing all of the information available to them, Watson and Crick designed a model of the structure of the DNA molecule.

The structure that Watson and Crick proposed is known as the **double-helix model** of DNA. Their model looks somewhat like a twisted ladder, with two polynucleotide strands running in opposite directions and winding about each other. The A-T and G-C base pairs form the "rungs" in the interior of the helix (figure 23.7). A purine (for example, adenine) on one strand is hydrogen-bonded with the complementary pyrimidine (in this case, thymine) on the opposite strand.

The helix is 2.0 nm in diameter, and it makes a full, spiraled turn every 3.4 nm (every ten base pairs). Thus, the helix rotates about its longitudinal axis 36° with each base pair or "rung"; the spiral winds through a full circle (360°) in the

length of helix occupied by ten base pairs. The DNA helix is stabilized by hydrogen bonding between A-T and G-C base pairs, by London dispersion forces (van der Waals forces) between the stacked bases, and through general hydrophobic interactions that favor a structure with bases to the inside and the sugar-phosphate backbone in contact with the surrounding water.

The overall structure of DNA, with its hydrogen-bonded base pairs, has important genetic implications because it suggests a mechanism by which genetic material can be replicated (copied). The two strands of a DNA molecule can separate and two new strands—each complementary to one of the original strands—can be made, so that two new, identical DNA molecules exist where there had been one. Thus, during cell reproduction, each one of two new cells can receive a complete and accurate copy of the DNA in the original cell. These implications, immediately recognized by Watson and Crick, were alluded to in their original paper, which was published in the journal *Nature* in 1953. They said: "It has not escaped our notice that the specific pairing we have postulated immediately suggests a possible copying mechanism for the genetic material."

The publication of Watson and Crick's brief paper describing their model of DNA structure also marked the point at which it was finally confirmed to nearly everyone's satisfaction that genetic information is carried by DNA molecules. Thus, the work of five scientists—Watson, Crick, Franklin, Wilkins, and Chargaff—transformed the study of biology by opening the door to new research on genetic mechanisms as well as on many functions of living cells that had not been accessible before.

Figure 23.7 DNA structure. All distances are in nanometers (nm). One nanometer equals 10^{-9} meter or 10^{-6} millimeter; that is, one nanometer equals 0.000001 mm. (a) The association of bases by hydrogen bonding. Hydrogen bonds are shown as colored dashes. (b) Diagrammatic structure of the DNA double helix. The DNA double helix can be visualized as a twisted ladder, where the sugar-phosphate "backbones" of the strands form the rails of the ladder and the paired bases are the rungs. The two strands of a DNA molecule run in opposite directions. This is shown by using the 3' and 5' carbons to indicate the directional orientation of the strands (see figure 23.5). (c) James Watson (*left*) and Francis Crick with their model of DNA.

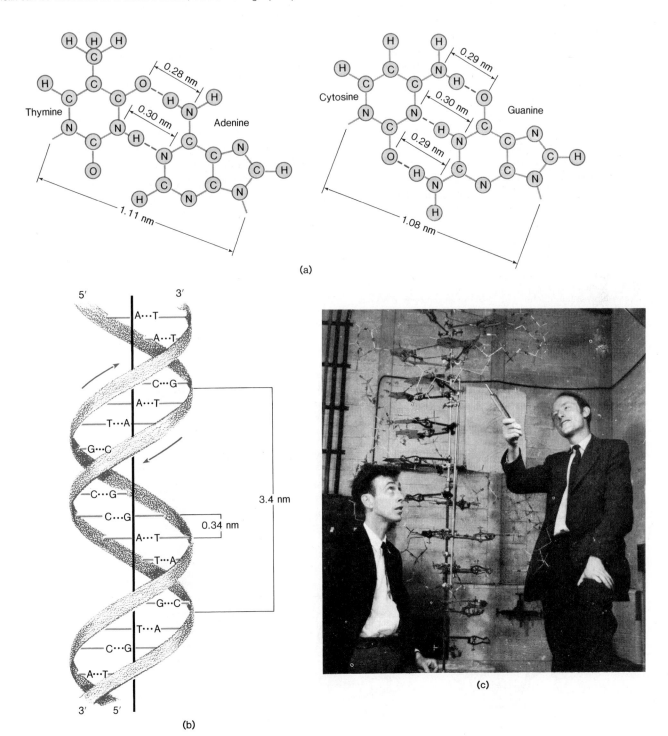

(a)

(b)

(c)

The Impact of the Watson and Crick Model of DNA

As a result of Watson and Crick's work, biologists realized that because DNA consists of two complementary strands of nucleotides arranged in the now-familiar double helix, each of these two strands could serve as a **template** (pattern) for the synthesis of a precise copy of the other strand. This would provide a mechanism for the duplication of genetic information during the S phase of the cell cycle. Also, it was pointed out that DNA molecules have a great deal of complexity and variety. The base pairing from one strand to another is restricted, since only complementary bases will pair; that is, adenine (A) always pairs with thymine (T), and guanine (G) always pairs with cytosine (C). However, there is no such structural restriction on the sequence of nucleotides within any one strand in the molecule. The sequence in which the purine- or pyrimidine-containing nucleotides are arranged determines the genetic coding contained in nucleic acids.

Nucleic Acid Diversity

For the great majority of living things, DNA is the primary genetic material. Yet there are some exceptions to this rule. Some viruses, for example, contain cores of RNA, rather than DNA, as their primary genetic material.

Moreover, DNA is not always found in the form known as the **B configuration (B-DNA),** which was described in the model that Watson and Crick proposed. In fact, there are several additional DNA configurations. **A-, C-,** and **D-DNA** configurations are all "right-handed" double-helical forms that differ from B-DNA in the spacing and angles of the base pairs. A radically different form, **Z-DNA,** was discovered by Alexander Rich and his colleagues in the late 1970s (figure 23.8). Z-DNA exists as a "left-handed" double helix. Although very different from B-DNA, the Z configuration has been found in short stretches in some eukaryotic organisms, and may be involved in gene regulation. In our examination of molecular genetics, however, we will focus on DNA-containing organisms and the B configuration.

Figure 23.8 Comparison of Z-DNA and B-DNA. Two slightly different forms of Z-DNA are shown here. Z-DNA is characterized by a left-handed helix and the zigzag configuration of the sugar-phosphate backbone (*the black line in each molecule*). B-DNA forms a smooth, right-handed helix with characteristic major and minor grooves as described by Watson and Crick.

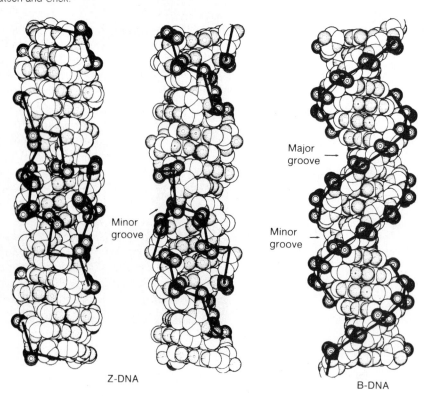

Z-DNA

B-DNA

DNA Replication

As we have said, the discovery of DNA structure led investigators to predict a mechanism of DNA replication based on *complementary base pairing*. It was proposed that a double-stranded molecule of DNA uncoils and that each single strand serves as a template for the formation of a new complementary strand (figure 23.9). Thus, where there is an A in the existing strand, a T would be incorporated into the forming strand; where there is a C in the existing strand, a G would be incorporated into the forming strand, and so forth. Two different experiments confirmed this pattern of replication in DNA molecules.

Figure 23.9 The two strands of a DNA molecule separate and each serves as a template for the formation of a new complementary strand. The strands of the original molecule are white. The new strands, which are synthesized by a group of enzymes acting in concert, are gray. This type of replication, resulting in two DNA molecules each of which is composed of one original strand and one new strand, is known as semiconservative replication. Note the complementary base pairing—adenine (A) always pairs with thymine (T), and guanine (G) always pairs with cytosine (C).

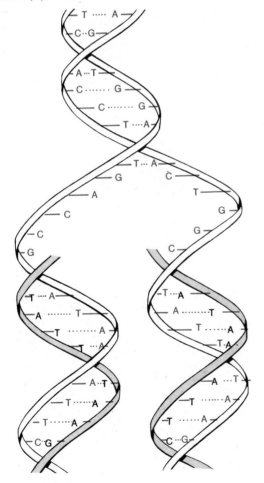

Experimental Proofs of Semiconservative Replication

The first conclusive experimental evidence concerning the nature of DNA replication was provided in 1958 by an experiment conducted by Matthew Meselson and Franklin Stahl. Meselson and Stahl were able to trace the mechanism of DNA replication by using strands of DNA with different densities. In order to produce DNA strands with different densities, they supplied growing bacteria with nucleotides (the building blocks of DNA strands) that contained different isotopes of nitrogen. During growth some bacteria were exposed to nucleotides containing the "light" (^{14}N) isotope, and others were grown in the presence of nucleotides with the "heavy" (^{15}N) isotope. Meselson and Stahl then measured the density of DNA synthesized by the bacteria by centrifuging it in tubes containing a cesium chloride (CsCl) gradient (figure 23.10). The DNA in the solution settled in a specific region of the gradient, that is, the point at which the density of the CsCl corresponded to the buoyant density of the DNA. As you can see in figure 23.10a and b, the heavy and light DNA formed distinct bands at different levels in the gradient, indicating that they had different densities.

Then Meselson and Stahl transferred bacterial cells that had been growing in the medium with nucleotides containing heavy nitrogen (^{15}N), and therefore had high-density DNA, to a medium with nucleotides containing light nitrogen (^{14}N). After the bacterial cells had gone through one cell cycle (one division), Meselson and Stahl found that the density of all of the DNA was intermediate between the density of light DNA (DNA containing only nucleotides with ^{14}N) and heavy DNA (DNA containing only nucleotides with ^{15}N) (figure 23.10c).

These results fit with the prediction that during DNA replication, the two strands of a DNA molecule separate and each serves as a template for formation of a new, complementary strand. These experiments demonstrated that the original double strand of heavy DNA had indeed split into two strands. Each of these strands served as the template for the formation of a new strand, into which were incorporated nucleotides containing the light nitrogen. Thus, two double-stranded DNA molecules resulted from the replication of each original DNA molecule, and each of these new DNA molecules had one light and one heavy strand. The density of this newly synthesized DNA, therefore, was intermediate between the density of DNA with two heavy strands and DNA with two light strands.

Meselson and Stahl next allowed some bacteria to remain in the medium with light, ^{14}N-containing nucleotides long enough for another division to occur. When they extracted and centrifuged the DNA, they found two bands in the CsCl gradient. You can see what happened in figure 23.10d. The original heavy strands paired with new light

Figure 23.10 Meselson and Stahl's experiments demonstrating semiconservative replication of DNA in *E. coli*. (a) When centrifuged in a tube containing a cesium chloride gradient of increasing density, light, 14N-containing DNA migrates a certain distance down the tube where it forms a single band. (b) When centrifuged under the same conditions, heavy, 15N-containing DNA migrates a greater distance and forms a single band further down the tube. (c) *E. coli* cells containing 15N DNA were placed in 14N medium. After one generation (one cycle of DNA replication), all of the DNA molecules had one heavy and one light strand. When centrifuged, this 15N/14N DNA migrated an intermediate distance in comparison with the heavy and light DNA. (d) After a second generation in 14N medium, *E. coli* cells possessed DNA that formed both an intermediate and a new light band after centrifugation. Meselson and Stahl used these results to support the semiconservative hypothesis of DNA replication predicted by the Watson–Crick model of DNA.

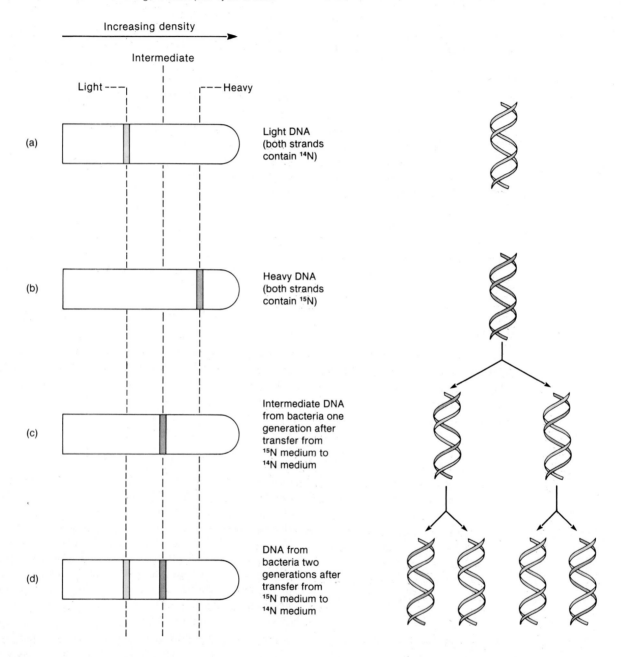

Increasing density

Intermediate

Light ─── Heavy

(a) Light DNA (both strands contain 14N)

(b) Heavy DNA (both strands contain 15N)

(c) Intermediate DNA from bacteria one generation after transfer from 15N medium to 14N medium

(d) DNA from bacteria two generations after transfer from 15N medium to 14N medium

strands, again producing DNA of intermediate density. The light strands from the previous generation served as templates for the assembly of new light strands, resulting in DNA molecules with light nitrogen in both strands. These DNA molecules with two light strands accounted for a low-density band in the CsCl gradient (figure 23.10d).

This type of replication, in which half of each new DNA molecule is carried over (conserved) from the previous generation and the other half is newly synthesized, is called **semiconservative replication.** Semiconservative replication was predicted after the publication of the Watson and Crick model of DNA structure and was thus confirmed by the experiments of Meselson and Stahl.*

A different type of experiment, conducted by Arthur Kornberg and his colleagues, showed clearly that a single strand of DNA can, by itself, act as a template for the formation of a new strand of nucleotides. Kornberg and his coworkers prepared a mixture of the four nucleotides found in DNA along with **DNA polymerase,** an enzyme that catalyzes the formation of new strands of DNA. To this mixture they added single strands of DNA. New, double-stranded DNA molecules resulted. The length of the new DNA molecules was determined by the length of the single strands of DNA that were added to the original mixture. When the nucleotide content of the new strands was analyzed, the new strands were found to be complementary to the original strands.

Molecular Events in DNA Replication

Any discussion of DNA replication must take into account the fact that DNA is in chromosomes. In prokaryotic organisms (bacteria), chromosomes are circular, while in eukaryotic cells chromosomes are linear. Let us take a closer look at DNA replication in prokaryotic and eukaryotic cells.

DNA Replication in Prokaryotic Cells

One type of chromosome replication in bacteria is shown in figure 23.11. In this process, known as **bidirectional replication,** replication is originated at a specific point (origin)

---

*The semiconservative model of DNA replication contrasts with two alternative models that had been suggested. The *conservative* model proposed that the two original DNA strands would remain together after directing the synthesis of two new strands, so that a completely new double helix would be formed. The *dispersive* model proposed that the DNA would fragment, be replicated, and then be reassembled. Meselson and Stahl's results proved that the semiconservative model of DNA replication is the correct one.

Figure 23.11 Bidirectional DNA replication of a bacterial (circular) chromosome. (a) The two strands of the helix separate at the origin (O), and replication proceeds away from this point in both directions. There are two replication forks, points at which the two original strands are separated and new, complementary strands are being assembled. The characteristic "theta" appearance is seen about midway through the process. (b) Electron micrograph of bidirectional replication occurring in the circular form of the chromosome of a virus, the bacteriophage lambda (phage λ). The two replication forks are indicated by arrows.

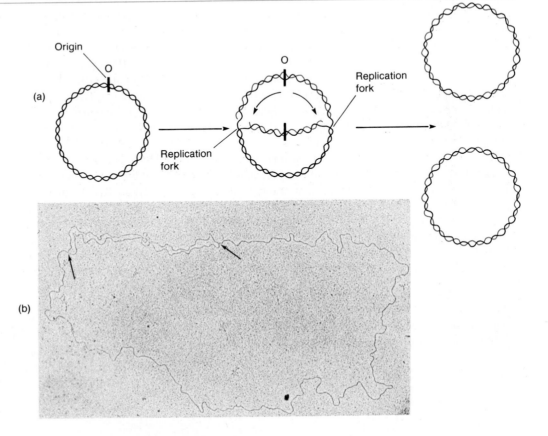

Figure 23.12 Rolling-circle (unidirectional) DNA replication. As one strand of the double helix is unwound from the circle, replication takes place on both strands. The process begins at the origin (*O*), and the single replication fork proceeds in one direction.

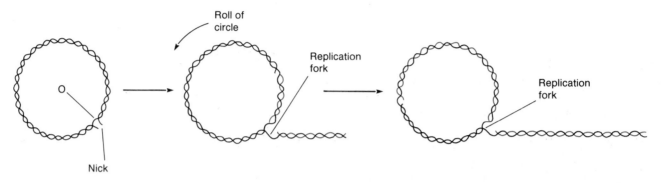

where the two strands of the circular DNA molecule are separated. Synthesis of a new complementary DNA strand begins along each original strand and proceeds away from the origin in both directions. The point where two strands of an existing DNA molecule are separated and new complementary strands are being assembled is known as a **replication fork.**

When viewed with the electron microscope, a partially replicated, circular chromosome has a characteristic appearance known as a "theta structure," since its shape resembles a somewhat twisted version of the Greek letter theta (*θ*). This process, which takes approximately forty minutes to complete in the bacterium *Escherichia coli,* is thought to be the most common way by which circular chromosomes of prokaryotic cells are replicated.

Another way that bacterial chromosomes, as well as some bacteriophage chromosomes and plasmids (see p. 595), may replicate is by **rolling circle (unidirectional) replication** (figure 23.12). Usually the bacterial chromosome undergoes this type of replication during conjugation, a form of bacterial mating (see p. 598). In this type of replication, one strand is cut (nicked) by an enzyme and separated from the other strand. Then synthesis of a new complementary strand begins along the other (uncut) strand that is now exposed. At the same time, the cut strand "rolls off" the circular chromosome, and synthesis of a complementary strand also occurs along it. Eventually, the new chromosome produced by replication of the cut strand closes up into a circle.

The Mechanism of DNA Replication
Now that you have a general idea of the process of DNA replication in bacterial chromosomes, let us examine in more detail some of the specific steps in this complex process.

What actually happens at a replication fork? Several different enzymes are required at the beginning of the replication process. The DNA strands are unwound and the helical twists of the molecule are relaxed by enzymes known as **DNA helicase** and **DNA gyrase.** At this point, then, individual strands are exposed for replication (figure 23.13).

As we noted earlier, DNA polymerases are the enzymes primarily responsible for the assembly of new strands that are complementary to existing single DNA strands. However, DNA replication is not simply a matter of DNA polymerase starting along each of the unwound, separated strands and adding complementary nucleotides as it goes. First of all, DNA polymerase, by itself, cannot *initiate* synthesis of a new DNA strand. Rather, a small segment of RNA, known as an **RNA primer,** must first be introduced through the action of an enzyme known as a **primase.** Once the RNA primer is in place, DNA polymerase begins the process of adding complementary nucleotides, resulting in the assembly of a new strand of DNA.

A problem arises, however, as a result of the way in which DNA polymerase functions. DNA polymerase can only assemble nucleotides into a strand that runs in a 5' to 3' direction. Recall, however, that the two strands of a DNA molecule run in opposite directions; one runs in a 3' to 5' direction and the other in a 5' to 3' direction (see figure 23.7). DNA polymerase has no problem in assembling a new strand that is complementary to the original 3' to 5' strand, but how can a new complementary strand be assembled for the other original strand, the one that runs in the opposite (5' to 3') direction?

The answer to this question was found by R. T. Okazaki and his colleagues. They discovered that production of the new strand which is complementary to the original 5' to 3' strand is accomplished by the assembly of relatively short

Figure 23.13 Diagrammatic model of DNA replication showing some of the currently known details of the process. DNA gyrase and helicase cause strand unwinding and helix relaxation so that the two strands are separated and can serve as templates for the synthesis of new strands.

After an RNA primer is in place, DNA polymerase catalyzes the addition of complementary nucleotides. Synthesis of the leading strand is continuous, but synthesis of the lagging strand is discontinuous. Along the lagging strand, DNA polymerase moves "backward," catalyzing the addition of groups of nucleotides (Okazaki fragments). These Okazaki fragments are subsequently joined when ligase closes the nicks between them.

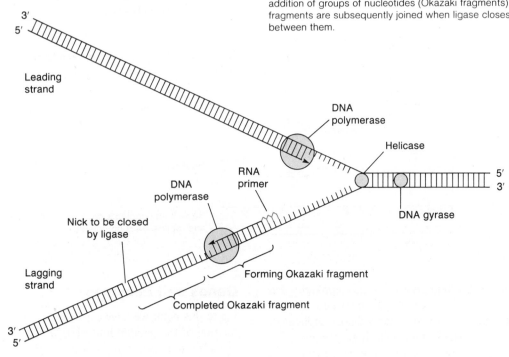

fragments of a complementary nucleotide strand, with DNA polymerase working opposite to the direction in which the replication fork is proceeding along the DNA molecule (see figure 23.13). These **Okazaki fragments,** as they are called, must then be linked together to make the polynucleotide chain of the new complementary strand continuous. Thus, one new strand, the **lagging strand,** is being synthesized discontinuously by the linking of Okazaki fragments, while the other strand, the **leading strand,** is being synthesized directly and continuously.

A special DNA polymerase removes the RNA primers and adds complementary nucleotides to complete the new DNA strand. But at these points and others, there are still nicks in individual polynucleotide strands—points at which the strands are not continuous. **Ligases,** enzymes that catalyze formation of covalent bonds between adjacent nucleotides, close up these nicks, making the strands continuous. Once all nicks are closed, each of the two replicated DNA molecules assumes the normal double helix configuration.

You can see, then, that DNA replication is a very complex process and involves many steps and enzymes. In addition to DNA polymerase (the enzyme that catalyzes the joining of nucleotides into strands), there are gyrases and helicases that assist in untwisting the double helix; primase, which inserts RNA primers; a polymerase that removes this RNA and fills the gaps, and ligases that close nicks. The description here is considerably simplified; you can find additional details of the process in some of the readings suggested at the end of this chapter.

DNA Replication in Eukaryotic Cells

The replication of chromosomal DNA in eukaryotic cells, is an even more complex process because each eukaryotic nucleus contains much more DNA than a prokaryotic cell does. Furthermore, the DNA of eukaryotic chromosomes is arranged in complex relationships with proteins (recall the structure of nucleosomes, p. 531). And yet, the time required for DNA replication in eukaryotic cells is not as great as you might expect. For example, a human nucleus contains about 2,000 times as much DNA as a bacterial cell, but the replication time of DNA in human cells is only about twenty times as long as in bacterial cells.

Figure 23.14 DNA replication in a eukaryotic chromosome. Replication begins simultaneously at a number of origins (*O*) along the chromosome and proceeds in both directions from each origin. The portion of a DNA molecule that is replicated following initiation at one origin is called a replicon. When the replicated stretches of DNA in all of the replicons have been joined, replication is complete.

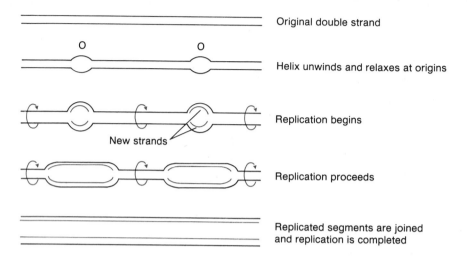

Original double strand

Helix unwinds and relaxes at origins

New strands

Replication begins

Replication proceeds

Replicated segments are joined and replication is completed

There are several explanations for this apparent discrepancy. First, the DNA in a human cell is distributed among 46 chromosomes, each of which is individually replicated, while the DNA in a bacterial cell is in a single chromosome. A second factor is that DNA replication in eukaryotic cells is initiated at hundreds of sites (origins), almost simultaneously, along each chromosome. Each origin is the point where the replication of a specific region of the chromosome begins. These chromosome regions are called **replicons.**

Just as in prokaryotes, DNA replication in eukaryotic cells is semiconservative, with a leading strand being replicated continuously and a lagging strand requiring formation and joining of Okazaki fragments. Moreover, the enzymes involved in replication are essentially the same. However, other enzymes are required to synthesize additional histones needed for new chromosomes.

In figure 23.14, you can see, in simplified form, a short segment from a eukaryotic cell chromosome showing replication of two replicons. What are the steps in the process? The double helix is unwound and relaxed much as in DNA replication in prokaryotic cells, but this takes place at multiple points called **origins (0).** Temporary nicks occur and function as swivel points as the strands of the double helix separate and unwind. Complementary strand synthesis begins at the origins and moves bidirectionally along each strand. As new complementary strands form along each of the previously existing strands, there is spontaneous rewinding and two double helices are formed where there had been one.

Genes and Enzymes

Up to this point we have considered some aspects of the replication of the genetic material, DNA. Now let us turn our attention to another important question in genetic research. How do genes determine the characteristics of living things? More precisely, how is the genetic information, coded as a sequence of bases in DNA molecules, expressed in the cells of living things?

Many types of genes function by providing coded information that directs the synthesis of enzymes or other proteins. Enzymes are the catalysts involved in virtually all biochemical reactions in living cells. Thus, many genes function by determining certain characteristics of these vital catalysts, thereby determining the nature of key functions in cells and organisms.

Garrod's Hypothesis Regarding Inherited Metabolic Deficiencies

The discovery of the relationship that exists between genes and enzymes has an interesting history. This relationship was first suggested not by a biologist working on experimental organisms in a research laboratory, but by the English physician Archibald Garrod, who studied human metabolic diseases in the early 1900s.

Figure 25.1 Transformation in bacteria. (a) Bacteria can be transformed by uptake and incorporation of DNA fragments. Stable transformation occurs only if the DNA fragment becomes incorporated into the bacterial cell's chromosome. In some cases, the DNA fragment is destroyed by cellular enzymes and it has no genetic effect on the cell. (b) Transformation may involve uptake of a plasmid. Because plasmids are autonomous and self-replicating, transformation by a plasmid does not require interaction with the cell's chromosome for the transformation to be stable.

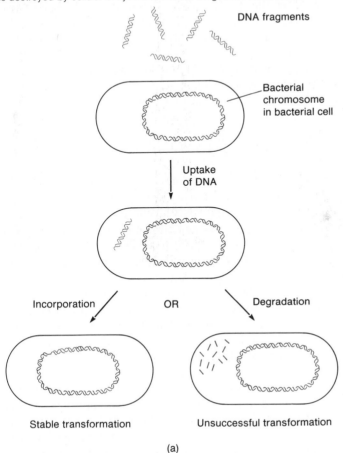

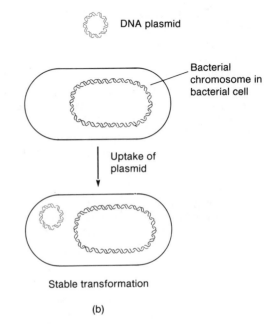

(a)

(b)

If one of the released DNA fragments contacts a cell of a species of bacteria that is capable of transformation, the DNA fragment may be bound to that cell (the **recipient**) and taken inside (figure 25.1). However, only **competent** (receptive) cells of bacteria such as *S. pneumonia* and *E. coli* are able to pick up such DNA fragments, and in nature only a small proportion of cells are competent. In many cases, though, competency can be enhanced in the laboratory by changing the ionic composition of the environment around the bacterial cells or by altering the temperature. It is interesting that the DNA taken up by competent bacteria need not originate from cells of the same species of bacteria. In fact, it is not even necessary that the DNA come from bacteria. Virtually any DNA fragment of natural or synthetic origin can be taken up by competent bacteria.

Once inside the recipient bacterium, the DNA fragment either will be broken down by enzymes into nucleotides, or it will be stably maintained within the cell. If the DNA fragment is broken down, then the uptake of that DNA will have no genetic effect. For transformation (effective gene transfer) to occur, the DNA fragment must be maintained in a stable condition. This can happen in one of two ways. One way is for the DNA fragment to be incorporated into the recipient cell's chromosome (figure 25.1a).

The other possibility is that the DNA that has been taken up is DNA that can exist inside a bacterial cell as an independently maintained piece of DNA, such as a circular **plasmid** (figure 25.1b). Plasmids are relatively small segments of extrachromosomal DNA found in many bacteria.

They consist of a circular double-stranded DNA molecule, and in most cases they are self-replicating entities; their replication is independent from that of the bacterial chromosome. Plasmids contain relatively few genes, less than 30 in most cases. Many plasmids contain genes for antibiotic resistance. The number of plasmids in single bacterial cells ranges from a few to many (up to 100). The natural roles of plasmids may be to provide multiple copies of genes (gene amplification) or to promote gene transfer in bacteria.

The process of transformation is very useful in the laboratory as a means of introducing engineered DNA molecules into bacterial cells. Furthermore, biologists have found that it is also possible for gene transfer by transformation to occur in eukaryotic cells. For instance, it is possible to introduce "foreign" DNA into cells of fungi and algae, and even into cultured mammalian cells. However, unlike bacterial transformation, transformation of these types of cells probably occurs only under laboratory conditions where complex procedures facilitate it, and it is much less efficient than transformation in bacteria. For example, in yeast, which was one of the first eukaryotic organisms to be transformed, the cell walls must be enzymatically removed prior to DNA uptake, and then allowed to grow back afterwards.

Transduction

As we saw in chapter 23, bacteriophages are viruses that infect bacteria by injecting their DNA into bacterial cells. The injected viral DNA either immediately or at some later time causes the bacterial cell to synthesize the various elements of new virus particles, which are then assembled and released. Each of these new particles can then infect another bacterial cell. In this process, viruses may serve as carriers for the transfer of bacterial genes from one cell to another. This mechanism of bacterial gene transfer is known as **transduction.**

Generalized Transduction

Transduction results when bacterial chromosome fragments are packaged in viral capsids (protein coats) during the assembly stage in the virus life cycle (see p. 914). When replicated viral chromosomes are being packed into protein capsids to make complete virus particles, random fragments of the broken-up bacterial chromosome may also be packaged. However, a viral capsid can contain only so much DNA. Therefore, if some bacterial DNA is packaged, some or even all of the viral chromosome will be left out. In many cases when this happens, the virus particles that are formed are defective. Such a virus particle is capable of injecting its DNA

into another cell, but it may have lost the ability to cause normal virus replication and the eventual breaking (lysis) of the host cell. The net effect, therefore, is that the virus has injected the bacterial genes that it carried into another bacterial cell (figure 25.2). Again, however, in order for the gene transfer to be stable, the injected DNA must be incorporated into the recipient cell's chromosome. This incorporation is very similar to what occurs in transformation. Because it is largely a matter of chance as to which bacterial genes are transferred in this manner, this form of transduction is called **generalized** (random) **transduction.**

Specialized Transduction

Instead of beginning reproduction immediately after entering a bacterial cell and destroying (lysing) the cell directly, viral DNA sometimes becomes integrated into the bacterial cell's chromosome. Then, when the bacterial DNA is replicated, the viral DNA is replicated right along with it, and the daughter cells contain the viral DNA as well. This process can go on for generations. Phages that interact with bacteria in this way are called **temperate phages,** and this type of infection cycle is known as a *lysogenic cycle* (as opposed to the *lytic cycle* we just described in which the bacterial cell is broken—lysed—directly). Bacteria in which this occurs are called **lysogenic bacteria,** and the virus DNA in the bacterial chromosome is called a **prophage** (pro = before), because it has the potential to eventually produce phage particles again.

If at some later time the prophage-carrying bacterial cells are exposed to certain environmental stimuli such as ultraviolet light, the prophage is induced to leave (*de-integrate* from) the bacterial chromosome. Then virus nucleic acids and proteins that are involved in the formation of new virus particles are synthesized, the particles are assembled, and the host cell is lysed.

Specialized (specific) **transduction** occurs during the lysogenic cycle of such temperate phages. Following induction by an environmental stimulus, when a prophage de-integrates from the bacterial chromosome and begins the steps that lead to virus assembly and release, some bacterial genes may be carried along with the prophage (figure 25.3). Thus, the phage particles that are produced carry bacterial genes, specifically those genes that were located adjacent to the prophage in the bacterial chromosome. After virus replication and assembly are completed, many phage particles are released, each of which contains the same bacterial genes. It is because all of the phages contain the same bacterial genes that this form of transduction is called specialized transduction.

Just as is true in generalized transduction, some of these phages may be defective, because while they carry some bacterial genes, they may lack certain virus genes that were left

Figure 25.2 Generalized transduction by a bacteriophage (phage). In some cases the bacterial chromosome is fragmented during viral infection of a bacterial cell. When virus capsids and DNA have been synthesized and virus particles are assembled, fragments of the bacterial chromosome are included in some virus particles on a random basis. Such phages may be defective because they lack some normal virus genes, but they are still able to transfer DNA to other bacterial cells. Just as with transformation, any inserted gene must combine with the recipient cell's chromosome to make the gene transfer stable.

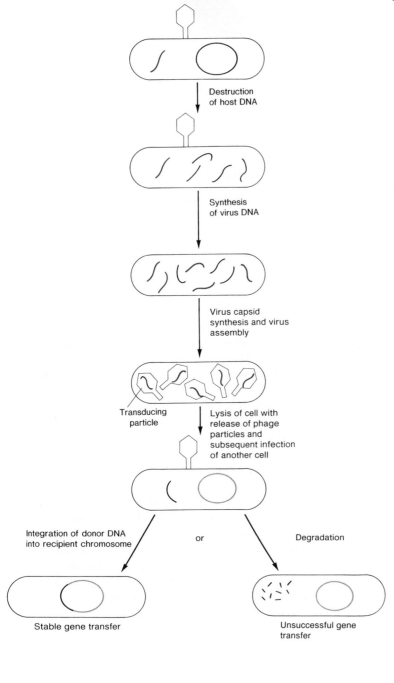

Destruction of host DNA

Synthesis of virus DNA

Virus capsid synthesis and virus assembly

Transducing particle

Lysis of cell with release of phage particles and subsequent infection of another cell

Integration of donor DNA into recipient chromosome

or

Degradation

Stable gene transfer

Unsuccessful gene transfer

Figure 25.3 Specialized transduction of a lysogenic bacterium by a temperate phage. In this form of transduction, specific genes located adjacent to a prophage are carried along with it when it de-integrates from the bacterial chromosome. These bacterial DNA fragments are replicated along with the virus DNA so that each virus particle eventually assembled will contain the same set of bacterial genes. The virus particles are then released by cell lysis. When the phage infects another cell, the genes will be injected and the transfer will be stable if the phage chromosome integrates into the host or if the bacterial genes introduced by the virus integrate by recombination.

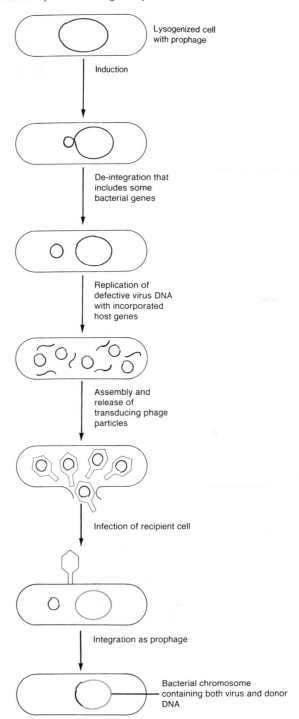

Lysogenized cell with prophage

Induction

De-integration that includes some bacterial genes

Replication of defective virus DNA with incorporated host genes

Assembly and release of transducing phage particles

Infection of recipient cell

Integration as prophage

Bacterial chromosome containing both virus and donor DNA

Figure 26.12 Fertilization membrane development. (a) A sand dollar egg just before fertilization. Note the pigment granules in the jelly that surrounds the egg. (b) Sand dollar zygote surrounded by a fertilization membrane.

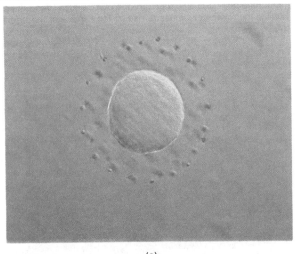

(a)

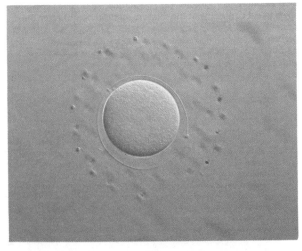

(b)

Material released from the cortical granules causes water to enter the space between the vitelline membrane and the plasma membrane by osmosis, thus expanding the vitelline membrane and elevating it off the plasma membrane. Then, structural proteins from the granules unite with the vitelline membrane to thicken it into a fully developed fertilization membrane. The development of the fertilization membrane begins where the first sperm makes contact, and spreads peripherally until it covers the egg's surface (figure 26.12). The fully developed fertilization membrane is a virtually impenetrable barrier to sperm cells and a permanent block to polyspermy.

Egg Activation

Fertilization is often defined only in genetic terms with emphasis on the fusion of the haploid sets of chromosomes in the gametes to produce the diploid zygote. You can see by now, however, that the egg-sperm interactions of fertilization begin long before the union of chromosomes—in fact, even before the gametes touch one another. Interactions continue through the time that egg and sperm cooperate to accomplish the sperm's entry into the egg. The egg does not wait passively for the motile sperm to burrow its way in; rather, the egg actively participates in the entry process once it has been contacted by the sperm.

Active participation in sperm entry is only the beginning of egg **activation** during fertilization. Activation involves many physical and biochemical changes in the egg cell. At the time of fertilization, the egg is a complex developmental system, poised and waiting for fertilization to stimulate it to develop further.

Ordinarily, the stimulus that causes egg activation is contact with a sperm. But egg activation is not entirely dependent upon sperm contact; **parthenogenesis** (development of an egg without fertilization) is also possible. Parthenogenesis occurs naturally in some species of animals (see p. 634), and it can be induced experimentally in the eggs of other species. Various physical treatments, such as chilling, heating, or pressure shocks, and chemical treatments that use weak acids or bases, can set off some of an egg cell's post-fertilization reactions. Sometimes a biological agent is used, as in the case of frog eggs that can be activated by pricking them with a needle dipped in frog blood. In some animal species, artificially activated eggs develop quite extensively. Over the years, observation of these responses to artificial activating agents and the ability of eggs to begin development without sperm contact has led biologists to conclude that egg cells contain essentially all of the information needed to direct early development, and modern biochemical studies have verified this conclusion. The egg seems to exert major control over early development; genetic information from the sperm becomes essential only after development is well under way.

Figure 26.15 Gastrulation and pluteus development. (a) At the beginning of gastrulation one side of the hollow blastula sinks in. Note the indented area on one side of the spherical embryo and the primary mesenchyme cells lying loose in its internal cavity. (b) Mid-gastrula stage of sand dollars. (c) Pluteus larva. The pluteus is the swimming, feeding larval stage of echinoderm development. The body is supported by skeletal spicules and has a complete digestive tract with mouth and anus.

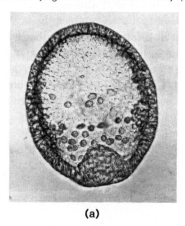

(a)

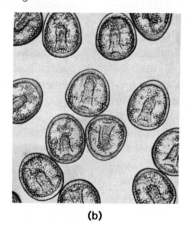

(b)

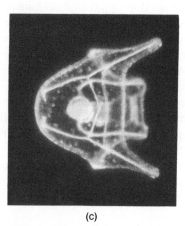

(c)

and the segregation of groups of cells. These processes are known collectively as **gastrulation.** Finally, cells become differentiated and organized into groups that will carry out specific functions in localized body areas. Let us now examine each of these phases of development in a little more detail, using photographs of developing sand dollar embryos.

Cleavage

Cleavage converts the unicellular zygote into a multicellular aggregate. During cleavage, each mitotic cell division produces two cells that divide again, and their daughter cells divide again, and so forth, all without growing between divisions (figure 26.14). This cycle of repeated cell divisions continues until a spherical mass of cells known as the **morula** is formed.

As cleavage divisions continue, cells begin to move apart so that spaces appear among the cells in the center of the mass. Cells keep pulling away from the central area until they have organized themselves into a single-layered, hollow ball surrounding a fluid-filled cavity known as the **blastocoel.** This hollow-sphere embryo, which develops at the end of cleavage, is called a **blastula.**

Development of the hollow blastula embryo requires that the embryo's cells behave and move in a specific way. This is only the first of several cell movement patterns that are essential steps in development. These specific cell movements are one-time events in the life history of the individual that must be conducted precisely if development is to proceed normally.

Through all of the cell divisions of cleavage and the cell rearrangements to form the blastula, some of the basic organization of the egg cytoplasm is preserved; that is, the various areas of egg cytoplasm are not displaced relative to one another. For example, a pigmented band present in an egg cell is represented by a pigmented band in the blastula. It is likely that specific kinds of cytoplasm, maintained in specific areas during cleavage, may determine the subsequent fates of cells in various parts of the blastula.

Gastrulation

Gastrulation converts the simple, single-layered sphere of the blastula into a several-layered body with structurally and functionally specialized areas. These changes in the embryo's organization, which are brought about by cell movements and cell shape changes, are part of **morphogenesis** ("development of form"). Morphogenesis is the progressive development of pattern and form of the developing embryo. Gastrulation in echinoderms begins when one side of the blastula wall sinks in to form a pit. You can visualize this process by imagining the indentation formed if you push your finger into one side of a balloon or a soft, hollow rubber ball. To cause this shape change, all of the cells in the area must change their shapes at the same time in a display of collective cell behavior (figure 26.15).

As this sinking process is beginning, some of the cells in the depressed area pull loose and come to lie free inside the blastocoel cavity. Soon, these cells move out to strategic locations in the embryo, where they position themselves and produce skeletal spicules that support the body of the developing larva. The cells that set off individually on this specialized developmental course are known as **primary mesenchyme cells.**

The blastula wall continues to sink in until the pit resembles the finger of a glove. This lengthening hollow structure is the developing **gut,** the future digestive tract. Cells at the tip of the developing gut send out long processes that contact the inside wall of the blastula and adhere to it. Then the processes contract, producing a cellular pulling force that draws along the developing gut. Finally, the gut contacts another point in the wall of the still-hollow embryo, where a breakthrough opens the gut to the outside. This newly produced opening becomes the **mouth.** The **blastopore,** the opening into the original pit on the side where the sinking process began, becomes the **anus,** the posterior exit from the larval digestive tract. Thus, as a result of gastrulation, the hollow sphere of the blastula stage embryo has been converted into a body that has several layers and includes a complete gut, beginning with a mouth and ending with an anus.

Cell Differentiation and Body Organization

The fundamental body layers, or **germ layers,** that characterize the basic body plan of the more complex, multicellular animals are fairly simply illustrated in the echinoderm embryo (figure 26.16).

Cells remaining on the surface after gastrulation make up a body surface layer known as **ectoderm.** Ectodermal cells form the "skin" of the echinoderm larva and they become ciliated, which permits the larva to swim.

Gut cells constitute the innermost body layer, the **endoderm.** Endoderm is the layer that produces the digestive system lining and its derivatives. The larval echinoderm gut quickly becomes functional, and the larva begins to feed on small organisms in the surrounding water.

Tissue between the ectoderm and the endoderm forms the middle body layer, the **mesoderm.** The mesoderm of the echinoderm embryo develops from primary mesenchyme cells and some additional tissue that also separates from the developing gut. Mesodermal cells produce the larval skeleton. Mesodermal tissue also lines the **coelom** (body cavity), the space between internal organs and the outer body wall. Eventually, in the adult echinoderm, mesoderm produces a number of structures that make up a considerable part of the body mass.

Table 26.1
Major Derivatives of the Three Germ Layers in Vertebrates

Ectoderm

Nervous tissue
Sense organs
Epidermis of skin

Mesoderm

Dermis of skin
Skeleton
Muscle
Circulatory system
Excretory system
Reproductive system
Connective tissue

Endoderm

Digestive system linings
Digestive glands
Lung and respiratory tract linings

Figure 26.16 Diagram of a fully developed pluteus larva. The mouth has formed, and the three basic body layers have differentiated. Skeletal spicules are produced by mesoderm (primary mesenchyme cells).

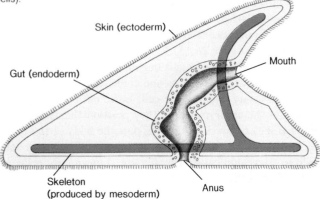

Skin (ectoderm)

Gut (endoderm)

Mouth

Skeleton
(produced by mesoderm)

Anus

Figure 26.17 Frog eggs surrounded by their jelly coats. Eggs of the wood frog at the beginning of the first cleavage division, showing the darkly pigmented animal pole and the light-colored, yolky, vegetal pole.

Vertebrate Development

The same three basic body layers (germ layers) that develop in echinoderms—ectoderm, endoderm, and mesoderm—also develop in vertebrate embryos (table 26.1). However, some noteworthy differences between early developmental processes in vertebrate and echinoderm embyros are apparent beginning with cleavage and gastrulation. An important factor in these differences is the relationship between the amount of stored nutrient material (yolk) in eggs and the eggs' developmental patterns.

Amphibian Development

Amphibian eggs contain considerable yolk, much of it concentrated in one hemisphere of the egg in an area known as the **vegetal pole.** The opposite side of the egg, called the **animal pole,** is darkly pigmented and has much less yolk (figure 26.17).

When an amphibian zygote divides, the cleavage furrow that cuts the cytoplasm in half moves quickly through the animal pole but is slowed by the bulky yolk in the vegetal pole. Even as the first division is being completed, a second set of cleavage furrows begins to divide each of the first two cells to produce four cells. At the third cleavage the division plane changes (figure 26.18). This third division yields two distinctly different sets of cells—four smaller, darkly pigmented cells lying above four larger, yolky cells. As cleavage continues, the smaller animal pole cells divide more rapidly, and they produce a population of small, uniformly shaped,

Figure 26.18 Scanning electron micrographs of cleavage stages in embryonic development of the clawed frog *Xenopus laevis..* (a) First cleavage division. The cleavage furrow cuts through the animal pole (*top*) quickly and then slows in the yolky vegetal pole (*bottom*). (b) The eight-cell stage. The size difference between the upper, animal pole cells and the lower, yolky vegetal pole cells is apparent. (c) Later cleavage stage.

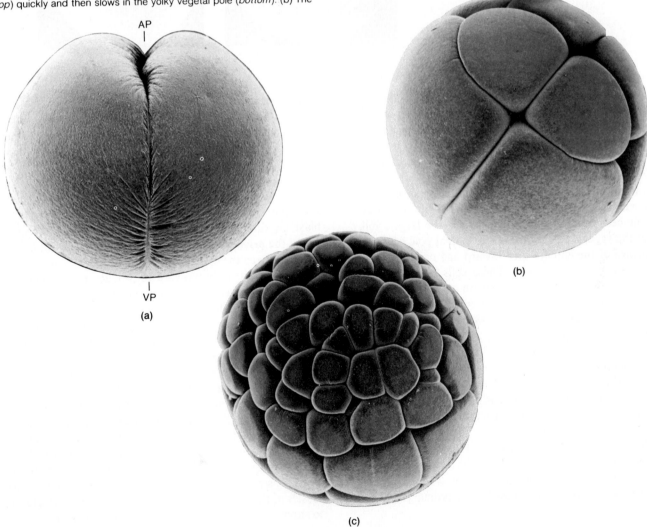

AP

VP

(a)

(b)

(c)

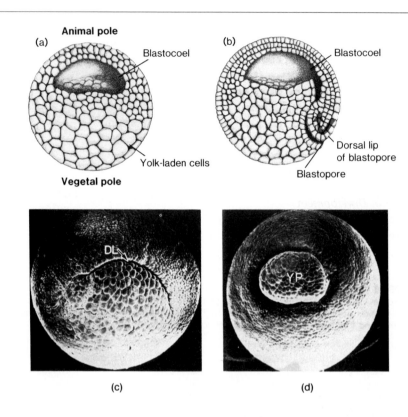

Figure 26.19 Amphibian gastrulation. (a) A section cut through a frog blastula-stage embryo showing the location of the blastocoel, which is displaced toward the animal pole side. (b) Section of a frog embryo at the beginning of gastrulation. Small animal pole cells move downward to surround and enclose the yolky vegetal pole cells. Some animal pole cells leave the surface and migrate to the interior through the blastopore (*colored arrow*). When they reach the interior, they migrate away from the blastopore and organize the embryonic mesoderm. (c) Scanning electron micrograph showing the surface of a frog embryo at the time cells begin to move into the blastopore. The first part of the blastopore to form is called the dorsal lip (*DL*) (magnification × 35). (d) Scanning electron micrograph of a later stage of frog gastrulation. As gastrulation proceeds, the blastopore becomes circular with cells migrating inward all around it. Small animal pole cells have completely enclosed the yolky vegetal pole cells except for the yolk plug (*YP*), made up of vegetal pole cells that lie in the center of the blastopore (magnification × 31).

pigmented cells. The larger vegetal pole cells divide more slowly and produce a population of larger, less regularly shaped cells. Differences between animal and vegetal pole portions of the embryo are also apparent when the embryonic cells organize into a blastula because the blastocoel is displaced toward the pigmented animal pole (figure 26.19).

Gastrulation in the amphibian embryo involves extensive movement by all of the cells in the blastula. Smaller animal pole cells move downward to envelop the larger, yolky cells of the vegetal pole. Some of the animal pole cells from the surface of the embryo turn inward and move to the interior through the blastopore. These cells form the mesoderm of the embryo. Cells that remain at the surface after these movements produce the ectoderm. Yolky vegetal pole cells rearrange themselves to produce the gut; they become the endoderm.

Bird Development

The avian (bird) egg proper is what we call "yolk" in everyday terminology. The albumen ("egg white"), the shell, and the shell membranes are accessory structures that surround and protect the developing embryo.

Patterns of cleavage and gastrulation in birds are very different from those of amphibians. The large mass of yolk is not cleaved during development of the avian embryo. In fact, it may be physically impossible for ordinary cell division processes to cleave that much yolk. Cleavage is restricted to a small quantity of developmentally active cytoplasm on one side of the yolk. The cleavage divisions there produce a flat disc of cells called the **blastoderm** (figure 26.20).

The cells in the blastoderm then sort into two distinct layers—an upper **epiblast** and a lower **hypoblast.** During subsequent gastrulation movements, certain cells of the epiblast migrate toward and down into a shallow depression called the **primitive streak.** After moving downward into the space between the epiblast and the hypoblast, many of these migrating cells move outward to produce the mesoderm of the avian embryo, located between the two previously existing layers. Epiblast cells that remain in the top layer constitute the ectoderm, while the original hypoblast plus some cells that come down from the primitive streak to join it constitute the endoderm.

Eventually the flat, three-layered avian embryo rolls up to produce a tubular body form similar to that of amphibian embryos and characteristic of all vertebrate bodies. During all of these changes, however, the large yolk mass of the avian egg is not directly involved in development of the embryo. As development proceeds, a **yolk sac,** which is attached to the embryo, encloses the yolk and absorbs nutrients from it. These absorbed substances are transported through blood vessels of the yolk sac into the embryo, where they are used to support the embryo's growth and development.

Figure 26.20 Bird cleavage and gastrulation. (a) The avian (bird) egg. The "yolk" is the egg proper, and development is restricted to an area of developmentally active cytoplasm at one side of the egg cell. Albumen, shell membranes, and shell are accessory structures that enclose and protect the developing embryo. (b) Cleavage divisions (1–4) in birds produce an embryonic disc (the blastoderm) at one side of the yolky egg. (c) Cells of the blastoderm (1–3) sort into two separate layers, an upper epiblast and a lower hypoblast. (d) Some of the cells of the epiblast migrate toward, down into, and away from the primitive streak to become the mesoderm of the avian embryo.

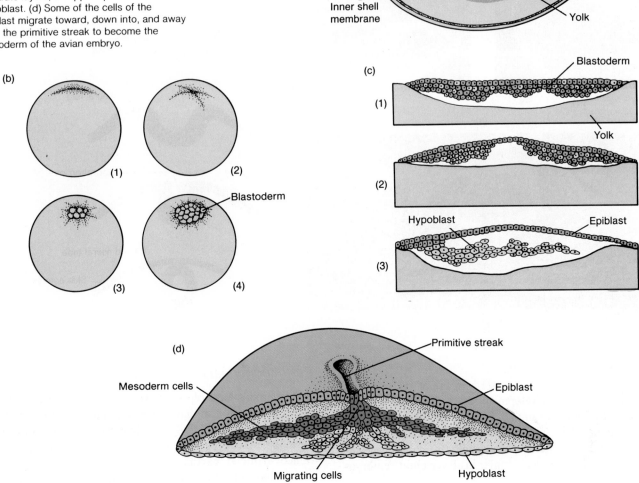

Development of Mammals

It is not surprising that reptiles, which also have large, yolky eggs, develop much as birds do. It is surprising, however, that early embryonic development in mammals also closely resembles the pattern seen in birds, despite the fact that the great majority of modern mammals produce eggs that contain very little yolk. The zygotes of mammals divide completely during cleavage, but the mammalian embryo still passes through a stage when it is a circular disc with relatively flat layers before it assumes a tubular form, just as the embryos of reptiles and birds do. This developmental similarity with reptiles and birds strongly suggests that mammals are evolutionary descendants of animals that had larger eggs containing much more yolk than do the eggs of present-day mammals.

We will learn more about development in mammals when we examine human development in chapter 27.

Figure 26.21 Nervous system development. (a) Scanning electron micrographs of frog embryos. In (1), the outlines of the neural plate can be seen as the edges of the plate begin to roll up as neural folds (magnification × 23). In (2), neural folds approach one another as neural tube formation is nearly completed (magnification × 30). (b) Scanning electron micrograph of a cross section of a chick embryo showing the shape of neural folds (magnification × 380). (c) A series of diagrammatic cross sections showing the process of neural tube formation in the chick embryo. Contraction of microfilaments strategically located in cells brings about shape changes in individual cells, and in entire sheets of cells, such as those involved in nervous system development.

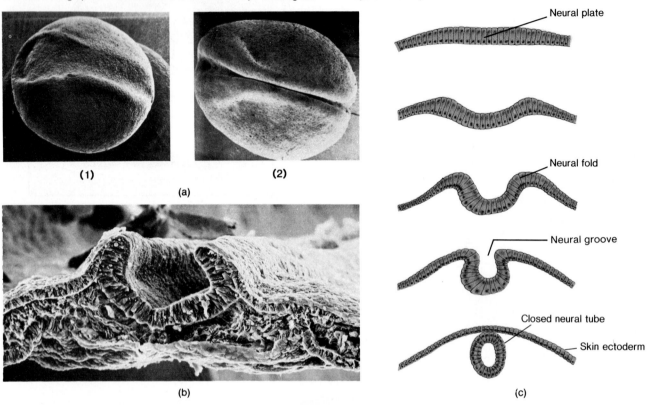

(1) (2)

(a)

(b)

(c)

Nervous System Development

Development of the nervous system is one of the most obvious aspects of early development in vertebrate embryos. It also provides a good example of the development of a vertebrate organ system. The morphogenetic processes involved in establishment of the nervous system are called **neurulation,** and neurulation is very similar in all vertebrate embryos (figure 26.21).

In neurulation, the ectoderm thickens along what will be the dorsal midline of the body to form a **neural plate.** Cells in the plate change shape due to contractions of microfilaments that are strategically located inside them. The collective action of these cell shape changes causes the neural plate to roll up, first into a pair of neural folds and then into a **neural tube.** The entire central nervous system (brain and spinal cord) of a vertebrate embryo develops from the neural tube.

Parthenogenesis

Some animals have reproductive processes that are very different from those we have seen thus far. One element that makes some animals' reproductive patterns exceptional is development by parthenogenesis. We have already mentioned artificial parthenogenesis, the experimental treatments that induce eggs to develop without fertilization. But parthenogenesis also occurs naturally in many animals, and it is a regular reproductive process in some of them.

For example, male honeybees develop by parthenogenesis. During mating on her nuptial flight, the queen bee receives a lifetime supply of sperm cells. She stores these sperm and controls fertilization of the eggs she lays. When she permits fertilization, the eggs that she lays develop into females, which become workers and a few young queens. When she

Figure 26.22 Parthenogenesis plays an important part in the life cycle of bees. Queens and workers (also females) develop from ordinary zygotes, but males develop parthenogenetically. Sperm are produced directly from haploid cells in testes.

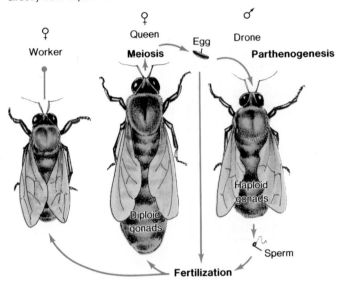

withholds sperm, thus preventing fertilization, the eggs develop into males (drones) (figure 26.22). These male bees retain the haploid chromosome number of the egg cell in all of their body cells, including the testis cells that eventually produce gametes. Therefore, meiosis is not required for sperm production in male honeybees. Haploid testis cells can be converted directly into sperm cells.

Parthenogenetic development, however, does not necessarily always produce a general haploid condition in all body cells, as it does in male honeybees. In some other parthenogenetically reproducing species, the second polar body is not formed, and two haploid sets of chromosomes are retained. In other animals, there is an initial chromosome duplication at the beginning of cleavage that is not followed by mitosis. Either of these means provides the new individual with a diploid chromosome number, but of course, it is a double set of maternal chromosomes.

Some organisms, such as the common, small aquatic animals known as rotifers (see p. 1007), produce eggs that develop directly by parthenogenesis at certain times of the year and then switch over to ordinary sexual reproduction with fertilization at other times of the year. Some gastrotrichs (tiny creatures similar to rotifers) seem to have no males at all and exist exclusively as populations of parthenogenetically reproducing females.

Sexual Reproduction and Plant Life Cycles

Plant life cycles are fundamentally different from most animal life cycles because, in plants, meiosis usually is not directly involved in gamete production. Meiosis in plants produces spores, and these haploid spores develop into haploid bodies (plants consisting entirely of haploid cells). Gametes are produced by direct differentiation of already-haploid cells within specialized areas of these bodies. Then, gametes fuse to produce diploid zygotes, which grow into diploid bodies (plants whose body cells are all diploid). The life cycle is completed when meiosis, occurring in specialized areas of these diploid bodies, produces haploid spores (see figure 26.23).

Since the haploid spores develop into multicellular haploid structures, plants have two body forms that alternate with one another. This is called **alternation of generations.** The diploid, spore-producing plant body is known as the **sporophyte generation,** and the haploid, gamete-producing plant body is known as the **gametophyte generation.**

While this general life history description can apply to all multicellular plants, there are many variations of the basic plan. In chapter 39, we will see how variations in reproductive strategies and details of life histories are related to the habitats and general biology of the various kinds of plants. In this section, we will focus only on the highly specialized flowering plants known as the **angiosperms.**

Flowers

The majority of angiosperms are terrestrial plants, and their reproductive strategies are tied to their habitat. The delicate, unicellular spores and gametes of angiosperms are enclosed and protected inside **flowers.**

A flower is a specialized shoot with a cluster of highly modified leaves, and it forms a protective envelope around the areas where cellular reproductive events take place. The modified leaves are arranged in concentric rings, or **whorls,** which are attached to a modified stem tip, the **receptacle.** The modified leaves in the lowest, outermost whorl are called **sepals,** and the next whorl inward consists of **petals.** Sepals may be colorful, but more frequently they are green and quite similar in appearance to ordinary foliage leaves. Petals often are large and colorful. Sepals and petals, although not directly involved in reproductive processes, attract insect or bird pollinators in those plant species that depend upon them for pollination.

The whorl inside the petals consists of **stamens,** and the innermost whorl is the **pistil.** Stamens are highly modified and not very leaflike in appearance. The pistil of most flowers consists of several very highly modified leaves that are fused into a single unit (figure 26.24).

Figure 26.23 Generalized plant life cycle (illustrated here by a fern life cycle) differs from the most common animal life cycles because most plants form a multicellular haploid structure. Meiosis is not directly involved in gamete formation.

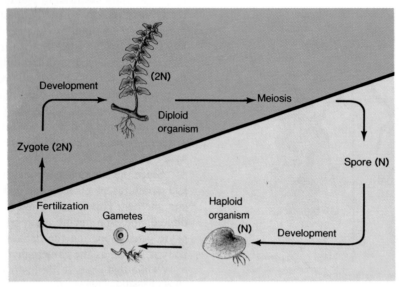

Figure 26.24 Structure of a flower.

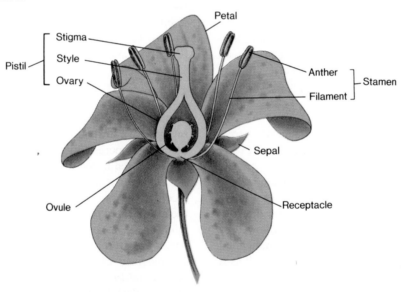

Many, but certainly not all, flowers contain both stamens and pistils (figure 26.24). In some plants, separate flowers with stamens but no pistils (**staminate flowers**) and flowers with pistils but no stamens (**pistillate flowers**) are borne on the same plant body. In other plants, staminate and pistillate flowers are borne on entirely separate plants.

Spores and Gametophytes

Flowering plants produce two different kinds of spores. **Megaspores** develop into female gametophytes within the pistil. **Microspores** become modified into **pollen grains.** Pollen grains are very small male gametophytes that are released and carried by the wind or by animal pollinators.

Figure 26.25 Development of female and male gametophytes, pollination, and double fertilization in an angiosperm.

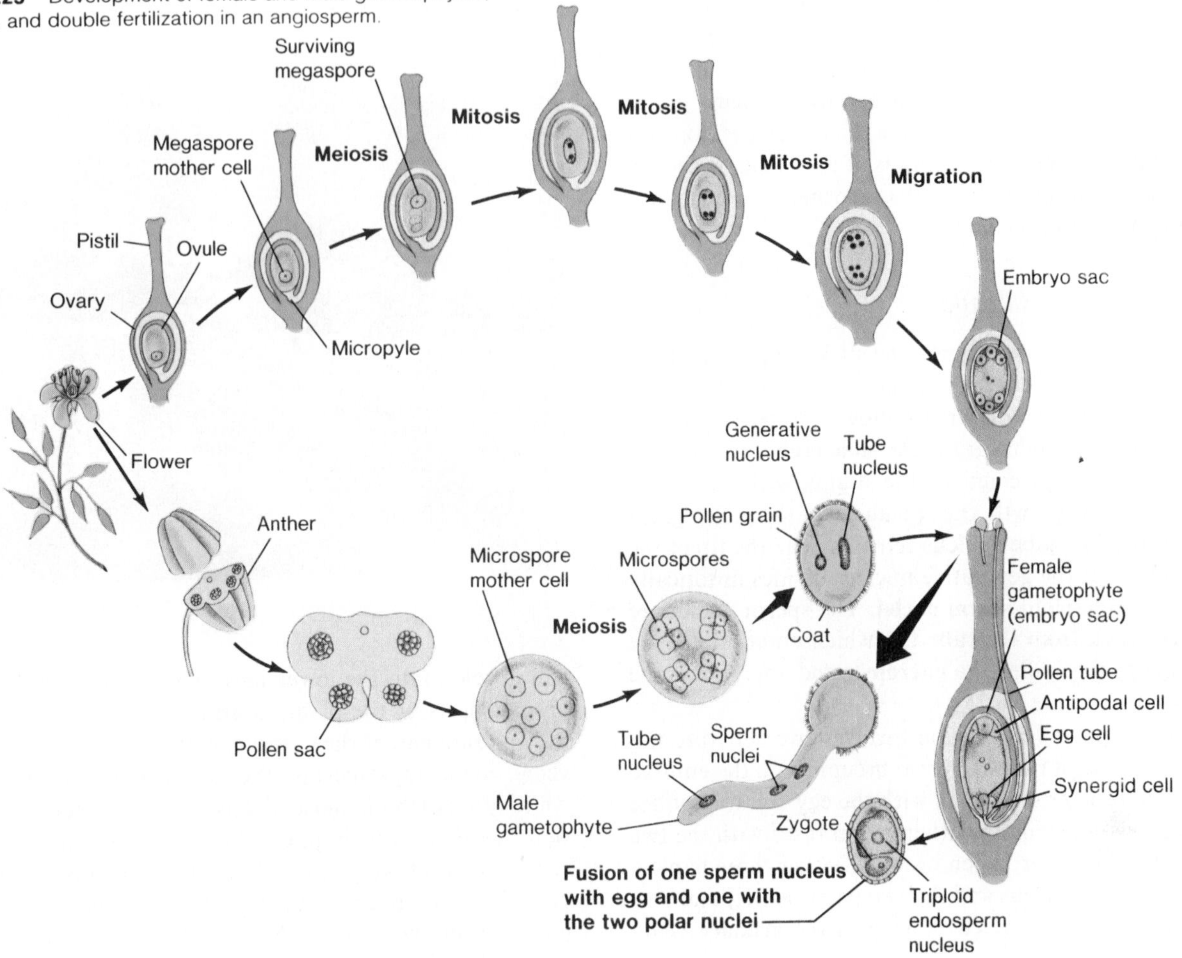

The enlarged base of the pistil, the **ovary,** contains from one to many **ovules** depending upon the plant species. Each ovule is the site of production of a functional megaspore. Megaspore production begins with a **megaspore mother cell** that divides meiotically to produce four haploid megaspores. In most flowering plants, only one megaspore is involved in producing the female gametophyte structure; the other three degenerate without developing further (figure 26.25).

Female gametophyte (**megagametophyte**) development also takes place inside the ovule where the megaspore was produced (figure 26.25). The functional megaspore cell expands, and three successive mitotic divisions produce eight nuclei in this one expanded cell. Three nuclei are clustered at each end of the cell, and the other two, the **polar nuclei,** are located in the middle of the cell. Cell walls develop around the nuclei at the ends to form two clusters of three small cells. The polar nuclei are left in the middle in a larger, seventh cell. This seven-celled structure is called the **embryo sac,** and

it is the fully developed female gametophyte. Thus, the haploid generation develops completely while still enclosed inside the ovule. Clearly, the haploid generation is a very reduced and inconspicuous part of the flowering plant's life history.

One of the cells at one end of the embryo sac becomes the functional egg cell. The other two cells at the egg-cell end (**synergid cells**) and the three cells at the opposite end (**antipodal cells**) all degenerate without playing any obvious role in reproduction. The large cell in the middle of the embryo sac with its two polar nuclei is involved in the formation of an important nutrient storage structure in the seed called the **endosperm.**

Microspores are produced in pollen sacs in the **anther,** which sits atop the **filament** at the tip of the stamen (figure 26.25; see also figure 26.24). Specialized **microspore mother cells** enter meiosis, and four microspores are produced per microspore mother cell. External ornamentation develops on the cell wall as the microspore becomes a functional pollen grain (figure 26.26). Internally, the haploid nucleus of the developing pollen grain divides mitotically to produce two nuclei, the **tube nucleus** and the **generative nucleus.** Pollen grains are released in this condition, and they develop further only if **pollination** occurs; that is, if they land on the appropriate part of the pistil. There they germinate and continue male gametophyte development.

Pollination and Fertilization

Pollen grains land on the **stigma,** the sticky upper tip of the pistil. Contact with the stigma induces pollen grains to germinate and begin a characteristic growth response. A long outgrowth from the pollen grain, the **pollen tube,** grows down through the **style** that connects the stigma with the ovary. During pollen tube growth, the two nuclei of the pollen grain enter the tube. The tube nucleus remains near the tip of the growing tube, but the generative nucleus divides mitotically to produce two haploid **sperm nuclei.** The sperm nuclei are located just back from the tube tip, which enters the ovule through an opening called the **micropyle** and approaches the embryo sac (see figure 26.25).

Flowering plant fertilization involves two separate nuclear fusions. One of the two sperm brought into the embryo sac by pollen tube growth fuses with the egg cell to produce the diploid zygote, while the other sperm fuses with the two polar nuclei. This latter fusion brings together three haploid nuclei (the sperm nucleus and the two polar nuclei) and produces a triploid (3N) nucleus known as the **primary endosperm nucleus.**

This primary endosperm nucleus will eventually divide repeatedly to establish a multicellular endosperm, which stores nutrients in the developing seed. The extent of endosperm growth and the endosperm's relative importance for food storage differ among angiosperms, but in many, the endosperm is the main storage site in the seed.

Early Zygote Development

Zygote development begins with a transverse cell division that produces two cells, a terminal cell and a larger basal cell near the micropyle. Several sequential transverse cell divisions follow an initial division of the basal cell. These divisions produce a linear chain of cells, the **suspensor,** which is an embryo attachment structure (figure 26.27).

Terminal cell divisions occur in several planes to produce the **embryo.** As shown in figure 26.27, the embryo body consists of several major areas. A **hypocotyl,** which eventually

Figure 26.26 Scanning electron micrograph of pollen grains from the prickly pear cactus (*Opuntia compressa*) (magnification × 338).

will develop into the lower part of the stem and the underground parts of the plant, is attached to the suspensor. At the opposite end of the embryo, **cotyledons** (seed leaves) develop. Some angiosperms (the dicots) have two cotyledons, while others (the monocots) have only one. As the description "seed leaves" implies, the cotyledons are temporary structures that will be lost later. In some angiosperms, especially those species having a large endosperm, the cotyledons are relatively small. But in others, the cotyledons may be very large and serve as the major nutrient storage sites in the seed. You have probably seen the large, fleshy cotyledons of bean seeds when beans split open while being soaked or cooked. Near the base of the cotyledons is the **shoot apex.** Although the shoot apex may seem a fairly inconspicuous part of the embryo, it is the source of most of the permanent aboveground parts of the plant.

Meristematic regions (centers of continuing cell division) are located in both the shoot apex and the **root apex** at the tip of the **radicle,** the portion of the hypocotyl that will form the root.

Seeds

The embryo, packed in with the endosperm, is surrounded by ovule tissue. Parts of the ovule harden to form tough, protective **seed coats.** Most seeds then dry out until their water content falls to very low levels (5 to 20 percent), and they maintain only a minimal maintenance level of metabolism. In many flowering plants, the remainder of the ovary develops into a **fruit** around the seed.

Figure 26.27 Stages in the development of a dicot embryo and seed.

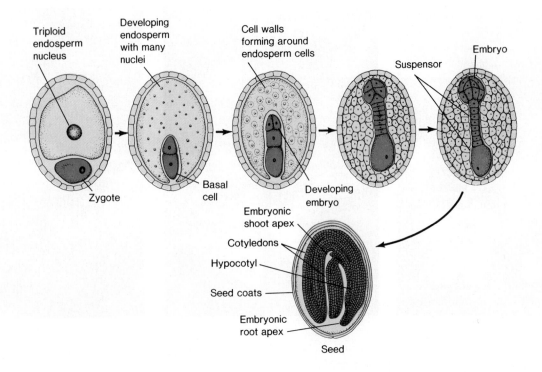

Fruits

Tissue develops around maturing seeds in diverse ways to produce various kinds of fruits (see table 26.2). The fruit tissue surrounding seeds that is derived from the ovary is called the **pericarp.** In addition to the pericarp, some fruits also contain accessory tissue derived from other flower parts or from the receptacle to which the flower is attached (figure 26.28). While there are a number of types of **simple fruits** (fruits that develop from a single ovary), other fruits are more complex. **Aggregate fruits,** such as raspberries, are produced from flowers that have several ovaries, and **multiple fruits,** such as pineapples, are derived from a number of flowers crowded together.

Germination and Seedling Growth

Seeds are dispersed in many ways. Some are blown by the wind or carried by running water; others are carried by animals. Still others are dispersed when animals eat fruits; fruit tissues are digested but seeds usually pass through the digestive tract intact. Because most plants produce many seeds, chances are good that at least some of them will reach locations favorable for seedling growth.

Embryos within seeds develop further only after seed **germination.** Germination occurs when a dry seed, in a favorable site, takes up water and swells. Then, metabolic activities of the embryo increase and rapid growth resumes. The seedling grows out of the seed and develops into a mature, diploid plant body (sporophyte).

Some kinds of seeds do not respond to conditions that generally induce seed germination because they are **dormant;** that is, they are not physiologically responsive, even when subjected to conditions that are fully appropriate for germination. Dormant seeds require exposure to a specific environmental stimulus (in many cases, a period of chilling) before they mature to the point where they can germinate.

When a seed germinates, active cell division resumes in the meristematic regions of the root apex and shoot apex. Cells produced in these regions are responsible for **primary growth,** the lengthening of the root and shoot. Later, other meristematic tissue produces **secondary growth,** growth in girth and lateral branching (see chapters 9 and 10). Continued growth and differentiation eventually produce a mature sporophyte plant (figure 26.29).

Figure 26.28 Fruit development. (a) Pear buds and flowers. (b) Early stage of fruit development. (c) Ripe pears.

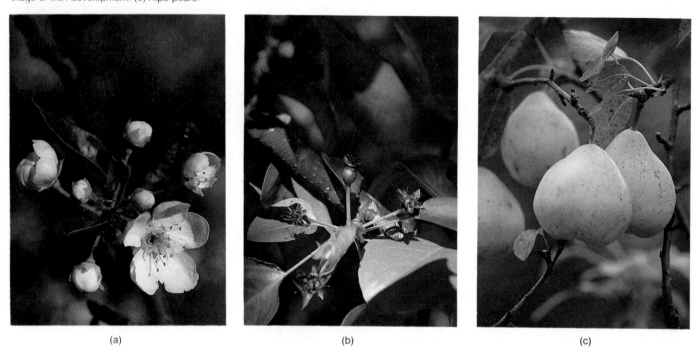

(a)　　　　　　　　　(b)　　　　　　　　　(c)

Figure 26.29 Early development of a bean seedling.

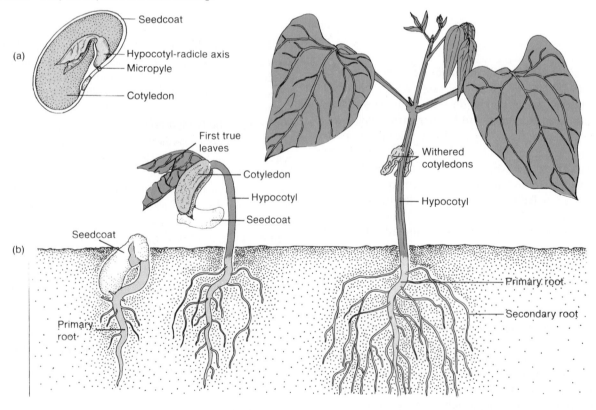

Table 26.2
Some Types of Fruits

Kind of Fruit	Description	Examples
Simple Fruits	Fruits that develop from a flower with a single pistil	
Fleshy Fruits		
Drupe	Outer part of the pericarp is fleshy; inner part is hardened to form a stony pit	peach, plum, cherry, apricot, olive
Berry	Entire pericarp is fleshy	tomato, grape, date, avocado
Hesperidium	A berry with a tough, leathery skin and numerous swollen, juice-containing growths derived from the ovary wall lining	lemon, lime, orange, grapefruit, tangerine
False berry	Contains accessory tissue in addition to the pericarp; entire wall is fleshy	gooseberry, blueberry, banana, cranberry
Pome	Contains accessory tissue in addition to pericarp; inner part of the pericarp leathery or papery	apple, pear
Dry Fruits		
Follicle	Dry fruit that splits open along one seam to release seeds	milkweed, peony, larkspur
Legume	Dry fruit that splits open along two seams	pea, bean, lentil, carob, mesquite, peanut
Achene	Dry fruit that does not split open at maturity; single seed attached to the pericarp only at its base	sunflower
Nut*	Dry fruit that does not split open at maturity; one-seeded fruits with thickened woody or stony walls	hazelnut, hickory nut, chestnut
Grain (Caryopsis)	Dry fruit that does not split open at maturity; single seed is fused to the pericarp and cannot be separated from it	wheat, rice, oats, barley, corn
Samara	Dry fruit that does not split open at maturity; pericarp extends out as a wing that aids in dispersal	ash, elm, maple (samaras in pairs in maples)
Aggregate Fruits	Derived from several to many pistils of a single flower; some have accessory tissue as well	raspberry, blackberry, strawberry
Multiple Fruits	Derived from several to many clustered flowers that grow together as fruits develop	pineapple, fig, mulberry

*Many fruits that commonly are called "nuts" are not nuts in the botanical sense. For example, peanuts are legumes, while walnuts, pecans, and cashews are drupes in which the fleshy part of the pericarp eventually dries and withers.

Summary

All organisms have the capacity to reproduce. Many organisms can reproduce asexually, but the majority of multicellular organisms reproduce sexually with specialized gametes fusing in fertilization to produce zygotes.

Most organisms reproduce seasonally. Their periods of reproductive activity are timed to occur when reproductive success is most likely.

Gametes are highly specialized cells that are well adapted for their reproductive functions. During animal gametogenesis, meiosis reduces the chromosome complement to the haploid number characteristic of gametes. Fertilization restores the diploid number in the zygote.

Sperm cells are specialized for motility. As they develop, sperm lose most of their cytoplasm, become streamlined in shape, and develop motile tails. They also develop an acrosome, which functions in fertilization. In oogenesis, unequal cytoplasmic divisions during meiosis concentrate cytoplasmic materials in the future egg cell. These materials are used during the early development of the zygote.

Animal fertilization is a complex set of interactions between egg and sperm, including membrane fusion that permits sperm entry into the egg. In most species, the egg becomes unresponsive to contact with additional sperm once it has been contacted by one sperm.

Fertilization causes metabolic activation of the egg, which initiates synthesis of nucleic acids and proteins required during cleavage and subsequent development. However, the earliest fertilization responses of the egg seem to be ionic changes, which precede changes in synthesis activity.

Relatively large, single-celled animal zygotes are converted into multicellular embyros by cleavage, a series of mitotic divisions without intervening growth. These cells become organized according to a basic body plan during gastrulation. Finally, cells become structurally and functionally differentiated to perform diverse, specialized functions in the body.

Some animals, such as honeybees, reproduce by parthenogenesis as well as by sexual reproduction with gamete fusion.

Plant life cycles include two separate generations—a haploid gametophyte generation and a diploid sporophyte generation. In flowering plants, gametophytes are several-celled structures, but they are reduced and inconspicuous. Gametes are produced by differentiation of already-haploid cells of the gametophytes.

Flowers are clusters of highly modified leaves that protect plant reproductive structures. Meiosis occurs during the development of spores. Megaspores develop into embryo sacs (female gametophytes), and microspores develop into pollen grains. Pollen grains complete their development into male gametophytes if pollination occurs. A pollen tube carries two sperm nuclei to the embryo sac where double fertilization occurs, producing a zygote and a primary endosperm cell.

Mitotic divisions of the zygote produce an embryo that is enclosed, along with the storage tissue of the endosperm, inside of seed coats. Fruits that enclose seeds develop from ovary tissue or from a combination of ovary and accessory tissues. Following seed maturation, embryo development pauses until the seed germinates. At germination, growth of the embryo resumes, and the growing seedling continues development as a sporophyte plant.

Questions for Review

1. Define the term clone.
2. Compare and contrast the cell divisions of meiosis in spermatogenesis with those of oogenesis.
3. What is polyspermy? How is polyspermy normally prevented?
4. Fertilization is important genetically because it brings together haploid paternal and maternal chromosome sets, but fertilization interactions begin long before fusion of the pronuclei. List some of these "early" interactions between egg and sperm in animals. Discuss egg activation.
5. How is the pattern of division without intervening growth related to the basic function of cleavage in animals?

6. Name and describe the development of the three germ layers in an animal embryo.
7. Define parthenogenesis.
8. How is spermatogenesis in male bees different from the typical situation in animals?
9. What do we mean by alternation of generations in plant life cycles?
10. Contrast animal and plant life cycles generally with reference to relationships of meiosis and gamete formation.
11. What are the two separate nuclear fusions that occur in fertilization in flowering plants?

Questions for Analysis and Discussion

1. Some biologists say that, for a species, reproduction is the ultimate homeostatic mechanism. Can you explain this point of view?
2. Suggest some advantages of sexual and asexual reproduction in terms of reproductive success of individual organisms and in terms of evolution.
3. What evidence can you give to counter the mistaken notion that fertilization depends upon a sperm burrowing actively into a passively waiting egg cell?

Suggested Readings

Books

Bewley, J. D., and Black, M. 1985. *Seeds.* New York: Plenum.

Cohen, J. 1977. *Reproduction.* London: Butterworths.

Johnson, L. G., and Volpe, E. P. 1973. *Patterns and experiments in developmental biology.* Dubuque, Iowa: Wm. C. Brown Publishers.

Meeuse, B., and Morris, S. 1984. *The sex life of flowers.* London: Faber and Faber.

Raven, P. H.; Evert, R. F.; and Eichhorn, S. 1986. *Biology of plants.* 4th ed. New York: Worth Publishers.

Saunders, J. W., Jr. 1982. *Developmental biology.* New York: Macmillan.

Stern, K. R. 1985. *Introductory plant biology.* 3d ed. Dubuque, Iowa: Wm. C. Brown Publishers.

Articles

Baskin, J. M., and Baskin, C. C. 1985. The annual dormancy cycle in buried weed seeds: A continuum. *BioScience* 35:492.

Epel, D. 1980. Fertilisation. *Endeavour* 4:26.

Gustafson, S. S. June 1985. The Methuselah bush. *Science 85.*

Meeuse, B. J. D. 1984. Pollination. *Carolina Biology Readers* no. 133. Burlington, N.C.: Carolina Biological Supply Co.

Schatten, G., and Schatten, H. September/October 1983. The energetic egg. *The Sciences.*

Stebbins, G. L. May/June 1984. The flowering of sex. *The Sciences.*

Tyler, M. H., and Carter, D. B. 1981. Oral birth of the young of the gastric brooding frog *Rheobatrachus silus. Animal Behaviour* 29:280.

Vasek, F. C. 1980. Creosote bush: Long-lived clones in the Mojave Desert. *American Journal of Botany* 67:246.

Human Reproduction and Development

27

Chapter Concepts

1. Human reproductive systems are specialized so that delicate reproductive cells and small, fragile, developing individuals are enclosed and protected within sheltered internal body environments.
2. The pituitary, the ovaries, and the uterus are all involved in the complex, monthly reproductive cycle of the human female.
3. Intricate hormonal interactions control reproductive processes and pregnancy responses.
4. Embryonic development establishes basic body organization, and growth and maturation proceed during fetal development.
5. The placental relationship sustains the infant throughout its development in the uterus.
6. After birth, a human infant is physiologically independent but still weak and helpless.
7. Fertility regulation and the increasing incidence of sexually transmitted diseases are worldwide social problems associated with human sexuality and reproduction.

Using ultrasound equipment, it is possible to visualize a human fetus inside its mother's uterus so clearly that an eye-blink can be detected. This ultrasound imaging, as it is called, allows quite close inspection of a fetus without harming or disturbing it. Ultrasound imaging and other techniques such as amniocentesis (p. 657) now allow us to learn a great deal about the developing human individual before its birth. Despite extensive research in the past and recent technical achievements, however, much remains to be learned about the development of a human infant. Most of us find this topic fascinating because each of us has come into being through these very developmental processes.

Furthermore, the study of reproductive processes goes far beyond scientific and medical questions into important personal and social issues. Human sexuality is much more than just a reproductive capacity; it is a powerful psychological and social force in human life. In this chapter, we will examine both the functional aspects of human reproduction and some related, socially important issues.

The basic design of the human reproductive (genital) system is a result of adaptation to life in a terrestrial environment. It is specialized so that delicate reproductive cells and small, fragile, developing individuals are enclosed and protected within moist internal body environments. For internal fertilization and development, sperm must be delivered to the inside of the female body. This internal delivery is achieved when the male **penis** deposits fluid containing sperm cells in the moist internal environment of the female **vagina.** Should fertilization occur, the developing zygote is maintained inside the female reproductive tract, and development proceeds inside the **uterus,** an organ highly specialized for maintenance of the developing embryo. The embryo becomes embedded in the uterine wall by a process known as **implantation,** and a **placenta** then develops in the implantation site. The placenta is a composite organ, made up of both maternal and embryonic tissues. It functions in the exchange of materials between the circulatory systems of mother and child during embryonic development and the long **fetal period** of growth and maturation until the time of birth.

The Male Reproductive System

The male gonads are paired **testes** that are suspended in a saclike structure, the **scrotum.** Testes have a dual function; in addition to producing sperm, they also synthesize and release male sex hormones.

Sperm production occurs in the **seminiferous tubules** of the testes. Each testis contains about one thousand of these small, highly coiled tubules. The total length of the seminiferous tubules in an average human testis is estimated to be as great as 250 meters. Spermatogenesis begins in the outer part of the seminiferous tubule, and as the process proceeds, the developing cells move from the periphery toward the center of the tubule. Mature sperm cells become detached and lie in the central cavity (**lumen**) of the tubule (figure 27.1).

Male sex hormones (chiefly, testosterone) are produced by the **interstitial cells** scattered in the spaces among the seminiferous tubules (figure 27.2). Testosterone is responsible for development of **secondary sex characteristics,** such as the general male body form and muscle development, male hair distribution, and voice deepening in maturing males. If the testes are surgically removed (an operation called castration) before the time of sexual maturation (**puberty**), male secondary sex characteristics do not develop.

Two hormones from the anterior lobe of the pituitary gland regulate testis functions. **Follicle stimulating hormone (FSH)** controls spermatogenesis, while **luteinizing hormone (LH)** stimulates testosterone production by interstitial cells. These hormones, along with several other pituitary hormones, are classified as tropic hormones (see chapter 15) because they stimulate secretion of other glands. Because they are tropic hormones associated with the gonads, they are called **gonadotropic hormones** or **gonadotropins.**

As with the secretion of other anterior pituitary hormones, gonadotropin secretion is controlled by releasing factors produced in the hypothalamus. Maturation of this control system is involved in the activation of reproductive function at the time of puberty. In adult human males, this hypothalamus–pituitary link is part of a feedback relationship that controls and balances the level of testosterone in the blood (figure 27.3).

Sperm Transport and Seminal Fluid

Sperm are moved along the seminiferous tubules, through small collecting ducts (the **vasa efferentia**), to the **epididymis,** a long, coiled tube lying on the surface of the testis. Sperm are stored there until they are propelled through the reproductive tract by peristaltic waves of muscular contractions that move through the system. These contractions are part of **ejaculation,** a reflex response to sexual intercourse or other sexual stimulation. The contractions move sperm out of the epididymis and on through a tubular transport tract beginning with the **vas deferens** (figures 27.1 and 27.4). The vas deferens runs out of the scrotum, passes into the body cavity,

Figure 27.1 Human testis organization and sperm production. (a) Schematic diagram of testis organization. Packed-in, coiled seminiferous tubules produce sperm, which are transported to the epididymis and through the vas deferens. (b) Diagrammatic cross section showing how sperm production proceeds as cells move from the periphery toward the lumen in the center of a tubule. Interstitial cells produce male sex hormones. (c) Stages of spermatogenesis in the small area outlined in (b). Mature sperm become detached in the lumen and are moved away toward the epididymis. Sertoli cells function in support and maintenance of developing sperm.

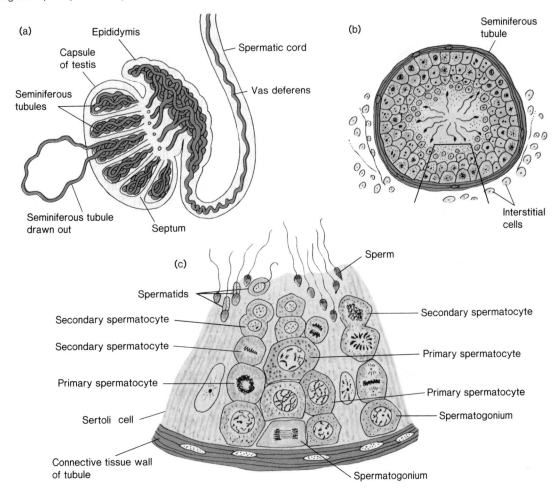

loops across the surface of the bladder, and then turns down toward the **prostate gland,** which surrounds the urethra just below the point where the urethra drains the urinary bladder. At the lower end of each vas deferens, a glandular **seminal vesicle** adds an alkaline mucous secretion to the sperm. The seminal vesicles have been mistakenly identified as sperm storage organs, but their function actually is secretory. Sperm, with this added secretion, then move into the urethra at a junction contained within the body of the prostate gland. At this point, the thinner, milky-colored secretion of the prostate is added to the sperm. From there on, the excretory and reproductive systems share a common duct, the urethra. Sperm and fluid move from the prostate area through the urethra toward the base of the penis. There, a third pair of

Figure 27.2 The structure of the steroid hormone testosterone. Carbons and hydrogens of the basic steroid molecular skeleton are not shown. Testosterone is the main male sex hormone (androgen), and it causes development of male secondary sex characteristics.

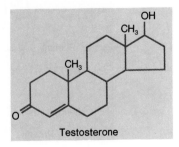

Testosterone

Figure 27.16 Embryonic development. (a) Drawing of an embryo near the end of the third week of development. Ectodermal folds (neural folds) are closing to produce the brain and spinal cord. Somites are also forming. (b) A human embryo at 40 days of development. Note the paddlelike limbs.

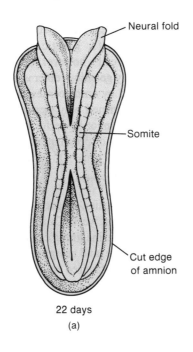

Neural fold

Somite

Cut edge of amnion

22 days

(a)

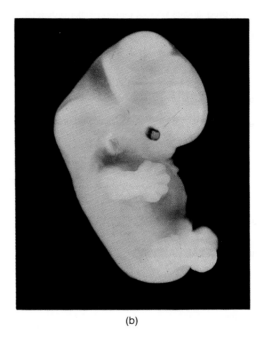

(b)

the embryo itself. Only later, during the second month, does blood cell formation begin inside the body, when the liver becomes the second temporary source of blood cells. Finally, the bone marrow matures and takes over as the third, and permanent, source of blood cells.

At the end of the first month, the embryo is still only about 5 or 6 mm in length, and the diameter of the chorion enclosing the whole developing system is only about 3 cm.

The Second Month

During the second month, rudiments of all the rest of the major body organs are produced in the embryo, which assumes an increasingly human appearance. The relatively simple digestive tube lengthens, and liver, gallbladder, and pancreas form as outgrowths from the embryonic gut. In addition, lung rudiments begin their development as pouches growing out near the anterior end of the gut. Paddlelike **limb buds** grow out from the sides of the body and gradually transform in shape into recognizable arms and legs (see figure 27.16b). Eventually, fingers and toes are produced, and the complex skeleton and musculature of the limbs develop. In the head, eye and ear development proceeds, and the face

differentiates. The embryo comes to look like a human being and can easily be distinguished from embryos of other vertebrates, which it earlier resembled very closely (figure 27.17).

Excretory and reproductive system elements develop extensively during the second month. At first, however, male and female embryos are indistinguishable since the developing reproductive system contains rudiments of both female and male structures. This "indifferent condition" of reproductive system development persists until about the seventh week of development, when signs of sex determination begin to appear.

The establishment of human appearance externally and all of the major organ rudiments internally marks the end of the embryonic period of development and the beginning of the fetal period. This is an important milestone in the development of the individual, and it has broader implications for the pregnancy, as well. The fetus generally is much less sensitive than the embryo to harmful external influences. Embryonic development includes a series of **sensitivity periods,** during which delicately balanced developmental processes involved in organ formation are subject to disruption by chemical agents or disease. Harmful agents still can have damaging effects on the fetus, but they are much less likely to cause gross developmental disturbances.

Figure 27.17 A fifty-six-day human fetus. From the end of the eighth week onward, it is proper to call the developing individual a fetus because rudiments of all major organs and systems are present. The fetus is clearly human in general body form.

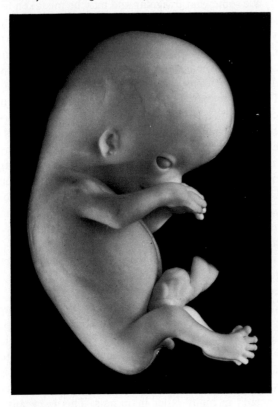

The Third Month

The third month of pregnancy is a relatively quiet period of development because the mother does not experience new symptoms, and fetal development involves mainly growth of structures already present. One notable feature, though, is that reproductive system development proceeds to the point that the sex of the fetus is externally apparent. Overall fetal growth is considerable because the fetus grows in length (crown of head to rump) from 2.5 cm at the beginning of the month to about 6 cm at the end of the month. Its weight increases from less than 2 gm to about 12 gm. While the body and limbs may move slightly, the fetus is so small that the movements are undetectable to the mother.

Four to Six Months (the Second Trimester)

During the second third of pregnancy (the second **trimester**), rapid fetal growth continues. For the first time, the mother can feel fetal movements. This "quickening" when the mother first feels movements within the uterus usually occurs late in the fourth month or during the fifth month. **Ossification** (calcified bone development) of the previously cartilaginous skeleton proceeds more rapidly during the second trimester than in the first, and general body form becomes increasingly human. Hair develops on the head, and a very fine hair, the **lanugo,** forms a downy covering over the body. Waxy, almost cheeselike, secretions cover the skin surface and apparently protect it during its long immersion in the watery amniotic fluid. The fetal brain develops impressively during the second trimester, and the characteristic ridges and furrows appear in the surface of the cerebral hemispheres. Sense organs become functional, and the fetus shows reflexes that clearly indicate it is becoming responsive to changes in its environment. The fetal heartbeat can be detected fairly readily with a stethoscope because of its distinctive sound and relatively rapid rate (120 to 150 beats per minute).

Especially during the sixth month, the fetus gains weight rapidly, and by the end of the second trimester, the average fetus weighs about 630 gm. Still, this is less than 20 percent of normal average birth weight. Some babies born prematurely at the end of the second trimester survive, but they invariably require intensive care under strictly controlled conditions. Infants born that early very frequently suffer from respiratory distress due to lung immaturity, and they must be carefully protected from environmental stresses such as temperature changes.

The Final Trimester

Fetal growth continues during the last trimester, with an average weight gain of about 25 gm per day. The mother's uterus continues to expand to contain the growing fetus and its membranes. At the end of gestation, the uterus weighs more than twenty times what it did before pregnancy began. The uterus also continues to move higher in the abdominal cavity, as it has done through much of pregnancy, until a couple of weeks before birth, when it moves downward toward the pelvic area again.

Nutritional research indicates that the third trimester is a critical period for brain development and that normal nervous system development is very dependent upon the mother's nutritional status. Protein deficiencies in the mother's diet, for example, can have adverse effects on nervous system development that may limit the infant's future mental capacities.

Within the ovaries of female fetuses, primary oocytes enter prophase of their first meiotic division and then pause until puberty. Testes of male fetuses begin to move during the seventh month, and they descend out of the abdominal cavity to their final location in the scrotum. This descent is necessary for eventual normal reproductive functioning that begins at puberty. Testes that remain in the abdominal area, a condition known as **cryptorchidism** (literally, "hidden testes"), are too warm to produce normal, functional sperm cells. Normal spermatogenesis occurs only at the slightly lower temperatures of the scrotum.

Other systems progress toward their full-term condition. Especially because of rapid lung maturation, the survival chances of prematurely born infants increase dramatically during the final trimester.

Maternal antibodies cross the placental barrier and enter the fetal circulation so that the newborn infant (**neonate**) carries immunities to bacteria, viruses, and other foreign materials. This passive immunity, "borrowed" from its mother, serves the infant only temporarily. Its own immune system begins to produce antibodies shortly after birth.

Birth

Human birth (**parturition**) occurs after an extended period of rhythmic uterine muscular contractions known as **labor.** Birth normally takes place about 280 days after the beginning of the last regular menstrual period (about 266 days after fertilization).

What determines the exact time of birth is obscure, but several factors, including changes in the condition of the uterus, may be involved. There may be a "readiness hormone"; that is, the fetus may release some chemical factor (possibly a pheromone?) that signals the mother when it has reached an appropriate stage of maturation.

Before birth, hormone levels in the mother's bloodstream change. Progesterone production decreases during the last two months of pregnancy. (As you recall, progesterone's role has been to keep the uterine lining intact.) Estrogen production increases gradually throughout the same period and then shows a sharp increase just before birth. (Estrogens increase the irritability and contractility of the uterine muscles.) The most direct hormonal stimulus, however, probably comes from **oxytocin,** an octapeptide hormone from the pituitary's posterior lobe that directly stimulates contractions of uterine muscles. The cause of the increase in oxytocin level and whether the increase is absolutely essential for normal birth are not known. But oxytocin's effect on uterine muscles is reliable enough that the hormone is used routinely to induce labor artificially when the birth process must be hurried along.

Labor is a progressive three-stage process that leads to the delivery of the fetus and the **afterbirth** (the placenta and the fetal membranes). Uterine contractions during early labor usually are somewhat irregular and may be mistaken for the gas pains or other intestinal discomforts typical of late pregnancy. However, when the contractions become regular and occur at intervals of twenty minutes or less, they signal the onset of labor, although they may still stop and resume some time later.

The first stage of labor involves important changes in the cervical portion of the uterus. A mucous plug that has blocked the cervical opening during pregnancy comes loose and is shed as the cervix begins to change shape. The cervix shortens and flattens so that it does not protrude as far into the vagina. The tiny cervical opening expands to a diameter of about 10 cm. Usually, the fetal membranes rupture early in labor, allowing the amniotic fluid to flow out through the vagina. By the end of the first stage of labor, uterine contractions have become much more forceful and frequent.

In the second stage of labor, expansion of the cervical opening is completed, and the infant is delivered through the opened uterine canal and the vagina (figure 27.18).

Continued uterine contractions during the third stage of labor dislodge the placenta, which normally is expelled through the birth canal about twenty minutes after delivery of the infant. The placenta is pulled loose from the uterine wall as a unit. The separation of the human placenta is accompanied by considerably more bleeding than is delivery of the placenta in many other mammals. This is because the bonds between fetal and maternal tissue in the placenta are looser in other mammals.

Newborn infants (neonates) must make rapid respiratory and circulatory adjustments during the transition from complete dependence on the placenta to independent life outside the uterus. Circulation to the placenta normally shuts down after delivery. This deprives the neonate of placental gas exchange and leads to a falling oxygen concentration and a rising CO_2 concentration in its blood. Brain respiratory centers respond to this increasing level of CO_2 by sending impulses to chest and abdominal muscles, thereby stimulating the contractions needed to initiate breathing if the infant has not already gasped or begun to cry in response to pressure changes or other stimuli experienced during and after delivery. Thus, the neonate makes the transition to independent gas exchange.

But still, a newborn human infant is helpless and immature in comparison to other neonatal mammals (figure 27.19). Baby pigs and calves, for example, stand up, walk around, and attempt to feed within minutes following birth. Human neonates are weak and only poorly able to regulate their body temperatures; they are completely dependent upon parental care.

Figure 27.18 Some stages of the birth process. (a) Side view of a fetus in the uterus just prior to the beginning of labor. (b) During the first stage of labor, the cervical opening of the uterus enlarges. Usually the fetal membranes rupture and the amniotic fluid flows out through the vagina. (c) During the second stage of labor, the infant moves into the birth canal. Later in the second stage, the head emerges. (d) In the third stage of labor, the placenta is dislodged and expelled about twenty minutes after delivery of the infant.

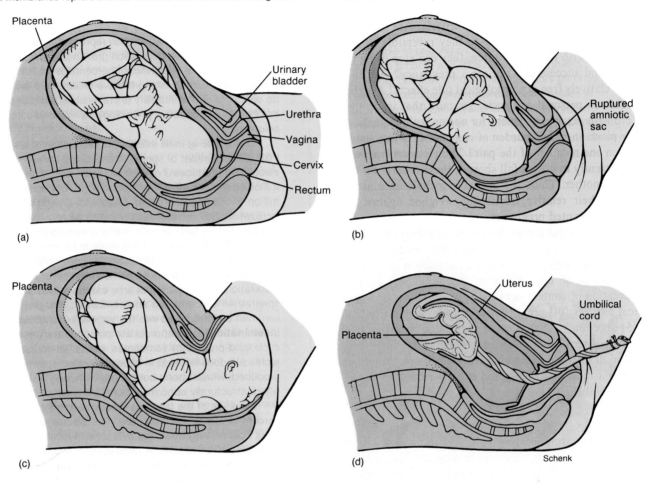

Placenta

Urinary bladder
Urethra
Vagina
Cervix
Rectum

(a)

Ruptured amniotic sac

(b)

Placenta

(c)

Uterus
Umbilical cord
Placenta

(d)

Schenk

Figure 27.19 A newborn human infant (neonate) just after delivery.

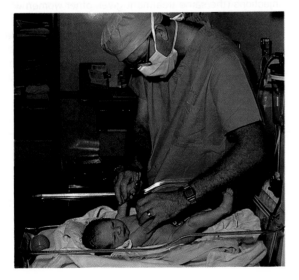

Regulating Human Fertility

While population growth rates are falling toward zero in the United States and several other industrialized countries, this is not the case in many underdeveloped and developing countries. Population growth either causes or aggravates practically all international social and political problems today. Effective means of dealing with population growth problems depend on development and distribution of safe and biologically effective birth control techniques.

As well as being an international social problem, regulation of fertility is a pressing individual concern. Family size can be an urgent personal and financial dilemma, and effective family planning depends upon adequate birth control technology and information.

Figure 28.3 Diagrammatic representation of the principle of differential gene expression in two different hypothetical cell types. "Housekeeping" activities are common to all cell types. "Specific proteins" are used for specific functions of specialized cell types. Different parts of the genome are expressed in different types of cells.

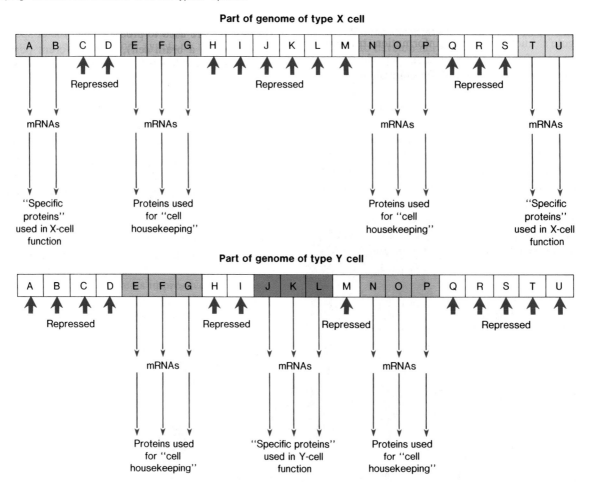

Part of genome of type X cell

| A | B | C | D | E | F | G | H | I | J | K | L | M | N | O | P | Q | R | S | T | U |

Repressed   Repressed   Repressed

mRNAs   mRNAs   mRNAs   mRNAs

"Specific proteins" used in X-cell function

Proteins used for "cell housekeeping"

Proteins used for "cell housekeeping"

"Specific proteins" used in X-cell function

Part of genome of type Y cell

| A | B | C | D | E | F | G | H | I | J | K | L | M | N | O | P | Q | R | S | T | U |

Repressed   Repressed   Repressed   Repressed

mRNAs   mRNAs   mRNAs

Proteins used for "cell housekeeping"

"Specific proteins" used in Y-cell function

Proteins used for "cell housekeeping"

Figure 28.4 Seventeenth-century biologists proposed that preformed bodies existed inside sperm cells. They made sketches such as this and called the miniature body a homunculus ("little man"). Absurd arguments arose over the possibility that still-smaller bodies of all future generations were inside the homunculus.

28.4). They proposed that the tiny, preformed body contained in the sperm could develop only after entry into the hospitable environment of the egg.

Then, in 1759, Caspar Friedrich Wolff published detailed observations of chick embryo development. In his studies, Wolff found no evidence whatsoever of a preformed body. Instead, he had seen "granules" (cells or nuclei?) that organized into layers which later folded into a body. Wolff concluded that the body is assembled from simpler, less organized material. This progressive organization of a body from less organized material was called the **theory of epigenesis.**

Figure 28.5 Results of early experiments on developmental potential of embryonic cells. (a) Roux's experiment. One frog embryo cell was killed with a hot needle. The surviving cell produced an abnormal half-embryo. (b) Driesch's experiment. Driesch separated the first two cells of a sea urchin embryo and found that each can produce a small, but normal and complete embryo.

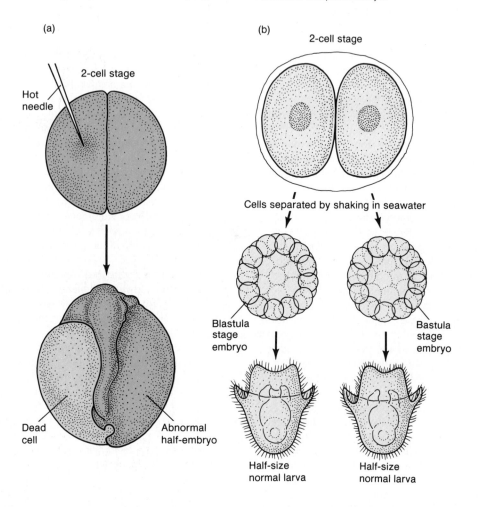

K. E. von Baer's work on animal development was another powerful blow to the idea of preformation. Von Baer's careful observations of progressive differentiation revealed no evidence that a preformed body was present at the beginning of development.

The First Experiments

Despite these findings, many biologists still argued in favor of preformation, and suggested that Wolff's granules might represent a preformed set of "determinants" for body parts that became separated and properly positioned by cleavage divisions. To test this idea, Wilhelm Roux experimented with two-cell-stage frog embryos. Roux destroyed one of the two cells with a red-hot needle. The surviving cell formed only half a body. On the basis of these results, Roux concluded that the surviving cell had produced an abnormal embryo because it received only half the determinants of the egg (figure 28.5a).

However, when Hans Driesch completely separated the first two cells of sea urchin embryos, each cell developed into a normal, but half-sized embryo (figure 28.5b). Driesch concluded that early development is not controlled by rigid preformation involving specific determinants, but rather that there is flexibility that permits adjustment of developmental processes. Years later, Hans Spemann and others showed that amphibian embryos, including those of frogs, could also show the same flexibility. If cells were completely separated, without being damaged, at the two-cell stage, each cell developed into a half-sized embryo and, eventually, into a miniature tadpole.

Box 28.1
Flexibility in Early Development: How Many Parents?

Hans Driesch showed long ago that the first cleavage cells of sea urchins have considerable developmental flexibility. If left to develop normally, the first two cleavage cells produce one pluteus larva. If separated experimentally, each of them develops into a half-sized, but normal pluteus. Since Driesch's time, biologists have demonstrated that the early cleavage cells of a number of organisms (but by no means all of those tested) have this ability to regulate their development following separation, just as Driesch's sea urchin cleavage cells did.

But what about developmental flexibility in the other direction? Instead of two embryos from one, can one embryo be obtained from two or more? Beatrice Mintz and her colleagues removed the zona pellucida, the membrane that encloses mammalian embryos during early development, from cleaving mouse embryos that had been taken from their mothers' oviducts. Two embryos at the eight-cell stage were placed together and allowed to fuse into a single cluster of cells. After fusion, the resulting embryo (for such a cell cluster does organize a single embryo) was placed into the uterus of a foster mother, where development proceeded normally. Mice that develop from two embryos are called tetraparental (four-parent) mice.

If the two embryos that are fused into one have different coat color genotypes, the results of fusion can be observed by examining the hair pattern of the mice after birth. The mice's coats are mosaics of patches of the two colors. This is clear proof that cells of both embryos survive the fusion technique and participate in the development of a single mouse.

In 1978, Clement Markert and Robert Petters fused three mouse embryos produced by three sets of parents, each of which had a different coat color genotype. These fused embryos were transferred to the uteruses of foster mothers, and some of the mice that were born had coats that were mosaics of three different coat colors. Thus, cells from all three of the original embryos had participated in the development of a single fused embryo. Mice that develop from three embryos are called hexaparental (six-parent) mice.

Beyond being an important demonstration of the flexibility of early development, these embryo-fusion techniques produce animals that are valuable subjects for research on other biological phenomena, such as the functioning of the immune system.

What seems even more amazing is that embryos of different species can be combined to produce a single organism. In 1984, two research teams—one in Cambridge, England, and the other in Giessen, Federal Republic of Germany—reported the birth of individual animals produced by combining sheep and goat embryos. After they were combined, the embryos were transferred to the uteruses of either female sheep or goats, where some of them developed to full term. Some of the offspring were obviously derived from cells of both animals. For example, they had patches of woolly sheep hair interspersed with patches of goat hair!

The secret of success in these experiments lies in how the membranes surrounding the combined embryos develop. For example, if the enclosing membranes consist of sheep cells, the embryo can implant successfully in a sheep's uterus, no matter what the cellular makeup of the embryo itself.

This discovery might someday enable ''foster mothers'' of other species to carry the embryos of endangered species or rare zoo animals, thereby producing more offspring than would result from natural processes. This could help considerably to enlarge and stabilize dwindling animal populations.

Box figure 28.1 Combined embryos. (a) A hexaparental (six-parent) mouse. This tricolored mouse was produced by aggregating three eight-cell embryos, each of which had a different coat color genotype. The mouse's coat is a mosaic of white, yellow, and black patches. (b) Sheep-goat combination produced by embryo manipulation. This animal has patches of both sheep hair and goat hair and has blood proteins of both species.

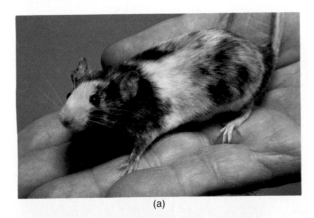

(a)

(b)

Nuclear Genetic Equivalence

Since each of the first cells of some animal embryos can, by itself, produce a small, but normal and complete body, each clearly has all the information required to direct the entire range of normal developmental processes. But might cell differentiation later in development result from a progressive series of restrictions of nuclear genetic capacity that leaves each nucleus capable of expressing only part of its original total genetic potential, only that part needed for development and function of a single cell type?

Nuclear Transplantation

Robert Briggs and T. J. King set out to test this hypothesis by transplanting nuclei taken from various embryonic cells of the frog *Rana pipiens* into eggs whose nuclei had been removed (**enucleated eggs**). These combinations tested the capacity of nuclei from cells at advanced stages of development to interact with mature egg cytoplasm to direct developmental processes (figure 28.6). Briggs and King found that nuclei from cells of blastula-stage embryos could interact with enucleated eggs to produce normal development. Thus, even in blastula embryos, which have 8,000 to 16,000 cells, it appears that nuclei have not lost any of their developmental capacity; they are still able to direct the entire range of developmental processes—from egg to tadpole and even on through metamorphosis to adulthood. In further experiments, however, Briggs and King found that if nuclei from cells of later stages of development were transplanted into enucleated eggs, very few normal tadpoles developed. They concluded at the time that as development proceeds beyond a certain point, nuclei of embryonic cells become restricted in their developmental potential, and that this nuclear restriction might be related to cell differentiation.

However, J. B. Gurdon obtained very different results when he did *two successive* nuclear transplantations using nuclei of the clawed frog *Xenopus*. Gurdon transplanted nuclei from advanced embryos into enucleated eggs. After these combinations reached the blastula stage, he transplanted nuclei from cells of these blastulae into other enucleated eggs

Figure 28.6 The nuclear transplantation technique used by Briggs and King. (The blastula embryo is enlarged to show detail.) Combinations of blastula nuclei with enucleated eggs developed into normal embryos in 60 to 80 percent of the cases, but they had less success with the nuclei taken from older embryos.

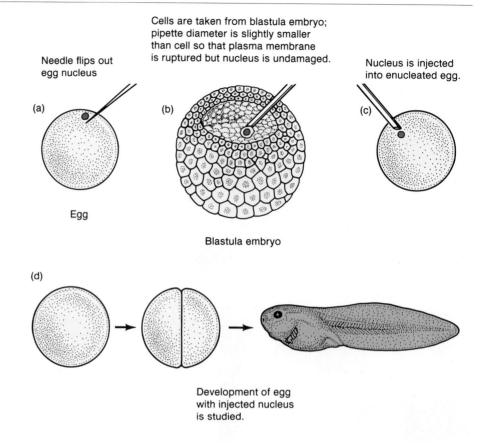

Needle flips out egg nucleus

(a)

Egg

Cells are taken from blastula embryo; pipette diameter is slightly smaller than cell so that plasma membrane is ruptured but nucleus is undamaged.

(b)

Blastula embryo

Nucleus is injected into enucleated egg.

(c)

(d)

Development of egg with injected nucleus is studied.

(figure 28.7). This several-step process apparently overcame a problem of timing incompatibility. Nuclei from slower-dividing cells of older embryos are not immediately compatible with enucleated eggs, which are geared up for the relatively rapid divisions of early cleavage. Using this serial transplantation technique, Gurdon found that even nuclei originally transplanted from differentiated gut cells of feeding tadpoles produced normal development with enucleated eggs. Thus, when tested under appropriate experimental conditions, nuclei from such differentiated cells of advanced embryos do not show signs of restricted developmental capacity. Eventually, Gurdon and his colleagues cultured adult frog skin cells and transplanted their nuclei into enucleated eggs (see figure 28.8). These combinations also produced development to the tadpole stage, although the tadpoles all died before they began to feed.

This work has attracted widespread interest because it suggests the possibility of cloning adult animals. Possibly even mammals, including humans, could be cloned. Experiments with mammals, however, pose great technical difficulties because mammalian eggs are much smaller than amphibian eggs and their nuclei are more sensitive to handling. In mammals, a membrane fusion technique produces better results than does mechanical injection of nuclei. Shortly after sperm entry, mouse zygotes are held in place while suction is applied to draw part of the plasma membrane along with the not-yet-fused egg and sperm pronuclei into a micropipette. Inside the pipette, the membrane fragment forms a sphere around the pronuclei. This sphere, with its enclosed pronuclei, can then be made to fuse (by treatment with Sendai virus) with another, similarly enucleated egg. This procedure introduces the pronuclei into the egg, and such combinations often produce normal mouse embryos. However, nuclear transfers from cleavage cells or later stages to enucleated eggs have been far less successful.

Figure 28.7 Gurdon's serial nuclear transplants using *Xenopus* embryos. After a second transfer, nuclei that are mitotically descended from an original gut cell nucleus can interact with an enucleated egg's cytoplasm to produce normal development. Tadpoles produced are a nuclear clone because they bear identical nuclear genetic information to one another and to the individual from which the first transplanted nucleus was taken.

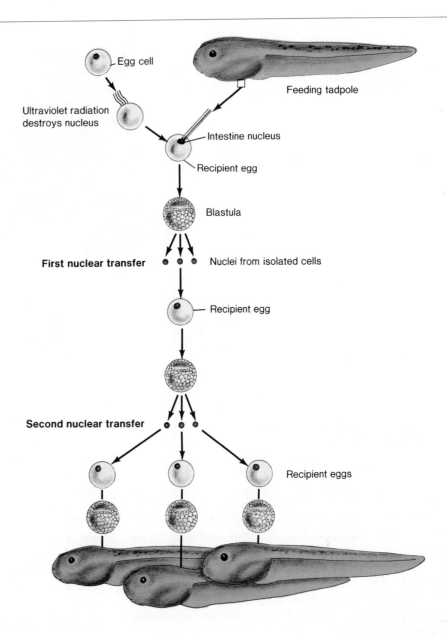

Totipotency of Plant Cells

Nuclei of even adult animal cells can express a wide range of developmental potentials, but testing their potential requires that they be placed in appropriate new cytoplasmic environments (enucleated eggs). In plants, however, the situation is quite different.

You already know that many plants can be propagated from cuttings of stems or leaves. Cuttings will eventually take root and produce whole growing plants. However, in plant tissue culture studies, whole plants can be grown from small clusters of cells broken off **calluses.** Calluses are masses of undifferentiated tissue that can be started from tissue taken from almost any part of a vascular plant. Thus, under appropriate conditions, entire plant bodies will develop from cells originally descended from only one part of a plant. This strongly indicates that during development, no permanent and irreversible restriction on the genetic capacity of plant cells occurs.

Figure 28.8 The technique for transplantation of an adult *Xenopus* skin cell nucleus into an enucleated egg. Cells taken from adult skin are placed in a culture dish, where they divide to produce a growing population of dividing cells. Then a nucleus from one of the cells in the culture is transplanted into an enucleated egg.

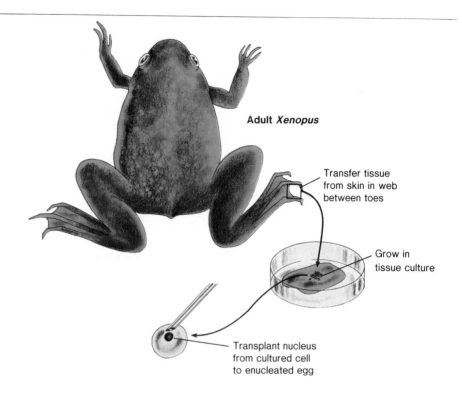

Adult *Xenopus*

Transfer tissue from skin in web between toes

Grow in tissue culture

Transplant nucleus from cultured cell to enucleated egg

If a small cluster of cells can grow into a whole plant, could a single cell also do so? F. C. Steward and his colleagues broke up callus cultures of carrot root phloem tissue into small pieces, some of which contained a few cells and some of which probably were individual cells. When these pieces were placed in a medium containing coconut milk, some began to divide and produce cell clusters that resembled early embryos. These clusters, called **embryoids,** developed into little carrot plants that when transferred to solid cultures, grew into whole plants (figure 28.9). Later, Vasil and Hildebrandt showed conclusively that individual tobacco plant cells isolated from calluses could divide to produce embryoids that eventually developed into completely normal tobacco plants.

Since then, embryoids have been obtained in cultures of cells from mature tissues of many species of plants. This ability of single plant cells to develop into whole plant bodies is known as **totipotency.**

As a result of these studies on totipotency, it is clear that nuclei of some differentiated plant cells retain the ability to express all of the genetic information needed to direct development of a complete plant. In other words, they can develop into an entire plant just as the original zygote did. This is an important difference between animal and plant cells. Individual plant cells can retreat from a commitment to being part of a differentiated multicellular body and divide to produce a number of cells that can each develop into an entire new plant. Totipotency of this sort has not been demonstrated for whole cells of animals.

Slime Molds: A Model of Differentiation

Because the cellular interactions that control differentiation in multicellular plants and animals are very complex, many developmental biologists prefer to study simpler systems (sometimes called "model systems") in the hope that fundamental developmental principles will be easier to discover there. An example of such a system is the life cycle of **cellular slime molds** such as *Dictyostelium discoideum,* a slime mold that has been studied in detail by K. B. Raper, J. T. Bonner, B. M. Shaffer, and others. Developmental changes in cellular slime molds parallel, in a simpler, more readily observable form, the changes in developing cells of more complex, multicellular organisms.

Figure 28.9 Steward's experiments on totipotency of carrot cells. Embryoids develop from small clusters of cells or single cells broken off cultured phloem tissue. Some embryoids can go on to produce whole normal plants.

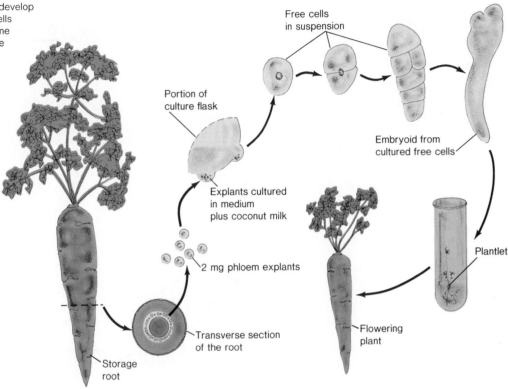

Free cells in suspension

Portion of culture flask

Embryoid from cultured free cells

Explants cultured in medium plus coconut milk

2 mg phloem explants

Plantlet

Transverse section of the root

Flowering plant

Storage root

Individual amoeboid *Dictyostelium* cells feed on bacteria, grow, and divide. During this period of feeding and cell multiplication, individual amoebae show no tendency to cluster; in fact, they repel one another when they get close together. This pattern of activity continues for generations, but when the food supply runs out, a change occurs. Amoebae swarm together and form a multicellular mass known as the **grex (slug)**. Grex cells secrete a slimy sheath around themselves, and the grex moves around in a sluglike fashion. Eventually, the grex develops into a **fruiting body** consisting of a stalk topped by a rounded, spore-forming structure. Spores with tough protective coverings and an ability to withstand dryness and other conditions that would kill the amoebae are produced in this spore-forming structure. Later, if the spores are distributed to favorable locations, they will germinate to release amoebae and the cycle will begin anew (figure 28.10).

Thus, in cellular slime molds, a population of identical, unspecialized cells goes through a series of specific differentiations, which result in aggregation, cellular cooperation within the moving grex, and finally, in development of a precisely patterned fruiting body.

The first change, initiation of aggregation, is a response to a chemical, **cyclic adenosine monophosphate (cyclic AMP),** that attracts cells. Cyclic AMP (figure 28.11) is secreted by cells forming an aggregation center. Waves of cyclic AMP secretion spread from these centers and cause amoebae to migrate toward the source and to develop specific cell surface adhesion sites that cause them to stick to one another. Cyclic AMP is also an important regulator substance in animals; it serves as a secondary chemical messenger in animal cell responses to several hormones (chapter 15).

After the grex has migrated for a time, it undergoes a second change by producing the fruiting body. Cells of the front one-third of the grex form the stalk. They secrete cellulose cell walls that help to hold the stalk upright. Cells of the back two-thirds of the grex form spores in a rounded mass at the top of the stalk.

Figure 28.10 Cellular slime molds.
(a) Asexual life cycle of the cellular slime mold
Dictyostelium discoideum. Different parts of
the cycle are drawn to different scales.
(b) Individual amoeboid cells moving toward
an aggregation center (magnification $\times$ 389).
(c) Streams of cells moving toward
aggregation centers (magnification $\times$ 12).
(d) Migrating grex. (e) Scanning electron
micrograph of a developing fruiting body
(magnification $\times$ 149).

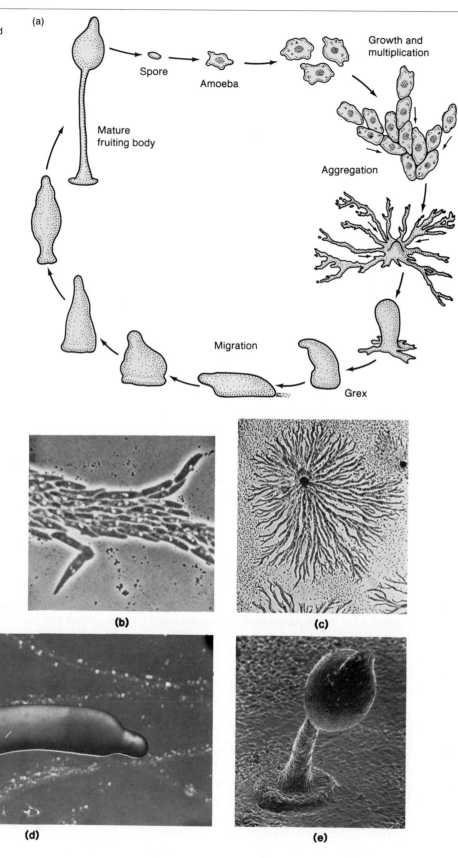

(a)

Spore

Amoeba

Growth and
multiplication

Aggregation

Mature
fruiting body

Migration

Grex

(b)

(c)

(d)

(e)

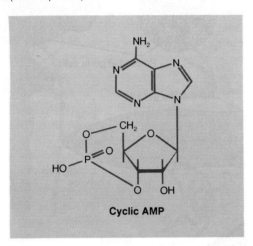

Cyclic AMP

A definite pattern of anterior cells destined to form stalk and posterior cells destined to produce spores is well established within the migrating grex, but experimental analysis has shown that individual grex cells are not irreversibly committed to their normal fate. If the grex is cut into parts, there is internal reorganization so that each part forms a small, but normal fruiting body (figure 28.12). Regardless of the size of the grex piece, the front one-third of its cells forms stalk, and the rear two-thirds forms spores. Thus, even within a subdivided grex, there is still a definite **polarity,** an end-to-end difference in developmental tendency.

As a model system, cellular slime molds show several interesting properties. Feeding amoebae change into aggregating cells that show specific movement behavior and develop new cell surface properties (figure 28.10). Within the grex, a definite pattern of polarity develops, and cellular commitments are expressed during development of the fruiting body. Cell movements shape the stalk and the spore-forming area, and cellulose is produced by stalk cells. Analysis of these various cellular developmental changes, as well as the control of these processes, have contributed much to our understanding of cell interactions in development of all multicellular organisms.

Determination and Differentiation

What control mechanisms cause structural and functional specialization of cells during development? In approaching this question, we must first understand that cells make a commitment to a specific course of differentiation some time before observable developmental changes occur. This commitment to undertake a particular differentiative pathway is

Figure 28.12   Results of cutting a *Dictyostelium* grex in half. Each half forms a fruiting body with stalk and spores.

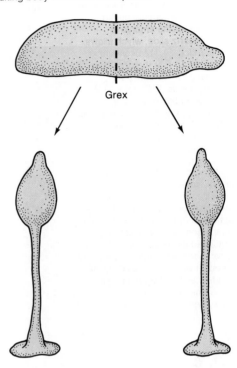

Grex

called **determination.** Some of the most challenging questions in developmental biology are concerned with determination and differentiation of cells in developing multicellular organisms. For example, what kinds of influences cause cells to become determined? And what specific cellular responses characterize cells that become developmentally determined?

Cytoplasmic Factors and Differentiation

As a cell divides, special parts of its cytoplasm may pass to only one of its two daughter cells. This daughter cell then develops quite differently from the cell that does not receive the special cytoplasmic material. For example, near the surface of the yolkiest part of a frog's egg is an area of cytoplasm that contains material long ago named the "germinal plasm." This material becomes enclosed in some relatively large cells situated, at the blastula stage of development, among cells destined to become part of the digestive tract. Unlike those surrounding cells, however, the cells containing the germinal plasm migrate from the developing gut area into the area of the developing gonads. These special cells, the **primordial germ cells,** settle in the gonads and become the ancestors of all of the eggs or sperm produced by the frog's gonads (figure 28.13a).

makes sense, considering that the environment of the developing mammal, the mother's body, is overwhelmingly female, at least in terms of hormones. (The small amounts of hormones produced by developing female gonads might not have much effect in such an environment.) But the situation is made more complex by the fact that developing testes also produce another hormone, called anti-Müllerian duct factor, that destroys the Müllerian ducts.

In a developing female embryo, where these two hormones are absent, the Müllerian ducts become female reproductive structures and the Wolffian ducts degenerate.

Pattern Analysis in Development

During the normal development of many structures, there are precisely ordered patterns of differentiation. Cells that are quite close to one another can develop in very different ways to produce the specifically arranged parts of complex structures. Cells develop as if they were responding to very specific positional information. Let us now examine some examples of precise pattern formation in development.

One- and Two-Dimensional Growth

A fern spore germinates and develops into a small, green, heart-shaped thallus that bears egg- and sperm-producing organs. Initially, a single cell from the spore divides to produce several green cells in an end-to-end chain called a **protonema.** Then the cell division plane changes, and the terminal cell of the chain divides to produce two side-by-side cells. This pattern change from one-dimensional to two-dimensional growth marks the beginning of the formation of the plate of cells making up the thallus (figure 28.17).

How does this change come about? The cells may somehow record the number of cell divisions it took to produce them. Then, at a certain "count," the change in division plane is triggered in the terminal cell. This cell has both a developmental history (a specific number of cell divisions have occurred) and positional information about its place in the pattern (it is contacted by another cell on only one side).

However, this normal developmental pattern of ferns is susceptible to change by external influences. For example, if kept under red light, protonemal cells grow long and thin, and one-dimensional growth continues indefinitely. Interfering with RNA synthesis also prevents the pattern change. This indicates that the developmental changes require specific genetic expression, but it is not yet clear how that genetic expression is normally controlled.

Chick Embryo Limbs

The idea of control of differentiation by an interaction of developmental history and specific positional information can also be applied to the problem of vertebrate limb development. In a vertebrate forelimb, for example, skeletal elements of the upper arm, the forearm, the wrist, and the hand all must develop in the proper sequence and in the proper relationships to each other. Then, a complex of muscles, blood vessels, and other tissues must develop in proper associations with the limb skeleton. This pattern has been thoroughly studied in the development of chick embryo wings.

Each limb begins its development as a small bulge, the **limb bud,** on the side of the embryonic body (figure 28.18). A limb bud consists of a core of mesoderm enclosed in a jacket of skin ectoderm. One strip of this ectodermal jacket is distinctly thickened to produce the **apical ectodermal ridge (AER).** The apical ectodermal ridge seems to be very important for normal limb development because surgical removal of the ridge terminates the outward growth of the bud. An incomplete limb produced by very early removal of the AER contains only **proximal** limb structures, that is, limb elements closer to the midline of the body. Slightly later removal of the AER allows development of a longer limb with more **distal** (further from the body's midline) skeletal parts present. Thus, there seems to be a progressive laying down of the limb's normal proximal-to-distal pattern that depends upon the presence of the apical ectodermal ridge.

Just under the apical ectodermal ridge there is an area in the mesoderm where many cell divisions take place. These cell divisions produce the bulk of the cells that are added to the length of the wing bud during its outgrowth. The active division zone is pushed further and further out as it leaves behind a lengthening limb. The cells that are left behind produce specific elements of the limb. But what makes cells at a given place form part of a humerus, as opposed to part of a radius, ulna, or even more distal element? What gives the cells positional information about their place in the proximal-to-distal pattern of the developing limb?

Lewis Wolpert and his colleagues in London have proposed a hypothesis for limb pattern formation based on experiments with developing chick embryo wings that suggests partial answers to these questions. They propose that the apical ectodermal ridge, together with the mesodermal area where cell divisions occur beneath it, constitute a "progress zone" in limb development and that this progress zone is the key to understanding how limb cells receive information about their proximal-to-distal position in the developing limb. They suggest that the cells left behind by the progress zone early in the limb's development form proximal structures (for example, the humerus). Cells left behind later form more distal structures (for example, the radius and ulna), and so forth,

Figure 28.17 Growth of the fern protonema from a spore. Rhizoids are pale cell extensions that function in nutrient absorption. (a) An end-to-end chain of cells switches to two-dimensional growth (3) and goes on to produce a heart-shaped gametophyte (4). (b) Cell shape is different when growth occurs in red light, and one-dimensional growth continues indefinitely. Note that the chloroplasts become clustered in one end of each cell.

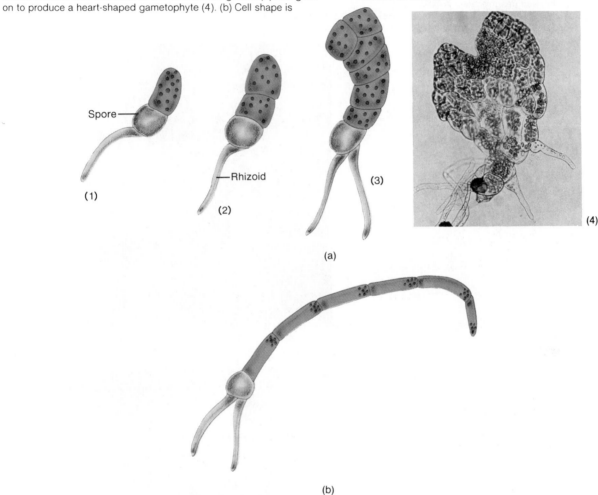

Spore

Rhizoid

(1)

(2)

(3)

(4)

(a)

(b)

as development continues. Somehow, cells have information about how long a time they spent in the progress zone before being left behind to differentiate. Possibly, these cells also "count" cell divisions.

Wolpert and his colleagues tested this hypothesis experimentally. If they took a progress zone from a wing bud at an early stage of its development and substituted it for the progress zone of a more advanced wing bud, the "young" progress zone proceeded with development according to its own developmental timer and added elements to the limb that repeated parts already there. If, on the other hand, they substituted a progress zone from a more advanced wing bud for that of a younger wing bud, a gap in development resulted, and some elements never formed (figure 28.19).

Other general hypotheses for wing bud development are consistent with these and other experimental data, but the Wolpert hypothesis regarding proximal-to-distal positional information in wing bud pattern formation is one of the most interesting and thought provoking.

Control of Pattern Formation

Another intriguing question concerning developmental pattern formation is how do similar cells that are located very close to one another differentiate in very different directions? For example, how do some cells within a particular part of a limb form bone while others form muscles, blood vessels, and so on? Some cells even have degeneration and death as their normal developmental fate. **Necrotic zones,** areas containing dying and degenerating cells, develop in limb buds

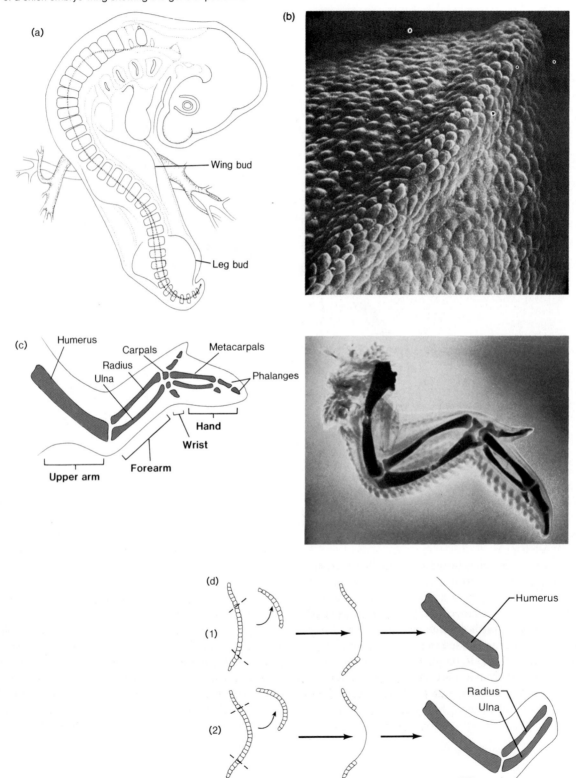

Figure 28.18 Chick embryo limb bud development. (a) Location of wing and leg buds of a chick embryo after about three days of incubation. (b) Scanning electron micrograph of the apical ectodermal ridge on a chick embryo's wing bud (magnification × 824). (c) The finished product of normal limb development. Simplified outline sketch and photo of a chick embryo wing showing the general pattern of relationships of parts in a vertebrate limb. ("Arm" terminology for areas is used because it is familiar.) (d) Results of surgical removal of the apical ectodermal ridge. (1) Early removal. (2) Later removal. These results are generalized from experiments of J. W. Saunders, Jr., and Edgar Zwilling.

Figure 28.19 Experiments of J. H. Lewis, D. Summerbell, and L. Wolpert in which progress zones (apical ectodermal ridge plus underlying mesoderm) were exchanged between wing buds at different stages of development. (a) The progress zone of a chick embryo limb bud. This scanning electron micrograph shows the apical ectodermal ridge and the underlying mesoderm cells. Note that cells in the ridge are taller than other ectodermal cells covering the rest of the bud. (The sphere beneath the ridge is a preparation artifact.) (b) Sketch of the composite wing produced when a progress zone of a younger wing bud is substituted for that of a more advanced wing bud. The host developed a normal humerus, radius, and ulna. The younger progress zone then added another humerus, radius, and ulna, as well as distal elements. (c) Sketch of a composite wing produced when a progress zone from a more advanced wing bud is substituted for that of a younger wing bud. A humerus was formed from host tissue, and the more mature progress zone added distal elements, but the radius and ulna are missing. (This sketch includes several shoulder elements not shown in the other figures.)

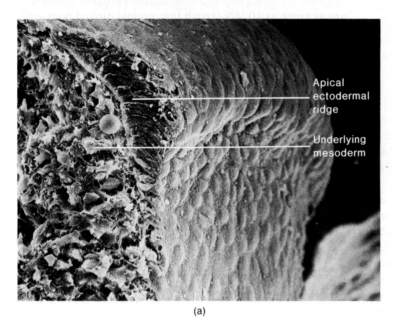

(a)

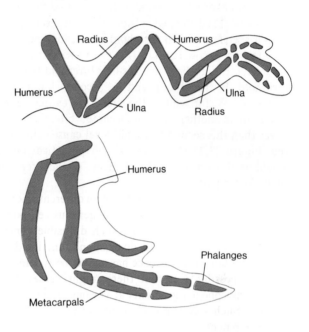

Figure 28.20 Distal part of a developing chick embryo's leg bud showing necrotic zones (*blue*) where morphogenetic cell death is occurring. The distal end of the bud initially is paddlelike, and morphogenetic cell death separates the toes. Similar morphogenetic cell death is involved in separating the digits of both hands and feet during human development.

(figure 28.20). The **morphogenetic cell death** that occurs in these necrotic zones is a necessary part of limb development; it produces the spaces that separate the digits in the originally paddlelike distal portion of limb buds, and gives the limbs their general shape.

How can such intricate patterns of differentiation occur? This kind of question arises again and again in the study of development. For such precise patterns to develop, uncommitted cells must receive and respond to specific information about their position relative to other cells. Lewis Wolpert has proposed a hypothetical model system that he calls the "French flag problem" to discuss the general concept of positional information.

The French flag consists of bands of three colors: blue, white, and red. Suppose that an imaginary row of cells can produce these colors and that the cells in the row can neither divide nor change position. How can they be made to form a French flag pattern? This is basically the same question asked about cells in many developing structures. How are these cells "notified" to produce different types of specialized tissues, depending on their location, and thus to assemble the normal pattern of tissues within a developing structure?

Figure 29.5 Complement. (a) Antibody binding on the cell surface makes possible a series of complement factor activations. C1 interacts with C2, C3, C4, and C5 in a complex set of reactions that yields active forms of C3 and C5. Active C3 promotes destruction of the foreign cell by phagocytosis. Active C5 sets off reactions that produce active forms of C6, C7, C8, and C9. C5 through C9 together form a complex that causes holes to develop in the foreign cell's surface. (b) Multiple lesions in a bacterial (*E. coli*) cell wall caused by complement activation (magnification × 176,000).

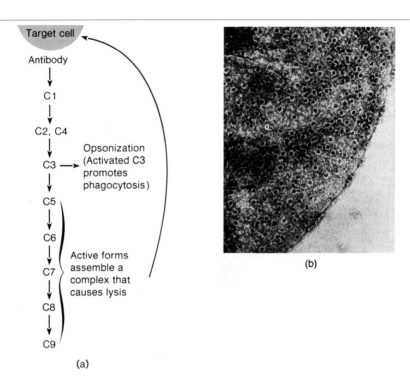

(a)

(b)

responses to infection. **Monocytes** are cells that regularly move freely between blood vessels and tissue spaces. In the tissue spaces they may enlarge and remain as **macrophages.** Macrophages are involved very early in responses to foreign antigens and generally are responsible for long-term phagocytic activities, such as cleaning up dead microorganisms and host cells at the end of a successful battle against infection.

After foreign cells have been taken in by phagocytic cells, they are digested by enzymes contained in lysosomes and eventually expelled from the cell (see figure 29.4b).

The Complement System

Certain blood serum enzymes known collectively as the **complement system** (or simply, **complement**) are always present in the bloodstream. The complement system is a series of about 20 proteins, the principal ones of which are designated C1, C2, C3, through C9. When activated, they work against invading bacterial cells.

The complement system's attacks on foreign cells depend on a series of reactions of the complement proteins. In one set of reactions, called the **classical pathway,** C1 combines with an antigen-antibody complex on the foreign cell and sets off a series of reactions involving the C2 through C5 proteins (figure 29.5a). This interaction "activates" complement proteins to fight infection in a number of ways. For example, the activated form of C3 serves as a chemical

attractant to phagocytic cells. This part of the complement reaction sequence, which makes foreign cells more susceptible to phagocytosis, is called **opsonization.**

When the elements C5 to C9 are activated, they combine into giant molecular complexes that form channels through the plasma membrane of an invading cell, thereby destroying the integrity of the membrane and causing cell lysis. Cytoplasm leaks out, and the cell dies (figure 29.5b).

The classical pathway of the complement system, therefore, is a set of general defenses that is effective against a wide range of foreign cells, but it depends upon specific immune interactions of antigens and antibodies. However, there is also an **alternative pathway** of complement activation that is set off directly by surface molecules of invading cells. This pathway serves as a first line of defense that does not need to wait for a specific immune response to develop.

Inflammation and Fever

Inflammation is a tissue and blood vessel response to infection or injury that is externally visible as a swollen, reddened, warm, and as you probably know from experience, often painful area. Inflammation involves movement of fluid, plasma proteins, and especially phagocytic cells out of the blood vessels into tissue spaces. Chemical regulators facilitate this mo ment by causing contractions of cells in the walls of blood vessels, which makes passage of materials easier (figure 29.6a).

Figure 29.6 Inflammation. (a) During inflammation, cells in vessel walls shrink, allowing easier movement of phagocytic cells (such as polymorphonuclear, PMN, leukocytes) into tissues, as well as increased movement of fluid and plasma proteins. (b) Histamine, a powerful inflammation promoter.

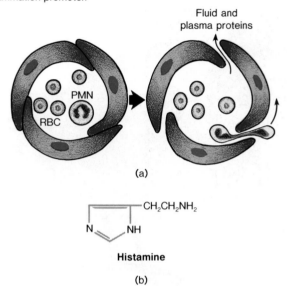

(a)

Histamine

(b)

One of the best known of these chemical regulators is **histamine** (figure 29.6b), a substance released from granules in **mast cells.** Complex inflammation responses are involved in several different types of normal reactions to foreign material, known generally as **hypersensitivity responses.** Some **allergy reactions,** however, can involve excessive and widespread expression of inflammatory-type responses. In some cases the response becomes so extreme that a decrease in arterial blood pressure, bronchial spasms, and even death result. Drugs known as **antihistamines** are used to combat excessive inflammatory responses.

Inflammatory responses elevate the temperature of a small, localized area, but sometimes **fever** develops in response to infection, and the temperature of the entire body rises above normal. White blood cells involved in defense reactions release a substance called **endogenous pyrogen** (literally "internal fire-maker"). Endogenous pyrogen causes resetting of the temperature regulating mechanism in the hypothalamus so that the body temperature becomes elevated. Sustained high fever can cause damage. Mild fever, however, aids in resisting infection in several ways. Certain metabolic processes in bacteria are inhibited by elevated temperatures. Endogenous pyrogen also strengthens the effect of interferon, and it stimulates production of some of the cells involved in defense mechanisms that depend upon specific recognition. We will discuss these mechanisms next.

Specific Recognition and Defense Mechanisms

Now that we have seen some general defense responses, let us look at specific mechanisms of defense.

It has been known for centuries that once people have recovered from certain diseases, they are not likely to contract them again; they have become **immune** (resistant) to subsequent infection. In the late eighteenth century, Edward Jenner proved that it is not necessary to contract a particular disease in order to acquire immunity to it. Jenner observed that farm workers who had been exposed to cows with cowpox did not contract smallpox, which at the time was a common and extremely serious disease. Jenner experimentally infected people with cowpox pus by scratching it into their skin. This caused a very mild disease with only one pock in humans, but such **vaccinated** (from the Latin *vacca,* meaning cow) people did not catch smallpox. Later, Louis Pasteur and a long line of other researchers developed vaccinations for many other infectious diseases. Through proper vaccination, some diseases have been almost eliminated in developed areas of the world. And in the late 1970s, smallpox, the first disease to be prevented by vaccination, became the first disease to be proclaimed completely eradicated by the World Health Organization.

Such specific immunities to diseases are only part of a much broader system of specific chemical recognition that normally makes distinctions between parts of the body ("self") and foreign cells and their cell products ("nonself"). For example, certain cells of vertebrate animals recognize foreign antigens, especially those on the surfaces of microbial cells, and respond with defense reactions aimed at removing or chemically neutralizing them.

Specific immune responses center on the activities of certain small white blood cells, the **lymphocytes.** Lymphocytes occur in the blood, in spaces among cells in tissues, and in the lymphatic system (see chapter 13). They circulate in lymph vessels and congregate in **lymph nodes.** Lymph nodes are enlarged, encapsulated masses of cells that occur along lymph vessels in many places in the body (figure 29.7).

Specific immune responses fall into two interrelated categories. One category involves specific interactions among immune system cells as well as direct attacks by lymphocytes on foreign cells. This is **cell-mediated immunity (CMI).**

The second category involves other lymphocytes, which specialize to become **plasma cells.** Plasma cells produce antibodies that combine specifically with antigens. This **humoral antibody synthesis** is important in body defense in general and absolutely vital for acquired immunity to specific disease-causing microorganisms.

Evolution

All living things are both products of and participants in a continuing evolutionary process.

Current concepts of evolution still rest on principles of natural selection proposed in the nineteenth century by Darwin and Wallace. Modern evolutionary theories have also incorporated the principles of population genetics. Modern evolutionary research takes advantage of techniques that help biologists determine the age of fossil organisms and make detailed molecular evolutionary studies of living things. Newly discovered fossils are providing more information on our human origins.

Chapters 30 through 32 deal with the origins of evolutionary theory, the evolutionary process, and human evolution.

Development of Evolutionary Theory

30

Chapter Concepts

1. The theory of uniformitarianism provided a conceptual framework for the explanation of earth's history and indicated that the earth was much older than had been estimated.
2. Charles Darwin's observations during his voyage around the world on H.M.S. *Beagle* led him to conclude that the environment plays a key role in the modification of existing species and the formation of new species.
3. The theory of natural selection suggests a mechanism for evolutionary change in natural populations.
4. The fossil record is physical evidence of the history of life on earth. When used with relative and absolute dating techniques, it provides considerable evidence about the evolutionary process.
5. Modern studies of evolution involve many scientific disciplines, ranging from biogeography to comparative anatomy.
6. Results of modern molecular evolution studies correlate well with conclusions drawn from other kinds of studies concerning lines of evolutionary descent.
7. It has been proposed that a process of chemical evolution led to the origin of life on earth.
8. Eukaryotic cells may have arisen from endosymbiotic associations among prokaryotic organisms.

Facing page   A sedge wren. These small birds live amid the tall, swaying vegetation of wet meadows. When disturbed, a sedge wren will often drop to the ground and scurry away like a mouse through the dense undergrowth. Like all other organisms, sedge wrens are adapted to life in their particular environment.

How old is the earth? When people asked that question early in the nineteenth century, most of the answers were probably in the range of five to ten thousand years. The most common belief at that time was that the earth and all living things on it had been created at one time and that they have remained essentially as they were created. By the end of that century, however, many scientists were convinced that there is good evidence to conclude that the earth is many millions of years old.

It had been suggested as early as the days of ancient Greece that modern living things might have arisen through **evolution,** a process of change over a long period of time. By the early 1800s a few scientists such as Buffon, Lamarck, and Diderot supported the idea that an evolutionary process had occurred. But there was the problem of time. It did not seem that the earth was old enough for such changes in living things to have occurred. Then geological discoveries made it clear that the earth was very much older than previously thought. These discoveries were to have a tremendous impact on biology.

With the new ideas about the age of the earth, it became possible to think seriously about the idea of evolutionary changes in living things over a very long period of time. These ideas were not widely accepted, however, until Charles Darwin provided the foundation upon which the modern **theory of organic evolution** has been built. In 1859 Darwin published *On the Origin of Species,* a thoughtful explanation of his views on the evolution of life on earth. In the more than 100 years that have elapsed since that time, the theory of organic evolution has gained almost unanimous acceptance among scientists and has been greatly strengthened by evidence obtained from a wide variety of scientific disciplines.

Charles Darwin and the *Beagle* Voyage

Charles Darwin (figure 30.1) was born in England into a wealthy family. His father was a prominent physician, and his mother was a Wedgwood, of the renowned china manufacturing family. Darwin himself married a Wedgwood, his cousin Emma Wedgwood. All his life Darwin loved nature. As a young man, he was an ardent naturalist; he roamed the fields and woodlands of the English countryside, studying plants and animals and observing rock formations. This interest continued into his college years, for although his father wanted him to become a physician, Darwin preferred to associate with professors of biology and geology. When it became obvious that Darwin would not become a doctor, his father attempted to guide him into a career in the clergy. That profession, however, held little more appeal for him than did medicine.

Figure 30.1 Charles Darwin (1809–1882) as a young man.

Finally, in 1831, Darwin was offered the opportunity to travel as a naturalist on H.M.S. *Beagle,* a ship commissioned to sail around the world on a survey mission. Darwin saw in this post an opportunity to see more of the natural world than was afforded most people of his time. In December of 1831 Darwin embarked on a voyage that was to change the history of biological thought (figure 30.2).

Lyell, Hutton, and Uniformitarianism

Darwin was not a good sailor; he spent much of his time in his bunk suffering from seasickness. During the early part of the voyage, he occupied his time by reading the first volume of Charles Lyell's *Principles of Geology,* in which Lyell further developed James Hutton's theory of **uniformitarianism** in geology. Hutton contended that the earth is not a static, unchanging sphere but that the present features of the earth's surface are the result of a continuous cycle of erosion and

Figure 30.2 Route of H.M.S. *Beagle* surveying trip around the world (1831–1836). Dashed lines indicate the return voyage.

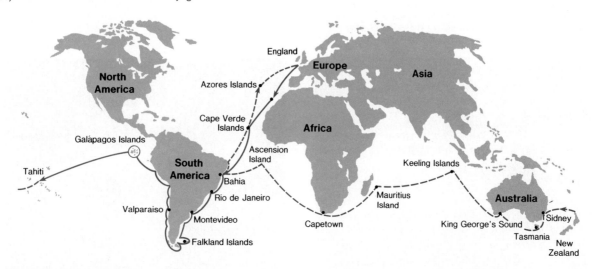

uplift. Hutton had seen evidence that the earth's continents have been worn away continually by the agents of erosion and that weathered rock debris is transported by rivers to the oceans, where the loose sediments are deposited in thick layers that are eventually converted into sedimentary rocks. Hutton also proposed that "subterranean forces" operate to uplift material from below sea level to form new land surfaces. Hutton's concept of uniformitarianism stated that these forces have acted at the same rate throughout the earth's history to produce similar geological events. (Hutton and Lyell's general ideas about continuing geological change are still accepted today, although modern geologists realize that rates of change have not always been the same.)

Darwin realized the implications of the uniformitarian theory for biology. If the earth itself is dynamic and its present features are the result of a long process of gradual change, Darwin reasoned, could it not be possible that the biological world is also dynamic rather than static?

Fossils of South America

Darwin found his first evidence of change in the biological world in South America. Along the east coast of Argentina, in the mud and silt of river deposits, Darwin discovered a large number of fossil bones literally sticking out of the loose sediment. Among these bones he identified the remains of a giant ground sloth, a huge hippopotamuslike animal, and a species of horse. All three of the fossil forms represented **extinct species,** species not represented by any living individuals. That some species had become extinct suggested to Darwin that not all species were immutable (unchanging).

Even more unsettling to Darwin, however, was the close resemblance of these fossil forms to living species of sloths, hippopotamuses, and horses. Was it possible, Darwin wondered, that these fossil forms might have been ancestral to species found on earth today? If so, the implication was that new species appear on earth as a result of descent from earlier species. Darwin later wrote that his first ideas of evolution began with his observations of these South American fossils.

Galápagos Islands

From Argentina, the *Beagle* slowly progressed southward, rounded Cape Horn, and worked its way northward along the coast of South America to a tiny group of islands called the Galápagos. In these islands Darwin discovered a veritable laboratory of evolution. The Galápagos were formed by volcanic eruptions and are relatively isolated, being separated from the South American mainland by some 950 km of ocean. Ancestors of the plants and animals that inhabit the islands came from the South American mainland. They may have been blown to the islands by strong offshore winds or rafted there on floating vegetation. The colonization of the islands by plants and animals was a slow process, but the weathered volcanic rock provided a good substrate for the growth of some plant species, and as seeds germinated and vegetation spread, food and shelter were available for animal species.

When Darwin arrived in the Galápagos in 1835, he found a variety of plants and animals, many of which are found nowhere else in the world (figure 30.3). Among the most interesting of the Galápagos creatures were the giant tortoises

Figure 30.3 Many species of plants and animals are unique to the Galápagos islands. (a) Giant treelike prickly pear cactus. (b) Marine iguanas are found only in the Galápagos. Note the large claws on its feet, which are well adapted to a life spent clinging to rocks along the shoreline. Galápagos iguanas eat seaweed. (c) The rare flightless cormorant is found only on two of the Galápagos islands.

(a)

(b)

(c)

Figure 30.4 Tortoises of the Galápagos islands. (a) Sketches of Galápagos tortoises from three different islands illustrating neck differences. Longer-necked races live in relatively dry areas and feed on the high-growing vegetation of cacti. Shorter-necked races live in moister regions with more abundant ground vegetation. (b) Map of the Galápagos islands with the islands' English and current Spanish names. (c) Photographs of tortoise shells illustrating the different shapes the islands' governor pointed out to Darwin. The round shell on the left was found on James island (Santiago) and the other on Hood island (Española).

(a)

Abingdon   Albemarle   Duncan

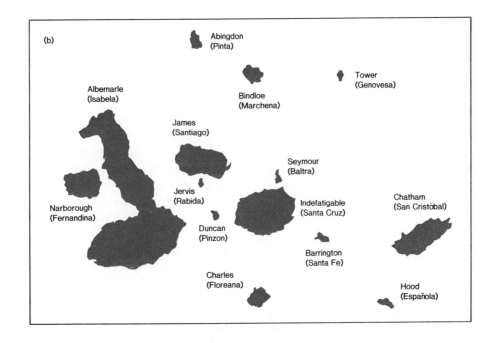

(b)

Abingdon (Pinta)

Tower (Genovesa)

Bindloe (Marchena)

Albemarle (Isabela)

James (Santiago)

Seymour (Baltra)

Jervis (Rabida)

Chatham (San Cristóbal)

Narborough (Fernandina)

Indefatigable (Santa Cruz)

Duncan (Pinzon)

Barrington (Santa Fe)

Charles (Floreana)

Hood (Española)

(c)

that gave the islands their name (*galápago* being Spanish for tortoise). Darwin's interest in these animals was aroused when the governor of the islands chanced to remark that he could tell from which island a tortoise had come by observing the shape of its shell. Upon closer inspection, Darwin noticed that not only did the shape of the shell vary from one island to another but that other features of the tortoise varied as well. For example, the necks of the tortoises living in dry areas were longer than the necks of those living in moist areas (see figure 30.4). Why, Darwin wondered, should so many distinct forms of the tortoise appear in such a limited geographical area?

Darwin later hypothesized that very few tortoises could have arrived at the Galápagos from the mainland, but once there they found little competition for food and no natural predators. Under such conditions, the tortoises became established on the islands and gradually increased in number, spreading out wherever conditions were favorable. Because the physical conditions for survival varied from island to island, the tortoise population of each island developed, over time, individual characteristics that distinguished it from the tortoise populations of neighboring islands. Long-necked tortoises inhabited dry areas where food was scarce, since a longer neck was helpful in reaching high-growing foliage. In moist regions with relatively abundant foliage, short-necked tortoises fared well. Thus, Darwin concluded that the major factor in development of interisland variation among the tortoises was the environmental variation that existed from one island to the next.

Adaptive Radiation

Darwin noted interisland variations among a number of other plants and animals in the Galápagos. For example, on each of the islands there are different species of the genus *Scalesia* (figure 30.5), a member of the sunflower family. Furthermore, in the Galápagos these plants are large, woody, and treelike. In contrast, members of this family that occur elsewhere are herbaceous and considerably smaller.

Darwin also took note of the range of variation among a group of small, drab birds that lived on the Galápagos islands, although he did not collect enough specimens to study them in detail. These relatively inconspicuous birds have been studied intensively, however, by biologists since that time and are now called "Darwin's finches." Like the tortoises, finches were introduced to the Galápagos from the mainland of South America. Also like the tortoises, finches found little competition for food and no natural predators. Under these conditions, the finch population quickly became established.

The original finches were seed eaters, and for a time there was plenty of food and they thrived. But the factors that fostered the initial success of the finches also led to a major problem. Finches are prolific breeders, and with an initial abundance of food and no predators to control population size, the population increased to the point that competition for food and nesting sites became severe. During this period of population increase, the finches spread to all of the Galápagos islands.

On the different islands, the finches underwent evolutionary modification in several directions. Although the finches were primarily seed eaters, they would also eat insects, and those finches that could not obtain sufficient seed to stay alive were compelled to turn to this alternate food source. By doing so, they decreased their competition with the seed eaters and increased their chances for survival and reproductive success. Over time, finches evolved that were insect eaters rather than seed eaters. An important factor in the success of this transition was the lack of competition from other bird species for the insect food supply. In most ecological situations, there are bird species that feed specifically on seeds, on insects, on fruits, and so on. Under such conditions it would be very difficult for members of a seed-eating species to make the transition to insect eating. However, because the Galápagos finches had little competition from other bird species over long periods of time, such evolutionary transitions were possible. Some of the finches even became specialized to other habits, such as cactus eating.

After periods of isolation on separate islands, various groups of finches came to be very different from one another. And these differences involved more than variations in their diet and appearance. These groups of birds, which were descended from common ancestors, had also changed genetically during the period of geographic isolation. They could no longer interbreed when they encountered one another later. They had become reproductively isolated from one another.

Thirteen distinct species of finches occupy the Galápagos islands and one species lives on Cocos Island 700 km to the northeast. Each species is adapted to a different lifestyle. In the Galápagos, finches have filled the ecological roles of warblers, parrots, woodpeckers, and other bird species that would normally occupy these niches in other geographical areas. There are distinct differences among the Galápagos finch species in such traits as plumage, nesting sites, body size, and size and shape of the beak (figure 30.6).

The populations of plants and animals of the Galápagos islands illustrate a process modern biologists call **adaptive radiation.** Adaptive radiation occurs when members of a single species enter environments in which there is little initial competition for resources. Under such conditions, after an

Figure 30.5 Members of the genus *Scalesia* on the Galápagos islands. In most cases, each species is restricted to a single island.

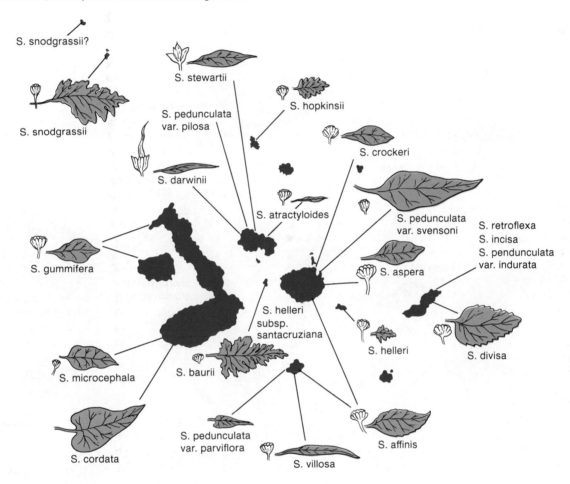

initial period of adjustment, population size increases rapidly. This leads to intense competition for resources, which in turn may result in fragmentation of the population. After a period of isolation a number of new species may develop; each of these new species occupies a distinct niche, as in the example of Darwin's finches.

The Impact of the Galápagos on Darwin's Thinking

Profoundly influenced by his observations in the Galápagos islands and elsewhere on his journey, Darwin concluded that species are not unchangeable and that new species are formed through the gradual modification of existing traits in response to the pressure of both the environment and biological competition.

Darwin had come to accept the idea of organic evolution—the concept that all species of plants and animals have arisen through descent with modification from previously existing species. However, not every species continues to interact successfully with its environment indefinitely, especially when there are changes in the environment. Change is the rule of nature, and the history of the earth has been filled with many geological and climatological changes. Thus, many species have become extinct. In fact, in the history of life on earth many more species have become extinct than presently inhabit our planet.

Figure 30.6 Galápagos finches. All of the species are descended from a common ancestor, but they differ markedly in appearance, habitat, and food sources. The most remarkable of the Galápagos finches is the woodpecker finch. It feeds on insects in tree bark as a woodpecker does, but it uses cactus spines as tools to draw insects out of holes in the bark.

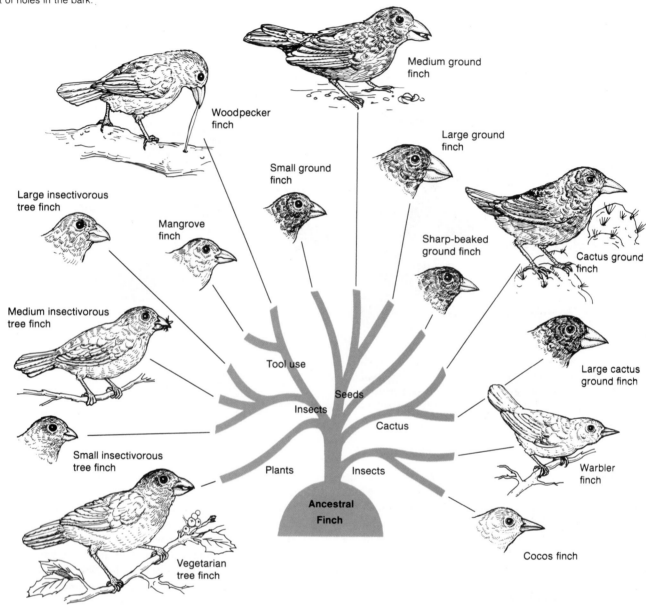

The Theory of Organic Evolution

After his return to England in 1836, Darwin married Emma Wedgwood and moved to a country house near the village of Downe in Kent. There he spent the next twenty years in research, reading, and developing the idea that living species are the products of series of gradual changes that have occurred through a long period of geological time. This is, in essence, the theory of organic evolution: that every species of plant and animal is the modified descendant of previously existing species. The concept of organic evolution was not original with Charles Darwin by any means, but he stated it clearly and directly, and systematically gathered evidence in support of the idea.

Darwin realized that this theory required a mechanism that would explain the ultimate source of biological variation and the means by which heritable traits were passed from generation to generation. Although Gregor Mendel was a contemporary of Darwin's, the gene theory of inheritance was not developed until the early years of the twentieth century. Thus, Darwin was unable to explain biological variation. Lacking an understanding of the mechanisms of genetics, Darwin was left with such vague and inadequate concepts as the "blending" theory of inheritance, which attributed the inheritance of traits to a mixing, or blending, of parents' traits. That theory was wholly inadequate to explain the wide range of variation found in any species and could not explain how some traits became more pronounced rather than being diluted generation after generation.

Darwin also realized that he needed an explanation for the mechanism by which new species were formed. The ideas of Jean Baptiste de Lamarck, though later proven wrong, provided Darwin with some food for thought.

Lamarck and Inheritance of Acquired Characteristics

Lamarck developed the concept of inheritance of acquired characteristics as part of his statement of the theory of organic evolution published in his *Philosophie Zoologique* in 1809. Lamarck accepted variation as part of the natural world and believed that the environment played a role in the origin of new species. He thought that new traits were acquired by an organism in response to a *need* imposed by the environment and that evolution proceeded toward a perfect form.

Lamarck stated that traits could be acquired through the use or disuse of an organ; that is, a body part that is not used by an organism atrophies and is lost to future generations, while a body part used extensively is amplified. For example, the Lamarckian explanation for the long neck of the giraffe is that elongation occurred because the ancestors of the giraffe constantly reached into the trees to feed on high-growing vegetation. This repeated stretching of the neck over time resulted in the modern giraffe with its very long neck. A similar argument would be that a son born into a family of weight lifters should have large muscles because his father, grandfather, and great-grandfather had developed bulging biceps by lifting weights.

Lamarck's concept of inheritance of acquired characteristics did not gain wide acceptance in scientific circles, partly because it was **teleological,** that is, it assumed a final goal toward which the process of evolution was directed. Furthermore, although his theory was an attempt to explain how traits are gained and lost, it lacked a basic mechanism of inheritance. And with the eventual rediscovery of Mendel's principles of inheritance, it became apparent that acquired characteristics, which are not genetically based, cannot be inherited. Nevertheless, Lamarck's concept of inheritance of acquired characteristics was one of the ideas about evolution that was considered by Darwin and other scientists of his time.

Artificial Selection

Because he lacked information about genetics, Darwin was not able to provide an adequate explanation for biological variation. He did, however, develop a clear explanation of a mechanism by which evolution of new species can occur.

An activity that has been practiced by the human race for thousands of years provided a clue in Darwin's search for an explanation of evolution. For centuries humans have selectively bred fruits and vegetables, cattle, chickens, and dogs in efforts to obtain greater yields of grain, better milk cows, more productive hens, or dogs more adept at sheepherding or rabbit hunting. Breeders observe the variations that appear in their plants and animals and select for breeding those individuals that best represent the properties they desire in future generations (figure 30.7). The success of this **artificial selection** is undeniable, and Darwin saw in artificial selection a model for change in the natural world. What puzzled him was how selection worked in nature. Human beings direct artificial selection, but what force directs selection in nature?

The Theory of Natural Selection

As essay on human population written by Thomas Malthus gave Darwin a clue to the solution to this problem. Malthus suggested that human populations have a general tendency to increase in size, but that such factors as fire, war, plague,

Figure 30.7 Darwin saw artificial selection, as practiced in the breeding of domesticated plants and animals, as a model for evolutionary change in nature. (a) These vegetables are all produced by plants belonging to the same species, *Brassica oleracea*. They have been artificially selected by breeders to emphasize different parts of the plant. (b) Variation among breeds of the domestic pigeon.

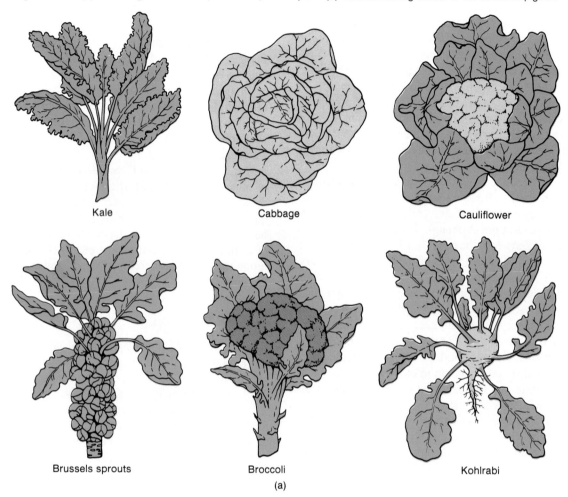

Kale

Cabbage

Cauliflower

Brussels sprouts

Broccoli

Kohlrabi

(a)

and famine serve to keep the human population of the earth more or less in check. If it were not for these factors of population control, the human population would increase explosively (a prediction that agricultural, medical, and technological advances of the last century have proven to be accurate).

Darwin applied the Malthusian concepts to the natural world and reasoned that *every* species produced many more young than could survive to reproductive maturity. Of the many new individuals produced by all species in nature each year, most are eaten by predators, succumb to disease, or are unable to obtain an adequate food supply or habitat in which to live. There is a constant struggle for survival. Only some members of any species live to reproduce. Thus, in any generation only the members that succeed in reproducing contribute to the evolutionary future of the species. Darwin

recognized that the same factors he had encountered in the Galápagos islands operated throughout the biological world: nature, in the form of environmental variation and biological competition, operates as the selective factor in natural populations.

From these hypotheses Darwin derived his major theory, the **theory of natural selection.** We can summarize the theory of natural selection as follows: All populations of organisms possess the potential to increase at a very rapid rate, but this rate of increase is very seldom observed among natural populations. Instead, the size of populations remains almost the same from one year to the next. Thus, Darwin reasoned, there is an intense, constant struggle for food and other resources among the many young that are born each generation. And those individuals with the most **adaptive** traits (traits that permit successful interaction with the environment) are the most likely to survive and reproduce. Thus, generation by

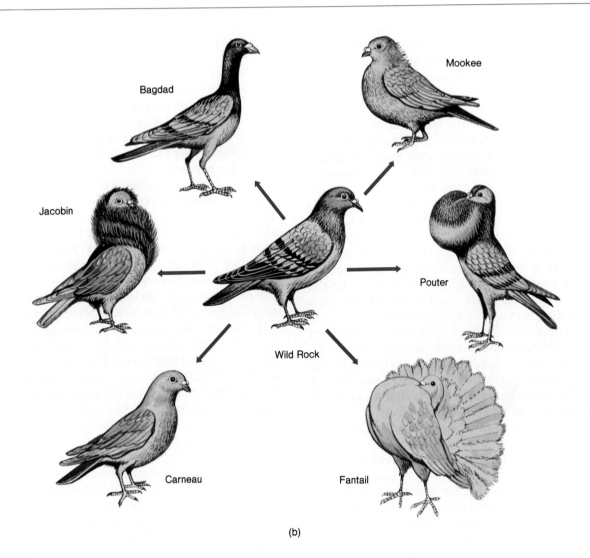

(b)

generation, adaptive traits become more frequent in the population, while nonadaptive traits tend to be eliminated.

Over long periods of time, natural selection produces species of organisms that are finely adapted to the set of environmental conditions in which they must compete. Each species is not only adapted to this physical environment, but to its biological interaction with other species as well.

Darwin's theory of natural selection also explains why some species perish while others survive, since survival is predicated on the adaptability of the species as a whole. Under changing environmental conditions, species with great biological variability are more likely to survive than species that are highly specialized to a given set of environmental conditions. Highly specialized species may have lost the flexibility required to survive in the face of changing conditions.

Alfred Russel Wallace

In 1858 Charles Darwin received an essay from the young English biologist Alfred Russel Wallace titled "On the Tendency of Varieties to Depart Indefinitely from the Original Type." Wallace wanted Darwin to review the essay and to decide whether it merited further critical reading by Charles Lyell. Darwin was startled to find in Wallace's short essay the basic elements of his own theory of natural selection. Wallace's ideas were so similar to his own that Darwin felt he should put off publication of his own work and let Wallace have credit for the idea of natural selection.

However, both Lyell and the botanist Sir Joseph Hooker knew of Darwin's long efforts in the development of his theory. They also knew that he had written his own ideas on the origin of species and natural selection in an abstract he had sent to Lyell in 1844 and in a letter to the famed American botanist Asa Gray in 1857. They persuaded Darwin to have these statements of his theory, along with Wallace's paper, presented to a meeting of the Linnean Society in the summer of 1858. Then Darwin hurried on with his preparation of *On the Origin of Species,* which contained his ideas of organic evolution and natural selection, and it was finally published in November 1859.

The similarity of Wallace's independently developed concept of species formation with that of Darwin might seem to be one of the great coincidences in the history of biology, but it is not particularly surprising when you compare the backgrounds of the two scientists. Wallace was a very talented person who had a number of experiences similar to those Darwin had as a young man. Wallace had taken an exploratory journey up the Amazon River and had extensive field experience in the East Indies. He was convinced that living things arose through an evolutionary process and, like Darwin, had pondered the mechanisms of evolution. He also had formulated his ideas about natural selection as a direct result of reading Malthus's essay on population.

Despite the similarity of their ideas, Wallace maintained that Darwin should receive primary recognition for the development and formalization of the theories of organic evolution and natural selection. Not only had Darwin's work preceded Wallace's chronologically, but it also was of considerably greater breadth and depth.

However, Darwin and Wallace came to disagree on the action of natural selection in the process of organic evolution. Wallace thought that every trait of an organism was the product of natural selection and therefore must possess an adaptive function. Darwin, on the other hand, maintained that selection was not the only means of species modification and that organisms are complex systems in which certain specific features may have adaptive value while others are nonadaptive. Darwin and Wallace did agree, however, that natural selection is the major force directing the process of organic evolution.

Evolutionary Evidence

New scientific theories often stimulate a great deal of controversy. This has a beneficial effect on scientific progress because scientists with differing views work intensively to find evidence to support their ideas. The theory of organic evolution and Darwin's theory of natural selection initiated such activity, as scientists sought data that would either support or refute these theories. The weight of scientific evidence over the past 100 plus years strongly supports Darwin's basic precepts, and such areas of biological specialization as biogeography, paleontology, comparative anatomy, comparative embryology, genetics, and molecular biology have made especially strong contributions to our modern concepts of evolution.

Biogeography

Biogeography is the study of the geographic distribution of plants and animals. It attempts to explain why species are distributed as they are and what their pattern of distribution reveals about prehistoric habitats and their climates. As Darwin toured the world on the H.M.S *Beagle,* he discovered that plant and animal species generally are not distributed as broadly as are their potential habitats. Studies in biogeography since the time of Darwin have confirmed this over and over again.

A fundamental conclusion drawn from biogeography is that a new species emerges in one place and then spreads outward from its point of origin. Some species become widely distributed, but they do not spread beyond natural barriers such as those that separate major biogeographic regions. Therefore, although virtually identical environments may exist in different biogeographical regions, they are seldom occupied by the same species. In fact, each major geographic region of the world (figure 30.8) has its own characteristic sets of plants and animals. For example, in Australia, marsupial (pouched) mammals fill the roles played by placental mammals in other biogeographical regions where marsupials are rare. In addition, the fossil record of each region reflects an evolutionary sequence of biological events distinct from that of all other regions. Within each sequence, the most recent fossil forms closely resemble the living species found in that particular region.

Such observational evidence reinforces the concept that natural selection applied by the environment is the major force that has molded modern species, adapting the life forms of each geographic region into species that are suited for survival within the topographic and climatic conditions of their surroundings. Darwin saw evidence for this when he discovered that species on particular islands are closely related to species on nearby islands, and that island species are generally related to species on the nearest mainland. On the other hand, there is no evidence that supports the existence of a set of "island species" that are characteristically found in island habitats everywhere around the world.

At a more specific level, biogeography provides numerous striking evidences of **convergent evolution.** Organisms in different biogeographical realms, although descended from very different ancestral stocks, possess remarkably similar adaptations to particular habitats. For example, plants of the cactus family (*Cactaceae*) are found in the deserts of southwestern North America and the high deserts of the Andes, but nowhere else. Comparable arid habitats in Africa are occupied by members of the spurge family (*Euphorbiaceae*). This clearly illustrates the molding power of natural selection which has shaped very similar sets of adaptations to similar environments in widely separated parts of the world (figure 30.9).

Figure 30.8 Major biogeographical regions of the world. Different biogeographical regions generally have distinctively different plants and animals. Some natural barriers that separate the regions are indicated in green. (1) The Sahara and Arabian deserts. (2) Very high mountain ranges, including the Himalayas and Nan Ling mountains. (3) Deep water marine channels among islands of the Malay Archipelago. (A. R. Wallace recognized and wrote about this barrier, and it has been called Wallace's line.) (4) The transition between highlands in southern Mexico and the lowland tropics of Central America.

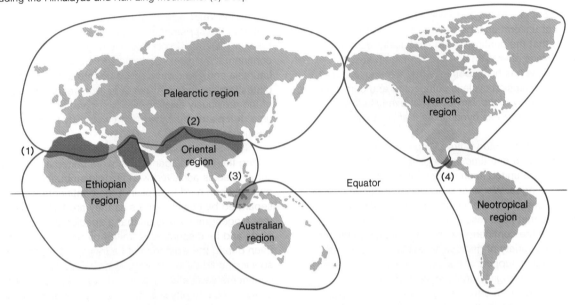

Figure 30.9 Convergent evolution of members of the Euphorbiaceae (a) and Cactaceae (b). These plants, descended from different ancestors, possess very similar adaptations to life in very dry environments. They have fleshy stems that can store water, very reduced leaves, and protective spines. This illustrates the power of natural selection in molding sets of adaptations to specific environments.

(a)

(b)

Figure 30.12 (a) Comparison of a modern bird and a representation of *Archeopteryx* based on fossil evidence. Like modern birds, *Archeopteryx* had feathers on its body and wings, but it had jaws with teeth, clawed digits at the margins of its wings, and various skeletal features intermediate between those of reptiles and modern birds. (b) A fossil of *Archeopteryx*.

(a)

(b)

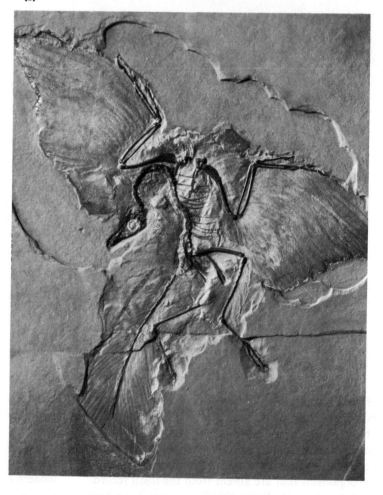

Another fossil form that had characteristics intermediate between major vertebrate groups is *Seymouria,* an amphibian whose skeletal structure shows features that are a mixture of reptilian and amphibian characteristics (see p. 1050). In fact, the skeleton of *Seymouria* is so intermediate in form that it has alternately been classified as a reptile and an amphibian by paleontologists. We will encounter more examples from the fossil record in chapters 37 through 40, in which we survey the diversity and evolution of life on earth.

We should mention at this point that there is some debate as to the tempo of evolutionary change. Many biologists think that the gradual transition in form represented by the fossil record of the horse is typical of evolutionary modification. If this view is correct, then eventually a number of intermediate forms should be found in the fossil record. However, other biologists think that there have been long periods of **stasis,** periods during which species underwent little structural change. According to this view, new species form very rapidly, during periods of environmental stress. Such a relatively rapid production of new forms would leave few transitional forms in the fossil record (see the discussion of **punctuated equilibrium,** chapter 31). This would make it highly unlikely that fossil "missing links" between certain groups would ever be found.

Relative Dating Techniques

To use the fossil record effectively to represent sequential changes that have occurred in the history of life on earth, we must be able to assign ages to fossil organisms. Paleontologists determine the age of fossil-bearing rock layers by two types of techniques: **relative dating** and **absolute dating.**

Relative dating techniques depend on the principle of **superposition,** which was established by Nicolaus Steno in the seventeenth century. This principle states that in any undisturbed sequence of sedimentary rocks, the bottommost layer is the oldest, and the topmost layer is the youngest. This interpretation follows from the manner in which sedimentary layers are formed. When a river reaches the ocean it deposits its sediment load on the ocean floor, and over a period of time hundreds of meters of sediments can accumulate. Eventually, these sediments may be converted into sedimentary rock, with the oldest sediments (and hence the oldest rock) being at the bottom of the rock sequence. Therefore, in any undisturbed sequence of sedimentary rock, fossils found in the bottommost layer are older than fossils found in the layers above.

Another method of determining the relative age of fossils was developed in the 1700s by William Smith, a British geologist, surveyor, and engineer. Smith's interest in geology was a practical one, since he was concerned with the physical properties of the rock layers he encountered in his road and canal construction work. Smith wanted to know whether the sandstone he encountered in one place was of the same origin, and thus possessed the same physical properties, as sandstone found in another area. With this goal in mind, he searched for a method to correlate strata from one geographic region with those of another.

Smith noted that in some sequences of sedimentary rock, the kinds of fossils in the bottom layers differed from those in higher layers. Because he had encountered similar fossil groupings in sedimentary layers occurring in widely separated parts of Britain, Smith proposed that all rocks containing similar fossils were of the same geological age. If so, then the fossil assemblage of a particular rock layer would provide the means of correlating the age of rock layers occurring in widely separated geographical regions.

Smith's idea proved to be a fruitful one and became known as the principle of **fossil correlation,** a valuable tool used by geologists to this day.

These two principles—superposition and fossil correlation—were the only dating techniques available to geologists before the twentieth century. They were used to painstakingly piece together the fossil record of life in the form of a set of correlations between contents of fossil-bearing rock layers (strata) and their relative ages. No one area on earth contains a continuous sequence of rock strata from the beginning of the earth to the present. Neither does any one location contain a continuous sequence of fossils. The task confronting geologists and paleontologists was to place these scattered rock layers into the proper time sequence. By carefully examining sedimentary layers in many parts of the world and correlating the fossils they contained, a continuous sequence of rock strata and fossil life was attained. The outlines of this record, known as the **geological time scale,** are presented in table 30.1.

But there still was no way to assign an absolute age to any single rock layer or fossil. With the discovery of radioactivity and the characteristics of radioactive decay, however, a means was devised to assign specific dates to particular rock layers.

Absolute Dating Techniques

Radioactive atoms have unstable atomic nuclei. When these nuclei break down (decay), they emit characteristic particles or rays. The end result of this radioactive decay is that another kind of atom is formed. For example, **potassium-argon dating** is a method used for determining the age of geological deposits. Radioactive potassium (^{40}K) decays to form ^{40}argon (^{40}Ar). When a sedimentary rock is first formed, it would

Box 30.2
Stable Isotopes and the Diets of Fossil Animals

Studies of radioactive isotopes have been used for some time to date fossils. However, newer techniques for studying naturally-occurring, stable (nonradioactive) isotopes have opened the door for other kinds of fossil research.

Mass spectrometers can separate quantitatively the common stable carbon isotope (^{12}carbon) from the much less abundant stable carbon isotope (^{13}carbon). This permits the relative abundance of the two isotopes in organisms to be calculated. This technique can be very informative because of the way in which the two isotopes enter living material. During photosynthesis, there is discrimination in favor of $^{12}CO_2$—as opposed to $^{13}CO_2$—by the enzyme systems that incorporate carbon dioxide into organic molecules. The degree of discrimination is different in the two major photosynthetic pathways for carbon fixation, the C_3 and C_4 pathways (chapter 7). Organic material produced by C_3 plants has a different $^{13}C/^{12}C$ ratio than material produced by C_4 plants.

Furthermore, the $^{13}C/^{12}C$ ratios in plant material seem to be perpetuated in the cells and tissues of the animals that eat the plants. Therefore, it is possible to use $^{13}C/^{12}C$ ratios in tissue, or in feces, to determine whether an animal has been eating plants that use the C_3 photosynthetic pathway, plants that use the C_4 pathway, or a combination of the two.

A number of researchers are using these techniques to study dietary patterns in modern animals, but other geologists and biologists—including M. J. DeNiro, S. Epstein, N. J. van der Merwe, and L. L. Tieszen—are using the techniques on fossil organisms. There is, for example, enough carbon preserved in some animal fossils to allow determination of $^{13}C/^{12}C$ ratios for those organisms. Such data can provide information concerning the organisms' food preferences or the food sources available to them. Although results of these studies can be difficult to interpret, biologists are able to partially analyze the diets of long-dead animals, including even members of extinct species.

Comparative anatomists compare many anatomical features of contemporary animals, but the comparative study of skeletal anatomy is most important to paleontologists because fossil evidences of anatomy consist almost entirely of skeletal material.

If all species of organisms have arisen from previously existing species, it would follow that those species that share a recent common ancestor would be more similar genetically. Over time, the number of genetic differences would increase. Thus, as you might expect, distantly related species differ more, both genetically and structurally, than do more closely related species.

Fundamental similarities in structure that have arisen through descent from a common genetic ancestor are called **homologies.** A good example of homology is the vertebrate forelimb (figure 30.13). All vertebrates are thought to have evolved from a common ancestor, and despite the fact that bird, whale, and human forelimbs serve different functions, they all possess the same fundamental bone structure.

Analogy is another important concept derived from comparative anatomy. Analogous structures are similar in function and appearance but differ in their fundamental structural plan. The wings of birds and insects are analogous structures. In both cases, the organs are used for flying, and each has a broad, flattened shape that is well adapted to this function. However, the wings of birds and insects are structured differently. The insect wing is a stiffened membrane supported by hard chitinous veins, while the bird wing has an internal bony skeleton covered by skin and feathers.

Analogous structures differ in basic form because they have arisen separately. The ability to fly evolved independently in birds and insects. Superficial similarities in the structures of their wings are due to the fact that a broad, flat surface is a prerequisite for flight. Thus, the ability to fly has appeared several times in the history of life. Analogous structures further illustrate convergent evolution, which, as you recall, is evolution from different ancestries toward a common adaptation to a similar life-style (figure 30.14).

Comparative anatomy also identifies **vestigial structures,** those structures that appear in a simple, poorly developed form in some organisms, but in a fully developed, functional

Figure 30.13 Homology in the vertebrate forelimb. All of these forelimbs are built on a fundamentally similar framework of the same set of homologous bones. This indicates that they share a common ancestry even though the limbs now look very different from one another and are adapted to a variety of functions. Homologous bones are indicated as follows: humerus (upper arm)—lavender, radius (forearm)—blue-green, ulna (forearm)—pink, carpals (wrist)—white, metacarpals (palm) and phalanges (digits)—tan. The digits are numbered, beginning with the first digit (thumb).

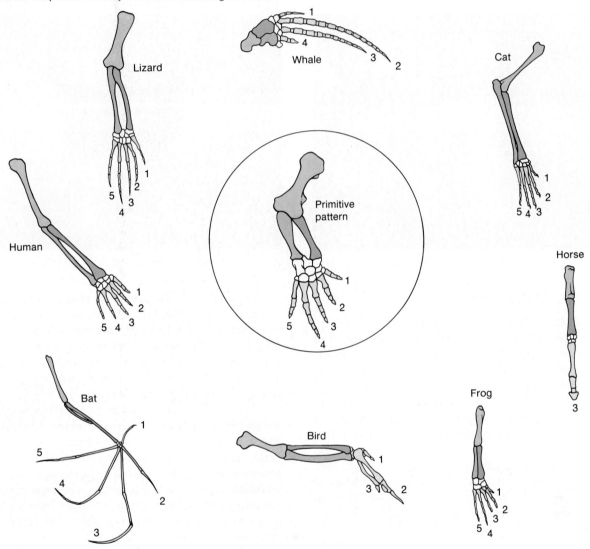

form in other, closely related animals; or that are poorly developed structures in a modern organism but thought to have been fully functional in its ancestors. Vestigial structures include the rudimentary, functionless eyes of cave-dwelling animals, the wisdom teeth of humans, and the three to five caudal vertebrae present in humans (remnants of the tail structure found in most other mammals).

One explanation for the decline of certain structures is that natural selection favors those individuals in which an unneeded structure is reduced because they expend less energy on its development and maintenance. For example, there is selection against full expression of the genes for eye development in cave-dwelling animals that live in complete darkness. Very slowly, over long periods of time, natural selection may completely remove such genes from populations.

However, it appears that the final instances of expression of certain genes may occur long before the genes are eliminated from the genome. Chickens, for example, have been found to retain genes for the development of tooth enamel, even though the genes are not expressed in modern chickens since they do not have teeth.

Figure 30.14 Analogous structures. A bird's wing (a) has feathers attached to skin covering a skeletal framework, and the wings of an insect (b) are stiff and membranous. Analogous structures have evolved from very different evolutionary precursors and have different basic structures, but they serve the same function. This independent evolution of wings as an adaptation for flight is another example of convergent evolution.

(a)

(b)

Comparative Embryology

Studies of the developmental stages of vertebrate animals reveal similarities that are not apparent in the adult stage. The early embryonic stages of fish, reptiles, birds, and mammals are surprisingly similar. In the past, these similarities were interpreted to mean that among vertebrate animals, "ontogeny recapitulates phylogeny"; that is, the stages of an organism's embryonic development (ontogeny) repeat the evolutionary history of the species (phylogeny). This notion has been discarded in modern evolutionary theory because embryos do not, in fact, pass through stages in which they resemble ancestral adults. Instead, developing embryos of each major vertebrate group pass through some stages in which they resemble the embryos of the other vertebrate groups.

Molecular Evolution

In addition to homologies found in anatomical structures, biochemists have found many homologies at the molecular level. Virtually all living things possess the same genetic material (DNA), use the same genetic code, and conduct energy conversions involving the same molecule (ATP).

The fact that practically all living things possess DNA as their genetic material implies that this trait developed very early in the evolution of life and has been passed down through the millenia to the organisms that have followed. It appears that since the DNA method of hereditary transmission and genetic control has proven to be efficient and effective, natural selection has not favored change in fundamental hereditary mechanisms.

Protein homologies have been of special interest to scientists studying evolution. Biochemists can determine the exact amino acid sequence of protein molecules. From this information, the makeup of genes can be inferred because amino acid sequences in proteins reflect nucleotide sequences in DNA molecules. This has made possible genetic studies of evolutionary relationships.

The evolutionary process, as revealed by study of protein molecules, has been very conservative in some cases, but much less so in others. The amino acid sequences of certain histones (proteins closely associated with DNA in chromosomes) have been highly conserved through the course of evolution. One of them has amino acid sequences that are virtually identical in molecules extracted from a wide range of organisms. For example, pea plants and cattle differ by only two of the 102 amino acids in these molecules. Apparently, the proper functioning of histones depends upon the maintenance of a very specific sequence of amino acids; amino acid substitutions in histone molecules that may have appeared as a result of mutation probably were eliminated by natural selection.

However, other proteins have not been so highly conserved, and we can use their amino acid sequences to estimate the closeness of evolutionary relationships among different species. If the amino acid sequences in proteins of two species are very similar, we can assume that these two species are closely related and have relatively recently evolved from a common ancestor. If, on the other hand, the proteins have very different amino acid sequences, the two species are distantly related, and have probably been separated from a common ancestor for a relatively long time.

Table 30.2
Evolution of Cytochrome *c*

Comparison between Different Organisms

Organisms	Number of Differences in Amino Acid Sequences
Cow and sheep	0
Cow and whale	2
Horse and cow	3
Rabbit and pig	4
Horse and rabbit	5
Whale and kangaroo	6
Rabbit and pigeon	7
Shark and tuna	19
Tuna and fruitfly	21
Tuna and moth	28
Yeast and mold	38
Wheat germ and yeast	40
Moth and yeast	44

Comparison with Human Cytochrome c

Organism	Number of Differences in Amino Acid Sequences
Chimpanzee	0
Rhesus monkey	1
Rabbit	9
Cow	10
Kangaroo	10
Duck	11
Pigeon	12
Rattlesnake	14
Bullfrog	18
Tuna	20
Fruitfly	24
Moth	26
Wheat germ	37
Mold (*Neurospora*)	40
Baker's yeast	42

From Volpe, E. Peter, *Understanding Evolution,* 5th ed. Compiled from studies by R. E. Dickerson. © 1967, 1970, 1977, 1981, 1985 Wm. C. Brown Publishers, Dubuque, Iowa. All Rights Reserved. Reprinted by permission.

Data from studies on cytochrome *c,* an electron carrier molecule in mitochondria, illustate this principle. The amino acid sequences in cytochrome *c* from humans and chimpanzees are identical, indicating relatively recent divergence of ancestors. Humans and chimpanzees differ from rhesus monkeys by only one amino acid of the 104 that constitute cytochrome *c,* but they differ from dogs by 11 amino acids, from rattlesnakes by 14, and from tuna by 20 (table 30.2).

Using this evidence about the accumulation of protein dissimilarities which have arisen through time by gene mutation, biologists can estimate the relative length of time since ancestors of present-day species diverged from one another.

Molecular Clocks

It is possible, however, that we can learn even more from accumulated changes in DNA nucleotide coding sequences. There is evidence that over long periods of time, gene mutations that alter nucleotide sequences occur at relatively steady rates. This permits biologists to use nucleotide sequence differences as **molecular clocks** that make it possible to calculate how long it has been since the ancestors of two present-day species diverged from a single common ancestor.

By studying the DNA nucleotide sequences of specific genes directly, biologists can analyze these molecular clocks of evolution. An important result of this analysis is the discovery that each gene has it own characteristic rate of nucleotide substitution. This means that, in a sense, there are a number of molecular clocks in an evolving genome, each ticking at its own rate. Nevertheless, phylogenies (interpreted patterns of evolutionary descent) based on molecular clocks generally agree with phylogenies developed by paleontologists using the fossil record and radiometric dating.

There is a cloud on the horizon, however, because of the recent discovery that nucleotide substitutions in certain genes occur at faster rates in some animals (rats and mice) than in others (humans). It is not yet clear what implications this finding has for future applications of the concept of molecular clocks in the study of evolutionary relationships.

Origins: Some Evolutionary Questions

Techniques such as radiometric dating, along with physical evidence from the fossil record and data concerning molecular evolution, provide us with concrete evidence about evolution and the time frame within which organisms have evolved. However, there are several fundamental aspects of the history of life on earth about which we have only circumstantial evidence: the origin of life and the evolution of eukaryotic cells.

The Origin of Life

When biologists survey the vast diversity of living things, they often wonder how life originated on earth. There is no way of knowing with certainty, but a number of ideas have been suggested.

The earth probably formed by condensation of dust and gases approximately 4,500 to 5,000 million years ago. The lighter gases, such as hydrogen and helium, were lost as the planet formed because it was too small to hold them by gravity. It is thought that a primitive atmosphere was produced later by volcanic action, but there is considerable disagreement about the exact nature of this primitive atmosphere. Harold Urey proposed in the early 1950s that

Figure 31.3 A cheetah cub. Cheetahs in South Africa are very homogeneous genetically, apparently as a consequence of a population bottleneck in the recent past. Loss of genetic diversity is a threat to endangered species that are reduced to small populations.

Certain other physical traits also were examined. Frequencies of some traits, such as middigital hair patterns, hyperextensibility of the thumb, and attached earlobes, were significantly lower among the Dunkers than in the surrounding population. On the other hand, Dunkers have essentially the same incidence of left-handedness as is found in the surrounding population.

We can attribute the special set of gene frequencies found among the Dunkers to the founder effect and genetic drift.

The Bottleneck Effect

Another special case of genetic drift is the **bottleneck effect,** a situation in which a population is drastically reduced by factors other than ordinary natural selection. For example, a particularly harsh winter in the temperate zone will greatly reduce the number of individuals in a population of organisms such as houseflies, that survive to reproduce during the next season. In this way, a large part of the variability in the

housefly population may be "squeezed out." Subsequently, relatively few individuals will give rise to all the flies produced during the next summer, and the population may become more homogeneous genetically than it was before.

The bottleneck effect has particular relevance for species that are near extinction. Even if a small number of protected individuals can reproduce successfully enough so that they increase in numbers, new problems may arise. Disease and environmental stress can have devastating effects if the remaining population is very homogeneous. The wider ranges of genetic variability normally found in larger natural populations usually permit at least some individuals to survive crisis situations.

An example of such an endangered species is the cheetah (figure 31.3). Genetic studies of cheetahs in South Africa reveal an extremely low level of genetic variability. Biologists hypothesize that this may be a result of a population bottleneck in the recent past. Cheetahs may face a dim future even if effective protection and conservation methods are applied.

Migration

Unless there are barriers that prohibit movement of individuals from one population to another, **gene flow** (gene migration) occurs as a result of migration and interbreeding. Gene flow adds to the variability of most natural populations by introducing new alleles. If migration is extensive it can significantly alter gene frequencies.

Mutation and Variability

As you know, variation in natural populations is a fundamental part of natural selection. While mutation is the source of *new* genes in gene pools, it is sexual reproduction that ensures that different alleles produced by mutation will be combined in many different ways to provide a rich variety of phenotypes in the population. In each generation, sexual reproduction segregates sets of parental genes during meiosis and recombines the genes in new combinations in the offspring.

Mutations can occur as modifications of chromosome structure. Most often, however, mutations are specific changes in individual genes that result from base substitutions in nucleic acid molecules (chapter 24). The frequency with which these changes occur varies because each gene locus has its own characteristic mutation frequency.

Mutation rates of different alleles for the same character are rarely in equilibrium; that is, the rate of **forward mutation**—mutation from the more common allele to the less common allele—is seldom the same as the rate of **reverse mutation**—mutation in the opposite direction. The difference between the two is a **mutation pressure** that tends to produce a very slow change in allele frequencies.

For some time biologists thought that mutation pressure was important in determining the direction of evolutionary change, but this no longer seems likely. It would take an enormous amount of time for mutation pressure alone to cause significant changes in allele frequencies. Furthermore, mutations are random events affecting individual genes, but they occur within the context of a whole constellation of genes in a gene pool. Natural selection acts on phenotypes that represent composites produced by the expression of these many genes.

The real importance of mutation is that it is the only mechanism by which new genetic material enters the gene pool, as *some* mutations produce new alleles.

Modern techniques reveal that mutations involving individual base substitutions in DNA molecules occur so often that it appears that most must be **neutral mutations,** mutations that are without effect. Neutral mutations might, for example, result in the substitution of amino acids in an enzyme molecule somewhere other than in the portion of the protein that makes up the enzyme's active site. Thus, the mutation would not affect the function of the gene's product.

However, of those mutations that are not neutral, a new allele is more likely to prove detrimental than beneficial when the gene is expressed. Since the gene pool of any population is the product of a long period of natural selection during which alleles producing more adaptive traits have increased in frequency and those producing less adaptive traits have decreased in frequency, it is not surprising that completely random mutational events usually produce alleles that are less adaptive than existing ones. However, most mutational events produce alleles that are recessive to the original allele. Such recessive mutant alleles are not expressed as phenotypes until they occur in the homozygous condition. Thus, new recessive mutant alleles are not immediately exposed to the effects of natural selection.

Maintenance of recessive alleles in a gene pool is important because this reservoir of genetic variability may prove advantageous at a later time. Should environmental conditions change, the adaptive value of a specific allele may also change.

An example of a detrimental mutant gene that can become beneficial under different conditions is seen in the water flea *Daphnia* (figure 31.4). Normally, *Daphnia* thrives at temperatures around 20° C but cannot survive temperatures of 27° C or more. There is, however, a mutant strain of *Daphnia* that requires temperatures between 25° C and 30° C, and cannot live at temperatures around 20° C. Thus, as is often the case, the adaptive value of this mutant condition is dependent on environmental conditions.

Balanced Polymorphism

Balanced polymorphism is an evolutionary situation in which several very different phenotypic expressions are maintained in a population without one increasing in frequency at the expense of the others.

Often the selective forces, if any, at work in balanced polymorphism are not evident. Human ABO blood types are one such example. The relative frequencies of the alleles that produce the various blood types are quite stable in populations, implying that the positive and negative selective forces acting on individuals with different phenotypes (blood types) must be in balance, thus producing little selective pressure for change in allele frequencies.

Figure 31.4 The microscopic water flea *Daphnia*, which normally lives and thrives at temperatures around 20°C and dies when temperatures rise above 27°C. A mutant strain of *Daphnia* lives at temperatures between 25 and 30°C, and cannot live at 20°C.

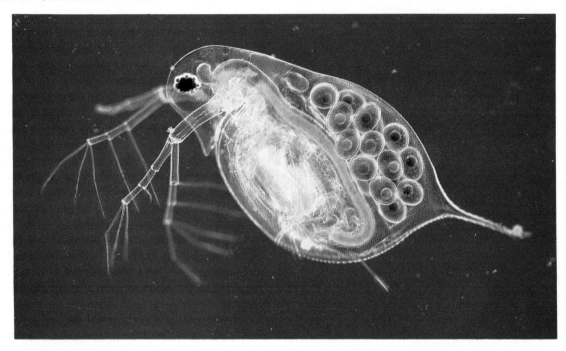

Heterozygote Superiority

Selective factors are understood in some cases of balanced polymorphism. One such case involves **heterozygote superiority**, a situation in which natural selection favors individuals that are heterozygous for a gene over either of the homozygous genotypes. Human sickle-cell anemia is a good illustration of heterozygote superiority, but it also shows that there can be differences in the survival value of a specific allele under different environmental conditions.

Sickle-cell anemia is an inherited disease that results from the formation of abnormal hemoglobin molecules in red blood cells (see page 52). Expression of the *S* allele produces normal hemoglobin, while expression of the *s* allele produces abnormal hemoglobin. Individuals with the *Ss* genotype have approximately 50 percent normal and 50 percent abnormal hemoglobin in their red blood cells (figure 31.5).

Individuals with the *ss* genotype suffer from sickle-cell anemia. When they are physically active, their red blood cells collapse and assume a characteristic sickle shape. These sickled cells can block small blood vessels and cut off circulation to a particular region of the body, which in turn can result in tissue necrosis (death). Individuals with the *ss* genotype have difficulty with even mild exercise and are generally weak. Many die before reaching reproductive maturity. Heterozygous (*Ss*) individuals sometimes experience problems when they exercise vigorously, but they are generally healthier than *ss* individuals.

You might expect that the *s* gene would have a very low frequency in the gene pool. Many *ss* individuals die young, and heterozygous (*Ss*) individuals also are seriously enough affected at times to be at a selective disadvantage to individuals who are homozygous for the normal hemoglobin allele (*SS*). However, when geneticists studied populations of black Africans, in which the disease is most common, they found that the recessive allele had a surprisingly high frequency (0.2 to as high as 0.4 in a few areas).

Further research showed that another important factor affects the frequencies of these alleles. In regions where malaria occurs, the heterozygous (*Ss*) genotype had an adaptive advantage over both the *SS* and the *ss* genotypes (figure 31.6). Because the malarial parasite, which enters red blood cells, does not inhabit cells that contain the abnormal form of hemoglobin produced by expression of the *s* gene, individuals with the *SS* genotype suffer a higher death rate from malaria than do individuals of the *Ss* genotype. And since a

Figure 31.5 Electrophoresis employs an electric current flowing through a gel to separate similar molecules from one another. Electrophoresis of hemoglobin extracted from red blood cells reveals differences between hemoglobin from normal individuals and those suffering from sickle-cell anemia. It also demonstrates that both types of hemoglobin are produced in heterozygous individuals.

Hemoglobin Electrophoretic Pattern

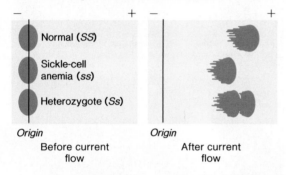

Figure 31.6 Geographic distribution of the sickle-cell condition, shown as percentages of the population afflicted with sickle-cell anemia. The frequency of the *s* gene is highest in areas where malaria is common.

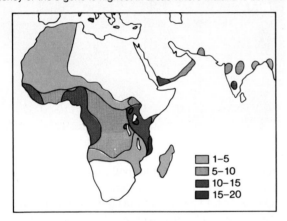

relatively high proportion of *Ss* individuals reproduce under these conditions, the frequency of the recessive gene remains greater than would be expected in the absence of the malaria factor. As you might expect, among American black populations, which are not exposed to malaria, the frequency of the *s* allele is lower than in black populations inhabiting malarial regions.

In the case of the sickle-cell phenomenon in Africa, heterozygotes have a strong selective advantage over either of the two types of homozygotes. Even though both homozygotes are selected against, neither gene is eliminated, since reproduction of the favored heterozygous individuals contributes both genes in equal quantities to subsequent generations. Such heterozygote superiority strongly favors balanced polymorphism.

Natural Selection

Natural selection is a key factor in any changes in gene frequencies within the gene pools of natural populations. Every other factor that causes evolutionary change must be viewed in the context of natural selection. This is because, as Charles Darwin correctly surmised, natural selection is a driving force that causes evolutionary change in living things.

Directional Selection

When the environment changes or when organisms migrate into new environments, natural selection operates to select those alleles that confer traits that are adaptive under the new set of environmental conditions. Because these changes represent a progressive adaptation to a changing environment, this type of selection is called **directional selection.**

Directional selection is clearly illustrated by **industrial melanism,** a progressive change in the color of moths that has occurred in industrial areas, particularly in Great Britain and continental Europe. The best-documented example of industrial melanism occurred in the population of peppered moths (*Biston betularia*) in England between the mid 1800s and the early 1900s.

In the mid 1800s, about 99 percent of the peppered moths were light-colored; a darker (melanic) form appeared only rarely. (Genetically, body color in the peppered moth can be treated as a simple dominant-recessive gene relationship, with the allele for dark body color being dominant to the allele for light body color.) Biologists proposed that the greater survival of the lighter-colored moths was due to the fact that it was more difficult for predatory birds to see these moths against light-colored vegetation, while the melanic forms stood out sharply and were quickly attacked and eaten.

The situation was different during the latter half of the nineteenth century, when great quantities of soft coal were burned to fuel the fires of industry. This produced a large quantity of air pollution, and soot settled over the areas surrounding industrial centers, progressively darkening the vegetation. Against the soot-darkened tree trunks, light-colored moths now stood out clearly, while the melanic form enjoyed the advantage of protective coloration. The adaptive advantage of the light-colored form of the peppered moth was lost, and natural selection favored the darker, melanic form (figure 31.7).

Figure 31.7 Dark and light forms of the peppered moth *Biston betularia*. Dark and light forms on a lichen-coated tree trunk in an unpolluted region (a) and on a soot-darkened tree trunk near Birmingham, England (b).

(a)

(b)

As a result, by 1900 the peppered moth population of industrialized areas of England was 90 percent melanic and 10 percent light colored. The frequency of the dominant allele had increased over this period of time, while the frequency of the recessive allele had decreased dramatically.

In order to test the hypothesis that the selecting factor in this situation involved predatory birds that feed on peppered moths and that moth body color influences prey selection, H. B. D. Kettlewell of Oxford University reared thousands of peppered moths in the laboratory. He then released equal numbers of the two color phases in two areas: (1) the nonindustrial, unpolluted Dorset area, and (2) the industrial and highly polluted Birmingham area. Each released moth was marked so that when recaptured, Kettlewell could distinguish those he had released from other moths that had inadvertently been captured.

From the unpolluted Dorset area Kettlewell recaptured 14.6 percent of the light-colored moths and 4.7 percent of the melanic form. From the polluted area around Birmingham 27.5 percent of the melanic form were recaptured as compared with 13 percent of the light-colored moths. Using blinds set up in each area, Kettlewell observed directly the selective predation of peppered moths by birds and even recorded the process on film.

Kettlewell's experiment demonstrated that a moth's body color is indeed a factor in its survival, with the selective advantage of body color varying with the color of the background vegetation. Industrial melanism clearly shows directional selection in which environmental change is accompanied by differential selection of phenotypes, resulting in changes in gene frequencies in the population.

Other Types of Selection

Directional selection generally results in change in one direction in response to changed environmental conditions. However, natural selection can have other effects on gene pools in other situations (figure 31.8).

Figure 31.8 Three types of selection. Each set of curves symbolizes the status of a particular set of characteristics in a population. Horizontal axes indicate the ranges of variability in the populations; vertical axes, the number of individuals distributed in different parts of the ranges. Directional selection acts against individuals possessing one of the phenotypic extremes. It tends to result in the elimination of one phenotype and a proportional increase in others. Disruptive selection favors extreme phenotypes and acts against intermediate ones. It produces two divergent subpopulations with very different gene pools. Stabilizing selection operates when conditions remain stable for long periods; it tends to make the gene pool more homogeneous.

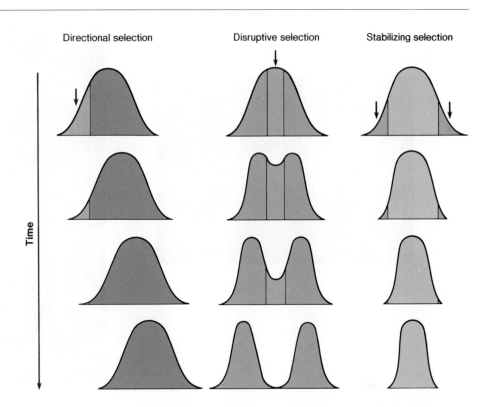

In **disruptive selection,** both categories of phenotypic extremes are favored over the average phenotype in a population. This can happen when a change in conditions removes the selective advantage of the average phenotype. Disruptive selection tends to divide a population into two contrasting subpopulations.

Another type of selection is **stabilizing selection,** which functions when the conditions under which a population is living remain constant over a period of time. Stabilizing selection works against phenotypic variation. It helps to conserve the adaptive fit of the population to its environment by selecting against phenotypes produced by expression of new genetic combinations. Thus, selection is not always an agent of change.

Adaptations

All biology texts, including this one, are books about adaptations, because all characteristics of living things must be examined in the context of adaptation. An **adaptation** is a characteristic of an organism that increases the organism's **fitness** for life in its environment. Fitness is a measure of the likelihood that an organism will live and succeed in reproducing. Fitness, therefore, determines the odds that an organism will make a genetic contribution to the next generation and is a critical measure of evolutionary success.

Each living thing possesses a battery of adaptations that contribute to its fitness. We can illustrate the complexity and diversity of adaptation by examining some examples of special types of adaptations, such as those involved in certain interactions between organisms of different species.

Cryptic Coloration and Mimicry

For most organisms, one of the most urgent and continuing threats to survival and eventual reproductive success is the danger of attack by predators. Many adaptations associated with predator avoidance are truly remarkable. For example, some organisms are camouflaged by **cryptic coloration** ("hidden coloration") that makes them virtually undetectable when they are in position against their normal background (figure 31.9). Other animals, like the well-known chameleon, undergo complex color changes that improve their "match" with the background as they move from one location to another.

Figure 31.9 Cryptic coloration. For cryptic coloration to provide effective concealment, cryptically colored organisms must be behaviorally adapted to remain motionless when danger threatens. (a) A leaf katydid from Brazil. (b) A stonefish.

(a)

(b)

Through **mimicry,** some organisms, instead of being hidden from the eyes of potential predators, present a showy but misleading appearance resembling that of another organism that predators normally avoid (figure 31.10). The similarity between the mimic and the model may be so close that predators avoid both to nearly the same extent. We will explore mimicry further in chapter 34.

Coevolution

Coevolution is a situation in which the mutual evolutionary interaction between two species is so intense that each exerts a strong selective influence on the other. We have already described some examples of adaptations resulting from coevolution. For example, rumen microorganisms and their grazing hosts (chapter 11) are products of a long coevolutionary process.

Some particularly striking products of coevolution are the relationships between certain flowering plants and the animals upon which they depend for pollen transport from flower to flower.

Most flowers pollinated by bees are showy and bright, and all of them are open during daytime hours, when worker bees do their foraging. Bee-pollinated flowers tend to be blue or yellow; few of them are red, since bees are blind to red colors. Furthermore, most of them have sweet, aromatic odors because bees depend on smell to locate nectar-containing flowers. Bee-pollinated flowers also have petal arrangements that provide bees a place to land (figure 31.11).

Hummingbirds, on the other hand, see red well but blue only poorly. Because hummingbirds have a poor sense of smell, aroma is not an important factor in attracting them to flowers. Like bees, hummingbirds forage in the daytime, but they require no landing perch. A hummingbird hovers in front of a flower and inserts its long beak and tongue, which is specialized for sucking, into the nectar-containing part of the flower. Characteristically, hummingbird-pollinated flowers are red or yellow and many of them are odorless.

Moths, such as sphinx moths, also hover as they feed on nectar from flowers, but they feed during evening or nighttime hours. Thus, the flowers that moths pollinate are ones that remain open at night, and in most cases are light colored.

The scarlet gilia (*Ipomopsis aggregata*), a plant that grows on mountainsides in the southwestern United States, changes flower color during its growing season and thus attracts the particular pollinators that are present at different times during its flowering period (figure 31.12). Early in the

Figure 31.10 Mimicry. A lacewing (*below*) that is a mimic of a wasp (*above*). The lacewing presents no threat to predators, but it, like the stinging wasp, is avoided.

Figure 31.11 Pollen that sticks to a bee as it pushes into a flower in search of nectar is carried to other flowers the bee subsequently visits. Bee-pollinated flowers tend to have landing perches for their pollinators.

Figure 31.12 Complex interactions between a plant and its pollinators. (a) A hummingbird feeding on the nectar of a red flower of the scarlet gilia (*Ipomopsis aggregata*). (b) Later in the season, when hummingbirds have left the area, scarlet gilias produce light-colored flowers that attract moth pollinators. A white-lined sphinx moth is shown here visiting these light-colored scarlet gilia flowers.

(a)

(b)

Figure 31.13 Clownfish among the tentacles of an anemone as an example of a symbiotic relationship. Clownfish deposit bits of food on the anemone's tentacles. In turn, anemones provide protection for the clownfish; their tentacles are covered with nematocysts that normally discharge and sting fish and other animals that contact them. Clownfish, however, swim safely among these formidable tentacles, apparently because mucus on the surface of a clownfish's body permits contact without triggering the stinging response.

season, when hummingbirds are present, most high-altitude gilias produce bright red flowers. Later in the season, after the hummingbirds have left the area, many of the plants produce lighter-colored flowers that attract moths, which continue to feed for several hours after dark. Ken Paige and Thomas Whitham, who studied this adaptation, reported that in late season, the percentage of gilia plants with light-colored flowers that set fruit (indicating successful pollination and fertilization) was significantly higher than that of red-flowered gilias.

Even behavioral adaptations of pollinating animals show evidence of the coevolutionary process. Bees, for example, tend to feed "faithfully" on one kind of flower at a time. This is adaptive for bees because plants of many species flower practically in unison, and by tending to seek more flowers like those in which they have found nectar, the bees' search for nectar is more efficient. If bees tend to move from plant to plant of the same species, pollination is effectively accomplished. A hovering moth likewise moves systematically from flower to flower in a flower bed. It pauses before a blossom long enough to unroll its long, sucking proboscis and extend it deep into the flower to feed. Then it rerolls its proboscis and moves on, repeating the process again and again until every flower has been visited.

Symbiosis

Some interactions between organisms that are products of coevolution are much more intensive and permanent than the temporary encounters between flowers and animal pollinators. **Symbiosis** (literally "living together") is a general category of relationships in which one organism lives in an intimate and continuing association with another. Symbiotic relationships range from the exploitative relationship of parasitism in which one member benefits at the other's expense to mutualistic relationships in which both organisms benefit from the association, as in the case of the algae and fungi that make up lichens (p. 954). In these and the many other symbiotic relationships in the living world, the impact of coevolution is apparent (figure 31.13).

Species and Speciation

The changes in gene frequency we have considered so far, such as the increasing incidence of antibiotic resistance in bacteria, are changes that occur within the gene pools of populations and constitute what is called **microevolution.** On occasion, however, sufficient genetic change has accumulated so that a new species has been formed. This is called **macroevolution.** Let us now turn our attention to the species concept and the mechanisms by which macroevolutionary change (the production of new species) takes place.

Species Defined

A **species** is a population of organisms that may display a range of phenotypic (and genotypic) variation, but that nonetheless represents a biological entity distinct from all other populations. One definition of species states rather simply that all members of a species are more like one another than they are like members of any other species.

However, the concept of **reproductive isolation** gives us a more satisfactory definition of a species. This concept defines a species as an interbreeding population that possesses a genetic identity because its members do not interbreed with members of any other species. Reproductive isolation may thus serve as a working criterion for species definition.

Reproductive Isolation

Certain factors prevent successful interbreeding between individuals of different species, and thus maintain reproductive (and genetic) isolation of the species. There are a variety of these **reproductive isolating mechanisms.**

In animals, certain *premating isolating mechanisms* operate to prevent mating between members of isolated populations. One such mechanism is **mechanical isolation.** For example, there may be differences in the reproductive structures of members of two populations that make copulation difficult, if not impossible. **Habitat isolation** occurs when two species inhabit the same general area, but occupy very different specific habitats. Thus, members of these two species seldom venture into each other's habitats. **Seasonal isolation** is a third premating mechanism. In this case, different species may become sexually active at different times of the year. A final premating isolating mechanism is **behavioral isolation.** The various songs, calls, display patterns, and courtship "dances" of birds and other animal species attract mates of that species but eliminate members of other species as possible mates, thus preventing gene flow between them.

Prefertilization reproductive isolation mechanisms function in plants also. For instance, when pollen grains of one species fall on the stigmas of another species, pollen tubes usually do not develop. Cross-pollinations among such species cannot result in fertilization.

Postmating isolating mechanisms operate after copulation has taken place. **Gamete mortality** occurs in matings between members of different animal species with internal fertilization. Sperm produced by males of one species may not be able to survive in the female reproductive tract of the other species, or the zygote may not be able to begin normal development. **Hybrid inviability** refers to the death, anytime before reproductive maturity, of hybrid offspring resulting from matings between members of different species. Even if hybrid offspring survive, there may be **hybrid sterility,** in which hybrid offspring are healthy and survive to reproductive age, but are incapable of producing offspring when mated with either parent type or with another hybrid. A familiar example of hybrid sterility is the mule, the hybrid offspring of a horse and a donkey. Mules are strong animals, well suited to certain work tasks, but despite scattered reports of their producing offspring, mules generally are sterile.

In situations characterized by hybrid inviability and hybrid sterility, members of two populations may mate successfully and produce offspring, but in neither case is a new genetic line of descent established. These isolating mechanisms function either because of significant genetic or chromosomal differences between the parents. We will note later, however, that barriers to hybridization between species are much less rigid in plants than in animals.

Figure 31.16 Phyletic gradualism and punctuated equilibrium. (a) Phyletic gradualism suggests that new species form by the gradual accumulation of different sets of adaptive traits over very long periods of time, and that a series of intermediate forms are part of the process. (b) According to the punctuated equilibrium hypothesis, speciation occurs by abrupt branching that takes place relatively rapidly. Thus, a series of intermediate forms would not be expected in the fossil record.

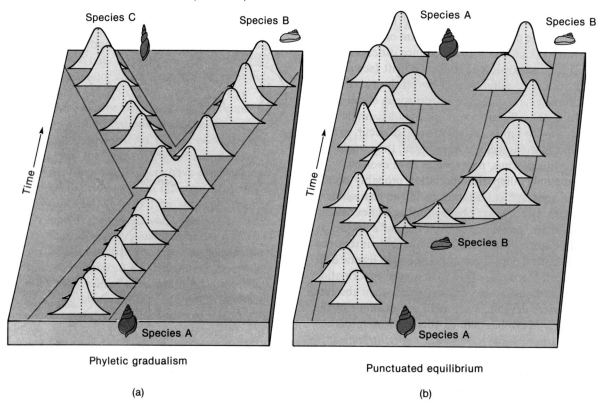

Phyletic gradualism

(a)

Punctuated equilibrium

(b)

Such chromosomal events occur relatively frequently in plants. In fact, about half of all flowering plant species in existence today arose by polyploidy. Many of these, including economically important plants such as wheat, originated as hybrid polyploids.

There is some debate, however, as to whether or not truly sympatric speciation occurs in animals. No habitat is entirely homogeneous, and at times, small subpopulations of an animal species may become highly specialized to localized environmental conditions. Can enough differences accumulate between these subpopulations to produce reproductive isolation, even though members of these subpopulations have the opportunity to interbreed? There is some evidence in certain insect species that specialization of subpopulations to utilize very different food sources might indicate that these subpopulations are on their way to becoming separate species. However, many questions about speciation within truly sympatric animal populations still remain.

The Tempo of Evolution

Early evolutionists often thought of evolution in terms of gradual change from one species to another. There is evidence of such straight-line descent, called **anagenesis,** in the fossil record. However, a more common pattern of speciation appears to have been **cladogenesis,** formation of a new species by branching from a previously existing species.

In recent years an interesting debate has developed over the tempo of this evolutionary branching. The traditional view has been that species are formed as a result of a slow, gradual accumulation of adaptive traits in populations through a selective process that requires many thousands or even millions of years for the separation of new species. This is the concept of **phyletic gradualism** (figure 31.16). Recently, however, a number of biologists, notably Stephen Jay Gould and Niles Eldredge, have advanced the hypothesis of **punctuated equilibrium,** which suggests that evolution has involved a series of relatively rapid speciation events. They argue that

Box 31.1
Evolutionary Controversies

The controversy between the phyletic gradualism and punctuated equilibrium viewpoints is by no means the first such controversy in evolutionary theory. Like many other scientific fields, evolutionary theory has had its share of controversy and disagreement. Even Darwin and Wallace, who together formulated the theory of natural selection, came to disagree about the actual role of natural selection in the evolutionary process.

Wallace favored a very strict application of the concepts of natural selection. He and others thought that each and every characteristic of every organism was a product of natural selection. Darwin, however, was convinced that natural selection was the main factor, but not the only factor, involved in evolutionary change. He viewed an organism as an integrated whole and thought that while natural selection might result in adaptive changes in one part, other parts might change in ways that had neither positive nor negative effects on the odds of reproductive success, which is the final test of adaptation according to the theory of natural selection.

Historically, a key part of evolutionary theory has been the idea of long periods of accumulating gradual change, usually associated with geographic isolation, and the eventual emergence of new and reproductively isolated species. The concept of punctuated equilibrium challenges this idea. It proposes instead that evolution of any particular line is characterized by long periods of relative stability punctuated by periods of abrupt change, change that occurs in periods of time that are relatively short compared to the enormity of the geologic time scale.

Disagreements such as the one that exists among evolutionary theorists regarding phyletic gradualism and punctuated equilibrium are not unusual in science. Unfortunately, however, people who do not think that life on earth has an evolutionary history cite such controversies as indicating some fundamental weakness in the theory of organic evolution, or that scientists are coming to doubt that an evolutionary process has occurred. These people have a mistaken view of science and the way scientists work. Such controversies are inherent in the nature of science and characterize most, if not all, fields of scientific endeavor. This vigorous debate is by no means a weakness in the theory of organic evolution. Rather it is a strength, an expected characteristic of an active field of scientific endeavor. Scientific progress is made through careful examination that leads to acceptance or rejection of competing conceptual schemes. The theory of organic evolution remains one of the central unifying themes of modern biology, and it provides one of the basic conceptual frameworks for interpreting biological phenomena.

Views of the history of life on earth, possibly more than any other set of scientific concepts, engender emotional responses and controversy of another sort. The great majority of scientists conclude that available evidence indicates a very long history of life on earth that is measured in thousands of millions of years and characterized by evolutionary descent. But some individuals believe that the history of life on earth is much shorter, possibly as short as 10,000 years, and that it is characterized by a series of divine creation events. Their view of life is essentially compatible with literal interpretation of the biblical creation story of the Judeo-Christian religious tradition.

Many scientists, however, do not find the idea of a long evolutionary process incompatible with their religious faith and experience. They do not feel that their faith is compromised because they interpret a creation story in terms of modern scientific understanding. They recognize that the biblical creation story was written in a form compatible with the experience of people living several thousand years ago. Possibly their view of faith and life can best be summarized with the words that Charles Darwin used following his summary of the theory of natural selection at the very end of the first edition of *On the Origin of Species:* "There is a grandeur in this view of life, with its several powers, having been originally breathed by the Creator into a few forms or into one; and that, whilst this planet has gone cycling on according to the fixed law of gravity, from so simple a beginning endless forms most beautiful and wonderful have been, and are being evolved."

throughout the greater part of its existence, a species displays very little change. This equilibrium is suddenly interrupted by rapid evolution of a small, local population at the periphery of the range of a species, and the process of branching into a separate new species occurs within a relatively short period of time.

Proponents of punctuated equilibrium suggest that the fossil record supports their interpretation of the evolutionary process because it shows long periods during which species remained essentially unchanged. They also point out the lack of transitional forms between taxonomic groups. Scientists who support phyletic gradualism have held that transitional forms are missing because relatively few organisms have been preserved in the fossil record, and it is not likely that, by chance, a large number of intermediate forms would be found. However, the supporters of punctuated equilibrium suggest that transitional forms are not found because there simply may not have been transitional forms. In other words, species formation occurs so rapidly that the long sequence of intermediate types predicted by phyletic gradualism were not a part of species formation.

One way that evolution could proceed by such leaps would be by the occurrence of a mutation, such as a major chromosomal reorganization that drastically alters early development so that the organism produced is extremely different from its parents. Under normal conditions, most such unusual individuals would probably die relatively young because they would not be well adapted to their environment. But occasionally, one or more would be very much better suited than ordinary members of the population for a rapidly changing or highly stressful environment.

Rapid speciation also might take place by other means, such as the occurrence of periods of very strenuous selection that favor rapid evolutionary change. Regardless of the mechanism by which it might occur, the concept of punctuated equilibrium has sparked a great deal of debate among biologists.

Unfortunately, some observers outside the field of biology misinterpret debates such as the one concerning phyletic gradualism and punctuated equilibrium. They mistakenly conclude that biologists "are beginning to have doubts about evolution." These biologists are *not* expressing doubts about the overwhelming evidence that indicates the occurrence of an evolutionary process. Rather, they are debating the way in which that evolutionary process proceeds and the mechanisms by which new species arise. Such debates actually stimulate biologists to reexamine the existing data and to seek new information regarding the process of evolution, thereby encouraging scientific progress.

Summary

Modern biologists explain the process of organic evolution in terms of changes in gene frequencies occurring in gene pools of populations. The Hardy-Weinberg principle provides a mathematical model of genetic equilibrium against which evolutionary changes in gene frequency can be measured. Gene frequency changes may be caused by genetic drift, migration, and mutation. Natural selection is a major force causing gene frequency change.

Each organism possesses a set of adaptations to its environment. Some adaptations are particularly striking because of their involvement in interactions between organisms of different species. Cryptic coloration and mimicry are adaptations that function in avoiding predation. In coevolution there is an intense mutual evolutionary interaction between two species in which each exerts a selective influence on the other.

A species is a population of organisms that is reproductively isolated from members of other populations. Species may also be defined on the basis of structural similarities and differences.

Allopatric speciation is a very important mechanism for species formation. A population is divided into smaller units and changes in gene frequencies accumulate within their individual gene pools until reproductive isolation is established, at which point new species have been formed.

It is also possible for species to form by sympatric speciation, as when the development of polyploidy reproductively isolates plants with different numbers of chromosomes.

There is disagreement among biologists concerning the tempo of species formation. Does speciation occur as a result of long-term, gradual accumulation of adaptive traits? Or are there relatively short periods of rapid species formation that punctuate longer periods of comparative equilibrium? Seeking answers to these questions helps to clarify the nature of the evolutionary process.

Questions for Review

1. Why does the synthetic theory of organic evolution deal with populations rather than individual organisms?

2. What is a gene pool?

3. Using either the Punnett square method or the algebraic method, demonstrate the genetic equilibrium described in the Hardy-Weinberg principle for a gene pool in which the alleles B and b have frequencies of 0.1 and 0.9, respectively. Assume that allele B is dominant to allele b.

4. What are some factors that cause changes in gene frequencies within a gene pool?

5. Define genetic drift.

6. Explain the founder effect.

7. What is the difference between stabilizing and directional selection?

8. Explain the significance of the bottleneck effect for the small remaining populations of endangered species.

9. Explain the importance of the concept of reproductive isolation in the definition of a species.

10. Distinguish the terms allopatric speciation and sympatric speciation.

11. What is cladogenesis?

12. How do proponents of the punctuated equilibrium hypothesis explain the scarcity of transitional forms in the fossil record? How does their explanation differ from that offered by proponents of phyletic gradualism?

Questions for Analysis and Discussion

1. Could you suggest some problems that might arise as a result of continuously feeding antibiotics to livestock and poultry to "keep them healthy and growing fast"?

2. For some people, phenylthiocarbamide (PTC) has a very bitter taste, but other people cannot taste it at all. The allele for tasting PTC is represented as T, and is dominant to the nontaster allele t. Both homozygous TT individuals and heterozygous (Tt) individuals are tasters. In a freshman biology lab, twelve of the twenty-four students could taste PTC and twelve could not. Use this student group as a population and calculate the frequencies of the genes T and t in the population's gene pool. What is the frequency of the genotype Tt? (Hint: The proportion of individuals who were nontasters (0.5) represents q^2 in the Hardy-Weinberg equation.)

3. What might Ernst Mayr's concept of peripatric speciation suggest about the reproductive isolation of Florida members of the species *Rana pipiens* from those in Vermont?

Suggested Readings

Books

Dobzhansky, T.; Ayala, F. J.; Stebbins, G. L.; and Valentine, J. W. 1977. *Evolution.* San Francisco: W. H. Freeman.

Grant, S. 1984. *Beauty and the beast: The coevolution of plants and animals.* New York: Charles Scribner's Sons.

Simpson, G. G. 1967. *The meaning of evolution.* rev. ed. New Haven, Conn.: Yale University Press.

Volpe, E. P. 1985. *Understanding evolution.* 5th ed. Dubuque, IA: Wm. C. Brown Publishers.

Articles

Ayala, F. J. 1983. Genetic variation and evolution. *Carolina Biology Readers* no. 126. Burlington, N.C.: Carolina Biological Supply Co.

Ayala, F. J. 1983. Origin of species. *Carolina Biology Readers* no. 69. Burlington, N.C.: Carolina Biological Supply Co.

Committee on Science and Creationism. 1984. Pamphlet. *Science and creationism: A view from the National Academy of Sciences.* Washington, D.C.: National Academy Press.

Cook, L. M.; Mani, G. S.; and Varley, M. E. 1986. Postindustrial melanism in the peppered moth. *Science* 231:611.

Friedman, M. J., and Trager, W. March 1981. The biochemistry of resistance to malaria. *Scientific American.*

Jones, J. S. 1981. An uncensored page of fossil history. *Nature* 293:427.

Lewontin, R. C. September 1978. Adaptation. *Scientific American.*

Meeuse, B. J. D. 1984. Pollination. *Carolina Biology Readers* no. 133. Burlington, N.C.: Carolina Biological Supply Co.

O'Brien, S. J., et al. 1985. Genetic basis for species vulnerability in the cheetah. *Science* 227:1428.

O'Brien, S. J.; Wildt, D. E.; and Bush, M. May 1986. The cheetah in genetic peril. *Scientific American.*

Paige, K. N., and Whitham, T. G. 1985. Individual and population shifts in flower color by scarlet gilia: A mechanism for pollinator tracking. *Science* 227:315.

Rensberger, B. April 1982. Evolution since Darwin. *Science 82.*

Wood, T. K., and Guttman, S. I. 1983. *Enchenopa binotata* complex: Sympatric speciation? *Science* 220:310.

Human Origins

32

Chapter Concepts

1. Humans belong to a relatively recently emerged species that is included in the order Primates of the class Mammalia.
2. The primates first evolved as arboreal animals, and certain characteristics of all primates reflect adaptation to tree dwelling.
3. Humans are descended from primates that became secondarily adapted to life on the ground.
4. Conclusions about the course of human evolution often are controversial because of disagreements concerning interpretation of the fossil record.
5. Cultural evolution is a key factor in the recent history of the human species.

In the introduction to his book *The Descent of Man,* Charles Darwin wrote: "During many years I collected notes on the origin or descent of man, without any intention of publishing on the subject, but rather with the determination not to publish, as I thought that I should thus add to the prejudices against my views." Darwin was right, as many of his contemporaries were hostile to the notion that humans are descended from other, earlier organisms and have not always been as they are now.

To a large extent, that original controversy over the evolution of human beings has passed; at least, it is not a significant issue among modern biologists. However, much still remains to be learned about our evolutionary history.

Taxonomic Relationships

We are mammals. Human beings belong to a group of vertebrate animals that are included in a taxonomic category called the **class Mammalia** (table 32.1). Mammals are homeothermic ("warm-blooded") animals that have body hair, feed their young milk produced by mammary glands, and possess a muscular diaphragm that separates their body cavity into thoracic and abdominal portions and functions in breathing movements.

The great majority of modern mammals are **viviparous;** that is, their offspring develop inside a specialized portion of the female reproductive tract, the uterus. Nutrients, gases, and other materials are exchanged between the bloodstreams of mother and developing offspring via a placenta. Generally speaking, placental mammals also provide their offspring with parental care, after birth, for a more extended time than do most other vertebrate animals.

The earliest fossil mammals have been found in rocks of the Jurassic period of the Mesozoic era (see table 32.3). These early mammals were small and inconspicuous compared to the reptiles that dominated that period of the earth's history. The rise of modern mammalian forms coincided with the mass extinction of the dinosaurs about 65 million years ago near the end of the Mesozoic era. During the early stages of the Cenozoic era, the mammals began an adaptive radiation that has resulted in their achieving dominance among terrestrial vertebrates.

Primate Evolution

Within the class Mammalia, humans are members of the **order Primates** (table 32.2). The primates first evolved as **arboreal** (tree-dwelling) animals, although not all modern primates live in trees. Most primate characteristics can be viewed as adaptations to arboreal habitats.

Table 32.1
Examples of Taxonomic Categories

Category	Human	Domestic Dog
Kingdom	Animalia	Animalia
Phylum	Chordata	Chordata
Subphylum	Vertebrata	Vertebrata
Class	Mammalia	Mammalia
Order	Primates	Carnivora
Family	Hominidae	Canidae
Genus	*Homo*	*Canis*
Species	*sapiens*	*familiaris*

Table 32.2
Classification of the Primates

```
Order Primates
    Suborder Prosimii (lemurs and tarsiers)
    Suborder Anthropoidea
        Superfamily Ceboidea
            Family Cebidae (New World monkeys)
        Superfamily Cercopithecoidea
            Family Cercopithecoidae (Old World monkeys and baboons)
        Superfamily Hominoidea
            Family Pongidae (apes)
            Family Hominidae (humans)
```

While many other mammals have undergone evolutionary reduction in the number of digits from the ancestral vertebrate five-digit condition, primates have retained five functional digits on their hands and feet. Their digits are very mobile, and in many primates, the thumbs are **opposable;** that is, they close to meet the fingertips and thus function quite efficiently in grasping. The tips of primates' digits have nails to protect them, instead of claws that extend beyond the digits, as is the case with many other mammals. Furthermore, a fleshy pad at the tip of each primate digit is very sensitive to touch, allowing these animals to grasp and manipulate objects with considerable dexterity.

In contrast to the shoulders of most other mammals, primate shoulder joints permit very extensive forelimb rotation, making possible climbing and foraging for food in the treetops. Other adaptations to life among the tree branches include relatively short snouts and eyes located at the front of the head. This placement results in excellent binocular stereoscopic (three-dimensional) vision that permits primates to make very accurate judgments about distance and position, judgments that are essential for animals that swing and leap from branch to branch. Paralleling this development of better depth perception has been evolutionary elaboration of anterior portions of the brain, especially the cerebral cortex. Compared with many other mammals, however, some primates have a relatively weak sense of smell.

Figure 32.1 A tree shrew. Modern tree shrews have many characteristics that are intermediate between those of modern insectivores (shrews and moles) and primates. Tree shrews may resemble the ancient ancestors of modern primates.

Table 32.3
The Cenozoic and Mesozoic Eras

Era	Period	Epoch	Millions of Years before Present*
	Quaternary	Recent	0.01
		Pleistocene	2.5
Cenozoic		Pliocene	7
	Tertiary	Miocene	26
		Oligocene	38
		Eocene	54
		Paleocene	65
	Cretaceous		130
Mesozoic	Jurasisc		180
	Triassic		230

*These are approximate dates of the beginnings of these intervals.

Most primates have a pair of mammary glands and produce only one offspring per pregnancy, whereas a great many other mammals produce litters, batches of several offspring born at the same time. Primates have relatively long infancies and develop strong, long-lasting mother-infant bonds.

Primates probably evolved from a group of small insect-eating animals that long ago entered the trees. In their treetop environment these animals found new food sources and were able to escape predators on the ground. The earliest fossil evidences of primate existence are a few teeth found in Montana, located in rocks from the Cretaceous period of the Mesozoic era. Teeth give important indications of an animal's diet, and these fossil teeth indicate that the earliest primates were insect eaters. Although fossil remains of the earliest primates are scarce, it generally is thought that they may have been quite similar to the living tree shrews of Southeast Asia (figure 32.1).

By 50 million years ago, however, tree-dwelling primates had evolved that differed considerably from their shrewlike ancestors. They were able to grasp limbs with their digits, as modern primates do. Instead of claws they had nails, and their eyes faced forward rather than to the sides.

By about 20 million years ago, the distribution of the evolving primates had changed. Climates had become cooler and drier, and the lush forests in which primates had thrived disappeared from many parts of the world. This reduction in the expanses of semitropical forests restricted primates to a part of Asia and the southern continents, Africa and South America, where their evolution continued.

The Primate order is divided into two suborders, the **Prosimii,** which includes the lemurs and tarsiers, and the **Anthropoidea,** which includes monkeys, apes, and humans (see table 32.2).

Figure 32.2 Two living prosimians. (a) A ring-tailed lemur. (b) A tarsier.

(a)

(b)

Prosimians

Fossils of prosimians are found in many parts of the world, including Europe and North America. But modern prosimians—which include lemurs (figure 32.2a), tarsiers (figure 32.2b), and other primitive primates with such exotic names as lorises, pottos, aye-ayes, and galagos—are very restricted in their ranges. Today, in fact, lemurs are found only on the island of Madagascar off the coast of Africa, and tarsiers only in the Philippines and the East Indies.

Monkeys

By the Oligocene epoch, which began about 38 million years ago (see table 32.3), the earliest members of the suborder Anthropoidea had diverged from the prosimians. Later two separate lines developed among the primitive anthropoids (figure 32.3). One group of anthropoids included ancestors of the **New World monkeys,** and the other group included ancestors of **Old World monkeys,** apes, and humans.

New World and Old World monkeys differ in several characteristic ways. New World monkeys have a long **prehensile** (grasping) **tail** that they regularly use almost as a fifth limb while swinging through the trees. Old World monkeys do not have prehensile tails. New World monkeys are sometimes called **platyrhines** ("flat-nosed") because their nostrils

are directed outward and separated by a broad, flat partition. In contrast, Old World monkeys, as well as apes and humans, are **catarrhines** ("downward-nosed"), because their nostrils are closer together and are directed downward. Many Old World monkeys, such as baboons, have brightly colored areas on their buttocks (called ischial callosities); New World monkeys do not.

New World monkeys are found in Central and South America. Some well-known New World monkeys are the spider monkey (figure 32.4a), the capuchin ("organ-grinder's monkey"), howler monkeys, and squirrel monkeys. Old World monkeys are found in Africa and southern Asia. Some Old World monkeys, such as colobus and mona monkeys are arboreal, but others are ground-dwelling descendants of arboreal ancestors. These ground-dwelling monkeys include colorful mandrills, baboons (figure 32.4b), and macaques, some of which are very widely used in medical research. All ground-dwelling monkeys walk on all fours.

Now we turn to the evolution of the **hominoids,** animals belonging to the superfamily Hominoidea. This superfamily includes two families, the **pongids** (the ape family) and the **hominids** (the human family). Note the slightly different spellings: homin*oid* is a broader term that includes both apes and humans, while homin*id* refers only to the human family.

Figure 32.3 A hypothetical "family tree" of the living primates.

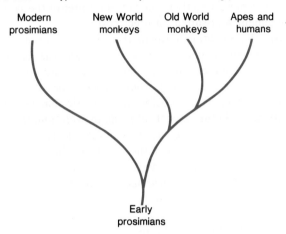

Early
prosimians

Figure 32.5 A chimpanzee "knuckle-walking."

Figure 32.4 Monkeys and baboons. (a) The spider monkey, a New World monkey. Note here how one monkey uses its prehensile tail to grasp a branch while being groomed. (b) Baboons, ground-dwelling Old World monkeys.

(a)

(b)

Dryopithecines

During the Miocene epoch, which began about 26 million years ago, apes became abundant and were widely distributed in Africa, Europe, and Asia. Among these Miocene apes, members of the genus *Dryopithecus* are of particular interest because some of them may have been the ancestors of modern apes (gibbons, orangutans, gorillas, and chimpanzees) and humans.

Dryopithecine apes were forest dwellers who probably spent most of their time in the trees, but their foot skeletons indicate that they also may have spent some time on the ground, probably "knuckle-walking" as modern apes do (figure 32.5). The dryopithecine skull had a low, rounded cranium and **supraorbital ridges** (bony ridges that protrude above the eyes). The face and jaws projected forward.

Later in the Miocene epoch, the climate continued to become progressively cooler and drier. The lush forests in which the dryopithecines thrived eventually dwindled, and vast **savannas** (grasslands with occasional clumps of trees) spread (figure 32.6). Some late Miocene relatives of the dryopithecines, however, were not restricted to the shrinking forests. These primates were able to live on the savannas, or at least at the edges of the savannas. Among these late Miocene primates, which came out of the trees, stood partially upright, and moved onto the savannas, were the ancestors of humans.

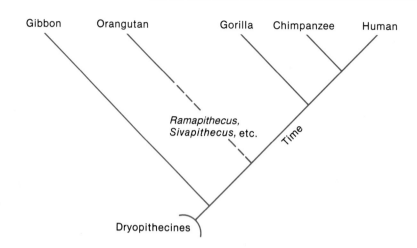

Figure 32.9 One interpretation of hominoid (ape and human) evolution based on structural features of living and extinct species and molecular clock data for living species. *Ramapithecus* and *Sivapithecus* are assigned to the orangutan branch on the basis of structural features. In this scheme, the degree of evolutionary relatedness among living species is inversely proportional to the estimated time since evolutionary ancestors diverged. For example, humans and chimpanzees would have a closer evolutionary relationship than would chimpanzees and gibbons based on this criterion.

The discovery in India of another fossil form called *Sivapithecus* has shed more light on the place of *Ramapithecus* in primate evolution. *Sivapithecus* and *Ramapithecus* fossils are very much alike, so much so, in fact, that some biologists suggest they should be in the same genus. However, far more skeletal parts have been found of *Sivapithecus* than of *Ramapithecus*. Examination of these additional bones strongly suggests that both of these organisms resembled the present-day orangutan, the only surviving Asian ape, much more than they resemble the earliest fossils that are classified as definitely hominid. Although not all biologists agree, it now seems to most of them that *Ramapithecus* was not an early hominid, or even an evolutionary ancestor of the hominids, but was part of a branch of primate evolution that produced the modern orangutan (figure 32.9).

This conclusion agrees fairly well with data from molecular clock studies of proteins and DNA (see page 757). These studies indicate that hominids and some of the pongids (gorillas and chimpanzees) did not diverge as separate evolutionary branches until after the time of *Ramapithecus*. This evidence also makes it seem very unlikely that *Ramapithecus* was an early hominid.

Removing *Ramapithecus* as a possible hominid ancestor further broadens an already large gap in the fossil record. About ten million years separate the dates assigned to fossils that may be those of possible ancestors of hominids from the earliest fossils that are recognizably hominid. This gap is one of the many mysteries concerning hominid evolution. Could there have been conditions during that long period of time that made fossil formation unlikely? This might account for

Table 32.4
A Classification of the Family Hominidae

Family Hominidae
    Genus *Australopithecus*
        A. afarensis
        A. africanus
        A. robustus
    Genus *Homo*
        H. habilis
        H. erectus
        H. sapiens neanderthalensis
        H. sapiens sapiens

the lack of fossil forms from that period. Or, are there undiscovered hominid fossils, formed during this long time interval, still waiting in the rocks for those paleontologists who may someday look in the right places? Further research and additional fossil discoveries may eventually answer these questions. Now let us leave *Ramapithecus* and turn our attention to the much more extensive fossil record of hominid evolution during the last four million years.

The Australopithecines

A small skull discovered in 1924 in a limestone quarry in South Africa was sent to the anatomist Raymond Dart in Johannesburg for study. Dart named the skull *Australopithecus africanus* ("southern ape of Africa") and proclaimed it to be a form intermediate between ape and human. At first, Dart's pronouncement created a great deal of controversy, but as more fossils were found in southern and eastern Africa, it became clear that *A. africanus* was indeed an upright-walking member of the family Hominidae (table 32.4).

Figure 32.10 Dental arcades of (a) a chimpanzee, (b) an australopithecine, and (c) a modern human. The two sides of the chimpanzee's jaw are roughly parallel, giving its dental arcade a rectangular shape. Its canine teeth are much larger than adjacent teeth, as they are in the jaws of all pongids (apes). The human and australopithecine jaws both curve gently to give a parabolic dental arcade. Note, however, that the large grinding molars (three rear teeth on each side) of australopithecines are much sturdier and broader than human molars.

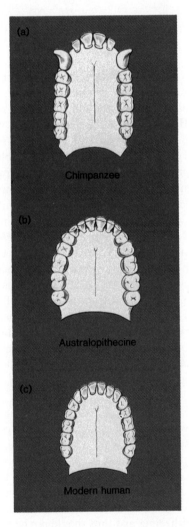

(a)

Chimpanzee

(b)

Australopithecine

(c)

Modern human

Figure 32.11 Skulls of *Australopithecus africanus* and *Australopithecus robustus*.

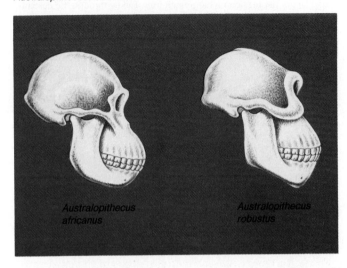

Australopithecus africanus

Australopithecus robustus

Many physical characteristics of *A. africanus* are transitional between apes and humans. For example, the brain capacity of *A. africanus* varied from just over 300 cc (cubic centimeters) to about 600 cc. The foramen magnum was farther forward than in apes but not centered as in humans. *A. africanus* had a generally parabolic (U-shaped) dental arcade and rather reduced canines (figure 32.10). Its face was flatter than that of apes, with a higher, rounder cranium. Furthermore, *A. africanus* had an S-shaped spine and a pelvis that was shorter than that of apes, which, along with other leg and pelvis characteristics, indicate that its locomotion was bipedal.

Another species of *Australopithecus, A. robustus,* has also been found in South Africa (figure 32.11). This hominid was larger than *A. africanus* and appeared somewhat later in the fossil record. Its large jaw and strong teeth suggest that its diet was of rougher vegetation. Both species lived during the period between 4 million years ago and 1.5 million years ago, although *A. africanus* is an older species and may have given rise to *A. robustus*. On the basis of additional fossil discoveries in eastern Africa, some biologists recognize yet another species, *A. boisei,* although other biologists contend that these still-larger and more robust fossil skeletons merely represent part of the range of variation in *A. robustus.*

In 1974, near Hadar in the Afar region of Ethiopia (see figure 32.12), Donald Johanson and his colleagues discovered a group of surprisingly complete fossil specimens, including the famous "Lucy" skeleton, that they named *Australopithecus afarensis.* Skeletal evidence indicates that this hominid walked fully erect. However, the skull capacity of *A. afarensis* was only 500 cc, and its dentition was persistently primitive, with a V-shaped dental arcade and relatively long canines. Some biologists have suggested that *A. afarensis* (if it does indeed represent a separate species) was ancestral to both *A. africanus* and *A. robustus* and that *A. africanus* gave rise, in turn, to *Homo,* the genus of modern humans. Johanson, however, thinks that *A. afarensis* may have given rise to two separate lines, one of which led to the other australopithecines and the other to the genus *Homo.*

However, some biologists disagree with the conclusions drawn about *A. afarensis* at an even more fundamental level. They contend that these fossils do not constitute a separate species at all, but should be included within the species *A.*

Figure 32.12 Location of the Rift valley of Africa. The Rift valley, which varies in width, lies within the lighter area of this map. It is part of a long line of depression running south from Turkey through the Jordan River area of Israel and the Red Sea, and on through eastern Africa. The rift apparently is the result of movement of continental plates (see p. 746). Sites of several very important hominid fossil discoveries are located in or near the Rift valley.

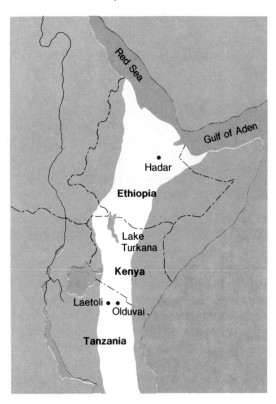

Table 32.5
Cranial Capacity of Hominids

Name	Cranial Capacity (in cc)
Australopithecus afarensis	500
A. africanus	300–600
A. robustus	300–600
Homo habilis	650
H. erectus	750–1,300
H. sapiens	1,400–1,700

The Leakeys have steadfastly maintained that the genus *Homo* is much older than previously thought, essentially as old as the australopithecines and, therefore, not descended from them. But there is much disagreement with this conclusion. Even though the skull of *Homo habilis* has a comparatively large (650 cc) cranial capacity (table 32.5), some researchers prefer to classify it with the australopithecines.

Arguments about relationships of the hominid fossils are far from settled and a number of hominid "family trees" have been suggested (figure 32.13). Strong support for the Leakey hypothesis of the early origin of the genus *Homo* came in 1972 when Richard Leakey (son of Mary and Louis Leakey) discovered an exceptionally interesting fragmented hominid skull in the Lake Turkana region of Kenya. This "skull 1470," as it is called, originally was dated at 2.6 million years ago (figure 32.14). Its large cranial capacity (nearly 800 cc), along with this early date, initially lent considerable support to the early origin of the genus *Homo*. However, some dispute arose concerning the original dating of "skull 1470" and it is now thought to be more recent, around 1.8 to 2 million years old.

In 1975, Mary Leakey and her colleagues found fossil teeth and jaws at Laetoli near Olduvai Gorge that closely resemble those of "skull 1470." The striking thing about these specimens is that they have been dated as being 3.6 to 3.8 million years old. Mary Leakey has classified these fossils as belonging to *Homo habilis*. Once again, however, disagreement has arisen, and other researchers feel that these fossils are australopithecines and thus should not be classified in the genus *Homo*.

Another very significant set of finds from eastern Africa, especially from Olduvai Gorge, includes a number of small stones that date to some 2.5 million years ago. These stones appear to have been intentionally chipped so that they could be used for cutting, pounding, and scraping; that is, they appear to have been made and used as tools (figure 32.15). It is also possible that the sharp stone flakes chipped off them were used as cutting tools.

africanus. In defense of their argument they cite dates now assigned to *A. afarensis* that may make it too recent to be a possible ancestor of both the other australopithecines and our genus, *Homo*.

The Leakey Hypothesis
There certainly are other interpretations. One of the most persuasive comes from the "first family" of hominid paleontology, the Leakeys, who have hunted hominid fossils at Olduvai Gorge and other sites in the Rift valley of Africa for many years (figure 32.12).

In addition to their many other important discoveries, Louis and Mary Leakey found an especially interesting hominid skull in 1964. The Leakeys considered this two-million-year-old skull to be much more advanced than any of its australopithecine contemporaries, and they named it *Homo habilis* ("handy man," or "man who is able to do or make").

Figure 32.13 "Family trees" of the hominids. (a) Version of hominid relationships in which *Australopithecus africanus,* or a similar form, gave rise to both the australopithecine and *Homo* lines. This version does not recognize *A. afarensis* as being separate from *A. africanus*. (b) Version that incorporates Donald Johanson's views regarding *A. afarensis* as an ancestor of all later hominids. (c) Version incorporating elements of the Leakey hypothesis, which proposes that the divergence of australopithecines and the genus *Homo* occurred earlier and that the two groups have descended as separate evolutionary lines.

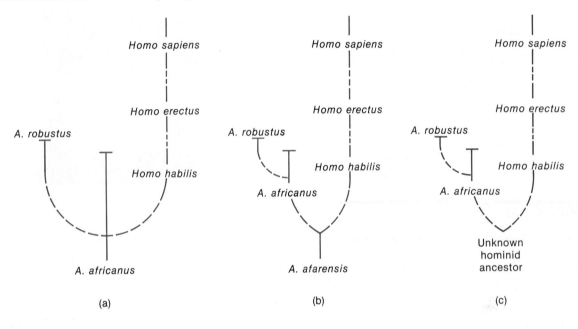

(a) (b) (c)

Figure 32.14 "Skull 1470," discovered on the eastern shore of Lake Turkana in 1972 by Bernard Ngeneo, a member of Richard Leakey's research team. This skull has been partially reconstructed.

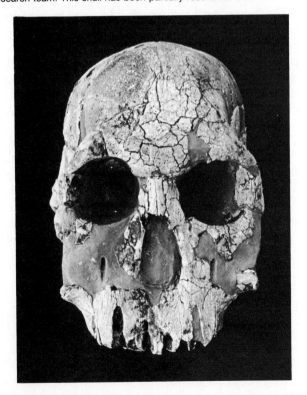

Figure 32.15 An example of the earliest tools, similar to those found by the Leakeys in Olduvai Gorge. Such tools were probably used as choppers in food preparation. Some scientists think that the sharp flakes chipped from these stones may have been used also, as tools for cutting.

Utilizing objects as tools is not unique to humans; in fact, it is relatively common among some primates. Chimpanzees and gorillas, for example, pick up and use objects, usually sticks and small stones, as tools and even weapons. What distinguishes the tools found at Olduvai Gorge is that they appear to have been consciously fashioned to perform a specific task. But who made these tools two and a half million years ago? One possibility is that the australopithecines might have made tools such as these. But another possibility is that while australopithecines may have found and used various objects as tools, the design and fabrication of tools is and has been a strictly human enterprise that was limited at that time to the work of *Homo habilis* or possibly other members of the genus *Homo*. The Leakeys and others use this argument to support the idea that the genus *Homo* is at least 2.5 million years old.

The Genus *Homo*

Many unanswered questions remain about *Homo habilis,* but there is general agreement among paleontologists about a series of fossils found in Africa, Asia, and Europe. These fossils are all included in the species *Homo erectus*.

Homo Erectus

The oldest of the *Homo erectus* fossils dates to about 1.6 million years ago and the youngest to about 300,000 years ago. Members of this species were bipedal and walked fully erect. Their cranial capacity (750 to 1,300 cc) was relatively large (see table 32.5). However, *Homo erectus* differed from modern humans in that it retained some primitive skull features. Along with rather thick cranial bones and large teeth, the face of *H. erectus* projected more than that of modern humans, and because the lower jaw sloped back, there was no distinct chin (figure 32.16).

The oldest fossil remains of *H. erectus* have been found in Africa, while those of later *H. erectus* individuals have been found in other parts of the world. This pattern suggests that the species originated in Africa and migrated outward to other suitable habitats—an event that took place in a relatively short period of time. Some of the sites where these fossils have been found yield considerable information about the life-style of *Homo erectus*.

Two sites explored near Torralba, Spain, provide evidence that these early humans hunted large animals. The bones of about 150 animals have been found there, including those of elephants, deer, horses, wild cattle, and rhinoceroses.

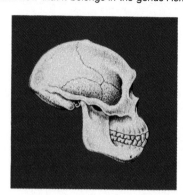

Figure 32.16 Skull of *Homo erectus. Homo erectus* was a widely distributed species, and many important human fossil finds (including ''Java man'' and ''Peking man'') have been assigned to this species. Earlier, the species was called *Pithecanthropus erectus,* but there is general agreement now that it belongs in the genus *Homo.*

The tools found at Torralba are much more elaborate than the older, more primitive tools found at Olduvai Gorge. Tools of *Homo erectus* included heavy, wedge-shaped choppers, small hand axes, and various small tools such as scrapers, borers, and engravers.

Another site, found during the excavation for an apartment complex at Terra Amata in Nice, France, has provided some insight into the daily life of *Homo erectus*. This site, dating to about 400,000 years ago, contained several temporary shelters approximately five meters by nine meters each. The shelters, built at different times, contained hearths, animal bones, and a fragment of a wooden bowl. Evidence clearly indicates that *H. erectus* used fire for cooking. These temporary shelters may have been used by members of a tribe as they followed the annual migration of the large animals they hunted. These evidences indicate that *Homo erectus* was a social animal, and that cooperative effort and some form of social structure contributed to group survival.

Early *Homo erectus* skulls had an average cranial capacity of about 940 cc, while the later ones averaged about 1,000 cc. This is a small difference, considering the total range of variability in cranial capacities of *H. erectus*, but it reflects an overall tendency toward increasing brain size during the course of human evolution (see table 32.5). This tendency continued in the evolutionary development of *Homo sapiens*.

Figure 32.17 Early humans. (a) Restored skulls of *Homo erectus,* a Neanderthal man, and a Cro-Magnon man. Significant differences are apparent in the shape and size of craniums, supraorbital ridges, and the structure of chins and jaws. (b) Artistic reconstruction of possible facial features of the three types.

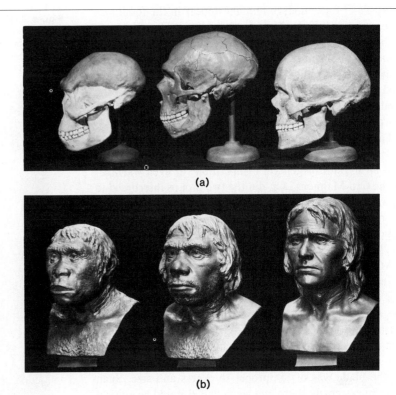

(a)

(b)

Homo Sapiens

Members of the species *Homo sapiens* display a number of differences in skull structure from *H. erectus*. These differences include an increase in cranial capacity from an average of 1,000 cc to an average of about 1,500 cc; a decrease in the size of the teeth and jaws; flattening of the plane of the face; rounding of the cranium; and lightening of the cranial bones.

How long ago did *Homo sapiens* emerge? Two specimens of early *H. sapiens,* one from Swanscombe, England (dated 250,000 years ago), and another from Steinheim, Germany (dated 200,000 years ago), resemble *Homo erectus* somewhat in their skull features but are different enough to be classified as *H. sapiens*.

About 100,000 years ago a distinctive group of humans called **Neanderthals** emerged. "Neanderthal" is German for Neander Valley, the area in Germany where the first fossil of this type was unearthed. Actually, the Neanderthals were widespread geographically; their fossils have been found in much of southern Europe, Asia (including China, Java, and Sumatra), and in northern, eastern, and southern Africa. During the first half of the tenure of the Neanderthals, the climate was temperate, but later a new glacial period developed, reaching its maximum late in the Neanderthal period, that lasted until about 40,000 years ago.

There is extensive evidence that the Neanderthals carried out ritualistic practices, which is strongly suggestive of abstract thought, and perhaps even a concept of religion and life after death. The arrangement of bear bones and rocks in many caves suggests that a bear cult existed among some Neanderthals. They were also known to bury their dead. A Neanderthal skeleton found in Shanidar Cave in Iraq was buried some 60,000 years ago amidst a profusion of flowers.

Usually, the Neanderthals are classified as *Homo sapiens neanderthalensis,* a variety of modern humans. The distinction is based on skeletal characteristics, including some persistently primitive skull features (a protruding forehead, no chin, and a flattish skull) (figure 32.17). The Neanderthal brain capacity was essentially the same as that of modern humans, however. Some researchers suggest that the Neanderthals gave rise to fully modern humans (*Homo sapiens sapiens*); others think that the Neanderthals were modern humans who simply displayed a normal degree of structural variation; still others propose that the Neanderthals became extinct and were replaced by humans who had a different ancestry.

Fossils of fully modern humans appear in the fossil record near the end of the Neanderthals' time. These humans have been designated **Cro-Magnon** people and are classified as *Homo sapiens sapiens* (as are present-day humans) to distinguish them from the Neanderthals. After a relatively brief period of coexistence with the Cro-Magnon people, the Neanderthals disappeared, and from about 40,000 years ago onward, no Neanderthal fossils occur in the fossil record.

The rapid (in terms of geological time) disappearance of the Neanderthals is somewhat of a mystery. Were they eliminated as a result of general competition (or even warfare) with the Cro-Magnons? Or, did the two groups interbreed to such an extent that distinctive Neanderthal characteristics were no longer recognizable, and the Neanderthals were simply incorporated into the general *Homo sapiens* population?

At about the time the Neanderthals disappeared, a new tool industry spread through the human population. The major characteristic of this industry was production of the **blade,** a stone tool with roughly parallel sides that is at least twice as long as it is wide (figure 32.18). Blades were produced from a long, relatively thin flake of stone. From this piece, smaller flakes were carefully removed from both sides until the blade was of the proper thickness and size. Such tools were extremely sharp and when dulled, could be quickly resharpened by simply removing another series of small flakes to produce a new cutting edge. The production of such beautifully worked stone tools indicates precise muscular control and considerable understanding of the fracture characteristics of different types of stone.

Many other tools also were made from a variety of materials, including bone, antler, wood, and ivory. These increasingly complex tools are physical evidences of the growth and spread of human **culture,** a body of information transmitted from generation to generation by means that did not depend on genetic mechanisms. People taught other people how to use and make tools. This and other information was transmitted by personal communication. Human evolution had entered a new and important phase, a phase involving cultural evolution, that has continued at an accelerating pace to the present time.

Cultural Evolution

In several important ways the ancient hominids and the living great apes differ very little from modern humans. For instance, humans, chimps, and gorillas possess over 90 percent of their genes in common. Furthermore, differences in skeletal anatomy of apes, early hominids, and modern humans are more a matter of degree than of absolute differences in

Figure 32.18 Blade produced by Cro-Magnon people. Some Cro-Magnon blades were as long as 30 cm and only 1/2 cm thick. Blades such as this represent the pinnacle of stone tool manufacture. (Compare this with the chopper in figure 32.15.) The next major step in human tool making occurred with the advent of metal working.

morphology, and the roots of human social behavior can be found in the behavior of other primates as well. The most significant differences between modern humans and ancient hominids—or between humans and apes for that matter—are the greater intelligence and cultural advancement of modern humans.

Culture is passed from one generation to the next through teaching and example, and it is dynamic, since it is continually modified as it is transmitted. Cultures do not evolve in the biological sense, but they do display an overall progressive change with occasional large-scale advances that take place in relatively short periods of time.

The earliest cultural stage of the hominid line was that of the hunter-gatherers, a stage in which hominids first learned to hunt small animals and later the larger, herd animals. The second major cultural stage was agriculture, a change in life-style that greatly altered not only the means by which humans provided for their physical existence but also marked a fundamental change in structure and values of human societies. The agricultural period extended from approximately 12,000 or 15,000 years ago until about 200 years ago, when a third major stage of human cultural development began with the advent of the Industrial Revolution, a technological and economic movement that was to have a profound influence not only on the growth and structure of cities, but on all areas of human endeavor.

Let us briefly examine each of these phases of cultural evolution.

The Hunter-Gatherers

From their earliest beginnings, humans were hunter-gatherers, and some contemporary peoples, such as the !Kung of the Kalahari and the Bushmen of Australia, still make their living in this way. In such cultures, the men hunt for game, and the women and children gather fruits, nuts, herbs, and roots. In the hunter-gatherer society, the women and children provide more than two-thirds of the tribe's food, but game is the favorite food, and hunting forms the central theme of these tribes in a ritualistic sense. This type of existence places a premium on communal effort. A single man is neither strong enough nor fast enough to track down and kill most game animals, but cooperative hunting efforts can be quite effective. Also, an individual woman cannot forage for food, care for an infant, and perform the myriad tasks required around the homesite, but a group of women of different ages can share the work effectively.

Tool making and use is important to hunter-gatherers. Their tools include digging sticks for unearthing edible roots, spears for hunting, scrapers for cleaning hides, and a variety of tools for cooking, clothesmaking, and so on. For the most part, hunter-gatherers used tools to obtain and prepare food and to make clothing, but they made objects for other purposes as well. Some artifacts (objects fashioned by humans) apparently were used in ritualistic practices, perhaps early efforts at obtaining some magical control over nature. Many cave paintings (see p. 4) suggest that the artists were attempting to guarantee success in the hunt. Small female figurines may have been fertility symbols used to guarantee successful reproduction. Burial of the dead dates to about 60,000 years ago. Because flowers, food, and implements were sometimes interred with the dead, there is at least some indication of belief in an afterlife.

The decline of the hunter-gatherer culture began some 20,000 to 15,000 years ago, perhaps as a result of a decrease in the number of large game animals. Possibly the ancient hunter-gatherers were the authors of their own decline. They had become efficient hunters, and possibly they began to kill many more animals than they needed for food and clothing.

Also, the decline of the game herds coincided with a period during which the human population increased rapidly. It has been estimated that there were between 1 and 3 million people on earth 30,000 years ago. By 12,000 years ago, the population is estimated to have increased to 6 to 10 million. The combination of decreased food supply and increased population probably provided a strong impetus for finding new, more dependable sources of food.

The Rise of Agriculture

Agriculture appeared independently in several places in Eurasia and the New World some 12,000 to 9,000 years ago, and there is evidence that people of the Nile Valley were milling grain as early as 15,000 years ago. It is thought that the first steps toward an agricultural existence were taken by some ancient hunter-gatherer tribes who settled close to a natural supply of grain or other food and gave up following the declining herds. At first, these people probably harvested crops without seeding for new ones and depended upon natural processes to provide a new crop each season, but eventually a variety of agricultural practices were developed that allowed the people to make a living by cultivating the soil. Many humans were no longer nomadic; instead they were tied to a specific geographic location.

By about 9,000 years ago, agricultural practices were well developed, and in the Middle East, a number of animals, including sheep, cattle, goats, pigs, and dogs, had been domesticated. Not only were these animals an important source of food and clothing, but some of them were also used as draft animals.

An entirely new array of tools accompanied the development of agriculture, and the origin of metallurgy some 8,000 years ago gave great impetus to the making of tools for agriculture and warfare. Once humans learned to produce temperatures hot enough to melt and cast copper, the use of other metals, such as lead, silver, and iron, as well as the production of metal alloys, followed. Bronze, a strong and durable alloy of copper and tin, proved very valuable and was widely used for tool making.

Apparently, as populations grew and people migrated to new areas, they cut down trees and burned off the surface vegetation, preparatory to planting crops. However, much of the farmland obtained in this way was not really suitable for long-term agricultural usage. This form of agriculture may have been responsible, at least in part, for the destruction of arable lands and the formation of deserts. The cutting of native forests and subsequent overfarming of the land has been shown to be responsible for the spread of deserts in India within historic times. The desertification of the Fertile Crescent of the Tigris and Euphrates Rivers may well have resulted from a similar sequence of events. And there is good evidence that the ancestors of the Egyptians lived in lush forests where only vast expanses of the Sahara spread today. Clearly, a new relationship between organism and environment was evolving because human agricultural activities caused large-scale changes on the earth.

The Industrial Revolution

The third major stage of human cultural development was ushered in during the eighteenth century with the arrival of the Industrial Revolution. During this period, there was a mass movement of people from agricultural areas to manufacturing centers. This movement of people from rural to urban societies continues even today.

Industrialization in some parts of the world has been accomplished using resources drawn from many other parts of the world. As a result of this unequal consumption of the earth's resources, the relative abundance of the Industrial Revolution has not been shared equally by all.

Industrialization also brought with it large-scale environmental degradation. Some early human activities had adverse effects on the environment, such as the decline and extinction of some species and the overfarming and desertification of large land areas. However, present human culture has the power for environmental disruption and destruction far exceeding that of any other stage of human development (see chapter 34). There is danger that cultural evolution, which represents the pinnacle of our humanness, could very well lead to our own destruction.

Summary

Humans are primates. Primates evolved as arboreal animals, and primate characteristics generally reflect adaptations for arboreal life. The oldest primates are the prosimians. Modern prosimians include lemurs and tarsiers. Monkeys, apes, and humans belong to the suborder Anthropoidea. Monkeys have diverged into New World monkeys, which have prehensile tails, and Old World monkeys.

Apes such as the dryopithecines were ancestors of modern apes and humans. They were forest dwellers who may have spent some of their time on the ground. At the end of the Miocene epoch, climatic changes led to a reduction of forested areas and produced vast savannas that provided new habitats for ground-dwelling hominid ancestors. However, some fossil forms that have been unearthed, such as *Ramapithecus*, probably do not represent those ancestors.

The australopithecines, which lived from about 4 million years ago until about 1.5 million years ago, definitely were hominids, and some of them were on, or near to, the line of human ancestry. *Homo habilis* was a contemporary of the late australopithecines, and its early emergence may indicate early divergence of australopithecines and the genus *Homo*. Hominids made simple tools more than 2 million years ago.

Homo erectus emerged at least 1.6 million years ago and spread rapidly over large areas of Africa, Europe, and Asia. *H. erectus* made more elaborate tools, hunted in groups, built shelters, and apparently had some social organization.

The modern human species, *Homo sapiens*, emerged between 250,000 and 200,000 years ago. From 100,000 years ago until about 40,000 years ago, Neanderthal humans lived in much of Europe and Asia. They were replaced (absorbed?) by fully modern humans, who have been dominant for the past 40,000 years.

Recent human evolution has been marked by rapid cultural evolution. Humans have progressed from hunter-gatherer and agricultural societies to the era of industrialization and the social development that has followed it.

Questions for Review

1. What human characteristics result in humans being classified as mammals?

2. Discuss those characteristics of primates that represent arboreal adaptations.

3. What are platyrhines and catarrhines? Discuss differences between them.

4. Describe some adaptive advantages of erect posture and bipedal locomotion, especially in a savanna environment.

5. What characteristics would you expect a gorilla jaw to have?

6. What are some differences between the skulls of *Australopithecus* and modern humans? Between those of *H. erectus* and modern humans?

7. Using the example of "skull 1470," explain why accurate dating is essential to understanding evolutionary relationships of early hominids.

8. What reasons might be suggested for the decline of the hunter-gatherer culture?

9. Discuss the possible relationship between population size and the beginning of agriculture.

Questions for Analysis and Discussion

1. In developing an understanding of human evolution, why is it useful to study chimpanzees, gorillas, and baboons in the wild?

2. Evidence of tool making 2.5 million years ago leads some scientists to think that australopithecines may have made tools. Explain how this evidence highlights differences between the proponents of the Leakey hypothesis and those who disagree with it.

3. In 1984, Kamoya Kimeu, a Kenyan colleague of Richard Leakey, discovered a *Homo erectus* skeleton on the western shore of Lake Turkana. The skeleton was excavated by other members of a research team led by Leakey, and Frank Brown, John Harris, and Alan Walker from the United States. The skeleton was identified as that of a twelve-year-old male and is the most complete *Homo erectus* fossil yet found—much of the skull *and* the remainder of the skeleton were present. Suggest how data gained from studying this specimen will aid in the study of other fossil hominids.

Suggested Readings

Books

Delson, E., ed. 1985 *Ancestors: The hard evidence.* New York: Alan R. Liss.

Fossey, D. 1983. *Gorillas in the mist.* Boston: Houghton Mifflin.

Johanson, D., and Edey, M. 1981. *Lucy: The beginnings of human kind.* New York: Warner Books.

Leakey, M. D. 1984. *Disclosing the past: An autobiography.* New York: Doubleday.

Leakey, R. E. 1981. *The making of mankind.* New York: E. P. Dutton.

Richard, A. F. 1985. *Primates in nature.* San Francisco: W. H. Freeman.

Articles

Day, M. H. 1984. The fossil history of man. *Carolina Biology Readers* no. 32. Burlington, N.C.: Carolina Biological Supply Co.

Hammond, A. L. November 1983. Tales of an elusive ancestor. *Science 83.*

Hay, R. L., and Leakey, M. D. February 1982. The fossil footprints of Laetoli. *Scientific American.*

Lewin, R. 1983. Fossil Lucy grows younger, again. *Science* 219:43.

Lewin, R. 1983. Unexpected anatomy in *Homo erectus. Science* 226:529.

Lewin, R. 1985. The Taung baby reaches sixty. *Science* 227:1188.

Lewin, R. 1986. When stones can be deceptive. *Science* 231:113.

Napier, J. R. 1977. Primates and their adaptations. *Carolina Biology Readers* no. 28. Burlington, N.C.: Carolina Biological Supply Co.

Pilbeam, D. March 1984. The descent of hominoids and hominids. *Scientific American.*

Rensberger, B. April 1984. Bones of our ancestors. *Science 84.*

Rensberger, B. January/February 1984. A new ape in our family tree. *Science 84.*

Stringer, C. December 1984. Fate of the Neanderthal. *Natural History.*

Weaver, K. F. November 1985. The search for our ancestors. *National Geographic* 168:560.

Behavior and Ecology

 All living things interact with their physical and biological environments. Animals show characteristic behavioral responses to members of their own species, as well as to members of other species. Such behavioral responses are important for maintaining homeostasis, reproducing successfully, obtaining food, avoiding predators, and other aspects of animal life.

On a larger scale, we can best understand living things in the context of populations, communities, and ecosystems. Several interacting populations constitute a community. A community, together with all the physical and chemical factors in the environment, make up an ecosystem, within which energy flows and materials cycle.

The growth of human populations and their impact on ecosystems around the world are important concerns of modern biology.

Behavior

33

Chapter Concepts

1. Behavior allows animals to respond rapidly to changes in their environment.
2. Organisms are genetically programmed through natural selection to behave in adaptive ways.
3. In some cases, there are rigid, relatively unvarying responses to a given situation, while in others behavior is highly modifiable through experience (learning).
4. The expression of certain behavior patterns requires that an organism be at a given state of morphological and physiological maturation.
5. Hormones control behavior by altering the internal motivational states of organisms.
6. Animal societies are composed of individuals of the same species that cooperate in an adaptive manner and interact with each other using a complex system of communication.
7. Sociobiology has provided a new way of looking at the genetics and evolution of social behavior, but the application of sociobiological principles to human behavior has caused some controversy.

Facing page   A male baboon engaged in a characteristic threat display. Such behavior often helps to resolve conflicts among individuals that might otherwise lead to fighting and possible injury. Behavioral patterns form an important part of the complex interactions between organisms and their environments.

Walking through a gull colony can be an exciting if somewhat baffling experience. Gulls fly overhead in all directions, calling, swooping down, sometimes buffeting your head with their feet. Gulls and their young are in constant danger from predators. Thus, it is not surprising that gulls attack humans just as they do any other potential marauders that wander near their nests (figure 33.1).

But if you hide in a blind, the gulls soon calm down and return to their normal activities. Some individuals stand alert by their nests or raise their heads in the raucous long-call characteristic of their species. At some nests, pairs are courting, jerking their heads up and down in a chokinglike motion, while at other nests the birds are fighting with their neighbors. Other gulls are settled on their nests, incubating their eggs, while their mates stand vigilantly nearby. Like houses in a suburban development, there is a regular spacing between neighboring nests. Each gull staunchly defends the area (**territory**) around its nest, allowing only its mate into the territory. The female lays several mottled, greenish brown eggs that are well camouflaged against the nest background. The two gulls take turns incubating the eggs for three weeks, and then they collaborate in the demanding task of feeding the young.

All of these actions are part of the gulls' behavior. **Behavior** is what an organism does (usually by movement) in response to various situations.

Analysis of Behavior

Shortly after the time of hatching, parent black-headed gulls systematically pick up empty eggshells from around the nest and remove them. If one of these discarded shells falls by the nest of a neighbor, the neighboring gull, in turn, also removes it.

This apparently trivial behavior intrigued Nobel Prize–winning ethologist Niko Tinbergen. **Ethologists** analyze behavior patterns in nature and are particularly interested in the adaptive significance of behavior. Why should the black-headed gull and other ground-nesting gulls take such pains to remove the shells?

There seemed to be several possibilities. For one, the sharp edges of the broken shell might be harmful to the chick or uncomfortable for the brooding parent. Yet many other birds, including cliff-nesting gulls called kittiwakes (figure 33.2), do not systematically remove eggshells from their nests. Another possibility was that the behavior might reduce the risk of predators attacking the chicks. Only the outside of the gull's eggshell is camouflaged. The empty shell with its jagged edge and white interior might flash like a beacon to any passing predators, especially to crows flying overhead.

Figure 33.1 A black-headed gull.

Figure 33.2 Kittiwakes are gulls that nest on steep cliffs. Thus, they suffer less predation than gulls that nest on flat ground. Kittiwakes do not systematically remove eggshells from their nests.

Tinbergen and his colleagues carried out a series of experiments to test these ideas about predation and eggshells. To test the hypothesis that the gulls were selectively picking out and removing the conspicuous eggshells, the scientists put dummy plaster eggshells in assorted colors, sizes, and shapes into the nests of wild gulls. They then hid in a blind nearby and observed gulls' responses to the shells. When a gull returned and settled down on its nest, it pecked at the dummy eggshell for a few moments and then picked it up and removed it from the nest. The more the dummy looked like the conspicuous inside of the gull's eggshell, the more likely the gull was to pick it up and carry it from the nest.

Figure 33.3 In an experiment with mock black-headed gull nests, predation was more likely if eggshell fragments were located close to the nest.

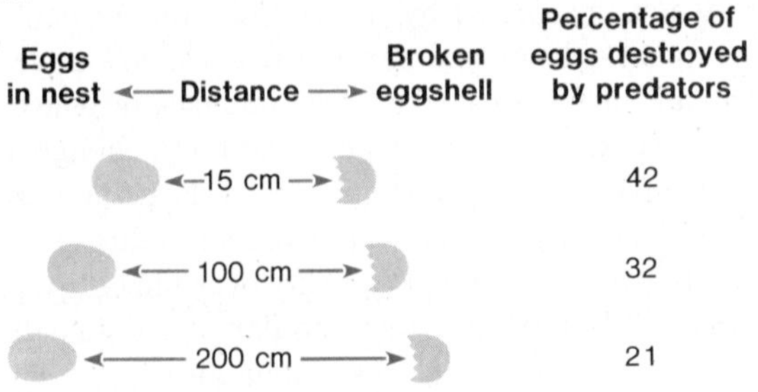

Eggs in nest ◄— Distance —► Broken eggshell	Percentage of eggs destroyed by predators
◄–15 cm —►	42
◄— 100 cm —►	32
◄— 200 cm —►	21

Thus, the results of these experiments tended to confirm the idea that conspicuous items that resemble broken eggshells were likely to be removed from the nest. But just how conspicuous are shells to would-be predators?

To answer this question, Tinbergen and his colleagues built artificial nests in various places around the edge of the gull colony and placed a clutch of eggs in each. Next they placed broken eggshells at varying distances from the nests. Exactly as their hypothesis had predicted, the nearer the eggshells were to the nest, the more likely it was that the eggs in the nest would be eaten (figure 33.3). It seems, then, that a parent gull saves its chicks from predation by removing the empty eggshells from around the nest.

However, one puzzling feature of the behavior remained. Despite the great risk of predation, parent gulls did not remove the eggshells immediately after their chicks hatched. If the nests were so predictably destroyed when eggshells were present, why did the gulls wait several hours after the chick had hatched before removing the broken shell? Tinbergen made a fascinating, if somewhat gruesome, observation. When parent gulls flew up and left a chick that had not yet dried off after hatching, the chick was very likely to be gobbled up by a neighboring gull. After a chick had had time to dry and gain a little strength, however, it could run and hide in the grass if the parents left the nest, and thus could escape cannibalistic neighbors if necessary. Tinbergen concluded that there are really two factors affecting eggshell removal. The parents need to remove the eggshells quickly, but as long as their chicks are wet, it is better that the parents not leave the nest, even for the few minutes required to dispose of a broken eggshell.

While it appears that parent gulls are thoughtfully considering the welfare of their offspring, they are simply responding mechanically and automatically to a **stimulus** in the environment. It is as if gulls are programmed by their genes to behave in this manner whenever they encounter eggshells or other conspicuous objects in or around their nests. In evolutionary terms, the complex combination of genes that causes gulls to remove objects from their nests is more likely to survive (and thus be passed on to future generations) than other genetic combinations that produce other behavior patterns. If a gull did not remove eggshells, all of its eggs might be eaten and its genes would not be passed on to another generation. There is natural selection for eggshell removal behavior, and selection pressure resulting from predation maintains the genes for the behavior in the population.

Simultaneously, however, there is selection against the gulls leaving wet chicks unattended, even for a few moments. This example illustrates the delicate balance between complex factors that results in the evolution of the finest details in the timing and form of behavioral responses.

Genes and Behavior

In discussing how a trait such as eggshell removal behavior might have evolved, we must assume that the trait was **inherited;** that is, that the offspring of parents with the behavioral trait would be more likely to show the behavior than offspring of parents without the trait. Somehow, the behavior must be programmed by genes, and those genes must be passed in an essentially unaltered form to the next generation. But how do genes program such complex behavior?

Genes act directly by controlling protein synthesis and thereby exert their effect through control of development and of physiological processes. However, such control is too slow and too indirect to be involved in animals' responses to immediate problems. Instead, animals' genes act to build bodies with nervous systems that make possible behavioral responses to changes in the environment. Behavior allows animals to respond quickly and adaptively. Furthermore, behavior is not limited to automatic, stereotyped responses that are genetically determined in advance. As we shall see later in this chapter, genes may also program animals to be able to learn to respond in particularly adaptive ways.

Thus, genes, and the behavior they program, are preserved because the individuals in which they occur are more likely to survive and pass their genes on to the next generation. Behavior patterns that clearly are inherited have been analyzed in organisms ranging from unicellular organisms such as the protozoan *Paramecium* to vertebrate animals.

Figure 33.4 The normal avoidance response of *Paramecium*. When the organism swims into an object (*1*), it backs up by momentarily reversing the direction of its beating cilia (*2*). After the *Paramecium* is positioned at a different angle (*3–5*), the cilia resume their normal movement and the organism once again moves forward (*6*).

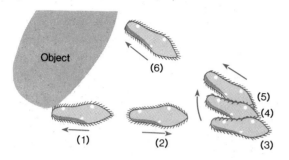

Swimming in Paramecium

A single-celled *Paramecium* moves through the water in a spiral fashion, propelled by its beating cilia. When it hits something, it backs up and moves away. We now understand much about the mechanism of this **avoidance response.** Running into an object causes a depolarization (a change in electrical charge) of the plasma membrane at the point of contact. This depolarization results from a sudden influx of calcium ions into the cell and spreads quickly over the entire plasma membrane. This process is very similar to the way that impulses pass along the nerve cells of animals (see chapter 16). The depolarization of the *Paramecium's* plasma membrane causes a reversal in the direction of movement of the cilia, and the *Paramecium* backs up (figure 33.4). Within a few

Figure 33.5 Inheritance of nest-building behavior in lovebirds. (a) Fisher's lovebird simply cuts a strip and carries it to the nest in her bill. (b) The peach-faced lovebird cuts a strip and then tucks it into her rump feathers. (c) A daughter of a cross between the two species at first acts completely confused. She succeeds in carrying a strip to the nest only when she carries it in her bill like Fisher's lovebirds. (d) The hybrid female eventually learns to carry the strip in her bill, but not without a turn of the head as though she were about to tuck it into her feathers. From "The Behavior of Lovebirds" by William C. Dilger. Copyright © 1962 by Scientific American, Inc. All rights reserved.

seconds, the membrane becomes repolarized by active transport of ions back across the membrane, and the *Paramecium* resumes its forward path.

A number of intriguing *Paramecium* mutants that differ in their swimming behavior have been found in the laboratory. One of these is a mutant named "paranoic." When a "paranoic" *Paramecium* runs into an object, it backs up as usual but continues swimming backward much longer than a normal *Paramecium* would. Because of a problem in the structure of the cell's plasma membrane, the normal process of repolarization fails. This "mistake" in membrane structure is caused by a change in a single gene, which, in turn, has a direct effect on the behavior of the organism.

Nest Building in Lovebirds

Lovebirds are small, green and pink African parrots that nest in tree hollows. There are several closely related species of lovebirds of the genus *Agapornis*. The species differ in the way they build their nests. A female Fisher's lovebird picks up a large leaf (or in the laboratory, a piece of paper) with her bill, perforates it with a series of bites along its length, and then cuts out long strips. She carries the strips, one at a time, to the nest in her bill (figure 33.5a) and weaves them in with others to make a deep cup. In contrast, a female peach-faced lovebird cuts somewhat shorter strips in a similar manner, but then she carries them to her nest in a very different way. She picks up each strip in her bill and inserts it deep into her rump feathers (figure 33.5b). In this way she can carry several strips on each trip she makes to the nest.

Crosses between these two species result in daughters that show intermediate behavior. When first beginning to nest, a hybrid female acts confused. She cuts strips of intermediate length and attempts to tuck them into her rump feathers. She does not push the strips far enough into her feathers, however, so that when she walks or flies, the strips come out (figure 33.5c).

Such hybrid behavior clearly indicates that the differences in the way the two species carry nesting material are the result of differences in their genotypes, and thus this is another illustration of how genes control behavior.

This research also illustrates another important point. Eventually, the hybrid bird's behavior changes so that she becomes more efficient (figure 33.5d). With time, she learns to fly to the nest with the cut strip in her bill without even trying to tuck it into her feathers, although she still makes at least a furtive turn of her head toward her rump before flying off.

Learned Components of Behavior

The improvement in the hybrid female lovebird's ability to carry nesting material suggests that there is an interaction between what the animal is programmed to do and the results of actually doing it. However, behavior is not limited to predictable automatic responses that are genetically programmed. Behavior can be altered by **learning,** the modification of behavioral responses by experience. Thus, adaptive behavior that results from an organism's experience with its environment is known as **learned behavior.**

When biologists analyze behavior, it is often difficult, or even impossible, to distinguish clearly between innate (predictable, automatic, and genetically determined) behavior and learned behavior. But some behavioral patterns do have recognizable learned components. Let us look at two examples.

Gull Chick Pecking

Parent gulls are faced with the arduous task of keeping up with the feeding demands of their chicks. The parent carries the food back to the nest in its crop (a sac in the first part of the bird's digestive tract). When an adult lands at the nest, the chick pecks at the parent's bill (figure 33.6). This behavior causes the parent to regurgitate the food it is carrying. All chicks, even newly hatched ones, peck at their parent's bill. This is an example of an innate, or **instinctive behavioral pattern.** The young gulls respond automatically to a **visual communication signal,** which in this case is the parent's bill.

Outside of the nest, newly hatched gull chicks will peck at a wooden model of a gull's head (figure 33.7a). By comparing the rates at which chicks peck at gull head models of different shapes and colors, biologists can obtain a reliable measure of how similar to the parent the chicks find each model.

The development and improvement in pecking behavior has been thoroughly studied in the American laughing gull. Newly hatched chicks are not very accurate in their pecking, but their aim improves markedly after only a couple of days in the nest. They also develop the ability to identify details of their parents' bills specifically. Eventually, they will peck only at heads that look like those of their own parents (figure 33.7b).

Box 33.1
Solar-Day and Lunar-Day Rhythms

Modification of the honey bees' waggle dance through the course of the day is only one of many behavioral patterns that change rhythmically with time. In fact, the total activity level of most animals is rhythmic; that is, animals are active at certain times of the day and inactive at others. Bats, rats, and moths are **nocturnal** (active at night), while dogs, butterflies, frogs, and lizards tend to be active in the daytime. Thus, members of certain animal species may never encounter members of certain other species even though they may live in the same area.

Solar-day (twenty-four-hour) **rhythms** are very widespread among living things (see chapter 19). Many such solar-day rhythms continue even when organisms are placed under constant conditions in a laboratory, where they are deprived of obvious information about the time of day.

However, solar-day rhythms are not the only rhythms expressed in animal behavior. Many animals that live in the intertidal zones along ocean shores display rhythmic changes in activity that correlate with the ebb and flow of tides.

Fiddler crabs of the genus *Uca,* for example, become active and feed during low tide periods, but plug their burrows from the inside and remain hidden during high tide periods. Since tides occur mainly in response to variations in the moon's gravitational pull on the earth during the 24.8-hour lunar day, tidal ebbs and flows occur about fifty minutes later each day than they did the preceding day. Because fiddler crabs emerge from their burrows and forage on exposed mud flats during each low tide, their activity periods begin fifty minutes later each day. Such tidal rhythms of alternating activity and inactivity are called **lunar-day** (moon day) **rhythms** because of the lunar periodism of the tides.

Frank A. Brown, Jr., and his colleagues discovered that these tide-related activity periods also persist under constant conditions in the laboratory. Fiddler crabs maintained in the laboratory are deprived of direct information about the ebb and flow of tides. Nevertheless, they are inactive during the times of high tide on their home beach, and they become active and move around when the tide is low on their home beach. Some tidal rhythms persist for many days in the laboratory, just as many solar-day rhythms do. In fact, in the absence of a tide table for fiddler crabs' home beach (which may be quite distant from the laboratory), it is possible to determine accurately the times of the tides there simply by observing the crabs' activity cycles!

Daily rhythms, as well as the underlying cellular clocks that time the rhythms, are important adaptations because rhythmic changes in physiology and behavior help to ensure continuing successful interactions with a fundamentally rhythmic environment.

Box figure 33.1   Fiddler crabs (genus *Uca*) show a definite tidal rhythm. At low tide, fiddler crabs feed along muddy beaches. During high tide, they remain inactive in their burrows.

the food source is on a straight line from the hive toward the sun (figure 33.11b). If the dance is straight down the comb, then workers know they can find food by flying away from the sun. For example, if the forager dances 20° to the left of straight up, the workers know they should fly 20° to the left of the direction of the sun, and so forth. The direction to the food source is thus symbolized by the dancer's movement with respect to gravity (called a **geotaxis**).

Even though the sun's position in the sky changes during the day, the bees can compensate for this even in the darkened interior of the hive because they possess an accurate **biological clock** (see chapter 19). A dancing bee inside a dark hive gradually changes the direction of its dance with the passage of time. In fact, the straight run of the dance rotates on the surface of the comb like the hands of a clock (only counterclockwise). Bees also use smell and visual cues, but the dance gives enough information to get the recruited workers very close to the food source.

This complex system of communication and location finding is adaptive because it communicates specific directions that permit more workers to reach food sources more quickly, thus enhancing the food-gathering efficiency of the hive.

This efficient communication system is also used when bees **swarm.** Swarming is a process by which a queen and a large number of worker bees leave an established hive and start a new hive. The bees fly out and settle in a mass on a tree branch or other surface (figure 33.12). Then individual workers fly out and scout the area for appropriate new hive sites. When a scout bee locates a potential hive site, she returns to the swarm and does a waggle dance to communicate the distance and direction to the site. This allows other bees to find and explore the site also. Eventually, a "decision" is reached, and the location for the new hive is chosen from among the sites being explored.

Again, use of the waggle dance is adaptive. Bees in swarms are vulnerable because they are exposed to predators and the weather. Thus, it is important that the new hive be established as quickly as possible, and communication by the waggle dance facilitates hive site location and selection.

The waggle dance is a form of symbolic communication because each dance is a miniaturized, symbolic version of the journey to a specific point. It is difficult to reconstruct the evolution of a complex behavior such as the waggle dance, but it is easy to understand that natural selection has favored the evolution of behavior that increases the efficiency of such vital processes as food gathering and location of appropriate living sites.

Figure 33.12 A swarm of honeybees on a fire hydrant in San Francisco. Swarming bees assemble in masses (usually on tree branches), while individual workers scout for a suitable new hive site.

The Seeing Ear

How do animals that are active in almost total darkness find their way? If they are not only active, but also are trying to catch elusive and fast-moving prey, they require a very accurate system for finding their way around. One solution has been the evolution of **sonar systems,** in which animals make noises and interpret the returning echoes. Using such a system requires that an animal be able to overcome many technical problems, but this is exactly what bats, dolphins, and a few other organisms have done in a variety of intriguing ways.

Donald Griffin clearly demonstrated that bats **echolocate.** He allowed them to fly in a room that was specially equipped with fine wires running from the ceiling to the floor. He soon learned that even blindfolded bats could find their way around these wires in total darkness. However, plugging a bat's ears or covering its mouth seriously affected its performance. Such a bat was very reluctant to fly at all. It would hang on its perch and groom its ears and mouth vigorously, trying to remove the plugs. If the bat was forced to fly, it flew "blindly," running into wires and crashing into walls.

Figure 33.16 Interactions between a bat and green lacewings recorded in photographs taken by Lee Miller and his colleagues. The camera shutter is kept open as bat and lacewing fly in the dark, and a series of brief flashes freeze their movements on film. (a) A bat captures a lacewing. The numbers indicate the position of each animal as each flash goes off. This bat swings its tail up, catches the lacewing against it, and flies off with the insect (*arrow*) in its mouth. (b) A lacewing evades the bat by folding its wings and diving before the bat reaches it. (c) The bat enters from the upper left, but the lacewing takes evasive action and the bat misses. This time, however, the bat swings around for another try.

(a)

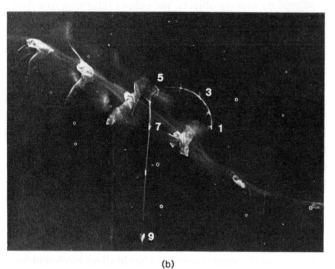

(b)

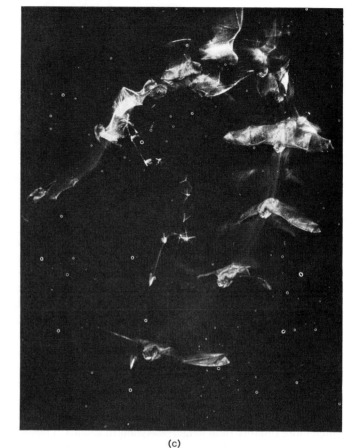

(c)

Ways of avoiding capture have evolved in bats' favorite prey. Some insects, including many of the common night-flying (noctuid) moths, have ears that are extremely sensitive to ultrasound. When moths hear batlike sounds, they respond quickly. If you jingle a set of keys (this makes some sounds in the ultrasonic range) near a moth that is clinging to a wall, the moth will let go and drop to the ground. It was just such a chance observation that focused Kenneth Roeder's attention on moths' responses to ultrasound.

For several summers, Roeder spent every warm evening in his floodlit backyard playing tape-recorded sounds he could not hear to moths that happened to be passing by. The moths' behavior was striking. When a moth was 20 to 30 meters from the loudspeaker, it turned and flew rapidly in the opposite direction. However, when a moth was very close to the sound, it either folded its wings and dropped to the ground, flew hard toward the ground in a power dive, or executed a series of loops and dives that generally took it downward. It was not possible to predict which one of these three courses of action the moth would take.

By watching interactions between moths and local bats, Roeder and his associates quickly accounted for the different kinds of moth behavior. A moth that is 20 meters away from a bat is still "invisible" to the bat; that is, it is out of the bat's sonar range. If the moth turns quickly, it can get completely out of the bat's range before it is detected. However, when a moth is only 10 meters or so away, it is within sonar range and it is in acute danger since bats can fly faster than moths. Only by executing evasive action can the moth escape.

If moths always used the same escape maneuver, however, bats would soon predict their behavior and intercept them regularly. Clearly, variable evasive maneuvers are adaptive for moths, or for any other animal escaping a predator for that matter. Therefore, natural selection favors moths that behave erratically, sometimes diving and sometimes going in for aerobatics. In experiments, moths that did not take evasive action were seven times more likely to be captured than those that did.

Behavior of predators and their prey usually includes measures and countermeasures, such as those seen in bats and various insects (figure 33.16). This is an example of co-evolution (see p. 776)—interactions between predator and prey are so intense that they have influenced each other's evolution. In most cases, a balance is eventually reached. The capture behavior of the predators is efficient enough so that a stable population of predators can obtain food. But the evasion and escape behavior of prey organisms is efficient enough so that a stable population of prey organisms is maintained.

Courtship Behavior

We have examined several examples of adaptive behavior used for gathering food. Behavioral mechanisms also are involved in various other aspects of animals' lives, such as reproduction.

In many animals, copulation is preceded by elaborate patterns of **courtship behavior.** Members of some species are normally aggressive, and courtship helps to resolve the conflict between the drive to be hostile and the drive to mate. In such species, courtship behavior often has both **aggressive** and **appeasing** components.

In many species, courtship behavior may be necessary to resolve conflicts involved in animals' touching each other, as they must do during mating. This is not a trivial problem since many adult animals respond very negatively to being touched by another animal. This general response is understandable if you consider that in most contexts, being touched can well mean being captured and killed. The negative response is so strong that it includes even members of the animal's own species. Courtship behavior resolves this conflict, and animals are able to come into close contact for mating and nesting or other parental care activities.

Figure 33.17 A male stickleback selects a nest site, excavates it by removing sand and gravel and then constructs a nest over it. The nest is made of bits of vegetation that the male glues together with a secretion from his kidney. The completed nest contains a tunnel in which the female will lay her eggs.

Another important function of courtship behavior is that it helps ensure that males attract females of their own species, and a variety of behavioral mechanisms that attract females have evolved.

Courtship of the Stickleback

The three-spined stickleback is a small, freshwater fish that lives in ponds or slowly moving streams throughout most of Europe and along both coasts of North America. The male stickleback defends a territory to which females come to lay their eggs. He is the sole guardian of the eggs. Each male competes with other males for possession of a territory with a good nesting site. Then he builds a nest by carrying mouthfuls of sand away from the nest site until a small pit is formed in the sandy bottom. Next he collects bits of vegetation in his mouth and spits them into the pit. When he has a little pile, the male glues the bits together with a secretion from his kidney. Finally, he forcefully wriggles and burrows his way through the pile of glued vegetation, making a tunnel (figure 33.17).

Box 33.2
Animal Compasses

One of the most remarkable aspects of animal behavior is that animals have what might be called a sense of place. Foraging animals move out, sometimes over great distances, from their homes and are able to find their way back. Some of these animals might use landmarks to find their way along familiar paths. However, many animals demonstrate far more impressive abilities to find their way from place to place, often traversing hundreds or even thousands of kilometers in the process of migration.

For example, young green sea turtles are found on the east coast of South America, where they feed and grow. Every two or three years adult females set off swimming eastward across the Atlantic. Their sea journey takes them more than 2,200 km to the beaches of tiny Ascension Island, which is only about 10 km across, where they lay their eggs. They leave the eggs buried in the sand and head out to sea for their return voyage. When hatchling turtles dig their way out of the sand, they scramble across the beach to the water. They apparently are carried westward by the current to the coast of South America, where the females live until they set out across the Atlantic on their own journey to Ascension Island.

Green sea turtles, however, do not hold the record for long-distance migration. Many common birds migrate across large parts of a continent or even from continent to continent. Probably the champion long-distance migrant is the arctic tern. In the autumn, the arctic terns that nest in the North American arctic regions migrate across the Atlantic and travel southward along the west coasts of Europe and Africa until they reach the tip of South Africa. Then they cross the South Atlantic to Antarctica, where they live along the shore during the Antarctic summer. In this journey they cover a distance of about 18,000 km.

Many other birds' long-distance migrations are spectacular because they involve nonstop flights requiring great endurance. For example, a flock of blue geese and snow geese was observed to migrate from James Bay, Canada, to Louisiana apparently without stopping. They covered this distance of 2,700 km in sixty hours.

Some species of birds migrate in flocks in which young birds and mature adults travel together, and thus young birds travel with individuals that have made the trip before. In other species, however, older birds migrate first, leaving young birds to make their first migratory journey without the company of experienced adults.

Birds also can find their way on long journeys that are not part of their normal migrations. For example, a Manx shearwater was captured on the west coast of England, banded, and transported by plane to Boston, Massachusetts, where it was released. The shearwater returned to its nest in England just twelve and one-half days later after crossing the Atlantic Ocean, a distance of about 5,300 km.

How do birds find their way on these long journeys? The use of familiar landmarks might be a factor near the beginning and end of a journey, but long-distance navigational capabilities must exist as well. Some birds use the sun and/or the stars for orientation. Such celestial navigation requires an accurate time sense because animals using the sun as a compass, for example, must continually change their orientation relative to the sun to stay on course in a single direction. Experiments have shown that birds do indeed have and use accurate time information in their celestial navigation.

However, the migratory flights of many birds do not stop on overcast days, when celestial navigation is not possible. Birds continue their journeys, sometimes even flying through clouds as they go. Early in the study of migration, some biologists suggested that migrating animals might use a magnetic sense for navigation, but the idea was dismissed because it was assumed that animals could not sense the low energy levels of the earth's magnetic field.

During the 1960s, Frank A. Brown, Jr., and his colleagues presented very persuasive data indicating that some animals show orientational responses to the earth's magnetic field. But again, despite the strong evidence presented, the idea of a magnetic sense for navigation was rejected by the majority of biologists for the same reasons it had been years before.

Finally, during the 1970s, the idea that animals can sense the earth's magnetic field and can use it in orientation and navigation became widely accepted. Some of the most convincing evidence came from studies of homing pigeons. Homing pigeons were carried some distance from their home loft and released to find their way home. Observers then recorded their vanishing bearing—the direction in which they disappeared as they flew away from the release point. The birds tended to orient themselves so that they disappeared in the direction of their home loft whether the day was clear or cloudy.

W. T. Keeton, Charles Walcott, and others have demonstrated, however, that a pigeon's homing ability on a cloudy day depends upon its being able to sense the earth's magnetic field. A pigeon's orientation is confused on cloudy days if it has a small bar magnet attached to its back or if it carries a small battery-powered coil that induces a uniform magnetic field in its head. Magnets attached to the pigeons do not disturb their orientation on clear days. Thus, it appears that use of a magnetic compass for orientation functions as a backup to the sun compass, which cannot be used on cloudy days.

Since the demonstration of a magnetic sense that can be used for orientation, some biologists have proposed that a magnetic sense may also be involved in animals' strong sense of place. Local and regional anomalies in the earth's magnetic field might provide "map" information to animals, but this proposal requires much more testing and evaluation.

How do organisms perceive magnetic fields? This question has been only partially answered. Some organisms have been found to possess grains of **magnetite** (Fe_3O_4). For example, bacteria that show specific orientation in magnetic fields contain chains of magnetite particles. Magnetite particles ("magnetosomes") also have been reported in a number of animals, including tuna, dolphins, whales, pigeons, and green turtles. In animals, movement of magnetite particles in response to magnetic fields might be detected by adjacent sensory receptors.

We now know that organisms can respond to relatively weak magnetic fields. Our next tasks are to determine how organisms respond to such magnetic fields and to investigate the adaptive significance of their responses.

The male vigorously defends his nest site from other stickleback males and other intruders, such as invertebrates or fish that are likely to eat eggs. A male stickleback recognizes another male by the bright red color of his belly (this is another example of a visual communication signal). When a male sees a spot of red in his territory, he charges out to expel it. It is easy to fool these aggressive little fish with a piece of cardboard of almost any shape, as long as it has red along the bottom side (figure 33.18).

If an intruder does not flee after an initial rush, the two males may circle, or they may stand on their heads in the water, jerking up and down rapidly in gestures of **threat**—rather like a dog baring its teeth in a snarl or a man shaking his fist in an opponent's face.

When the territories of two males are side by side, their fighting may take on a rather bizarre appearance. Each male is **dominant** in his own territory, but **subordinate** in the territory of another. Therefore, a stickleback that is fearlessly pursuing a rival will suddenly turn tail and flee when he discovers that he has crossed the territorial boundary.

When a female enters the male stickleback's territory, his behavior is very different. He recognizes a female ready to lay eggs by her swollen belly. (Again, it is easy to fool him; he will respond to any object of about the right size and shape.) He launches into an elaborate course of zigzag swimming in front of the gravid female, interrupted now and then by a quick dash back toward his nest (figure 33.19). If she follows him, he leads her back to his nest and sticks his snout into the entrance. The female may then enter and wriggle into the nest tunnel. If the female does enter the nest, the male nuzzles her tail, and as he does so, she spawns her eggs and then swims out the other side of the tunnel. The male then immediately enters and sheds sperm on the newly laid eggs.

The female stickleback may break off this **chain of behavior** at any point and swim from the territory. This only causes a renewed course of zigzag swimming by the male. A neighboring male also may court the female and attempt to lure her to his own nest. He may even sneak in behind a spawning female and shed sperm on eggs in another male's nest!

After fertilization, the male's behavior changes dramatically. He chases the female and drives her from his territory. After arranging the egg mass on the floor of the nest, he repairs any damage done to the nest, and also lengthens it to make room for additional batches of eggs. Within an hour he again courts any female that swims into his territory.

Figure 33.18 Male sticklebacks attack models of various shapes such as these, so long as the "belly" is red.

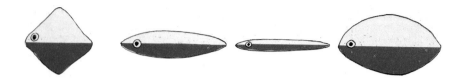

Figure 33.19 Sexual behavior of the three-spined stickleback. When a female enters a male's territory, he courts her with zigzag swimming. If she responds, he leads her to his nest, where he adopts a special posture with his snout in the entrance. Then if the female enters the nest, the male nuzzles her tail, and as he does so, she spawns her eggs (*inset*). After she spawns, the female swims out of the nest and the male enters and sheds sperm on the eggs.

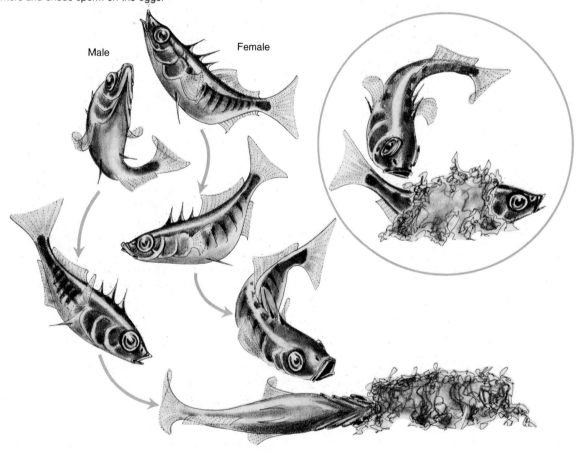

Male

Female

As the embryos develop, however, they require more of his time, and he is less likely to court. He swims near the nest and beats his fins, forcing a current of water through the nest. This fanning behavior is essential because it makes adequate gas exchange possible for the developing embryos. During this period of tending the young, the male stickleback loses much of his bright red coloration. This probably further protects the nest from predators.

When the young fish begin to hatch, the male continues to hover over them for several days. If any should stray away from the nest, he will suck them up into his mouth and spit them back into the nest. Finally, when they are large enough, they swim off, or the male simply deserts both them and the nest.

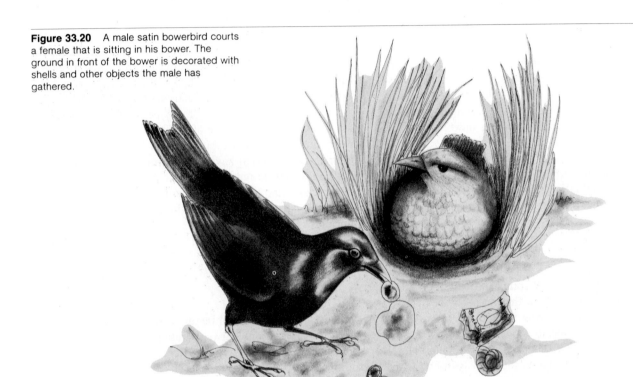

Figure 33.20 A male satin bowerbird courts a female that is sitting in his bower. The ground in front of the bower is decorated with shells and other objects the male has gathered.

The reproductive behavior of this little fish illustrates a number of important ethological concepts. First, the male stickleback is not responding generally to the female or a rival male. Only a specific part of the overall stimulus (for example, the red belly of another male) elicits a response. A specific stimulus such as another male's red belly is called a **sign stimulus,** and it is said to **release** the male's aggressive behavior. Sign stimuli are found throughout the animal kingdom. For example, the parent gull's bill is a sign stimulus that elicits begging movements in newly hatched chicks.

Second, there are progressive changes in the male's **motivation,** or likelihood to behave in a certain way. For example, just prior to mating he vigorously courts the female, and just afterwards he chases her from the territory as though she were a trespasser. Experiments have demonstrated that it is the presence of freshly laid eggs in his nest that causes him to drive away all other fish, including the female, and to tend the eggs closely. Understanding the bases of motivational changes such as these is an important part of understanding and predicting animal behavior.

Courtship of the Bowerbird
In many species, courtship interactions involve specific responses of females to combinations of coloration and display by males. But the situation can be more complex, such as it is in the case of bowerbirds. Male bowerbirds court by collecting objects and displaying them for females.

Bowerbirds are chickenlike birds that live in the forests of Australia and New Guinea. Unlike the males of many other bird species, the male bowerbird is drably colored, much like the female. However, he makes up for it in a most surprising way—he constructs a "bower."

The male satin bowerbird builds a walled avenue of sticks (figure 33.20). He collects distinctively colored objects and displays them at one end of the bower. Satin bowerbird males show a preference for anything that is blue, including blue feathers, blue berries, and freshly picked blue flowers. When a male establishes his territory near human habitation, he may collect and display blue crockery, rags, bus tickets, candy wrappers, and so forth. These males also make a paste of blue fruit pulp and saliva, which they paint on the inner walls of their bowers with the aid of a piece of bark that they use like a paintbrush.

Other species of bowerbirds choose different kinds of display objects. The spotted bowerbird, for example, uses white objects, such as bleached bones, dried white snail shells, and so forth. When they are nesting near humans, male spotted bowerbirds collect an astonishing array of miscellaneous objects including silverware, coins, keys, and jewelry. One of these birds even hopped through an open window and stole a gentleman's glass eye from his bedside!

When a female comes to a male's bower, he **displays** by picking up feathers, flowers, or other objects in his bill. As he does this, he produces a rhythmic whirring, not unlike the sound of a mechanical toy. He arches his tail, stiffens his wings, and hops around the bower with his display objects. In most cases the female leaves before courtship is consummated, but occasionally a female crouches down and copulates with the male in the bower.

In these bowerbird species, females apparently choose mates on the basis of their ability to assemble materials in a bower. Here mate selection focuses on the display objects that the males collect, rather than on a display of their plumage.

Foster Parents

Many kinds of animals expend a great deal of energy on the care and feeding of their young offspring, activities that often expose parents to predators. Nevertheless, behavior that assures the care of helpless offspring is adaptive because it is essential for reproductive success, and thus it forms an important part of the behavior of many adult animals. However, some animals avoid the burdens of parental care altogether—they trick other animals into caring for their young.

Some birds, referred to as **brood parasites,** lay their eggs in the nests of other species. In many cases, the birds of the **host species** then care for the parasite's eggs and nestlings as if they were their own. The behavior of brood parasites is a lesson in deception.

Female cuckoos spend much time patiently watching the nest-building activities of a prospective host species. Their egg-laying is timed to coincide with that of the host. When the female cuckoo is ready to lay an egg, she quietly flies to the host's nest, moves one of the host's eggs, slips onto the nest, and lays one of her own in its place. In a matter of seconds she is gone, taking a stolen host egg with her. Each female cuckoo lays her eggs in the nests of one particular species of host, and it is thought that she chooses the same species by which she herself was raised. The cuckoo's eggs closely resemble those of the host species—eggs that are less likely to arouse the host's suspicions are more likely to be accepted and cared for.

The incubation period of the cuckoo egg is slightly shorter than that of the host's eggs, and as a result, the nestling cuckoo generally hatches first. A day after hatching, the seemingly helpless cuckoo becomes totally intolerant of any objects in the vicinity and begins the strenuous task of evicting everything around it from the nest. Supporting an egg or nestling on its back and holding it with its wings, the young cuckoo pushes it along the rim of the nest and out over the edge (figure 33.21a). Soon the cuckoo is the sole occupant of the nest, and the foster parents devote all of their attention to it. The baby cuckoo grows quickly, and in some cases, soon becomes larger than its foster parents (figure 33.21b).

There is an evolutionary "race" between brood parasites and their hosts. In the case of the cuckoo, natural selection favors hosts that avoid taking care of cuckoo young. Hosts often desert nests that have been disturbed by cuckoos or remove eggs that do not look like their own. Host birds often mob cuckoos and try to chase them away whenever they see them. But natural selection also favors cuckoos that become more successful at getting hosts to accept their eggs. Thus, this is a balanced relationship, and neither host nor parasite wins the "race."

Societies

We have considered a number of examples of **social behavior,** adaptive interactions between individuals of the same species, but thus far we have said little about the organization of animal societies.

A **society,** or social group, is a group of individuals of the same species that interact with each other and influence each other's behavior in definite ways. This behavioral interaction is a key characteristic of a society and distinguishes a society from a simple **aggregation,** a group of animals that may be together but do not interact behaviorally. Insects flitting around a light on a summer evening or huge flocks of birds of many species roosting together in trees in the autumn are aggregations, not societies.

Insect Societies

Organization of a society may be very rigid, as it is in many insect societies. In a honeybee colony, for example, individuals belong to **castes,** groups of individuals whose body forms and behavioral roles are predetermined by chemical and nutritional controls over development. A female larva that is destined to become a queen (a reproductively active female) is fed a special diet. Other females that do not receive the special diet become sterile workers.

Worker bees do all of the work in the colony. During their brief lives, they perform a series of tasks that change as they mature. In their first couple weeks they care for the young of the colony (figure 33.22). Later they build comb and guard the nest. Finally, when they are about three weeks old, they begin to search for food and transport it to the nest. They continue this demanding work until they die at about six weeks of age.

Figure 33.21 Brood parasitism by cuckoos. (a) A newly hatched cuckoo pushes its host's eggs out of the nest. (b) A hedge sparrow feeds a young cuckoo that is much larger than its foster parent.

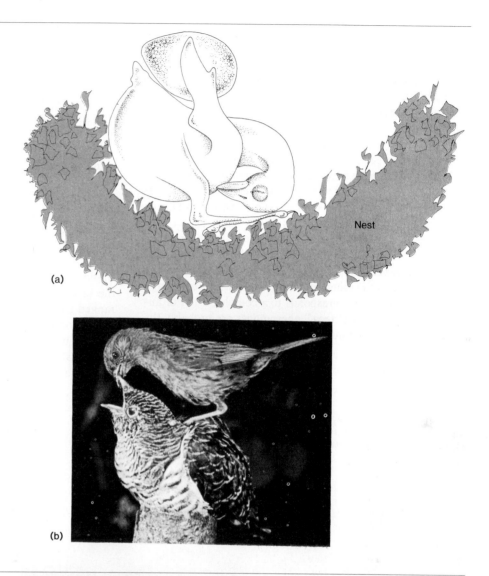

Nest

(a)

(b)

Figure 33.22 Honey bee workers constructing a queen cell in which a larva destined to become a queen will develop.

Each bee automatically fulfills its function in the regular life of the colony, participating in complex behavior such as communicating information about food sources by waggle dancing (see p. 817). Even during specialized activities such as swarming (p. 819), behavioral responses of bees are stereotyped and predictable.

The complex social organization of honeybees, as well as that of many species of ants and termites, is maintained to a large degree through the action of pheromones (see p. 379). A **pheromone** is a chemical, or combination of chemicals, that is released by an individual and in some way influences the behavior (and in some cases even the development) of other individuals of the same species. The queen honeybee, for example, secretes the pheromone 9-oxodec-*trans*-2-enoic acid (a compound originally named "queen substance") that strongly influences a wide variety of behavior in all the other bees in the colony. Although each member of the hive also responds to a complex array of auditory and tactile "messages," pheromones are probably the primary mode of communication and control.

Vertebrate Societies

In vertebrate societies, behavioral interactions are far less rigid and stereotyped than in insect societies. Vertebrate individuals do not belong to developmentally determined castes, and reproduction is not biologically restricted to a very small number of individuals, as it is in insect societies. Furthermore, the organization of vertebrate societies depends on the ability of society members to recognize each other as individuals.

A particularly important aspect of vertebrate social organization is the existence of **dominance hierarchies (pecking orders)**, in which some individuals always dominate others and receive preferential access to desirable living sites, food sources, and mates (figure 33.23). In some cases the dominance hierarchy is a linear chain of individuals in which a single animal at the top dominates every other member of the group, a second individual dominates all but the top-ranked individual, and so on down.

A dominance hierarchy emerges through a series of hostile encounters, which can involve bloody fights between individuals in some species. In many species, however, once winners and losers have been determined, the fighting largely

Figure 33.23 A wolf of low rank in a wolf pack shows submissive behavior to dominant individuals by assuming a vulnerable position on its back.

ends and stable relationships persist. This stability is adaptive because it reduces conflict among individuals that would otherwise waste energy and be a continuing source of injury to group members.

The stability of a dominance hierarchy is maintained by a set of signals that pass between individuals. In an African baboon troop, for example, a subordinate (lower-ranking) male demonstrates submission to the dominant male in the troop by "presenting"—he assumes the same posture that female baboons do when they are ready to mate. The dominant male responds by placing a paw on the subordinate male's back. Occasionally, if a subordinate does not demonstrate submissiveness, the dominant male will threaten him. The subordinate male usually responds quickly with the appropriate submissive gesture and further conflict is avoided (figure 33.24).

In this case, dominance is very important in terms of reproduction. Subordinate males do mate with females that are in estrus, a state of sexual receptiveness around the time of ovulation. However, when a female is in that phase of estrus during which ovulation occurs, the dominant male is the only male that mates with her. Thus, during the time that he is dominant, this one male is likely to father the majority of offspring produced in the troop.

Figure 33.24 Dominance hierarchies in animal societies are maintained by individuals exchanging signals. In this encounter between two male baboons in Tanzania a subordinate male "presents." A dominate male places a paw on the subordinate male's back as the subordinate male crouches and gives a fear grin.

Sociobiology: A New Look at the Evolution of Social Behavior

During the 1970s, a new emphasis developed in behavioral research. This new emphasis, **sociobiology,** focuses on certain genetic and evolutionary explanations for social behaviors. Advocates of sociobiology maintain that many complex social behaviors, including human social behavior, could be better understood if they were reexamined in the context of sociobiology. Because many biologists disagree with some of its basic assumptions, a great deal of controversy surrounds sociobiology. Let us look at an example of a sociobiological analysis and interpretation of behavior patterns.

Helpers at the Nest

Gray-breasted (Mexican) jays are sociable birds that live in permanent flocks of five to fifteen birds in the Arizona mountains. Only one or two pairs of birds in the flock build nests at any one time. All of the eggs in a given nest are laid by one female and incubated by a single pair of jays. The unusual thing about the jays is that after the nestlings have hatched, all of the birds in the flock feed the young, even the parents from the other active nest. It is a sort of bird commune. Although the nestlings' parents supply more food than other individuals in the flock do, they provide only about half the food eaten by the hungry nestlings (figure 33.25).

This cooperative behavior in caring for offspring seems unusual, but it is by no means unique. Many female mammals, such as lions and wild pigs, are known to nurse another female's offspring now and then, and African hunting dogs will help feed another female's pups (figure 33.26).

Figure 33.25 A gray-breasted (Mexican) jay feeding young in a nest.

Figure 33.26 Two female lions, probably sisters. One is nursing both her own cubs and those belonging to the other female.

Such seemingly unselfish interest in the welfare of others is called **altruism**, and behavioral biologists call this **altruistic behavior.** How did this behavior evolve? How could natural selection favor taking care of another individual's offspring with the same care that would be given to one's own?

It appears to be very difficult for young gray-breasted jays to establish themselves in a new flock. Therefore, they usually remain in the same flock throughout their unusually long lifetime. Young jays do not begin breeding until they are three years old, so a flock usually contains several young birds. Because most jays remain within the same flock, most of the individuals in a flock are likely to be very close relatives. The flock actually is an extended family unit, and thus, if an animal helps another member of the flock, it is almost certainly helping a close relative. Many examples of altruistic behavior are found, like this one, to be cases of animals helping their relatives. But the question still remains, how could such behavior evolve?

In the early 1960s, W.D. Hamilton, then a graduate student at London University, developed a hypothesis that has generated much interest and controversy among biologists who study behavior. Hamilton pointed out that a parent caring for its young is just a special case of the more general phenomenon of animals helping genetic relatives. Close genetic relatives, in addition to parents and offspring, have genes in common. Thus, an animal showing altruistic behavior toward a close relative is promoting survival and transmission of at least some genes that are the same as its own. Hamilton proposed that natural selection favors such altruistic behavior toward relatives.

Does Hamilton's hypothesis provide any insight into the evolution of helpers at the nest in gray-breasted jays? Since members of a flock are almost always close relatives, helping to feed another individual's nestlings means that a jay is helping a close relative. Since individuals often cannot breed for several years, the best way to promote one's own genes is by helping relatives in the flock that are able to breed and transmit those same genes.

This is not to suggest that animals "know" in any conscious sense who their relatives are. However, animals that behave in this way have more surviving relatives that share and transmit the same genes, including genes for altruistic behavior, than those that do not. Even altruistic behavior can therefore be self-serving in an evolutionary sense if the altruism is directed toward close relatives with whom the altruists have genes in common. The frequency of genes of organisms involved in such **kin selection** is increased in subsequent generations through the reproductive success of their close relatives.

Sociobiology and Human Behavior

Sociobiology assumes that individuals within social groups behave in ways that improve the chances for passing on their own genes. Put another way, sociobiologists think that individuals in a social group have been programmed through evolution to behave as self-serving opportunists. Because the adaptiveness of behavior is measured ultimately in terms of effects of the behavior on successful transmission of genes, some sociobiologists go so far as to say that the basic "selfish" units are genes rather than individuals.

Sociobiologists insist that these ideas apply to human behavior as well as to the behavior of other animals. It is on this point that sociobiology has generated the greatest amount of controversy. People who disagree with some of its basic assumptions argue that sociobiology as applied to human beings promotes a kind of **biological determinism.** They feel that sociobiology leads to the conclusion that much of human behavior, including some of its most negative aspects, is at least in part genetically determined and thus is inevitable and unalterable. Opponents of sociobiology feel that this concept of determinism can lead people to an acceptance of the social status quo that can condone racism, sexism, and class determinism. They argue that culture and the social environment, not genes, are of primary importance in shaping human behavior.

Sociobiologists answer this criticism by saying that it is not dehumanizing to investigate the roles of genes in establishing human behavior, and that the prospect of discovering that human behavior patterns are more a part of human biological nature than originally thought should not make people uneasy. They argue that it does not detract from one's humanity to discover that humans are different from one another in terms of behavior and that some of these differences have genetic roots deep in our evolutionary past.

The questions raised by sociobiologists are intriguing, and we can certainly expect to hear more about these arguments in the future.

Summary

Adaptive behavior increases the probability that animals, or their offspring, live to reproduce successfully so that their genes are perpetuated in future generations.

The innate (genetically determined) nature of certain behavior is clear. Behavioral mutants show altered behavior patterns that are inherited by offspring of the mutant organisms. The genetic basis of behavior is also demonstrated when hybrid individuals show behavior patterns that are intermediate between the two different behavior patterns of their parents.

Learned behavior results from an organism's experience with its environment, and the influence of learning on behavior patterns begins very early in life. Even obviously instinctive behavior patterns, such as the pecking response in the feeding behavior of gull chicks, is modified by learning through experience.

Many animals learn to recognize members of their own species and distinguish them from other organisms because they become imprinted on their parents at an early critical age. All behavior results from a combination of animals' genes and interactions with their environments (learning).

Behavioral expression depends on internal motivational states. Physiological changes, such as changing hormone levels during reproductive cycles, result in changing motivation. Other motivational changes depend on completion of specific sequences of behavioral expressions.

Adaptive behavior patterns play roles in diverse aspects of animals' lives, including such vital activities as obtaining food and finding proper living places. Bee communication through the waggle dance increases the efficiency of food gathering and facilitates location of new nest sites during swarming. Bats have highly specialized sensory and behavioral mechanisms that permit them to fly safely at night and to locate and capture flying prey.

Courtship behavior is adaptive because it brings animals together with prospective mates of their own species and resolves potential behavioral conflicts between males and females. In many cases, courtship displays are part of a chain of behavior leading to parental behavior that provides care for helpless offspring. Brood parasites such as cuckoos shift a large part of the burden of their reproduction to members of other species that they trick into incubating their eggs and feeding their young.

Societies are groups of individuals of the same species within which behavioral interactions occur among individuals. Insect societies are based on developmentally established caste systems. Vertebrate societies involve more flexible behavioral interactions among individuals.

Sociobiology focuses on genetic and evolutionary explanations for social behavior. Among other things, sociobiology attempts to explain how altruism directed toward close relatives promotes transmission of at least some genes that are the same as those of the altruist.

When the concepts of sociobiology have been used to explain certain aspects of human social behavior, considerable controversy has arisen.

Questions for Review

1. Use the example of ground-nesting gulls removing eggshells from their nests to discuss the evolution of behavioral patterns.
2. What does an ethologist do?
3. What evidence is there that nest-building behavior in some lovebirds has a strong genetic component?
4. How does imprinting affect the behavior of ducks, geese, and some other animals throughout their lifetimes?
5. Discuss the development of chaffinch singing as an example of the interaction of genes and learning in the development of a behavior.
6. What do behavioral biologists mean by motivational state?
7. How does the waggle dance of bees communicate information about direction and distance?
8. How does the use of ultrasound make the bat echolocation system more efficient than would the use of lower-frequency sounds?
9. What is a sign stimulus? Give two examples.
10. Courtship behavior produces several important results. List and describe three of them.
11. What is the adaptive value of dominance hierarchies for vertebrate societies?
12. Explain the sociobiological reasoning leading to the conclusion that altruistic behavior can be adaptive.

Questions for Analysis and Discussion

1. Although brood parasitism seems to be highly adaptive for some animals, there are only a few species in which it is part of normal reproductive behavior. Why do you suppose this is true?
2. Consider two small flocks of chickens. One flock is left intact with the same set of individuals kept together more or less permanently. The composition of the second flock is altered repeatedly by removing several individuals and then adding several new individuals to replace them. What might be some differences between these two flocks in terms of behavior and general well-being?
3. Many people feel that great care must be taken in applying the principles of sociobiological analysis to human behavior. Explain their concern about biological determinism in terms of making future social and political decisions.

Suggested Readings

Books

Alcock. J. 1984. *Animal behavior: An evolutionary approach.* 3d ed. Sunderland, Mass.: Sinauer Associates.

Gould, J. L. 1981. *Ethology.* New York: Norton.

Hart, B. L. 1985. *Behavior of domestic animals.* New York: W. H. Freeman.

Lorenz, K. 1961. *King Solomon's ring.* New York: Crowell.

Manning, A. 1979. *An introduction to animal behavior.* 3d ed. New York: Addison-Wesley.

Wilson, E. O. 1980. *Sociobiology: The abridged edition.* Cambridge, Mass.: Belknap Press.

Articles

Bekoff, M., and Wells, M. C. April 1980. The social ecology of coyotes. *Scientific American.*

Carr, A. 1986. Rips, FADS, and little loggerheads. *BioScience* 36:92.

Dilger, W. C. January 1962. The behavior of lovebirds. *Scientific American.*

Fenton, M. B., and Fullard, J. H. 1981. Moth hearing and the feeding strategies of bats. *American Scientist* 69:266.

Goodenough, J. 1984. Animal communication. *Carolina Biology Readers* no. 143. Burlington, N.C.: Carolina Biological Supply Co.

Gould, J. L. 1984. Magnetic field sensitivity in animals. *Annual Reviews of Physiology* 46:585.

Gwinner, E. April 1986. Internal rhythms in bird migration. *Scientific American.*

Hailman, J. P. December 1969. How an instinct is learned. *Scientific American.*

Johnson, C. H., and Hastings, J. W. 1986. The elusive mechanism of the circadian clock. *American Scientist* 74:29.

Jolly, A. 1985. The evolution of primate behavior. *American Scientist* 73:230.

Keeton, W. T. December 1974. The mystery of pigeon homing. *Scientific American.*

Palmer, J. D. 1984. Biological rhythms and living clocks. *Carolina Biology Readers,* no. 92. Burlington, N.C.: Carolina Biological Supply Co.

Scheller, R. H., and Axel, R. March 1984. How genes control an innate behavior. *Scientific American.*

Silver, R. 1978. The parental behavior of ring doves. *American Scientist* 66:209.

Ecology
Organisms and Populations

34

Chapter Concepts

1. Ecology is the study of relationships that exist between organisms and their environments.
2. Organisms respond to physical factors, such as temperature, in a variety of ways.
3. Populations have certain properties and characteristics, and they respond in predictable ways to various factors.
4. When resources in an environment are nonlimiting, populations grow exponentially; as resources become depleted, population growth slows and eventually stops.
5. Population size may be influenced by forces within or outside the population.
6. Two species with identical ecological requirements cannot coexist indefinitely.
7. Predator and prey interactions are the outcome of coevolution.
8. Several kinds of symbiotic relationships exist between pairs of species.

Imagine a field trip to a prairie grassland in which you observe the living and nonliving elements of the grassy plain. While most of the plants are grasses, here and there are broadleaf plants growing singly or in small clusters. Except for a few trees lining a streambank in a distant ravine, nothing grows much taller than waist-high for as far as the eye can see.

As you look more closely, you see many insects feeding on the grasses and other plants. Some scurry and some crawl along leaves and stems, but others hop, sometimes great distances, from plant to plant. Down on the ground, mice, rabbits, lizards, and other small animals live and feed on the lower portions of the plants, on insects, and on seeds that fall from above. Many birds wing their way over the tall grass, singing as they fly, now and then alighting briefly to feed on plants or on plant-eating insects.

High overhead you occasionally spot a hawk or an eagle soaring silently through the sky. One of these birds may suddenly plummet earthward to capture one of the small animals on the ground. Such birds are not the only hunters of the grassland, however. Foxes, coyotes, and weasels move stealthily through the gently swaying vegetation in search of prey (figure 34.1).

As you bend down to examine the soil surface, you see a layer of dead and decaying plant material that has fallen during several growing seasons. Close examination of this litter reveals many kinds of tiny insects, quite different from those feeding above on green plant parts. You may also see worms and other invertebrates that are also common consumers of the litter material. Both in the litter and in the plants above, spiders spin their delicate webs and patiently wait for unwary insects to become fatally ensnared.

If you examined a sample of the litter under a microscope, you would find it is also populated by countless millions of bacteria and fungi. These organisms play an indispensible role in the decay processes that break down litter constituents and add new layers to the fertile soil of the grassland.

Finally, as you survey the landscape stretching out before you in all directions, you notice the wind, an ever-present physical factor of the prairie. Wind buffets the prairie almost continuously, causing the tall grasses to ripple almost like waves in an ocean. The effects of wind and very modest rainfall make prairie grasslands relatively dry places. But prairie organisms, living together in complex, interdependent networks, are well-adapted to life in this dry and often harsh environment.

Thousands of acres of our once-vast prairies have been converted to agricultural lands. Fortunately, a few native prairie grasslands still remain in North America. Within these protected areas, biologists study the fascinating interrelationships that exist among the plant and animal inhabitants of the prairie.

Figure 34.1   A coyote partially hidden in the tall prairie grass.

Ecology

The study of such relationships is the province of the science of **ecology.** The word ecology comes from the Greek root *oikos,* meaning "house" or "household." Modern ecology emphasizes the study of **ecosystems,** the assemblage of organisms in a particular area along with their physical environment. Ecologists analyze the complex relationships among all elements of an ecosystem, such as the prairie grassland in our example. The factors that influence an individual organism within an ecosystem include the **abiotic** (nonliving) components of the physical environment, such as moisture, temperature, light, gravity, and mineral nutrients (especially important for plants). The **biotic** (living) components of an organism's environment include other individuals of its own kind and members of other species.

As we begin our consideration of ecology, we need to introduce two basic ecological concepts: niche and habitat. An organism's **niche** is the particular way in which it interacts with its environment. How does the organism obtain nutrients? What other organisms might obtain nutrients by consuming the organism? How is the organism affected by changing physical factors in its environment? These interactions are specific to each kind of organism and have their origins in its evolutionary history. They enable that organism to survive and reproduce in its environment.

It is important not to confuse the concept of niche with that of **habitat.** An organism's habitat basically describes where that organism is to be found, the kinds of plants and animals with which it lives, and the array of physical factors with which it must deal. Thus, to say that a mallard duck lives in a marsh describes its habitat, and to say that it eats submerged vegetation describes one aspect of its niche.

Physiological Ecology

The branch of ecology that deals with the interactions between a single organism and physical factors in its environment is called **physiological ecology** or **ecophysiology.** We have already discussed a number of physical factors that are important components of an organism's environment. For example, we examined the importance of light in photosynthesis (chapter 7), photoperiodism (chapter 19), and sensory perception (chapter 17). We considered water relationships of plants in chapters 7, 9, and 10 and of animals in chapter 14. Now we will examine another physical factor: temperature.

Temperature

Organisms respond to changes in environmental temperature in different ways. As we saw in chapter 8, only about 40 percent of the energy available in organic nutrients is used for ATP production in the cells of living things. Most of the remaining 60 percent is emitted as heat, which, in all plants and most animals, is quickly lost to the environment. The body temperatures of these organisms are influenced by the temperature of their environment. When the environmental temperature rises, their body temperatures tend to rise, and when the environmental temperature falls, their body temperatures tend to fall. Organisms with environmentally influenced, variable body temperatures are called **poikilothermic,** which means "changeable temperature" (the Greek word *poikilos* means changeable).

However, some animals, specifically birds and mammals, conserve body heat and regulate their metabolism so that they maintain stable, relatively constant body temperatures. These animals are called **homeothermic** ("same temperature").

Metabolism is the sum total of all cellular activities in an organism's body, and the **metabolic rate** is a collective measure of these processes. Sometimes metabolic rate is measured as heat released from the body, but in organisms using aerobic metabolism it is often measured as oxygen consumption or CO_2 release or both.

Metabolic rates change with an organism's activity level. In humans, for example, the rate of oxygen consumption during heavy work can be 20 times the resting rate. When a butterfly is flying it consumes 100 times as much oxygen as when it is resting. Metabolic rates thus depend on measurement conditions. In laboratory studies, the resting metabolism of an organism carrying on only life maintenance functions is called **standard metabolism.** Standard metabolism is comparable to **basal metabolic rate** in humans, which is determined in midmorning, when a person is relaxed but awake and has eaten nothing since the preceding evening. Our discussion of temperature relationships and metabolism will focus on temperature effects on standard metabolism.

Poikilotherms

Because the rates at which chemical reactions proceed in living organisms are temperature dependent, environmental temperature changes and subsequent body temperature changes have a significant effect on the metabolic rates of poikilothermic organisms.

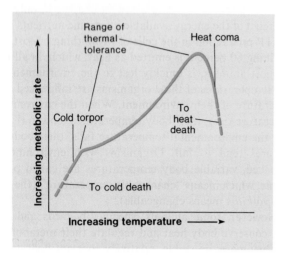

Figure 34.2 The relationship between temperature and the metabolism of a poikilothermic organism.

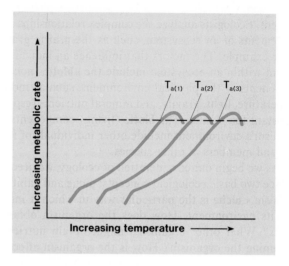

Figure 34.3 Results of acclimation to different storage (acclimation) temperatures: $T_{a(1)}$, $T_{a(2)}$, and $T_{a(3)}$. The broken line intersects each R:T curve at the acclimation temperature of that group of animals. This demonstrates that acclimatory compensation has occurred, making rates measured at the acclimation temperature the same in each case, even though measurement temperatures are different.

Figure 34.2 shows a rate:temperature (R:T) curve that illustrates how changing temperature affects the metabolic rate of a hypothetical poikilothermic organism. As you can see, there is a temperature range across which the organism is able to exist without serious difficulties, even though its metabolic rate changes as the temperature changes. This is known as the **range of thermal tolerance.**

Beyond the lower and upper extremes of this acceptable temperature range, some key processes are adversely affected and slow to below a critical rate, depressing the activity and metabolism of the whole organism. **Cold torpor** and **heat coma** are names given to the inactive states at the lower and upper extremes of the range of thermal tolerance. **Cold death** or **heat death** will occur when temperatures are extreme enough to cause extensive, irreversible damage.

However, not all poikilotherms respond to temperature change in the same way. Different temperature responses have evolved in various organisms. Consider, for example, the difference in the effects of various water temperatures on an arctic fish and a tropical fish. In water at 4° C, an arctic fish thrives and carries out its life activities normally. However, a tropical fish placed in water at 4° C would be paralyzed or worse by the cold. On the other hand, a tropical fish normally lives in 20° C water. Yet an arctic fish placed in 20° C water would suffer heat death.

These different responses to environmental temperature are due to genetic differences that have evolved over long periods of time. But how do organisms that are subjected to significant short-term temperature changes, such as the seasonal changes in freshwater lakes and ponds, survive?

Acclimatization

Many organisms that live in environments with changeable temperatures make physiological adjustments that permit them to continue to function normally even though their environmental temperature goes through a definite annual temperature cycle. This physiological adjustment made under natural conditions is called **acclimatization.**

In the laboratory, responses to changing thermal environments are studied experimentally by keeping all conditions except temperature constant. In contrast to acclimatization, metabolic adjustment under such experimental conditions to changes in just one factor (rather than to the full range of environmental influences) is known as **acclimation.** Let us look at how an acclimation experiment is carried out.

In an acclimation experiment, a group of organisms is divided into several samples, each of which is held at a different temperature for a period of time. Then, metabolic rates, or the rates of other responses, are measured. The results of one such experiment are shown in figure 34.3. The rate:temperature curves for the metabolism of organisms from the three groups differ according to the temperatures at which they had been held. As you can see, the metabolic rate:temperature relationships of some poikilothermic organisms became adjusted—these animals were able to function efficiently at the environmental temperatures they experienced.

Figure 34.4 An eastern collared lizard basking in the Texas sunshine.

Behavioral Thermoregulation

Are poikilothermic animals completely at the mercy of their thermal environment, or are some forms of adjustment available to these organisms that face hourly temperature changes on a daily basis? Some poikilotherms behave in ways that adjust their body temperature. This is called **behavioral thermoregulation.**

On cool mornings, snakes and lizards that are made very sluggish by low temperatures bask in the sunshine until their body temperatures rise to the point where they can carry out their normal activities (figure 34.4). Later in the day, if their body temperatures rise above a certain critical point, they seek shelter in cool burrows.

Another type of behavioral thermoregulation is the warm-up period that some insects go through on cool mornings. Butterflies spread their wings and vibrate them slowly until the heat generated by the metabolism of their muscle cells raises their body temperature. When their temperature is high enough to permit the intense muscular activity needed for flight, they take off. Such forms of behavioral thermoregulation help poikilothermic animals cope with changing environmental temperatures.

Homeotherms

As we mentioned earlier, a different strategy for dealing with environmental temperature fluctuations is to maintain a constant internal body temperature. Animals that do this are called homeotherms. Temperatures in areas near the body surface may vary, but the temperature deep inside a homeotherm's body, the **core temperature,** remains relatively constant. Thus, major internal organs function in an environment where the temperature changes very little.

Homeotherms (that is, birds and mammals) have relatively higher metabolic rates than poikilothermic organisms, and many of them also have insulating layers of fat, feathers, or fur to help conserve body heat. Furthermore, they have nervous and hormonal mechanisms that regulate their metabolic rates in response to body temperature changes. The temperature-maintenance processes of homeothermic animals operate like "a furnace rather than a refrigerator"—body temperature is set and kept near the high end of the organism's normal environmental temperature range. The reason for this seems to be that regulation by heat conservation and generation is very efficient, while animals' cooling capacities are usually much more limited. Normal body temperatures of mammals usually range from 36° to 39° C, while birds' body temperatures are generally somewhat higher, in the range of 40° to 43° C.

For a homeotherm, a balance between heat loss and the heat produced by normal resting metabolism occurs in a range of environmental temperatures known as the **thermoneutral range.** Within that range, an animal regulates its body temperature by altering the amount of blood flow near body surfaces. At temperatures low in the thermoneutral range, blood vessels near the body surface are constricted and heat loss to the environment is reduced. At temperatures in the upper part of the range, the vessels are dilated, resulting in greater blood flow near the body surface and greater heat loss to the environment.

The human thermoneutral range is rather small, extending only from 26° or 27° to 31° C. When environmental temperature falls to a point such that adjustments in blood flow are no longer effective in maintaining body temperature, the nervous system responds by initiating **shivering.** Shivering is a series of involuntary muscle contractions that produce heat that helps to compensate for heat loss. For example, when you become very chilled, shivering can be quite vigorous, even to the point that your teeth "chatter" as jaw muscles become involved. On the other hand, when the environmental temperature rises above a certain value, your body initiates **sweating,** which dissipates a great deal of body heat by evaporative cooling. Both shivering and sweating require energy expenditures and raise the metabolic rate (figure 34.5).

Early in this century the ornithologist Joseph Grinnell formalized the concept of niche. Grinnell considered the niche to be a "habitat space," the smallest subdivision of a habitat that can be occupied by one species. He further stated that "no two species of bird or mammal will be found to occupy exactly the same niche." In his 1927 book *Animal Ecology,* the English ecologist Charles Elton described the niche as an animal's place in the community. He looked upon the role of the animal in its community as its "profession." Elton thought of the niche largely in terms of how animals utilized food resources.

Later, the American ecologist G. E. Hutchinson visualized the niche as an abstract multidimensional hypervolume. The axes of this hypervolume are described by all of the environmental variables important to a particular species. We can illustrate Hutchinson's view of the niche by considering a limited number of environmental variables. For example, an important environmental variable for an animal species might be food item size. We can plot the range of all food item sizes that could potentially be utilized by the species as a line in one dimension (figure 35.3a). A second variable might be foraging height (distance from the ground at which the animal feeds), and we can plot this as a line in a second dimension (figure 35.3b). These two lines would then enclose a rectangle that would represent a two-dimensional niche. We can take our example one step further by adding a third dimension for another variable—humidity. The three lines would then describe a three-dimensional niche (figure 35.3c).

Hutchinson used the term **fundamental niche** to describe such a multidimensional niche. A fundamental niche includes the full range of environmental conditions under which an organism can live in the absence of interference from other species. Because most species occur together with other species, however, the fundamental niches of the species in a community often overlap. When this takes place, species compete for that part of the environmental resources represented by the overlap. There are two possible outcomes of this competition: either the region of overlap is incorporated into the niche of one species or the overlap area is divided

Figure 35.2 Prairie dogs, like all other oganisms, occupy a niche; they interact in a particular way with other species in the community. Prairie dogs eat grass and are in turn eaten by coyotes, black-footed ferrets, and other prairie carnivores. Prairie dog activity produces bare soil areas where seeds can germinate and grow without being shaded or crowded out. The dirt that prairie dogs pile on the ground surface provides a place for bison "wallows," areas where bison roll vigorously in the dirt taking dust baths. Abandoned prairie dog holes provide nests for burrowing owls.

Figure 35.3 Niche dimensions. (a) A one-dimensional niche, the range of food size that a species can exploit. (b) A two-dimensional niche, in which a second axis, the height of foliage over which the species forages, is included. The enclosed rectangular space represents two dimensions in the ecological niche of the species. (c) The two-dimensional niche has been expanded to include a third axis representing humidity. The enclosed volume now represents a three-dimensional fundamental niche.

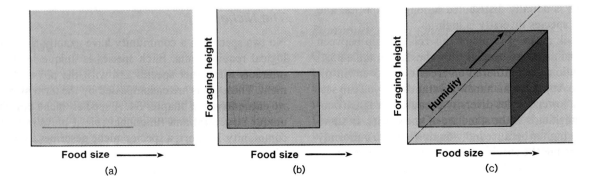

among the species competing for it (see the discussion of species guilds on p. 853). In either case, the fundamental niche is reduced to what is called the **realized niche,** the part of the fundamental niche that is actually occupied under natural conditions.

Figure 35.4 Primary succession. (a) Succession on a sand dune. The beach grass shown here is a pioneer species. Shrubs and trees replace the grass on older dunes after the dunes become stabilized and some organic matter accumulates. (b) Succession on volcanic ash on Mount St. Helens in Washington. Scattered clumps of fireweed have become established.

(a)

(b)

Succession

Natural communities are constantly changing. Old fields become forests, ponds fill in to become meadows, and forest trees die and are replaced by others. As the vegetational components of a community change, so do its animal components. In this way, communities undergo a predictable series of changes in structure and function that eventually result in the emergence of a relatively stable community known as a **mature (climax) community.** We call the process by which a mature community is achieved **succession.**

Ecologists distinguish between two kinds of succession in terrestrial communities. **Primary succession** occurs in areas that are devoid of soil and have not previously supported a community, such as sand dunes, lava flows, glacial tills, and rock outcrops (figure 35.4). On the other hand, **secondary succession** occurs in areas that were previously occupied and where soil is present. Secondary succession occurs on such disturbed areas as abandoned agricultural fields (figure 35.5), burned-over land, and areas where vegetation has been cut and removed.

Figure 35.5 Secondary succession on an old field site. The shrubs and invading trees indicate that succession in this field will eventually produce a mature forest community.

Primary Succession

An isolated sandy beach by the ocean often provides a first-hand look at primary succession. As one walks into the dunes beyond the beach, the vegetation changes. Surface temperatures on the sand dunes during the summer are high, and moisture is in short supply. These and other conditions make it difficult for organisms to colonize the dunes, but some do succeed. One of the most successful pioneering plants on sand dunes is marram (beach grass). Once this hardy grass becomes established and stabilizes the dunes against movement by the wind, similar plants invade the area. These are followed by mat-forming shrubs, which in turn may be succeeded by trees—first pines and then deciduous trees such as oaks and maples.

In primary succession, changes in vegetation come about in part because the plants themselves make the environment more favorable for invasion by other species. The pioneer species build up organic matter, shade the ground from the sun, reduce evaporation, and improve moisture conditions. In this way, grass species such as marram pave the way for shrubs, and shrubs in turn improve conditions for trees. When pines become established, they eventually shade out the shrubs. Since pine seedlings are unable to grow in the shade of mature pines, however, the pines eventually give way to deciduous trees.

Secondary Succession

We can use a study of old-field succession in North Carolina as an example of secondary succession. When cornfields are abandoned after the harvest in the fall, annual crabgrass, an opportunistic species (see p. 851), takes over. The seeds of another plant species, horseweed, also germinate, and by early winter horseweed forms clusters of growth in the disturbed soil. The following spring, horseweed gets a headstart on crabgrass and crowds it out. During that summer another plant, white aster, germinates and begins to crowd out the horseweed. Dying horseweed opens up sites for the establishment of broomsedge, a bunchgrass. Eventually, broomsedge dominates the field, but because broomsedge grows in clumps, open ground still exists between plants. These moist, lightly shaded, and plant-free spots provide an ideal place for pine seeds to germinate. If pine seedlings become established, a stand of pines develops within a few years and shades out the broomsedge. And since pine seedlings cannot grow in the shade of adult pine trees, after some years the pines eventually die and are replaced by oaks and hickories, whose seedlings can grow successfully in the shade. At this point succession will have reached the relatively stable mature condition.

Figure 35.6 A burn area several years after a forest fire. Fire can be a powerful influence on vegetational development because it influences species composition and shapes the character of the community.

This is not to imply, however, that mature communities are static and unchanging. Although a mature community is relatively permanent and self-sustaining, small-scale changes, such as the replacement of old, dying trees by young vigorous ones, occur on a regular basis here and there within the boundaries of the community.

Factors Affecting Succession

A number of community characteristics change predictably as a community undergoes succession. In a community that is moving toward a mature (climax) forest community, one such change is that of increased stratification. As we saw earlier, a mature forest may have as many as five or six layers (see figure 35.1). Other changes involve community biomass and species diversity. As succession continues, biomass increases. Species diversity also increases, in part because of the greater number of habitats available due to the more highly stratified vegetation.

Figure 35.7 One example of the effects of overgrazing on a community. Wild burros have overgrazed the area to the left of this fence in Death Valley National Monument.

The direction in which succession will proceed is affected by a number of physical factors, such as soil, slope, nutrient and moisture availability, fire, and grazing. Fire is of major importance because it sets back succession, shapes the character of a community, and influences the community's future species composition (figure 35.6). Although some plant species are eliminated by fire, others depend on fire for their persistence. Some entire communities are fire-controlled—their growth is renewed by periodic burning. For example, if fire is prevented in a fire-adapted community such as a native North American prairie, trees and shrubs soon invade the prairie and grow up among the grasses. Native prairie species generally survive fire but the invading shrubs and trees do not. Thus, fire is an important management tool used to maintain native prairies in those areas where they are preserved.

Both domestic and wild animals can arrest succession or influence the species composition of a community by grazing and browsing (figure 35.7). For example, in parts of eastern North America there are large populations of white-tailed deer. In some areas the deer eliminate all of the seedlings of certain plant species (pin cherry, black cherry, and sugar maple) but ignore others (beech, birch, and certain oaks). As a result of their feeding habits the deer influence the species composition of entire forests. In other areas, deer inhibit the reestablishment of forest plant species on burned or cutover sites. In these areas woody vegetation is replaced by grasses and other herbaceous plants.

Ecosystems

As you know, an ecosystem consists of the community of organisms in a given area, together with the abiotic (nonliving) components of their environment. The term ecosystem was coined in 1935 by the British ecologist A. G. Tansley, and today it is applied to both small and large ecological systems. Thus, we might consider a single small pond as an ecosystem, or, on a much larger scale, we can examine vast reaches of a continent or huge areas of an ocean as ecosystems.

Ecosystems have both structure and function. The structure of an ecosystem is determined by the components that make up the system, while ecosystem function is determined by the manner in which these components interact in a complementary way. Let us examine these characteristics in more detail.

Ecosystem Structure

All ecosystems possess both biotic and abiotic structural components. Biotic components include all of the organisms in the system, while abiotic components include such things as soil, water, light, inorganic nutrients, and weather variables.

We can categorize the biotic components of an ecosystem on the basis of how they obtain energy and nutrients. The **producer** organisms in any ecosystem are chiefly green plants that use photosynthesis to convert radiant energy from the sun into the chemical energy of carbohydrates, fats, and proteins. In terrestrial ecosystems the producers are predominately herbaceous and woody plants, while in freshwater and marine ecosystems the dominant producers are various species of algae. From a nutritional point of view, producer organisms are **autotrophic;** that is, they can synthesize organic compounds using only inorganic raw materials and energy from an external source (usually sunlight).

Consumer organisms obtain energy and nutrients by ingesting producer organisms or other consumers. **Primary consumers (herbivores)** eat green plants, while **secondary consumers (carnivores)** eat primary consumers. Both types of consumers are **heterotrophic**—they obtain organic nutrients by utilizing other organisms.

In using the word *consumer,* we do not necessarily imply that the material being eaten was alive. In any ecosystem there are heterotrophic organisms that ingest dead organic matter, thereby reducing large masses to smaller chunks that are more susceptible to the decomposing action of bacteria and fungi. Dead plant and animal tissue is called **detritus,** and organisms that ingest this material, such as earthworms and soil arthropods, are called **detritivores.**

Production stored as wood forms a nutrient pool unavailable for short-term cycling, but some short-term nutrient cycling does take place through the decomposition of roots and litter. Nutrient cycling in deciduous forests is strongly influenced by the geology of the site on which the forest grows. For example, forests growing in soils over nutrient-poor bedrock store and cycle smaller quantities of nutrients than do forests growing in soils over rocks rich in calcium and magnesium.

Vertical stratification is well developed throughout the deciduous forest biome. The forest supports a great diversity of consumer organisms, the most numerous of which are insects. Insect populations seldom grow excessively large, but there is a notable exception. The gypsy moth, which was accidentally introduced into the northeastern United States, feeds on a wide variety of hardwood trees, especially oaks.

A major large herbivore of deciduous forests is the white-tailed deer. In areas of high population, white-tailed deer can influence the structure and development of a forest by overutilizing certain species of forest tree seedlings and other sprouts. Because of human activity, deer predators such as wolves and mountain lions have virtually disappeared from some areas.

Chaparral

In regions with hot, dry summers and mild, damp winters, the dominant vegetation is known as **chaparral** (figure 36.18). Chaparral is characterized by low, shrubby vegetation with tough, waxy-coated leaves that are resistant to drought. Net production averages about 3,000 kcal/m² a year, with much of the production tied up in woody growth and litter. Periodic fires that sweep through the chaparral are important for nutrient cycling and stimulating new vegetative growth. Chaparral occurs in the southwestern United States, in Mexico, around the Mediterranean Sea, and in parts of southern Australia, South Africa, and South America.

Tropical Rain Forest

Tropical rain forest is located in the equatorial regions of Central America, central and northern South America, western Africa, and Indonesia (figure 36.19a). Tropical rain forests are found in areas where average temperatures are high (between 20° and 25° C) and precipitation is heavy (in excess of 200 cm per year). Temperatures fluctuate little, and the rainfall is evenly distributed through the year. The forest is dominated by broadleaf, nondeciduous trees, which are part of a very rich diversity of plant species. The canopy of the

Figure 36.18 Chaparral occurs in the semiarid areas of western North America, the Mediterranean region, Australia, South America, western India, and Central Asia, where summers are hot and winters wet. Chaparral is characterized by communities of broadleaf shrubs and dwarf trees. They are highly susceptible to fire because they possess volatile and flammable compounds in their leaves. Chaparral is a fire-dominated ecosystem; burning renews growth and recycles nutrients.

forest is highly stratified and it provides habitats for a tremendous variety of animal species (figure 36.19b). Because of the dense shade created by the towering trees, however, there is little undergrowth.

In such a warm, moist environment, organic matter decays rapidly, and only a small amount of litter accumulates on the forest floor. Nutrients are cycled directly from the litter to plant roots by way of mycorrhizae. Thus, most nutrients are stored in trees and nutrient reserves in the soil are small. Yet although the soil itself is relatively infertile, net production in a rain forest is very high because of high temperatures, a twelve-month growing season, and the rapid recycling of nutrients from the litter.

Figure 36.19 Tropical rain forest. (a) The tropical rain forest forms a worldwide belt around the equator. The largest continuous rain forest is found in the Amazon basin, where its existence is threatened by land clearing. The tree species number in the thousands, climbing plants are common, and the interior is dark and moist. (b) A colorful toucan from the tropical rain forest of Brazil.

(a)

(b)

Even though such soils are poorly suited for agriculture, tropical rain forests are being cleared and converted to agricultural use at an alarming rate. However, soil nutrients are so scarce in these areas that the soil seldom supports productive agriculture for long, and the cleared areas are usually abandoned after only a few years. If the cleared areas are small (less than 2–3 hectares), the rain forest will reoccupy them in time. Larger clearing operations, however, whether done for lumbering or agricultural purposes, are threatening to permanently destroy large areas of tropical rain forest in many parts of the world.

Bordering the tropical rain forest is the **tropical seasonal forest.** Not as lush as rain forests, tropical seasonal forests grow in regions where rainfall is seasonal with pronounced wet and dry seasons. Many trees in these forests are deciduous and lose their leaves during the dry season.

Tropical Savanna

Also associated with tropical regions is the **tropical savanna,** which is dominated by tall grasses and also supports scattered trees. In tropical Africa these trees are often flat-topped, thorny *Acacia* trees, while in South America they are mostly palms. Tropical savanna occurs in regions where temperatures are high and where rainy and dry seasons are pronounced. During the rainy season, precipitation ranges between 90 and 150 cm. The heavy rains of the wet season cause extensive leaching and result in nutrient-poor soils. Nevertheless, annual net production ranges between 800 and 8,000 kcal/m².

Figure 36.20 Wildebeests migrating across the Serengeti plain of East Africa, a tropical savanna. Tropical savannas are found over large areas in the interior of continents. The best known are the African savannas, characterized by grassy plains with scattered, flat-topped *Acacia* trees. Savannas support large populations of herbivores such as wildebeest, zebra, and many species of antelope.

Figure 36.21 Bison herd on the grassland of Wind Cave National Park, South Dakota. Before settlement by Europeans, the interior of the North American continent was one of the great grasslands of the world. These great expanses of grass once supported vast herds of bison and other herbivores. Today much of the native grassland is under cultivation.

African savanna (figure 36.20) supports a rich diversity of grazing animals whose cycles of reproduction and migration are correlated with the rainy seasons. Tropical savanna has been greatly disturbed by human activity, particularly by agricultural development and overgrazing by domestic animals. For example, the large herds of native grazing animals that once covered the African savanna, such as wildebeest, zebra, and many species of antelope, are rapidly disappearing, as are many of the predators, including lions, leopards, cheetahs, and hunting dogs that prey on them.

Grassland

Grassland is found on the temperate plains areas of North America, Eurasia, and South America. Rainfall is variable; in many of these areas the rainfall could support tree growth, but the encroachment of trees is halted by fires and periodic drought. In North American prairies, along an east to west gradient of decreasing moisture, there is a transition from tall-grass, to mid-grass, and finally to short-grass grassland.

Compared to forest ecosystems, grasslands exhibit little biomass accumulation, and much of the annual production dies each year. Thus, the turnover of biomass and nutrients is rapid. Furthermore, because of relatively low rainfall and high evaporation rates, leaching is not excessive, and nutrients tend to accumulate in the lower part of the soil. At the same time, organic matter accumulates in the upper part of the soil. As a result, grasslands have very fertile soil and they have been exploited worldwide for agriculture. Most of the native temperate grasslands of the world have been converted to agricultural usage.

Grasslands, like tropical savannas, can support large populations of herbivores (figure 36.21). Insect herbivores, notably grasshoppers and ants, are very common. In North America, millions of bison once roamed the native grasslands. There were also large numbers of pronghorn antelope and deer; their numbers were kept under control by predators such as coyotes and wolves.

Today, however, most native grassland herbivores have been largely replaced by cattle, sheep, and goats, and in many places overgrazing has resulted in severe deterioration of the grassland ecosystem.

Figure 36.22 Deserts occur in two distinct belts around the earth, one near the Tropic of Cancer, the other near the Tropic of Capricorn. In North America there are two types of desert. One is the cool desert of central and northwestern North America, which lies in the Great Basin in the rain shadow of the Sierra Nevada and Cascade Mountains. This desert is dominated by sagebrush. Hot desert is found in the southwest, in Arizona (pictured here), New Mexico, and southern California. This desert is dominated by cactus, creosote bush, paloverde, and ocotillo.

Desert

Desert forms a worldwide belt at about 30° N and 30° S latitude. Desert is associated with the rainshadows of mountains, coastal areas next to cold ocean currents, and the deep interiors of continents. It is characterized by rainfall of less than 25 cm per year, low humidity, a high evaporation rate, and high daytime temperatures and low nighttime temperatures. Desert plant life consists of widely scattered shrubs that have short, waxy, drought-resistant leaves. The soil is high in inorganic salts and low in organic matter. Because of lack of water and extreme temperatures, annual net production is very low (below 200 kcal/m²).

Deserts may be hot or cold. In North America, hot deserts, dominated by cactus and desert shrubs, are found in the southwestern United States and Mexico (figure 36.22). Northern cold deserts are dominated by sagebrush.

Even deserts are subject to extensive human disturbance. Irrigation has made it possible to utilize deserts for agriculture. However, water for irrigation is often pumped to the surface from deep wells, and it is usually removed faster than it can be replaced by natural processes. Also, as a result of the evaporation of irrigation water, salts accumulate in surface soils. Finally, there are grave questions concerning the future availability of water for direct human use in areas such as the southwestern United States, where the human population is growing very rapidly.

Environmental Concerns

The world has serious environmental problems, and many natural ecosystems are being threatened. Generally, most ecosystems can recover from minor disturbances, but there is a point beyond which an ecosystem loses its ability to return to normal. This usually occurs when it is subjected to a very large disturbance or to continuing disruptive influences over a long period of time, or when key species become extinct.

Since an ecosystem consists of a complex array of interconnected parts, a disruption of even one of its aspects eventually affects the entire system. Despite our sometimes arrogant attitudes, we humans are an integral part of ecological systems, and our activities very often have negative effects on the environment. A large percentage of the environmental problems to which we contribute arise simply because there are so many of us, and it looks as if the total human population of the world will continue to grow until at least well into the next century (see p. 846).

Pollution

To begin with, world population growth contributes to environmental pollution problems. **Pollution** is an undesirable change in the characteristics of an ecosystem, and pollution can be very harmful to all living things in the ecosystem, including humans.

Water Pollution

Toxic wastes from industries and sewage systems can poison living things in aquatic ecosystems directly. But the release of toxic wastes can be detected and, at least to a degree, controlled. Moreover, there is general support for the enforcement of laws that prevent the direct disposal of poisonous substances and other pollutants into streams, rivers, lakes, and oceans.

There are more subtle water pollution problems, however. Even very efficient sewage treatment plants release nutrients such as nitrates and phosphates into bodies of water. Nutrient molecules are also carried off farmland by rainwater runoff. The resulting nutrient enrichment of aquatic ecosystems causes abnormally dense growth of photosynthesizing aquatic organisms, especially algae in lakes. Such algal "blooms" cause the water to become murky and often result in the death of at least some organisms in the ecosystem (figure 36.23). Over time, nutrient enrichment leads to rapid eutrophication (see p. 886) and causes lakes to fill in much more rapidly than they would in the absence of the pollution.

Power plants often use water from rivers and lakes for cooling purposes and then release the warmed water directly back into the bodies of water from which it came. This **thermal pollution** also encourages excessive growth of certain organisms and produces some of the same results as nutrient pollution in aquatic environments. The warm water may accelerate the growth and reproductive rates of some species, but others are less tolerant of increased temperature and are inhibited or even killed by the change in their environment.

Air Pollution

Industries, automobiles, and even home heating systems all release pollutants into the air. Some of these pollutants are particles of soot and dust that settle on vegetation or are inhaled by humans and other animals. Certain chemical pollutants, such as lead and carbon monoxide from automobile exhaust, are harmful directly. Other pollutants may be less dangerous initially, but they react with substances in the air to produce very harmful compounds.

Air pollution that occurs relatively near the source of the pollutants often takes the form of **smog**. Smog is produced when pollutants react with water vapor in the air.

However, other pollutants entering the atmosphere are carried great distances and eventually return to the earth's surface with rain. In this way they contribute to a phenomenon known as **acid precipitation (acid rain)**. Under natural, unpolluted conditions, the pH of rainwater is slightly acidic (5.6) because of reactions with CO_2 in the atmosphere. But

Figure 36.23 A fish kill caused by oxygen depletion resulting from decay of algae produced in an algal bloom.

pollutants such as sulfur oxides and nitrogen oxides can change the normal pH of rain drastically. They react with water to produce sulfuric and nitric acids, respectively. When this water eventually falls to the earth as rain (often hundreds of kilometers from the source of the original pollutants) it has a pH well below normal. In fact, the pH of acid rain can be as low as 3.0, which makes it as acidic as vinegar!

Acid rain overcomes the buffering systems of natural waterways such as lakes and ponds and lowers their pH, often with devastating results on a wide range of living things. Lowered pH kills some adult organisms and prevents successful reproduction by others. It also increases the quantities of certain metals such as aluminum that dissolve in water and that may prove toxic to many organisms in the ecosystem.

There is also good evidence that acid rain negatively affects plant health and growth. Because acid rain often does its damage far from the pollution source, it has been the topic of considerable debate between regions in the United States, as well as between countries such as Canada and the U.S.

As with water pollution, it is possible to recognize and regulate obvious causes of air pollution if we choose to make the effort. But as is also the case with water pollution, there are subtle forms of air pollution that may have far-reaching effects. For example, carbon dioxide produced by the burning of huge quantities of fossil fuels is accumulating in the atmosphere and could have a significant impact on world climate. Increased CO_2 in the atmosphere may contribute to the **greenhouse effect** (see p. 878) and lead to an overall warming of the earth's surface.

Box 36.1
Will We Have a Nuclear Winter?

By now everyone is aware of the devastating immediate effects of the explosion of nuclear weapons. Many school children can tell you that the average thermonuclear warhead has a yield of two megatons, the equivalent of two million tons of TNT. Our minds have become almost numb to the realization that one such explosion would be more powerful than the total explosive force of all weapons used in World War II. We are aware of the blast power and the intense, short-term lethal effects of radiation released by nuclear explosions. There are now enough nuclear warheads to destroy all military installations, most concentrations of industry, and *every* city with a population of more than 50,000 in both the U.S.S.R. and the United States.

For months after nuclear explosions, radioactive material would continue to fall from the upper atmosphere. The world received a frightening lesson about radioactive fallout when a nuclear reactor at Chernobyl in the U.S.S.R. exploded on April 26, 1986, releasing radioactive material into the atmosphere. Yet this accident was miniscule compared with the worldwide radioactive fallout that would follow a nuclear war.

Furthermore, scientists from many countries including the United States and the U.S.S.R., have developed a model that suggests that dust and smoke raised by nuclear explosions of a moderately large war (with 5,000 megaton yield) might create yet another set of problems. Dust and smoke could, within a few days, reduce the amount of sunlight reaching the earth to just a small percentage of the normal amount, far too little to support the photosynthesis conducted by primary producers in all ecosystems.

Dust and smoke could also cause a chilling "nuclear winter," no matter what time of year the war occurred. Temperatures could fall to $-20°$ C and remain there for several months. The combination of radiation, cold, and inhibited photosynthesis would probably destroy most ecosystems in the Northern Hemisphere and kill the majority of organisms living there. It is also likely that these problems would, in time, spread to the Southern Hemisphere, though they would probably not be as severe there.

When the atmosphere cleared as the dust was washed down by rain, the effects of chemical disturbance of the ozone layer would become severe. All remaining organisms would be exposed to harsh, unfiltered, ultraviolet radiation that could seriously burn human skin in minutes. Any extended exposure would greatly increase the incidence of skin cancer.

Some scientists dispute the nuclear winter model, but to most biologists, the potential effects of massive exchanges of nuclear warheads on ecosystems are horrifying, even if a nuclear winter did not develop exactly as this model predicts. Normal patterns of ecosystem structure and function would be radically disturbed, leaving a bleak, unpredictable future for any organisms that might accidentally survive. Thus, the danger of nuclear war is not just another modern environmental problem; it is *THE* problem of our time, not just for human ecology, but for all life on our planet.

Pesticides

Pesticides represent a special case with reference to the general concept of pollution. Pesticides are chemical substances that are deliberately introduced into the environment to kill organisms that are considered, for one reason or another, to be undesirable. Included within the pesticide arsenal are **herbicides** (for killing plants) and **insecticides** (for killing insects). Let us look at some of the effects of insecticide use.

When insecticides are used, the targeted insect pest is usually a threat to humans because it transmits disease or destroys crops. Insecticides have been effective in many such situations; they have reduced populations of dangerous or damaging insect species. However, most chemical insecticides are nerve poisons that kill a variety of insects when they are applied. Thus, along with the targeted pests, many harmless and even beneficial insects are killed. Sometimes the predators that prey on the target species are among the organisms killed, thus reducing the effectiveness of natural controls on the target species. Furthermore, these toxic substances become incorporated into food chains and food webs and do considerable damage to a variety of organisms.

Some insecticides such as DDT are very stable and can persist in the environment for up to fifteen years. Other insecticides such as malathion and parathion decompose to harmless products in a relatively short time. However, during their application these less-persistent insecticides are more directly toxic to birds and mammals (including humans) than are some persistent insecticides, and thus pose a greater threat to many ecosystem inhabitants.

Figure 36.24 An example of severe soil erosion at the edge of a Tennessee corn field.

Not unexpectedly, resistance has also become a major problem. Widespread use of insecticides is a form of artificial selection that dramatically favors those insects that are less susceptible to the chemical substance. Increasing insecticide resistance has usually been countered by applying greater quantities of an insecticide or using new formulas. There is some indication that, environmental dangers aside, the escalating cost of chemical insect control in certain agricultural situations may eventually exceed the economic loss caused by the insects.

Resource Depletion

Growing human populations place diverse demands on the environment. Many minerals are already in short supply worldwide, and efforts to recycle scarce minerals have not yet received the attention or government support needed to make them truly effective. Sooner or later our supply of fossil fuels, which are **nonrenewable** energy sources, will dwindle. Progress in the development of alternative, **renewable** energy sources such as solar and wind power has been painfully slow.

There are many other forms of resource depletion. For example, soil erosion, aggravated by modern large-scale agricultural processes, is rapidly reducing the productivity of cropland in many countries (figure 36.24). In other areas, overgrazing of grasslands and deforestation without subsequent replanting are contributing to the spread of deserts, a process called **desertification.** Thus, as the world's human population grows, the land available for food production is decreasing.

Formerly it was thought that the world's oceans could produce a virtually unlimited supply of food, once modern technology was applied in the fishing industry. The annual world fish catch did indeed triple between 1950 and 1970, but there were only small increases between 1970 and 1979. More recently, it has become clear that if overfishing practices continue, some fish species will become seriously depleted, if not pushed to the brink of "commercial extinction"; that is, they will be so scarce that it will no longer be profitable to fish for them.

The Future

One view of the future is that the depletion of natural resources and energy reserves will soon bring human population growth to a screeching halt, no matter what we do about population control. Another view is that new technologies will expand human possibilities and allow for a much larger world population, though members of that population will need to abandon their desires for solitude and access to undisturbed natural areas.

Faith in the development of new technologies has faltered somewhat, however, as it has become apparent that "food from the sea," nuclear power, and even the "green revolution" in agriculture all have definite limitations. Thus, an acceptable human future depends upon our decision to act now to manage population size and to cut natural resource consumption (especially in the United States, the world's largest per capita resource consumer). Are we willing to make the hard and perhaps even sacrificial decisions that would help to assure a decent future for the generations to come?

Summary

The earth supports a number of recognizable ecosystem types. Some of these are aquatic (water-based) systems, while others are terrestrial (land-based) systems.

Aquatic ecosystems may be fresh water or salt water (marine). Lotic freshwater ecosystems are characterized by flowing water and include brooks, streams, and rivers. Lentic freshwater systems are characterized by still water and include lakes, ponds, and marshes. In many lentic systems there is both vertical stratification and horizontal zonation. Some marine ecosystems are the open sea, intertidal zones, coral reefs, and estuaries.

Major terrestrial ecosystems are categorized by the type of mature plant communities that dominate them. These community types reflect differences in climate and soil. Each of the major terrestrial ecosystems (biomes) is inhabited by characteristic groups of organisms.

North of 60° N latitude is a vast, treeless region called the arctic tundra. South of the tundra lies a worldwide belt of coniferous forest, the taiga. Still further south in areas of moderate climate is the temperate deciduous forest. Regions of the world with hot, dry summers and mild, wet winters support the chaparral biome type. Tropical rain forest occurs in equatorial regions with high temperatures and heavy rainfall. Tropical savanna also is found in tropical regions and is an extensive area of tall grass with scattered trees, in many cases populated by large herds of grazing herbivores. Grassland is found in temperate areas where rainfall is limited and variable and evaporation is high. Desert is located at about 30° N and 30° S latitudes where rainfall is very low. Deserts may be either hot or cold.

Worldwide population growth has created grave environmental problems for humans and many other species. Depletion of natural resources, water and air pollution, the use of pesticides, habitat destruction, and the looming threat of nuclear war are only a few of the problems facing us today.

Questions for Review

1. Distinguish between aquatic and terrestrial ecosystems.
2. What are some characteristics of a lotic freshwater ecosystem? of a lentic freshwater ecosystem?
3. How does vertical temperature stratification develop in temperate lakes?
4. What are benthic organisms?
5. Distinguish between oligotrophic and eutrophic lakes.
6. Explain why environmental conditions are relatively severe on rocky shores and sandy beaches.
7. How are oceans zoned horizontally? What are some characteristics of these zones?
8. Explain the role of zooxanthellae in coral reefs.
9. What is an estuary?
10. What is meant by the term biome?
11. Characterize each of the following as to location, general climate conditions, and dominant mature plant communities: arctic tundra, alpine tundra, taiga, temperate deciduous forest, chaparral, tropical rain forest, tropical savanna, grassland, desert.

Questions for Analysis and Discussion

1. Explain why an isolated lake in the coniferous forest of Manitoba remains clear throughout the summer, while a lake located in the agricultural area of southern Minnesota and lined with cottages becomes green and "soupy" by early August.
2. In general, soils are rather nutrient poor in both coniferous forests and tropical rain forests, and nutrients tend to be accumulated in living plants. Compare and contrast mechanisms of nutrient cycling in these two ecosystems.
3. Discuss the ecological significance of the nonselectivity of many insecticides.

Suggested Readings

Books

Alcock, J. 1985. *Sonoran desert spring*. Chicago: University of Chicago Press.

Attenborough, D. 1984. *The living planet*. Boston: Little, Brown and Co.

Ayensu, E. S.; Heywood, V. J.; Lucas, G. L.; and Defillips, R. A. 1984. *Our green and living world*. Washington, D.C.: Smithsonian Institution Press.

Huxley, E., and van Lawick, H. 1984. *Last days in Eden*. London: Harvill.

Miller, G. R., Jr. 1986. *Environmental science: An introduction*. Belmont, Calif.: Wadsworth.

Odum, E. P. 1983. *Basic ecology*. Philadelphia: W. B. Saunders.

Valiela, A. 1984. *Marine ecological processes*. New York: Springer-Verlag.

Articles

After nuclear war. October 1985 (entire issue). *BioScience* 35:9.

Brown, L.; Chandler, W. U.; Flavin, C.; Pollock, C.; Postel, S.; and Wolf, E. C. April 1985. State of the earth 1985. *Natural History*.

Cooper, C. F. April 1961. The ecology of fire. *Scientific American*.

Diamond, J. M. 1982. Man the exterminator. *Nature* 298:787.

Koehl, M. A. R. December 1982. The interaction of moving water and sessile organisms. *Scientific American*.

Laws, R. M. 1985. The ecology of the southern ocean. *American Scientist* 73:26.

Moore, J. A. 1985. Science as a way of knowing—Human ecology. *American Zoologist* 25:483.

Norman, C. 1986. Hazy picture of Chernobyl emerging. *Science* 232:1331.

Perry, D. R. November 1984. The canopy of the tropical rain forest. *Scientific American*.

Pimentel, D., and Levitan, L. 1986. Pesticides: Amounts applied and amounts reaching pests. *BioScience* 36:86.

Porter, K. G. 1977. The plant-animal interface in freshwater ecosystems. *American Scientist* 65:159.

Richards, P. W. December 1973. The tropical rain forest. *Scientific American*.

Salati, E., and Vose, P. B. 1984. Amazon basin: A system in equilibrium. *Science* 225:129.

Schindler, D. W., et al. 1985. Long-term ecosystem stress: The effects of years of experimental acidification on a small lake. *Science* 228:1395.

Van Cleve, K., et al. 1983. Taiga ecosystems in interior Alaska. *BioScience* 33:39.

The Diversity of Life

The most striking characteristic of the world of life is its diversity. Organisms of innumerable types possess a multitude of specializations that permit various living things to interact successfully with virtually every type of environment on earth. How do biologists make sense of this sometimes bewildering diversity?

Biologists divide living things into categories (taxonomic groups) on the basis of certain ranges of shared characteristics. No single classification scheme is, or probably ever will be, ideal. However, generally accepted taxonomic schemes are essential tools for biologists.

Taxonomic grouping aids in recognition and study of organisms. It also permits scientists of many countries to communicate clearly and directly with one another about the organisms they study. Furthermore, the systematic process of classifying organisms into taxonomic groups helps to clarify evolutionary relationships.

In chapters 37 through 40 we will sample the fantastic diversity of life on earth within the context of one of the most widely used general taxonomic schemes.

Taxonomic Principles, Viruses, and Monerans

37

Chapter Concepts

1. Taxonomy is the branch of biology concerned with classifying living things.
2. All organisms can be placed in one of five kingdoms on the basis of their mode of nutrition, whether their cells are prokaryotic or eukaryotic, and their general organizational complexity.
3. Viruses are infectious particles that can reproduce only within living cells.
4. Bacteria are all prokaryotes; they occur in a wide range of shapes and sizes and display considerable metabolic diversity.
5. Although many bacteria cause disease, the vast majority are beneficial and are indispensable ecologically.
6. Cyanobacteria (blue-green algae) are prokaryotic cells whose photosynthetic mechanisms more closely resemble those of algae than those of other photosynthetic bacteria.
7. Cyanobacteria participate in a number of symbiotic associations with other organisms.

Facing page   Newly emerged adult mayflies. Mayflies emerge by the millions from lakes, ponds and streams where they develop. As adults, mayflies are highly specialized for reproduction. In fact, they lack functional mouthparts and thus never feed during their brief period of adulthood, which lasts only a few hours, or at most, a few days. These insects illustrate one of the striking adaptations encountered among the almost unimaginable diversity of living things.

As a young man, Charles Darwin delighted in studying a great variety of living things. Once while on a collecting trip, Darwin tore off some old bark from the side of a tree and immediately spotted two rare kinds of beetles. Just as he seized one in each hand, however, he spied yet a third kind of beetle. Darwin could not bear to let this third specimen escape, but rather than release one of those already captured, he popped the beetle he held in his right hand into his mouth and proceeded to collect the third!

Darwin, like many other people throughout the ages, was prodded by intense curiosity and a strong desire to understand the world about him. Undoubtedly, even before the beginning of recorded history, people were observing and mentally cataloging plants and animals. However, such cataloging does more than just satisfy our human curiosity. The process of naming living things and relating them to each other in an organized system is very important in biology because it allows scientists from around the world to communicate using universally understood names. Such a system also facilitates the efficient organization of an enormous amount of information.

In this chapter, we begin a survey of the living world, starting with a brief consideration of the principles by which living things are named and categorized.

Principles of Classification

A fundamental principle of biology is that organisms possess biological individuality. Even identical twins, though they are alike genotypically, often are slightly different phenotypically. For this reason, the task of organizing or grouping individual organisms is extremely difficult.

The part of biology devoted to describing and naming organisms and arranging them in a system of classification is **taxonomy** (from the Greek *taxis* meaning "arrangement"). Taxonomists arrange organisms in a system of classification that takes similarities, differences, and evidences of evolutionary relationships into account. Such a task is enormous. Life arose at least 3,000 million years ago, and organisms have been evolving ever since with increasing complexity and diversity. At the present time, approximately 400,000 plant species and 1,200,000 animal species have been described, with perhaps many more yet to be found.

The first recorded attempts to categorize organisms systematically were made by Greek philosophers. Aristotle (384–322 B.C.) grouped plants into herbs, shrubs, and trees. He used medicinal value and size as criteria. In general, early taxonomic schemes were based on concrete and practical criteria such as size, habitat, medicinal value, and ability to harm humans.

Modern classification stems from the work of John Ray (1627–1705), an English naturalist. He developed a classification scheme for his *Historia Plantarum* and was the first to outline the species concept; that is, that there are small groups of organisms with similar morphological characteristics that can be distinguished from other closely related organisms. This scheme was followed for the most part by Carolus Linnaeus (1707–1778), a Swedish physician and botanist whom we consider to be the father of modern taxonomy.

Linnaeus revolutionized biological **nomenclature** (naming process) in 1758 with the publication of the tenth edition of his *Systema Naturae*. Rather than using a polynomial ("many-name") system that included long, descriptive phrases to identify organisms, as others had done previously, Linnaeus used only two names to identify a species, and his **binomial** ("two-name") system is still in use.

However, species names alone are not adequate to help us make sense of the world of life because there are so many species. What is needed is a classification hierarchy in which each species is placed in a succession of ever-broader taxonomic categories. Linnaeus suggested such a scheme in which he recognized four **taxa** (groupings) of organisms: class, order, genus, and species. Linnaeus placed organisms in these taxa mainly on the basis of structural similarities and differences. Modern taxonomists have added only two major taxa to his scheme: phylum (or division) and family. Let us take a closer look at nomenclature and the classification hierarchy as they are used in modern biology.

Nomenclature

Science is a truly international enterprise; scientists of all nations must be able to communicate with one another, and the use of universally understood names for organisms helps to make this communication possible. Using an organism's common name is often ineffectual because common names, as you might imagine, are different in different languages. They even vary considerably within an individual country. For example, in the United States, the common sparrow (*Passer domesticus*) is also known as the house sparrow and as the English sparrow. The European white water lily is a more extreme example; it has 245 different common names distributed among four languages!

To be useful, a scientific nomenclature (naming) system must have at least three properties. (1) Each organism must have a unique name. (2) This name must be universal; it must be used by scientists regardless of their native tongue. Finally, (3) the name must be as unchanging as possible. Once an organism has been properly named, its name should not be changed without a compelling reason.

Figure 37.1 This hypothetical example shows how a number of species (A–U) might be arranged taxonomically to form genera, families, and orders within a single class. The darkest-colored area is a genus containing three species (A, B, C). The family containing this genus is composed of two genera and seven species; the order, of two families and three genera (the second family has only one genus containing the species H, I, and J).

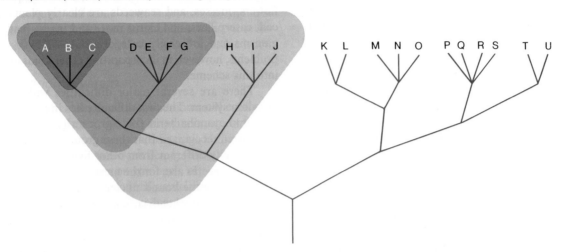

Before Linnaeus's time there was little order or consistency in the way species were named. Now, each organism is given a two-word, Latin or latinized name. The first word is the name of the genus to which the organism belongs; the second word indicates the particular species. The initial letter of the genus name is capitalized, and the name is italicized in print or underlined when written or typed. For example, consider *Paramecium*, the common ciliated protozoan encountered at one time or another by nearly all biology students. *Paramecium* is the genus name, and within this genus are a number of species—*Paramecium aurelia, Paramecium bursaria*, and so forth.

To ensure the uniqueness, universality, and stability of each name, all new names must meet the criteria set down in one of three special international codes: the International Rules of Botanical Nomenclature, the International Code of Zoological Nomenclature, or the International Bacteriological Code of Nomenclature.

The Hierarchy of Taxa

As we noted earlier, so many species inhabit the earth that the situation would be chaotic without further grouping or classification. A hierarchy of taxa (groups of organisms) has been developed to organize species in a coherent, meaningful way. Species form the smallest unit in this hierarchy. A species has usually been defined as a population of morphologically similar organisms that can sexually interbreed but that are reproductively isolated from other organisms. However,

Table 37.1
Examples of Taxonomic Categories

Category	Human	Domestic Dog
Kingdom	Animalia	Animalia
Phylum	Chordata	Chordata
Subphylum	Vertebrata	Vertebrata
Class	Mammalia	Mammalia
Order	Primates	Carnivora
Family	Hominidae	Canidae
Genus	*Homo*	*Canis*
Species	*sapiens*	*familiaris*

this definition is not adequate for organisms such as bacteria that generally do not reproduce sexually. In such cases, individuals are placed in the same species on the basis of overall similarity.

Each species is placed in a succession of ever-broader taxonomic categories. The major taxonomic ranks, from broadest to most specific, are: **kingdom, phylum** or **division, class, order, family, genus,** and **species.** Usually, every rank in the hierarchy contains several groups in the level below it (for example, a particular family normally is composed of several genera, and each genus may have several species within it). Part of such a hierarchical classification system is shown in figure 37.1. Table 37.1 illustrates the use of the system in classifying two kinds of organisms, human beings and domestic dogs.

Figure 37.5 The structure of a bacteriophage. The virus attaches to a bacterial cell by its base plate and tail fibers. Then an enzyme makes a hole in the cell wall and the virus inserts its double-stranded DNA into the bacterium. In (a) the tail sheath is relaxed; in (b) it has contracted as it injects its DNA into the bacterial cell.

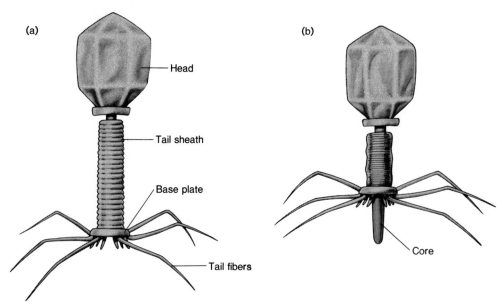

(a)

Head

Tail sheath

Base plate

Tail fibers

(b)

Core

Although living cells contain both DNA and RNA, virions have either DNA or RNA, but not both. The nucleic acids range in size from a few thousand nucleotide base pairs (fewer than five genes) to around 250,000 base pairs (several hundred genes). Depending upon the type of virus, the nucleic acid can be single- or double-stranded. In most instances, each virion has a single nucleic acid molecule. However, some RNA viruses have their genome in pieces; for example, the influenza virus has eight or nine pieces of RNA.

Viral Reproduction

Probably the most-studied virus reproductive cycle is that of the large, complex viruses that infect the bacterium *Escherichia coli*. Since these viruses attack bacteria, they are referred to as **bacteriophages,** or simply **phages** for short (figure 37.5). Most is known about the reproduction of a group of bacteriophages known as "T-even phages," and we can use them to illustrate how a virus infects a living cell, takes control of it, and causes it to reproduce virus particles.

The phage's life cycle begins with a chance collision between the virus and an *E. coli* cell (figure 37.6). The base plate and tail fibers attach the phage to the bacterium at a specific surface receptor site. Next, a phage enzyme digests a portion of the bacterium's cell wall. Then the tail sheath

contracts, injecting the core of viral DNA into the bacterium. The virus genetic material takes control of the host cell within a few minutes and initiates the destruction of host DNA and inhibition of the synthesis of host cell proteins. Messenger RNA is synthesized, and it directs the production of virus proteins and any enzymes required for the manufacture of new virions. The host bacterium provides all the required energy and building blocks for virion synthesis. As soon as all virus components have been prepared, the virus particles are assembled. Finally, the mature virions—in the case of T2 phages, about 100 per bacterial cell—are released when **lysozyme,** an enzyme synthesized under virus direction, disrupts the cell wall. Each of the new virus particles can then attack a neighboring *E. coli* cell.

The reproductive cycle we just described is called a **lytic cycle** because the host cell is lysed (broken open) by the phage. In contrast, some bacteriophages do not always immediately destroy their hosts, but instead are sometimes reproduced along with the bacterium to generate more infected bacteria. This relationship is called **lysogeny.** Infected bacterial cells capable of producing phage virions are said to be **lysogenic,** and viruses that can enter into this relationship are **temperate phages.** When a temperate phage such as the phage lambda (λ) infects a bacterium, it directs the synthesis of a special repressor protein. If this repressor reaches a high concentration fairly quickly, it inhibits phage reproduction. Instead of lysing the bacterial cell, the lambda DNA is

Figure 37.6 Infection of a bacterial cell (*Escherichia coli*) by a T-even bacteriophage. Capsids are shown only in outline so that DNA (red) is visible in all cases.

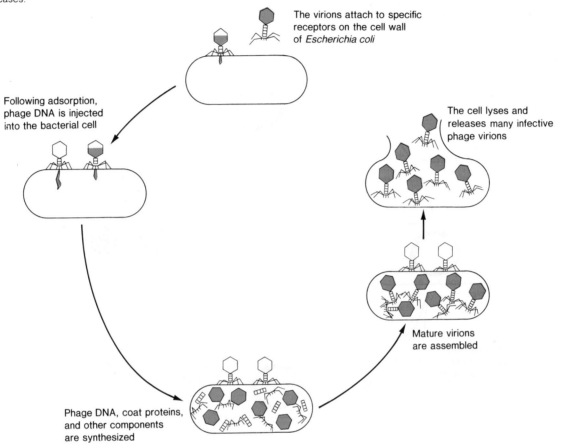

The virions attach to specific receptors on the cell wall of *Escherichia coli*

Following adsorption, phage DNA is injected into the bacterial cell

The cell lyses and releases many infective phage virions

Mature virions are assembled

Phage DNA, coat proteins, and other components are synthesized

integrated into the bacterial DNA and becomes a **prophage** (see figure 37.7). The prophage is replicated along with the bacterial chromosome, and all of the lysogenic cell's progeny will carry the prophage but not be lysed. If at some time lysogenic cells are exposed to harsh environmental factors (such as ultraviolet light), repressor levels drop and **induction** occurs. As a result of induction, the prophage leaves the bacterial chromosome, virus nucleic acids and proteins are synthesized, new virions are constructed, and the host cell is lysed.

Before we leave the topic of viral reproduction, we must emphasize that not all viruses infect cells in the same way that T-even bacteriophages do. For example, many animal viruses seem to be taken into cells by phagocytosis and their capsid removed by host cell digestive enzymes in a process referred to as **uncoating.** The mechanism of virus release also varies. Some bacterial viruses and a number of animal viruses are released through the host cell plasma membrane without lysis. Indeed, in infections by some types of viruses,

the host cell plasma membrane is used in the formation of the virion envelope when the virions are released by a kind of cell budding process.

We mentioned earlier that a number of viruses differ from living cells in that they have RNA as their primary genetic material. Some viruses have double-stranded RNA but function much as if they had DNA. A special **RNA transcriptase** copies one strand to produce mRNA, which can then direct protein synthesis or be copied to produce new double-stranded RNA. Thus, genetic expression depends on RNA→RNA information flow, rather than the more familiar DNA→RNA flow. The primary genetic material in some other viruses is single-stranded RNA. These single-stranded viruses operate in a different way. An **RNA replicase** (coded for by the virus genome) copies the single strand to form a double-stranded RNA replicative form. This replicative form, which is present only during virus reproduction, then directs the synthesis of single-stranded RNA for new virions. Polio virus is one example of a virus that reproduces in this way.

(mumps). In addition, viruses inflict extensive damage in other animals (swine influenza, rabies, hog cholera) and in plants (mosaic diseases, leaf curls).

Cell and tissue damage can result in a number of ways. Some viruses, particularly enveloped viruses, may possess antigens that stimulate the host to form antibodies against some of its own constituents. A number of virus-induced proteins are actually toxic for host cells and cause morphological damage or block cellular biosynthetic processes. Lysosomes may be disrupted so that the cells digest themselves. Virions can even directly disrupt cell structure through the formation of **inclusion bodies,** large collections of incomplete or mature virions that accumulate in the nucleus or the cytoplasm. The destruction of a particular tissue may result from the simultaneous operation of more than one of these processes.

Attempts to prevent viral diseases in humans and domesticated animals have met with mixed success. Nearly 200 years after Edward Jenner first suggested the possibility of immunization against smallpox, the disease was finally eradicated in the 1970s through the efforts of the World Health Organization. The smallpox virus now exists only in a few highly controlled research laboratories (and possibly in some secret biological warfare facilities).

Despite some successes in the control of viral diseases, however, other disease-causing viruses present very difficult challenges. For example, the viruses that cause influenza are very changeable (mutable), especially in the makeup of their coats. New influenza strains with new sets of coat proteins are constantly being produced. This means that vaccines that are quite effective against one particular strain of influenza virus can prove nearly useless against newly developed strains. Viruses that cause the common cold are similar in this respect. Unfortunately, the HTLV-III virus associated with acquired immune deficiency syndrome (AIDS) presently appears to behave in a similar fashion. If this is indeed true, it may prove very difficult to prevent and control this dreaded disease.

Viroids

A number of plant diseases—potato spindle-tuber disease, exocortosis of citrus trees, chrysanthemum stunt disease, and others—are caused by a class of infectious agents called **viroids.** These are very short strands of RNA that are not enclosed in coats and can be transmitted between plants through mechanical means or by way of pollen.

Viroids are smaller than even the smallest viruses. The potato spindle-tuber disease agent has been studied intensively. Its RNA is about 130,000 molecular weight, much smaller than the nucleic acids of viruses (figure 37.8). It can exist as either a linear RNA strand or a closed circle collapsed into a rodlike shape due to intrastrand base pairing.

Viroids are found mainly in the nuclei of infected cells. They do not serve as messengers to direct protein synthesis, and it is not yet clear how they are reproduced or cause disease symptoms.

Although the viroids so far discovered cause plant diseases, it is possible that similar agents are responsible for certain animal diseases as well. There are a number of nonbacterial, neurological diseases (scrapie of sheep and goats;

Figure 37.8 Illustration comparing *Escherichia coli*, several viruses, and the potato spindle-tuber viroid with respect to size and the amount of nucleic acid in the genome. All dimensions are enlarged about 40,000 times. Bacteriophage f2 is one of the smallest known viruses (the capsid is about 20–25 nm in diameter).

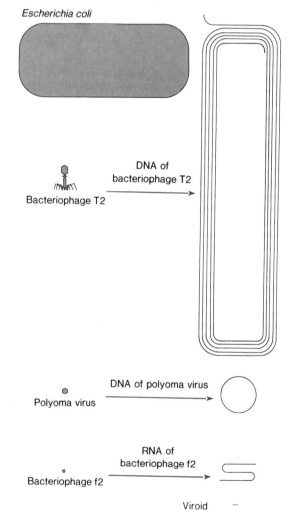

Part 7

Diversity of Life

kuru and Creutzfeld-Jacob disease in humans) in which it has not yet been possible to isolate viruses from diseased victims. Some scientists think that infectious nucleic acids might be involved.

Finally, there are some very mysterious particles called **prions** that also are suspected of causing infectious diseases. The puzzling thing about these tiny particles, which are much smaller than even viroids, is that they are proteins and do not seem to include any nucleic acid. At this point, we can only tell you to be on the lookout for news about clues in the prion mystery.

The Kingdom Monera (Prokaryotae)

The kingdom Monera includes two groups of organisms that are prokaryotic cells, the bacteria and the cyanobacteria (blue-green algae).

The Bacteria

Bacteria were discovered by Anton van Leeuwenhoek, who published drawings of a wide variety of them in 1684. Bacteria are prokaryotes that range in size from small spheres 0.2 or 0.3 μm in diameter to rod-shaped bacteria hundreds of micrometers in length. The average cell is 1 to 5 μm in size. A bacterium may be spherical (a **coccus**), rod-shaped (a **bacillus**), or helical (a **spirillum**) (figure 37.9). The rods

Figure 37.9 Scanning electron micrographs showing the three most common bacterial shapes. (a) Cocci. (b) Bacilli (these bacilli have flagella). (c) A spirillum.

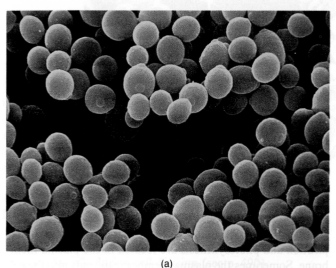

(a)

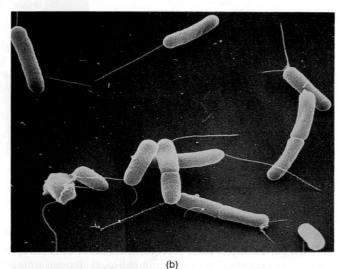

(b)

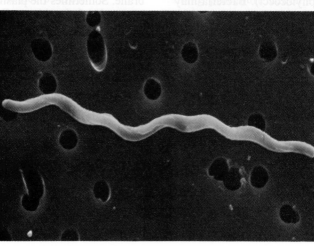

(c)

Bacterial Reproduction

Bacteria normally reproduce asexually by cell division (**binary fission**) (see figure 37.11). After the DNA has been replicated, the plasma membrane pushes inward to form a central transverse septum. Next, the cell wall grows inward within the transverse septum and eventually divides the cell in two. The newly duplicated genetic material is attached to the plasma membrane. Since the older cell walls are elongating as the new walls are being formed, the two chromosomes are separated and pulled into the daughter cells by cell wall and plasma membrane expansion.

Although it takes several hours or even days for most eukaryotic cells to reproduce, bacteria can divide much more rapidly. For example, under optimal laboratory conditions, *E. coli* populations double about once every twenty minutes. If it were possible for this rate of division to continue for a day or so, the mass of *E. coli* formed would be many times the weight of the earth! However, such maximal growth rates are not sustained for very long. Growth rates decrease due to exhaustion of nutrient supplies or accumulation of toxic wastes. In nature, where growth conditions are not so favorable, a population can take days to double in number. For example, *E. coli* has a doubling time of twelve hours even in the intestinal tract (a fairly nutrient-rich environment). This is much slower than the twenty-minute doubling time seen in culture media that provide ideal growth conditions for the bacteria.

Bacterial Nutrition and Metabolism

One of the most remarkable qualities of bacteria is their vast metabolic diversity. There probably is no natural organic molecule that cannot be degraded by at least one bacterial species. Indeed, some bacteria are chemical omnivores. *Pseudomonas multivorans,* for example, can use more than ninety different organic molecules as its source of carbon and energy. Bacteria that are quite fastidious also can be isolated—methane-oxidizing bacteria, for example, metabolize only methane, methanol, and a few related substrates. Metabolic diversity makes bacteria extremely important components of the ecosystem because they play a critical part in the decomposition and recycling of organic matter, including many man-made materials.

The majority of bacteria are heterotrophs. Most require oxygen for growth; that is, they are **aerobic.** Oxygen serves as an electron acceptor for the electron transport system during aerobic respiration. Many bacteria possess the Embden-Meyerhof pathway, the Krebs cycle, and an electron transport system (although their electron transport system differs from that seen in most eukaryotes).

Other bacteria are **facultative anaerobes.** They do not require oxygen but grow better when it is present. *E. coli* is a well-known facultative anaerobe. It grows adequately under anaerobic conditions in the human intestine but flourishes when incubated in the aerobic atmosphere of a laboratory incubator. Under anaerobic conditions, *E. coli* switches to the process of fermentation for its source of energy, oxidizing a sugar such as glucose to pyruvate and producing ATP. To eliminate the excess electrons generated during this process, it reduces pyruvate to lactate (see chapter 8 for more details on respiration and fermentation). Here an organic molecule, not oxygen, is serving as the electron acceptor. Many other bacteria carry on other forms of fermentation.

Not only do **anaerobic bacteria** not require oxygen for growth; they can be killed by it. They lack special enzymes that protect aerobic and facultative organisms from toxic oxygen derivatives like hydrogen peroxide. Members of the genus *Clostridium*—the causative agents of botulism, tetanus, and other serious diseases—are **obligate anaerobes;** that is, they can grow only under anaerobic conditions. This explains why tetanus normally develops only from deep puncture wounds and not from surface scratches.

Specific identification of bacteria has always been a difficult task because many of them lack complex, distinctive morphological features. Therefore, microbiologists make extensive use of physiological tests for identifying bacteria, particularly potential pathogens. In practice, the nutrient sources used and the products released by growing bacteria can be identified by using color reactions (figure 37.15). A hospital laboratory, for example, can often make a tentative identification of a type of bacteria within a few hours using commercially available color test systems.

In order to obtain energy in the form of ATP, animals like humans must oxidize organic molecules such as sugars and fatty acids. Many bacteria, however, are not limited to the use of organic molecules as an energy source. **Chemoautotrophic** bacteria can extract electrons from hydrogen sulfide, elemental sulfur, iron, and other inorganic substances and use these electrons to generate ATP by electron transport. They then use the ATP to provide energy for synthesis of organic compounds from simple, inorganic precursors.

These bacteria are quite important ecologically because they oxidize large quantities of sulfur and ferrous iron to sulfuric acid and insoluble ferric iron (figure 37.16). They can render a mine drainage stream (which contains large quantities of sulfur and iron) barren and lifeless by their metabolic activities.

Figure 37.15 Examples of physiological tests used in bacterial identification. Results of some tests are determined on the basis of color changes in the medium. Some other tests are based on the presence or absence of gas production.

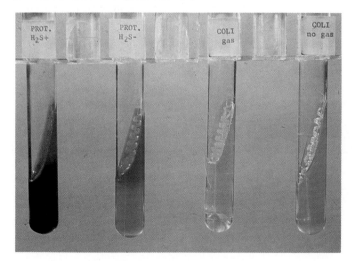

Purple and green bacteria are **photoautotrophs.** They use light as their energy source and produce reduced carbon compounds using carbon dioxide as a precursor. They lack the chloroplasts possessed by eukaryotes and their photosynthetic membranes are spread throughout the cytoplasm. Their chlorophylls and other photosynthetic pigments also differ from those of plants. Plant chloroplasts have two functional photosystems and yield oxygen from water oxidation (see chapter 7). In contrast, photosynthetic bacteria have only one photosystem (Photosystem I) and do not oxidize water to O_2. Instead, they use light to produce ATP and extract electrons from inorganic molecules like H_2S in order to reduce NAD^+ or $NADP^+$. As a result, granules of elemental sulfur are products of photosynthesis rather than O_2. Purple and green bacteria are usually found a few meters below the surface of lakes, where there is sufficient light and a supply of H_2S (figure 37.17).

Bacteria as Pathogens

When most people think of bacteria, they think of "germs" and disease. This is quite natural because **pathogenic** (disease-causing) bacteria have had a dramatic effect on the human race (table 37.3). In fact, modern microbiology developed in the nineteenth century primarily in response to the need for disease control.

Figure 37.16 Yellow precipitate of salts produced as a result of oxidation by bacteria in a stream draining a coal mining area.

Table 37.3
Some Bacteria That Cause Disease in Humans

Bacterium	Result of Infection
Gram-Negative Rods:	
Escherichia coli	Normally present in intestine; can cause diarrhea and urinary tract infections
Pseudomonas aeruginosa	Opportunistic pathogen, usually infecting hospitalized persons, especially those with severe burns
Salmonella typhi	Typhoid fever; other species of *Salmonella* cause food poisoning
Shigella dysenteriae	Dysentery
Vibrio cholerae	Cholera
Yersinia pestis	Bubonic plague
Gram-Negative Cocci:	
Neisseria gonorrhoeae	Gonorrhea
Gram-Positive Rods:	
Bacillus anthracis	Anthrax
Clostridium botulinum	Botulism
Clostridium tetani	Tetanus
Clostridium perfringens	Gas gangrene
Gram-Positive Cocci:	
Staphylococcus aureus	Boils, pimples, toxic shock syndrome, and staphylococcal food poisoning
Streptococcus pneumoniae	Pneumonia
Streptococcus pyogenes	Scarlet fever, tonsillitis, "strep throat"
Spirochetes:	
Treponema pallidum	Syphilis
Borrelia recurrentis	Relapsing fever
Actinomycetes:	
Mycobacterium tuberculosis	Tuberculosis
Corynebacterium diphtheriae	Diphtheria
Rickettsias:	
Rickettsia prowazekii	Typhus
Rickettsia rickettsii	Rocky Mountain spotted fever
Chlamidias:	
Chlamydia trachomatis	Various strains cause venereal disease, eye diseases, and infant pneumonia
Mycoplasmas:	
Mycoplasma mycoides	Oral, respiratory, and kidney infections

Figure 37.17 Masses of purple sulfur bacteria in a small sulpher spring in Indiana. The cells form a pink/purple mat over the leaf litter in the pond.

One of the foundations for the **germ theory of disease** resulted from the research of Louis Pasteur on spontaneous generation—the notion that living things spontaneously arise from nonliving material whenever proper conditions exist. During the course of his studies, which indicated that spontaneous generation does not occur, Pasteur showed that the air is filled with microorganisms that can cause decay unless they are removed or destroyed. The British surgeon Joseph Lister was very impressed by Pasteur's work. He worked to protect wounds from airborne microbes by sterilizing surgical instruments, dressings, and the operating room. His success in reducing infection provided indirect evidence in support of the belief that "germs" cause disease.

In 1876, the German physician Robert Koch published an exhaustive study of anthrax, a disease of cattle that can be transmitted to humans. He proved conclusively that anthrax is caused by the spore-forming bacterium *Bacillus anthracis*. We can express his approach and method of proof in terms of the criteria now known as **Koch's postulates.** Koch maintained that one can prove a bacterium causes a particular disease if four conditions are met:

1. The microorganism must be present in all diseased animals and not in healthy ones.
2. The organism must be isolated from a diseased animal and grown in pure culture.
3. When the isolated microorganisms are injected into healthy animals, precisely the same disease must result.
4. It must then be possible to reisolate the suspected pathogen from the experimentally infected host and show it to be the same as the one first isolated.

Koch's work not only established rigorous criteria for the study of disease agents, but it also stimulated the use of **pure culture** (culture containing only one kind of bacterium) and the laboratory cultivation of bacteria. Furthermore, Koch pioneered the use of solid nutrient media, particularly agar. This made the isolation of single colonies derived from individual bacteria much easier. A mixture of bacteria could simply be streaked out on agar to such an extent that each cell reproduced to form an isolated colony (figure 37.18).

These technical and conceptual breakthroughs led to a vigorous spurt of activity in medical microbiology beginning in the 1870s. During the following thirty years bacterial pathogens were discovered for many of the most deadly diseases: anthrax, gonorrhea, typhoid fever, tuberculosis, cholera, diphtheria, tetanus, pneumonia, meningitis, gas gangrene, plague, botulism, dysentery, whooping cough, and many more. These discoveries made possible the subsequent progress in disease control that has transformed human lives.

Figure 37.18 A streak plate. If bacteria are spread adequately, some small colonies contain only descendants of a single original cell.

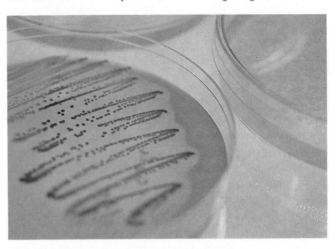

How do bacteria actually harm their hosts? The ability of bacteria to cause disease is called **pathogenicity.** The relative degree of pathogenicity, called **virulence,** is basically a function of two major factors—invasiveness and toxigenicity. **Invasiveness** refers to the ability of a pathogen to proliferate in its host. **Toxigenicity** is the capability of producing a chemical substance, a **toxin,** that can damage the host and lead to disease. A pathogen may be virulent either because of high invasiveness or great toxigenicity. An excellent example of invasiveness is *Streptococcus pneumoniae,* the bacterium that causes pneumonia. *S. pneumoniae* does not appear to produce a toxin, yet in the lungs it reaches such a high population level that the lungs fill with serum and white blood cells in response to its presence. In contrast, *Clostridium tetani,* the bacterium that causes tetanus, rarely leaves the wound in which it lives. Nevertheless, its toxin is so potent that the victim may well die of the paralysis caused by the toxin.

There are two types of bacterial toxins, each with different effects. **Exotoxins** are protein toxins released, in most cases, by living Gram-positive bacteria. Botulism, tetanus, gas gangrene, diphtheria, and cholera are all caused by exotoxins. For example, diphtheria toxin inhibits protein synthesis; one of the gas gangrene toxins attacks the host cell membrane phospholipid, lecithin; and cholera toxin disrupts ion transport in the intestinal epithelium and thus interferes with the ability of the host to maintain normal ionic composition of body fluids and normal body fluid volume. **Endotoxins** are derived from the outer portion of the Gram-negative bacterial cell wall. They are very complex in composition—containing lipid, polysaccharide, and protein—and normally seem to be released only upon bacterial cell death.

Common responses to endotoxins include shock, diarrhea, hemorrhage, and fever.

The disease process is very complex and depends on the interaction of host and pathogen. Some bacterial pathogens cause disease symptoms by triggering allergic responses rather than by producing a toxin; as a result, the infected host is harmed by its own defense mechanisms.

Beneficial Activities of Bacteria

Pathogenic bacteria have such a widespread impact that we often tend to consider all bacteria dangerous and harmful. In truth, beneficial bacteria vastly outnumber the pathogens and they are so important that the ecosystem could not function without them. For example, bacteria play an indispensable role in both the carbon and nitrogen cycles (chapter 35). They help degrade dead organic material, mostly soil humus, to CO_2, which makes the carbon available for photosynthetic incorporation by plants. Organic nitrogen in decaying material is released as ammonium ions by bacterial activity. Nitrifying bacteria can oxidize this ammonia to nitrate, a form of nitrogen readily used by plants.

One of the most important ecological contributions of bacteria is **nitrogen fixation,** the utilization of N_2 (atmospheric molecular nitrogen) as a source of nitrogen to be incorporated into organic molecules. Because most organisms cannot use molecular nitrogen, the amount of available nitrogen limits productivity in many environments. Around 85 percent of nitrogen fixation is biological; the remainder results from lightning or industrial activity. *Rhizobium* is a bacterium that fixes nitrogen when in the root nodules of legumes such as soybeans, clover, and alfalfa (see p. 209). Several free-living bacterial species also fix nitrogen and thus contribute to soil fertility.

Many bacteria in our bodies are beneficial rather than harmful. Their importance has been demonstrated in studies on **germ-free animals** grown in special isolation units. These germ-free animals have poorly developed immune systems and are therefore very susceptible to pathogens. The absence of normal bacteria also lowers these animals' resistance to disease since an invading pathogen does not encounter any competition from nonpathogenic bacteria that would normally be present. Intestinal bacteria also seem to be necessary for proper intestinal development; the intestinal walls of germ-free animals are thin and underdeveloped. Finally, intestinal bacteria produce and release vitamins such as vitamin K, which their hosts can absorb.

Bacteria are also indispensable to many industries. Lactic acid bacteria are used by the dairy industry in the manufacture of yogurt, cottage cheese, and cheese. The distinctive flavors of cheddar, Swiss, Parmesan, and Limburger cheeses are the result of bacterial products. Vinegars are made by allowing acetic acid bacteria to oxidize the alcohol in wine, apple cider, or malt. Bacteria also are used to synthesize amino acids (glutamic acid, lysine), organic acids (lactic acid, butyric acid), steroids, and enzymes. Most of the important antibiotics, with the exception of penicillin and ampicillin, are produced by bacteria.

The future of industrial microbiology looks even more promising. It is quite likely that methane for use as fuel can be manufactured from agricultural organic wastes by methanogenic bacteria. Several types of bacteria are already being grown for use as biological insecticides. For example, a toxin from *Bacillus thuringensis* kills a dozen or more common insect pests (cabbage worm, gypsy moth, and others). More intensive use of biological control agents is almost certain in the future.

Finally, some countries are already growing bacteria as a source of protein. The utilization of bacteria as a food source is especially attractive because bacteria can grow rapidly on many indigestible materials, such as wastes from the petroleum industry.

Cyanobacteria

The cyanobacteria (blue-green algae) are a large, complex group of prokaryotic microorganisms that are similar to eukaryotic algae in terms of photosynthesis and general appearance. However, it is probably best to call them bacteria since they are genuine prokaryotes; that is, their cells lack a true nucleus and other complex, membranous organelles found in eukaryotic cells (figure 37.19). Even their cell walls are similar to those of Gram-negative bacteria.

The group is very diverse morphologically, ranging from single cells to colonies and filamentous forms (figure 37.20). There is similar diversity in size. A number of species coat themselves with a gelatinous sheath. The smallest cyanobacteria are about 1 μm in diameter, while *Oscillatoria princeps,* a giant among prokaryotes, can reach a diameter of 60 μm.

Although a number of cyanobacteria are nonmotile, many possess gliding motility. When a cell or filament is in contact with a solid surface, it glides along by some presently unknown mechanism. Some of them even rotate and flex as they move. Phototaxis and chemotaxis have also been observed. Finally, some cyanobacteria can swim. Yet these

Figure 37.19 An electron micrograph of the cyanobacterium *Anabaena azollae*. Note the prokaryotic structure and the extensive photosynthetic membranes (lamellae). Polyhedral bodies are structures commonly found in cyanobacterial cells that seem to be involved in some photosynthetic functions.

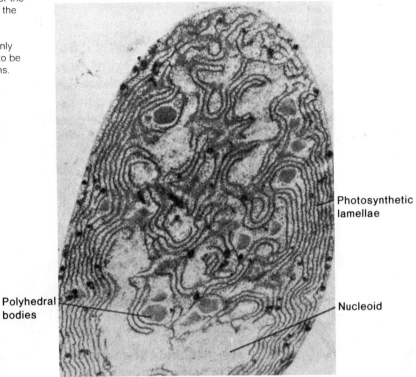

Photosynthetic lamellae

Polyhedral bodies

Nucleoid

swimming cells do not seem to have flagella, and thus are not exceptions to the generalization that cyanobacteria have never been found to have flagella. The mechanism by which they swim, however, remains to be determined.

As with other prokaryotes, cyanobacteria reproduce asexually by binary fission. Sometimes, filamentous forms break off fragments (**hormogonia**), which then glide away as separate filaments. Cyanobacteria may also form thick-walled resting spores called **akinetes** that can survive periods of drying or cold and then germinate under more favorable conditions (see figure 37.20).

Cyanobacteria are photosynthetic and usually have very simple nutritional requirements—CO_2, nitrate or ammonia, and some inorganic ions. Their photosynthetic apparatus, even though it is not contained in a chloroplast, is more similar to that of red algae than it is to the systems of photosynthesizing green and purple bacteria. Cyanobacteria have chlorophyll *a* and the accessory pigments **phycocyanin** (blue in color) and **phycoerythrin** (red). Because of these pigments, cyanobacteria are not always blue-green but may be red, yellow, purple or even brown. Unlike other photosynthetic bacteria, cyanobacteria do possess Photosystem II and produce O_2 photosynthetically. Many also have **gas vesicles** that allow them to float close to the surface, where light intensity is greater.

A number of cyanobacteria can fix nitrogen. When nitrogen gas (N_2) is the only available source of nitrogen, unique structures called **heterocysts** develop (figures 37.20 and 37.21). Heterocysts are specialized, thick-walled cells that have lost their nuclei and been transformed into nitrogen fixation centers. **Nitrogenase,** the enzyme system responsible for N_2 fixation, is inactivated by oxygen, and therefore can function only in an anaerobic environment. When the heterocyst develops, Photosystem II (the photosystem required for O_2 generation) is lost. The mature heterocyst still contains Photosystem I and can generate ATP, but without producing O_2. Thus, the heterocyst provides just the proper anaerobic environment for nitrogen fixation. Cyanobacteria fix nitrogen so efficiently that in certain rice-growing regions of the world, Southeast Asia in particular, nitrogen fertilizers are unnecessary because cyanobacteria abound on the surface water of rice paddies. Consequently, rice can be grown on the same land year after year without the addition of fertilizers.

Figure 37.20 Common cyanobacteria (blue-green algae). (a) *Lyngbya.* (b) *Gomphosphaeria.* (c) *Chamaesiphon.* (d) *Anabaena.* (e) *Spirulina.* (f) *Oscillatoria.* (g) *Aphanocapsa.* (h) *Gloeocapsa.* (i) *Nostoc.* (j) *Merismopedia.* Akinetes are thick-walled resting spores. Heterocysts are specialized, thick-walled cells in which nitrogen fixation occurs.

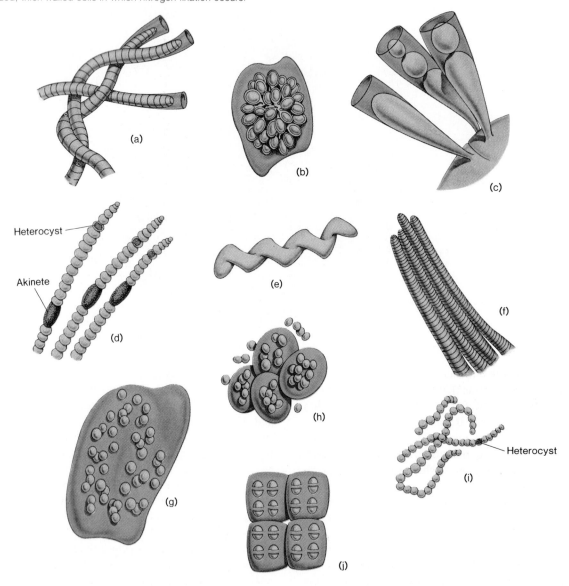

With their resilience and simple nutritional requirements, cyanobacteria are capable of colonizing areas unable to sustain eukaryotic algae (hot springs and deserts, for example) and are found in almost all other environments as well. These organisms may have been the first colonizers of land during the course of evolution. Once established, their mass could have provided a physical as well as chemical substrate for the ultimate attachment and growth of plants.

Cyanobacteria also participate in a variety of symbiotic associations. For example, they are associated with liverworts and ferns, and with corals and other invertebrates. One of the most interesting of these associations is the **lichen,** a symbiotic association of a fungus with either green algae or cyanobacteria (see p. 954). In a lichen, the cyanobacterium provides organic nutrients for the fungus, while the latter protects the cyanobacterium and furnishes the inorganic nutrients required by its photosynthetic partner.

Figure 37.21 Filaments of the cyanobacterium *Anabaena* showing heterocysts, which form when N$_2$ is the only available nitrogen source.

Heterocyst

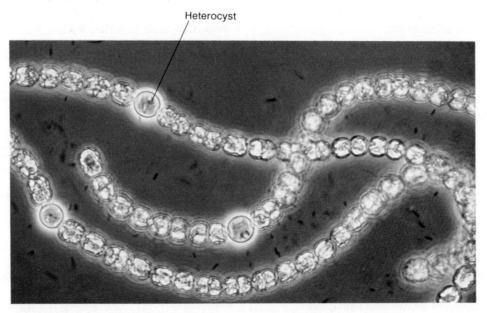

Although exceedingly beneficial and ecologically important, cyanobacteria also can become a nuisance. They thrive in environments that are high in phosphates and nitrates. If care is not taken in the disposal of industrial, agricultural, and human wastes, phosphates and nitrates drain into lakes and ponds, resulting in a "bloom" of cyanobacteria. The surface of the water becomes cloudy, and as a result, light available for photosynthesis by aquatic algae is reduced. Furthermore, the toxic by-products of the cyanobacteria can kill fish and other aquatic animals.

Summary

Taxonomy is concerned with naming organisms and arranging them in a system of classification based on their similarities and probable evolutionary relationships. A species is the basic taxonomic unit and is represented by a Latin binomial (two-word name) consisting of a genus name and a species name. Species are arranged within a succession of ever-broader taxonomic ranks. These ranks, in decreasing order of breadth, are: kingdom, phylum or division, class, order, family, genus, and species.

The five-kingdom classification system developed by R. H. Whittaker and others uses cellular structure, general level of organizational complexity, and modes of nutrition (photosynthetic, absorptive, ingestive) to group organisms into one of five kingdoms: Monera, Protista, Fungi, Animalia, and Plantae.

Viruses are not considered to be living in the usual sense of the word. They do contain nucleic acids and protein, the basic chemicals of living systems, but in order to reproduce, they must enter host cells and cause them to replicate viral nucleic acids and synthesize virus capsids. Cells are often destroyed in this process, and viral diseases can develop.

Bacteria are unicellular prokaryotes. Prokaryotic cells differ from eukaryotic cells in that they do not possess a true nucleus, their DNA is a circular molecule not bound to histone proteins, and they lack membrane-bound organelles such as mitochondria and chloroplasts. Their reproduction is asexual, usually by binary fission.

Cyanobacteria (blue-green algae) are also prokaryotic cells that range from single cells to colonies and filaments. Cyanobacteria differ from other photosynthetic bacteria because they possess Photosystem II and release oxygen as a by-product of photosynthesis.

Both bacteria and cyanobacteria are of great importance to human beings and the ecosystem as a whole. They are critical components of the nitrogen and carbon cycles, since they are involved in nitrogen fixation and the decomposition of dead organic matter. Bacteria are useful industrially in the manufacture of food, antibiotics, and other important products. Some bacteria are pathogenic for humans and the plants and animals on which humans depend.

Questions for Review

1. Briefly describe the five-kingdom system and its advantages over the older two-kingdom system.
2. How are T-even bacteriophages reproduced?
3. What is lysogeny?
4. How do RNA viruses reproduce?
5. What are viroids?
6. In what ways do bacteria differ from eukaryotic microorganisms?
7. Define these terms: aerobe, facultative anaerobe, and anaerobe.
8. What are Koch's postulates and what is their importance?
9. Define pathogenicity and virulence. What factors determine virulence?
10. Describe four major ways in which bacteria benefit humans or the human ecosystem.
11. What are cyanobacteria, and how do they differ from other bacteria? Briefly describe their morphology and reproduction.
12. What are heterocysts, and what is their significance?

Questions for Analysis and Discussion

1. Both lizards and human beings (as well as other primates) have five digits at the ends of their limbs. What implications does this have for taxonomists using phenetic methods?
2. Taking antibiotics orally often results in digestive upsets. Suggest an explanation.
3. The tiny water fern, *Azolla,* has small cavities that contain the nitrogen-fixing cyanobacterium *Anabaena azollae.* Of what value is this relationship to the fern? Explain why *Azolla* is grown in rice paddies in Asia.

Suggested Readings

Books

Brock, T. D.; Smith, D. W.; and Madigan, M. T. 1984. *Biology of microorganisms.* 4th ed. Englewood Cliffs, N.J.: Prentice-Hall.

Goodfield, J. 1985. *Quest for the killers.* Boston: Birkhauser.

Pelczar, M. J., Jr.; Chan, E. C. S.; and Krieg, N. R. 1986. *Microbiology.* New York: McGraw-Hill.

Stanier, R. Y.; Ingraham, J. L.; Wheelis, M. L.; and Painter, P. R. 1986. *The microbial world.* Englewood Cliffs, N.J.: Prentice-Hall.

Articles

Brock, T. D. 1985. Life at high temperatures. *Science* 230:132.

Clark, P. H. 1985. Microbial physiology and biotechnology. *Endeavour* 9:144.

Demain, A. L., and Solomon, N. A. September 1981. Industrial microbiology. *Scientific American.*

Diener, T. O. 1983. The viroid—A subviral pathogen. *American Scientist* 71:481.

Goodfield, J. October 1985. The last days of smallpox. *Science 85.*

Gould, S. J. June 1976. The five kingdoms. *Natural History.*

Janzen, D. H. 1977. Why fruits rot, seeds mold, and meat spoils. *The American Naturalist* 111:691.

Krogmann, D. W. 1981. Cyanobacteria (blue-green algae)—Their evolution and relation to other photosynthetic organisms. *BioScience* 31:121.

Mayr, E. 1981. Biological classification: Toward a synthesis of opposing methodologies. *Science* 214:510.

Miller, L. K.; Lingg, A. J.; and Bulla, L. A., Jr. 1983. Bacterial, viral, and fungal insecticides. *Science* 219:715.

Prusiner, S. B. October 1984. Prions. *Scientific American.*

Sanders, F. K. 1981. Interferons: An example of communication. *Carolina Biology Readers* no. 88. Burlington, N.C.: Carolina Biological Supply Co.

Waterbury, J. B.; Willey, J. M.; Franks, D. G.; Valois, F. W.; and Watson, S. W. 1985. A cyanobacterium capable of swimming motility. *Science* 230:74.

Woese, C. R. June 1981. Archaebacteria. *Scientific American.*

38

Protists and Fungi

Chapter Concepts

1. Members of the kingdom Protista are mainly unicellular or colonial eukaryotes. This kingdom includes protozoa, unicellular algae, and funguslike protists.
2. Both autotrophs and heterotrophs are found among the protists.
3. Many protists are major primary producers and of great ecological importance. Other protists cause serious diseases in humans, their crops, and their domesticated animals.
4. Fungi are multicellular heterotrophic eukaryotes that absorb nutrients after secreting extracellular digestive enzymes.
5. Fungi are important decomposers in ecosystems.
6. Although many fungi cause diseases in both plants and animals, others are edible or are used in the production of many kinds of food products.

Many people living in industrialized countries tend to view malaria as a disease of the past. Yet in tropical and subtropical areas of the world, malaria is still a very severe health problem. Nearly 2,000 million people live under the constant threat of malaria, and 150 to 300 million people are afflicted with this dread disease each year. Furthermore, it is estimated that malaria causes between two and four million deaths each year. All of this human misery and death is caused by tiny, single-celled protozoans that are members of the **kingdom Protista.**

Although certain other protists also cause a wide variety of human and animal diseases, not all members of the kingdom Protista have such a negative impact on other living things. Many photosynthesizing protists are so numerous that they rank as the major producers of nutrients in aquatic ecosystems and thus form the critical first link in aquatic food chains.

Certain members of the **kingdom Fungi,** the other main group of organisms we will consider in this chapter, also bring about considerable disease and destruction. For example, it is estimated that 70 to 80 percent of all major crop diseases are caused by fungi, and that in the United States alone, fungal diseases cause several thousand million dollars' worth of damage to crops each year. However, the overall impact of fungi on other organisms and the environment is actually quite positive; many fungi play key roles as decomposers that help to make the nutrients in dead organisms and organic wastes available for use by other living things.

In this chapter, we will survey these two kingdoms of life, the kingdom Protista and the kingdom Fungi.

The Kingdom Protista

In many ways, the kingdom Protista is the most heterogeneous and fascinating of the five kingdoms. This kingdom contains many different kinds of unicellular and simple colonial eukaryotic organisms (table 38.1).

The kingdom Protista encompasses a tremendous diversity of organisms, but the borderlines between the protists and other kingdoms are not entirely clear. For example, as we mentioned in chapter 37, the green algae (chlorophyta) contain unicellular, colonial, and multicellular forms, and some biologists consider them to be protists while others recognize them as plants. (We will do the latter and defer our discussion of the green algae to chapter 39.) Furthermore, the distribution of protists among the categories protozoa, unicellular algae, and funguslike protists is not as clear and unambiguous as implied in table 38.1. For example, some biologists consider euglenoids, dinoflagellates, and slime molds all to be protozoa.

Table 38.1
The Kingdom Protista

Protozoa

Mastigophora	Flagellated protozoa
Sarcodina	Pseudopodial protozoa
Sporozoa	Spore-forming protozoa
Ciliophora	Ciliated protozoa

Unicellular Algae

Euglenophyta	*Euglena* and relatives
Pyrrophyta	Dinoflagellates
Chrysophyta	Diatoms and others

Funguslike Protists

Chytridiomycota	Chytrids with a single, posterior flagellum
Oomycota	Water molds with biflagellate zoospores
Gymnomycota	Slime molds

Despite these problems, however, there is a unifying theme among the protists: they all must cope with their environments without the benefit of the complex multicellular organization found in plants and animals.

The Protozoa

The **protozoa** are single-celled, heterotrophic eukaryotic organisms that do not form funguslike fruiting bodies. Although they are unicellular, protozoans certainly are not uniformly simple. Many are highly specialized cells in which organelles take the functional roles that organ systems play in more complex multicellular organisms.

There are over 65,000 known species of protozoa, with more being discovered regularly. Protozoa are found in a wide variety of environments. The majority are free-living and inhabit freshwater or marine environments. There are, however, a number of terrestrial protozoa, and even beach sand has it own particular protozoan population. Other protozoa are parasitic and cause some widespread diseases such as malaria.

Each protozoan maintains homeostasis through a complex set of interactions with its environment. Protozoa take in nutrients in a variety of ways: by simple diffusion, by active transport, and by pinocytosis and phagocytosis. Usually, nutrients are digested in food vacuoles. Gas exchange takes place through the plasma membrane. The contractile vacuoles of freshwater protozoa eliminate excess water acquired from the hypotonic environment (see p. 328). Protozoa eliminate nitrogen wastes (usually ammonia) by simple diffusion through the plasma membrane.

Figure 38.1 Trypanosomes. (a) A scanning electron micrograph showing a trypanosome among red blood cells. (b) Structure of the intermediate bloodstream form of *T. rhodesiense* as revealed by the transmission electron microscope. The pellicle is a flexible, proteinaceous covering over the cell.

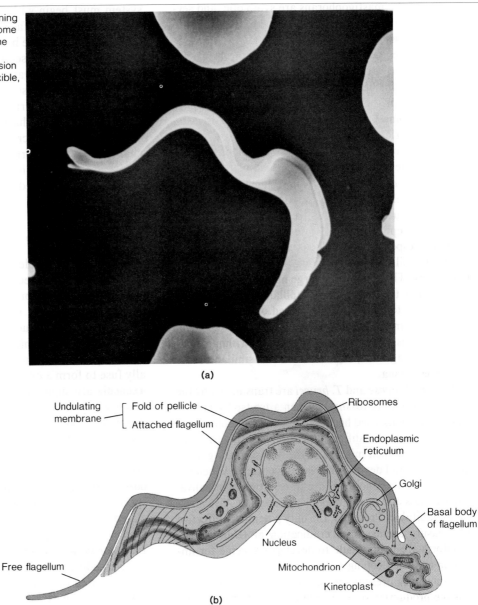

(a)

Undulating membrane — Fold of pellicle — Ribosomes — Attached flagellum — Endoplasmic reticulum — Golgi — Basal body of flagellum — Free flagellum — Nucleus — Mitochondrion — Kinetoplast

(b)

Protozoa move by several different means and often are quite responsive to changes in their environment. They also show considerable variety in reproduction, and while asexual reproduction is the rule, many protozoa can also reproduce sexually during some part of their life cycle.

Protozoan classification is quite complex, but for the sake of simplicity, we will use a traditional system of four phyla: **Mastigophora, Sarcodina, Sporozoa,** and **Ciliophora** (see table 38.1). These four groups are distinguished from one another on the basis of their locomotor organelles and their modes of reproduction.

Phylum Mastigophora

Mastigophorans have one or more flagella which have the characteristic 9 + 2 microtubular structure found in eukaryotic cells (see p. 80). Evolutionarily, mastigophorans are considered to be quite ancient, and it is very likely that the other groups of protozoa arose from the flagellates.

Most mastigophorans are uninucleate, but some possess an extranuclear accumulation of DNA, a mitochondrial structure called the **kinetoplast** (figure 38.1). Asexual reproduction by way of longitudinal binary fission is most common.

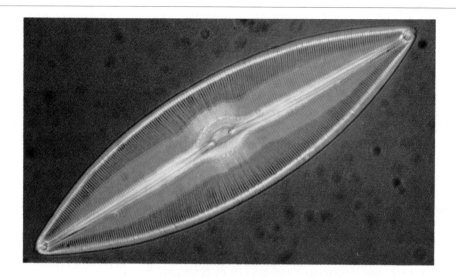

Figure 38.12 A diatom. Diatoms are major producer organisms in aquatic food chains. Their silica-containing shells appear glassy.

Funguslike Protists

A number of organisms that formerly were classified as fungi can be placed with the protists because of their simple organization. At least three protist groups can be classified as funguslike: the chytrids (**division Chytridiomycota**), the water molds (**division Oomycota**), and the slime molds (**division Gymnomycota**).

Division Chytridiomycota

The **chytrids** are normally aquatic, although some live in moist soil. Many have very simple morphology, often consisting of a spherical cell that penetrates a host with colorless, rootlike **rhizoids** (figure 38.13). Their cell walls contain chitin. At some point in their life cycle all chytrids produce motile cells, each with a single, posterior, whiplike flagellum.

Many of the aquatic chytrids are parasitic on algae, but some parasitize aquatic plants and animals, and a few parasitize terrestrial plants. Other species are **saprophytic**—they grow on decaying plant and animal remains. Species of *Allomyces* and *Blastocladiella* are valuable research organisms in the study of morphogenesis, but on the whole, chytrids seem to have little direct effect on human welfare.

Division Oomycota

The **water molds** are very common aquatic organisms and are often found as cottony masses on sick or dead insects and fish. They are also widespread in the soil. Their structure usually consists of a network of branching tubes called **hyphae** (singular: **hypha**) that grow together to form a mass of

Figure 38.13 Growth of a typical chytrid parasite on algae. Rhizoids penetrate the host cells. The chytrid on the right is releasing flagellated spores (zoospores) that will eventually develop into new adults.

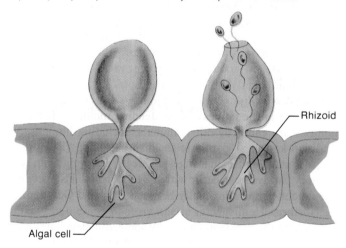

Rhizoid

Algal cell

hyphae known as a **mycelium.** Hyphae lack cross walls and thus are **syncytial** (**coenocytic**); that is, they are not divided into individual cells. Their cell walls contain cellulose and other glucose polymers. Most water molds produce biflagellated cells at some point in their life cycle.

Most members of the Oomycota are harmless saprophytes, living on decaying organic matter. However, several parasitic forms are of great economic importance, and a soil-dwelling member of this group dramatically shaped history. *Phytophthora infestans* causes potato blight and was responsible for the 1840s potato famine in Ireland. The entire potato crop was destroyed in one week during the summer

Figure 38.14 Plasmodium of the slime mold *Physarum.*

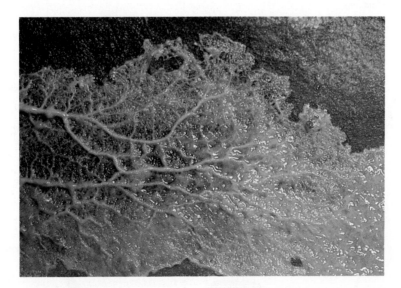

of 1846, resulting in a famine of frightening proportions. The mass migration of the Irish to the United States before the turn of the century was at least partially due to this organism.

Interest in protists as disease agents was also stimulated by the loss of French wine grapes during the latter half of the last century to *Plasmopara viticola,* the causative agent of grape downy mildew. In only a few years' time, *P. viticola* almost destroyed the wine industry in France. This disaster led to the development of the first fungicide, the "Bordeaux mixture," a combination of copper sulfate and lime.

Saprolegnia, another water mold, parasitizes fish, including those kept in aquariums. As this water mold grows, a mycelium forms over the fins and eventually covers the entire fish, first paralyzing and then killing it.

Division Gymnomycota

The **slime molds** are an unusual and varied collection of amoeboid organisms that possess both plant- and animal-like characteristics. One of the two best-known slime mold groups, the **cellular slime molds** (Acrasiomycetes), has been studied very intensely by developmental biologists; we discussed their differentiation in chapter 28. Here we will focus our attention on the other major group, the plasmodial slime molds.

There are 450 to 500 species of **plasmodial (true) slime molds** (Myxomycetes), and they are found worldwide in moist, dark areas, such as under the bark of decaying trees or in layers of decomposing leaves. The active, growing vegetative form of the organism is a **plasmodium** that moves over surfaces, feeding on bacteria, protozoa, and other organisms (figure 38.14). As it grows, its diploid nuclei undergo mitosis, but there is no division of the cytoplasm. Thus, a plasmodium is a large syncytial mass with many nuclei. Although some plasmodia are small, many can become rather large (30 cm or more across) and quite colorful. The sudden appearance of a large, bright-yellow plasmodium on someone's lawn has more than once caused excitement in a neighborhood.

When the plasmodium has matured, it creeps out into a lighted area and forms **fruiting bodies** (light is often required for fruiting) (figure 38.15). Slime mold fruiting bodies, though only a few millimeters tall, are varied and quite beautiful (figure 38.16). As the spores develop within a structure called the **sporangium,** meiosis takes place so that the mature spores are haploid. These spores can survive for years under unfavorable environmental conditions and then germinate in the presence of moisture to release **myxamoebae** or flagellated **swarm cells** that feed and divide. Eventually, haploid cells fuse to form a diploid zygote. The zygote then feeds, grows, and multiplies its nuclei through synchronous mitotic divisions. Finally, a mature plasmodium develops, and the life cycle has made a full turn.

Although plasmodial slime molds have little direct economic importance, they are fascinating organisms and have proven useful in basic biological research. For example, the slime mold *Physarum polycephalum* and its close relatives can be cultured in the laboratory and have been used extensively for research in cell physiology and molecular biology.

Figure 38.15 The life cycle of a plasmodial slime mold. (Different parts are drawn at different magnifications.)

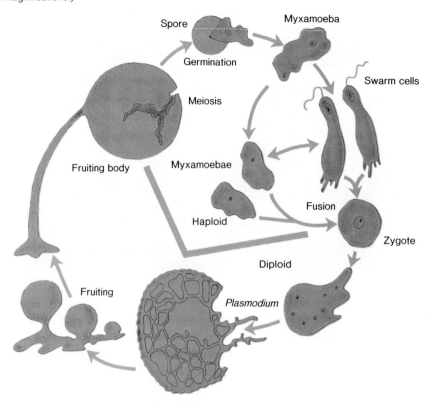

Spore

Germination

Myxamoeba

Meiosis

Swarm cells

Fruiting body

Myxamoebae

Haploid

Fusion

Zygote

Diploid

Fruiting

Plasmodium

Figure 38.16 Plasmodial slime mold fruiting bodies. (a) *Hemitrichia.* (b) *Stemonitis.*

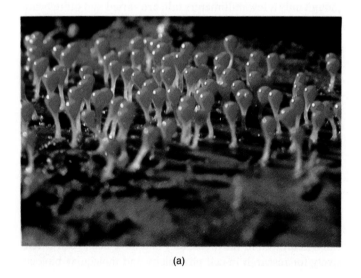

(a)

(b)

The Fungi

Most **fungi** (singular: **fungus**) are nonmotile organisms composed of masses of hyphae. As with water molds, the mass of hyphae in a single fungus forms a mycelium.

Like plant cells, fungal hyphae have rigid cell walls, but fungal cell walls usually are formed of chitin, a substance seldom found in plants. (Chitin is, however, a major component of the hard exoskeletons that cover the bodies of insects, spiders, crabs, and many other animals.) Some hyphae lack cross walls (septa) almost entirely, except where a reproductive organ is formed. Other fungi have septa with pores so that the hyphal cells are partially separated from one another.

Fungi are heterotrophic; they must absorb nutrients from their environment. Thus, they secrete digestive enzymes and then absorb the soluble digestion products. Fungi are particularly important as decomposers, aiding in decomposition of dead matter and the subsequent recycling of inorganic and organic molecules in an ecosystem.

Fungi reproduce both asexually and sexually. Asexual reproduction can be by the simple fragmentation of a mycelium to produce new individuals, but many fungi also produce asexual spores. In some of the less complex fungi, asexual spores develop on specialized hyphae, called **sporangiophores,** within a saclike sporangium. In the more complex fungi, spores develop in the tips of specialized hyphae, the **conidiophores;** these minute spores are referred to as **conidia** (from the Greek, meaning "dust").

The pattern of sexual reproduction differs among the major groups of fungi, and we can separate them into four divisions—Zygomycota, Ascomycota, Basidiomycota, and Deuteromycota—on the basis of reproductive characteristics.

Division Zygomycota

Members of the **division Zygomycota** are widespread and include some of the familiar molds often found growing on foodstuffs such as bread and fruit. They are terrestrial and therefore lack flagellated spores or flagellated gametes.

A commonly encountered member of this division is *Rhizopus stolonifer,* the black bread mold, so-called because mature sporangia turn black. The life history of *R. stolonifer* begins when a spore germinates on bread (figure 38.17). Certain hyphae called **stolons** extend laterally over the surface of the bread, growing with amazing rapidity. Once

Figure 38.17 The life cycle of the black bread mold, *Rhizopus stolonifer.* Both asexual and sexual processes are shown.

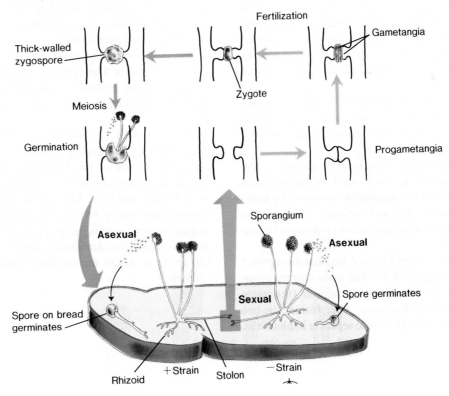

Figure 39.5 *Ulothrix* life history. The zygote (*darker green*) is the only diploid cell.

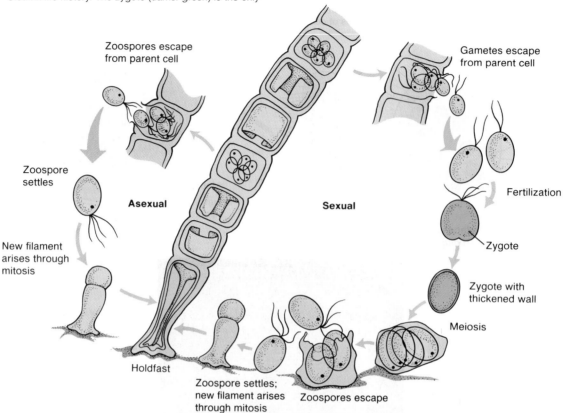

Zoospores escape
from parent cell

Gametes escape
from parent cell

Zoospore
settles

Fertilization

Asexual

Sexual

Zygote

New filament
arises through
mitosis

Zygote with
thickened wall

Meiosis

Holdfast

Zoospore settles;
new filament arises
through mitosis

Zoospores escape

survive adverse conditions. Later, the zygote divides by meiosis to produce four haploid zoospores, each of which can develop into a new *Ulothrix* filament (figure 39.5).

Spirogyra is a very common, freshwater, filamentous green alga. It is often an important part of "pond scum" when algal blooms cover water surfaces. *Spirogyra* cells have one or more large helical chloroplasts with numerous pyrenoids.

Sexual reproduction in *Spirogyra* occurs by a special fusion process called **conjugation.** Usually in the autumn, two filaments come to lie side by side. Protuberances that form on the cells enlarge, meet, and fuse to form **conjugation tubes.** Once cells are paired in this way, one of the cells pulls away from its cell wall, rounds up, squeezes through the conjugation tube, and fuses with the other cell. Thus, individual vegetative cells function as gametes. This happens all along the paired filaments and results in the production of zygotes in one filament and a series of empty cell walls in the other filament. Each zygote forms a thick, resistant wall. Before germination, meiosis occurs, and four haploid nuclei are produced. Three of them degenerate. The fourth becomes the functional nucleus of the cell and grows out of the broken zygote wall to begin the development of a new filament (figure 39.6).

Just as in *Chlamydomonas* and *Ulothrix,* all *Spirogyra* vegetative cells are haploid. The zygote is the only diploid cell in *Spirogyra*'s entire life history.

Oedogonium, another common freshwater alga, has a markedly different reproductive pattern. *Oedogonium* has two distinctly different types of gametes. Small, motile sperm cells fuse with large, nonmotile egg cells. Sexual reproduction involving fusion of such unlike gametes is called **heterogamy,** because these are **heterogametes** (*hetero* = different). Any vegetative cell can differentiate into an egg-forming cell (**oögonium**) or divide to produce several small sperm-forming cells (**antheridia;** singular: **antheridium**). An antheridium produces two swimming sperm cells, each of which has a circle of flagella.

Sperm are released and swim to eggs, apparently attracted by some chemical released by the eggs. As in other filamentous algae, the zygote secretes a thick wall and can withstand adverse conditions. The zygote eventually undergoes meiosis to produce four haploid zoospores. Each of them can initiate formation of a new *Oedogonium* filament (figure 39.7).

Figure 39.6 *Spirogyra* conjugation. (a) Vegetative cell. (b) Stages of conjugation. Zygotes form in one filament of each pair, leaving empty cell walls in the other filament.

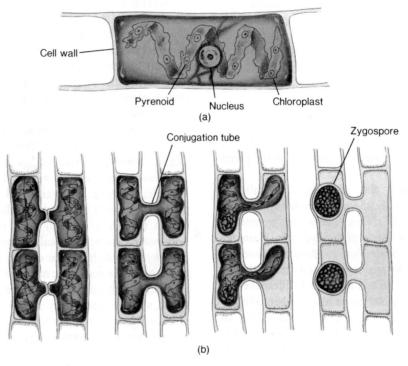

(a)

(b)

Figure 39.7 Reproductive processes in *Oedogonium*.

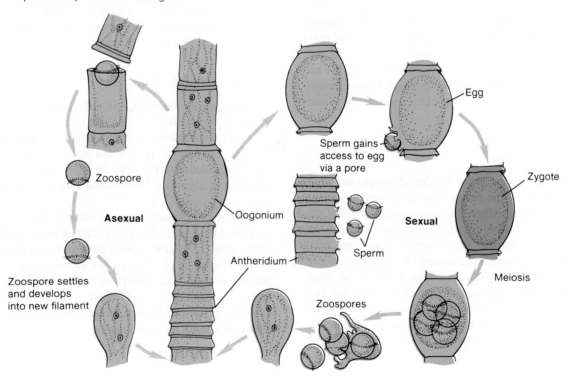

Figure 39.22 Conifers. (a) A giant sequoia (*Sequoiadendron*). (b) Bristlecone pine (*Pinus aristata*). Some living bristlecone pines are known to be over 4,000 years old. What may have been the oldest living thing, a 4,900-year-old bristlecone pine, unfortunately was cut in 1965 so that its age could be determined. (c) Pine needle cross section. The hypodermis below the epidermis is a further barrier to surface water loss. Injury to the needle causes resin ducts to release resin, which closes wounds.

(a)

(b)

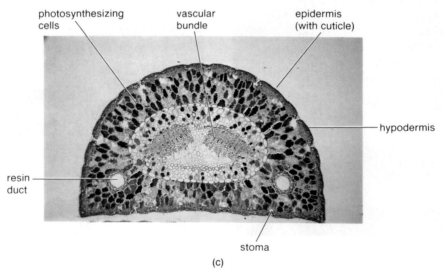

photosynthesizing cells

vascular bundle

epidermis (with cuticle)

hypodermis

resin duct

stoma

(c)

Figure 40.32 (a) A chiton, a member of the class Amphineura. (b) The giant clam *Tridacna*, a member of the class Pelecypoda. *Tridacna* shells can reach lengths of well over 1 meter. Note the open siphon and the coral colonies growing on the clam's two shells.

(a)

(b)

Nautiluses are enclosed in shells, but the shells of squids are reduced and internal. Octopuses lack shells entirely (figure 40.34). These cephalopods can squeeze their mantle cavities so that water is forced out and thus propel themselves rapidly backwards by a sort of jet propulsion. Cephalopods also possess ink sacs from which they can squirt out a cloud of brown or black ink. This action often leaves a would-be predator completely confused.

Cephalopods use the ring of long **tentacles (arms)** around their mouth to capture prey, which they then tear up with a sharp **beak.** Experienced divers fear the beak of an octopus much more than its tentacles.

Although giant octopuses are a standard part of science fiction, octopuses seldom grow to be very large. Some giant squids, however, are enormous; they are by far the largest invertebrate animals. There are accurate records of the capture of a squid that was 18 meters long from tentacle tip to tail and weighed at least two tons. Fairly reliable evidence indicates also that even larger squids may exist in the ocean depths, and there are reports that these giant squids actually fight back when they are attacked by sperm whales.

Figure 40.48 Mites and ticks. (a) Scanning electron micrograph of a mite. Note the body hairs. Many mites live as scavengers on the surfaces of plant and animal hosts, but others, such as chiggers, suck blood. Some mites are carriers (vectors) of disease-causing microorganisms, which they transmit from one host to another (magnification × 42). (b) The Rocky Mountain wood tick (*Dermacentor andersoni*). This tick carries the rickettsia (small bacterium) that causes Rocky Mountain spotted fever in humans. All ticks are parasites. Some cause little direct damage themselves, but a number of tick species are vectors of serious animal and human diseases. (a) from Kessel, R. G. and Shih, C. Y.: *Scanning Electron Microscopy in Biology.* © 1976 Springer-Verlag.

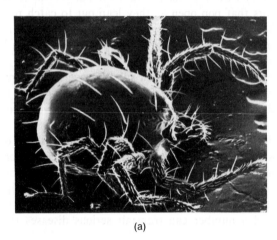

(a)

(b)

Mandibulate Arthropods

The second subphylum of the phylum Arthropoda, the subphylum Mandibulata, includes crustaceans, insects, millipedes, and centipedes. In contrast to the chelicerates, mandibulate arthropods have one or two pairs of antennae, and they have **mandibles** rather than chelicerae (figure 40.49). Mandibles usually are modified for biting or chewing, but they never are pincerlike. Most mandibulate arthropods have two additional pairs of mouthparts called **maxillae.**

Class Crustacea

Because they are edible, some members of the **class Crustacea,** such as **lobsters, shrimps, crabs,** and **crayfish,** are quite familiar to most people. However, **pill bugs,** which are terrestrial, **water fleas, brine shrimp,** and **barnacles,** as well as a host of tiny animals that live as part of the plankton of the world's oceans, also are crustaceans.

All crustaceans have two pairs of antennae and the three pairs of characteristic mandibulate arthropod mouthparts. Crustaceans also have appendages on their abdomens. But beyond these characteristics, it is hard to generalize about the class Crustacea because it is a very diverse group (figure 40.50).

Figure 40.49 Head appendages of a hermit crab, an example of a mandibulate arthropod. Mandibulates have mandibles, rather than chelicerae, as their first mouthparts.

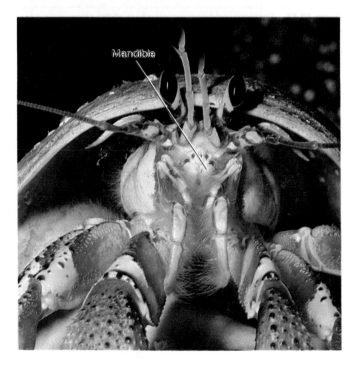

Figure 40.50 Crustaceans small and large. The class Crustacea is a very diverse group of animals. (a) The water flea, *Daphnia,* a small freshwater crustacean, photographed by dark-field microscopy. The average *Daphnia* is 1 to 2 mm long. (b) *Homarus americanus,* the commercially important lobster that occurs along the northeast coast of the United States. (c) The King crab *Paralithodes.* This tasty crab occurs in the north Pacific and is sometimes called the Alaskan King crab. (d) Pill bugs (also called sow bugs or wood lice) are terrestrial crustaceans that live in moist places. (e) Krill. These animals grow to only about 6 cm in length, but are so numerous that they provide abundant food for large animals such as blue whales.

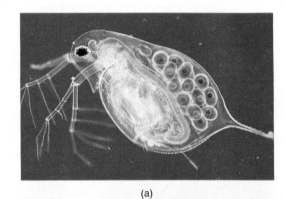

(a)

(b)

(c)

(d)

(e)

Figure 40.51 Barnacles. Barnacle larvae are free-swimming, but adults live in shells attached to substrates. (a) Drawing of the organization of an adult acorn barnacle, *Balanus.* Cirri sweep food out of the water toward the mouth during feeding. Each cirrus is covered with stiff bristles (setae). Note the presence of a penis, which can be protruded out of the body and into an adjacent individual. A barnacle that cannot "reach" another barnacle from its attached position cannot mate. (b) *Lepas,* one of the stalked barnacles (also called goose barnacles or gooseneck barnacles). They are attached to the substrate by a flexible stalk.

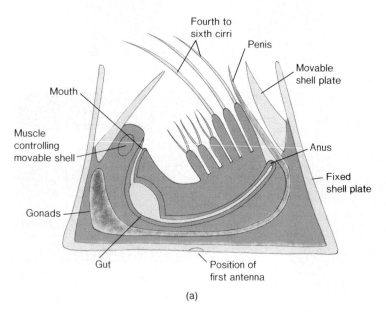

(a)

(b)

The large, familiar, bottom-dwelling crustaceans, such as crabs, lobsters, and crayfish, all belong to a group (order) known as **decapod** ("ten-foot") Crustacea. They are called decapods because they have five pairs of walking legs. Decapod crustaceans have rigid exoskeletons that are heavily impregnated with calcium carbonate.

Many crustaceans, however, are tiny, soft, delicate creatures, and some of them are extremely important links in aquatic food chains. For example, small shrimplike crustaceans known as **krill** provide food for many oceanic animals. Krill occur in great swarms that can cover areas of the ocean as large as several city blocks in layers 5 meters thick. Within such a swarm, there can be more than 60,000 individuals per cubic meter. Despite the small size of the individuals, such huge masses can provide food for even very large animals.

For example, blue whales swim into the swarms and filter huge quantities of krill out of the water. A blue whale can consume up to a ton of krill at a single feeding, and it may feed four times a day. Krill are now being considered as a possible source of human food, and Russian and Japanese fisheries are experimenting with methods for harvesting krill.

Most crustaceans are free-swimming throughout their lives, but adults of one group, the barnacles, live attached to a substrate. Initially, barnacle larvae are free-swimming, but eventually, a young barnacle settles headfirst on a surface, attaches, and secretes a calcium-containing shell around its body. Some of the barnacle's legs develop into delicate, feathery sweepers (**cirri**) that strain food out of the surrounding water and kick it into the animal's mouth (figure 40.51).

Barnacles attach to rocks, other organisms (for example, snails, crabs, and even whales), wharf pilings, buoys, and ship bottoms. Ships with barnacle-fouled bottoms are slowed considerably because the barnacles interfere with the smooth flow of water over their hulls. In order to remove these persistent crustaceans, ship bottoms must be scraped periodically.

Classes Chilopoda and Diplopoda (Centipedes and Millipedes)

Both **centipedes** ("hundred legs") and **millipedes** ("thousand legs") are terrestrial and have elongate, wormlike bodies with a head and a trunk consisting of many segments. Both also use tracheal tubes in gas exchange and Malpighian tubules in excretion, yet they are different enough to be placed in two separate arthropod classes.

Centipedes (**class Chilopoda**) are active, fast-moving, carnivorous animals that use their first pair of legs, which are modified as poison claws, to kill their prey. Centipedes usually eat insects and other small invertebrates, but larger centipedes have been known to feed on snakes, mice, and frogs (figure 40.52).

A centipede has antennae, mandibles, and two pairs of maxillae on its head. Its body is flattened, and each segment bears one pair of walking legs. Actually, the great majority of centipedes have far less than one hundred legs. Centipedes' reproductive ducts open at the posterior end of the body.

Millipedes (**class Diplopoda**) do not have nearly a thousand legs, but each body segment behind the first four or five does have two pairs of legs (see figure 40.42). These legs appear to move in waves as a millipede slowly moves along.

Millipedes do not have poison claws and they are herbivores or scavengers that feed on decaying material. These secretive animals typically live beneath leaves, stones, or logs, and avoid trouble by rolling up and feigning death.

Millipedes differ from centipedes in still other ways. Millipedes have rounded, rather than flattened, bodies. They have two pairs of spiracles leading to tracheal tubes in each body segment; centipedes have only one pair per segment. And a millipede's reproductive ducts open near the anterior end of its body.

Class Insecta

The **class Insecta** includes more species than any other class of organisms, and insects are regarded as one of the most successful (possibly *the* most successful) groups of organisms that have ever lived. Ninety percent of the million or so species of arthropods are insects.

Insects occupy almost every kind of freshwater and terrestrial habitat in the world, and they are variously specialized to utilize a tremendous variety of food sources. Some are parasites on animal bodies; some are scavengers that eat dead, decaying organisms; and others are specialized in countless ways to eat specific parts of plants. For example, the remarkably specialized mouthparts of aphids and treehoppers can penetrate plants and pierce individual phloem cells without causing an injury response in the normally sensitive cells (figure 40.53). Thus, they can feed on a continuing supply of phloem sap.

Figure 40.52 A giant South American centipede (class Chilopoda) attacking a frog. Centipedes have paired antennae, poison claws, and flattened bodies with a single pair of walking legs attached to each trunk segment. Compare with the pictures of a millipede in figure 40.42.

Figure 40.53 A treehopper. Treehoppers, like aphids, have specialized mouthparts that permit them to feed on phloem sap of plants.

Table 40.1
Some Major Orders of the Class Insecta

Order Thysanura (~700 species)	Silverfish and bristletails. Fast-running, primitive, wingless insects; chewing mouthparts; long antennae; simple eyes; two or three long, taillike appendages on rear of abdomen. Common in moist environments; some species in houses, particularly bathrooms and basements. Incomplete metamorphosis.
Order Odonata (~5,000 species)	Dragonflies and damselflies. Rapid-flying, predaceous insects; two pairs of long, narrow, net-veined wings; large, highly developed compound eyes; chewing mouthparts. Incomplete metamorphosis with immature stages (nymphs) in fresh water.
Order Orthoptera (~23,000 species)	Grasshoppers, locusts, crickets, mantids, walking-sticks, roaches. Large-headed insects with chewing mouthparts, compound eyes, two or three simple eyes; usually two pairs of wings; membranous hindwings folded beneath narrower, leathery forewings at rest. Incomplete metamorphosis.
Order Isoptera (~1,800 species)	Termites. Soft-bodied, highly social insects with winged and wingless individuals composing the colony; membranous forewings and hindwings of equal size; simple or compound eyes; chewing mouthparts. Live underground or in wood. Incomplete metamorphosis.
Order Hemiptera (~40,000 species)	True bugs. Highly variable body; usually two pairs of wings; forewings thick and leathery at base, membranous at tip; hindwings membranous; piercing/sucking mouthparts; herbivorous or predaceous. Terrestrial or aquatic. Incomplete metamorphosis.
Order Homoptera (~20,000 species)	Cicadas, leaf hoppers, and aphids. Closely related to the Hemiptera. Wings, if present, membranous and held in a tentlike position over the abdomen; mouthparts as a sucking beak. Incomplete metamorphosis.
Order Anoplura (~200 species)	Sucking lice. Ectoparasites of birds and mammals. Flattened bodies; wingless; eyes reduced or absent; piercing/sucking mouthparts; legs and claws used for clinging to host. Incomplete metamorphosis.
Order Coleoptera (~300,000 species)	Beetles and weevils. Largest order of insects. Hard bodies; chewing mouthparts; forewings form hard protective covering for membranous hindwings. Herbivorous or predaceous. Complete metamorphosis.
Order Lepidoptera (~110,000 species)	Butterflies and moths. Long, soft bodies; two pairs of large wings covered with pigmented scales; compound eyes and antennae well developed. Larvae (caterpillars) have chewing mouthparts and eat plants. Adult mouthparts modified as a coiled proboscis used for sucking flower nectar. Complete metamorphosis.
Order Diptera (~85,000 species)	True flies: mosquitoes, gnats, midges, horseflies, houseflies. One pair of functional, membranous front wings; reduced, knoblike hind wings (halteres) act as balancing organs; well-developed eyes; piercing, sucking, or sponging mouthparts. Complete metamorphosis.
Order Siphonaptera (~1,100 species)	Fleas. Intermittent ectoparasites. Very small wingless insects with laterally flattened bodies; long legs well adapted for jumping; piercing and sucking mouthparts used for feeding on the blood of birds and mammals. Complete metamorphosis.
Order Hymenoptera (~100,000 species)	Wasps, bees, ants, and sawflies. Chewing or chewing/lapping mouthparts; wings membranous when present; some adult females with stinger or piercing ovipositor. Complete metamorphosis.

Some insects are less than 0.25 mm long (smaller than large protozoans), while others have wingspans up to 30 cm. Fossils indicate that some extinct insects were several times that large. These larger insects of the past may have competed much more directly with insects' chief rivals in the terrestrial environment, the vertebrate animals. Smaller insects, on the other hand, generally are able to occupy habitat niches that do not have sufficient resources to support most vertebrates. Thus, natural selection may have favored the evolution of many species of small insects rather than very large insects.

Insects are so numerous and diverse that the study of this one class is a major speciality in biology, and this science, **entomology,** occupies whole departments in many universities. Table 40.1 and figure 40.54 provide a brief introduction to insect diversity and include some of the major orders of insects.

The Insect Body
A typical adult insect body is divided into a head, a thorax, and an abdomen. The head bears well-developed sense organs, including one pair of antennae, **compound eyes** (eyes with many separate focusing units), and sometimes simple eyes as well. Insect mouthparts include a pair of mandibles, a pair of maxillae, and a **labium,** which is a sort of lower lip that has evolved through the fusion of a second pair of maxillae (figure 40.55). There are, however, literally thousands

Figure 40.54 Insect diversity. Representatives of some of the orders of insects described in table 40.1 (drawn to different scales). (a) Silverfish (order Thysanura). (b) Dragonfly (order Odonata). (c) Camel cricket (order Orthoptera). (d) Termite soldier (order Isoptera). (e) Cinch bug (order Hemiptera). (f) Human body louse (order Anoplura). (g) Beetle (order Coleoptera). (h) Royal walnut butterfly and its caterpillar (order Lepidoptera). (i) Gall gnat (order Diptera): (j) Buffalo treehopper (order Homoptera). (k) Flea (order Siphonaptera). (l) Wasp (order Hymenoptera).

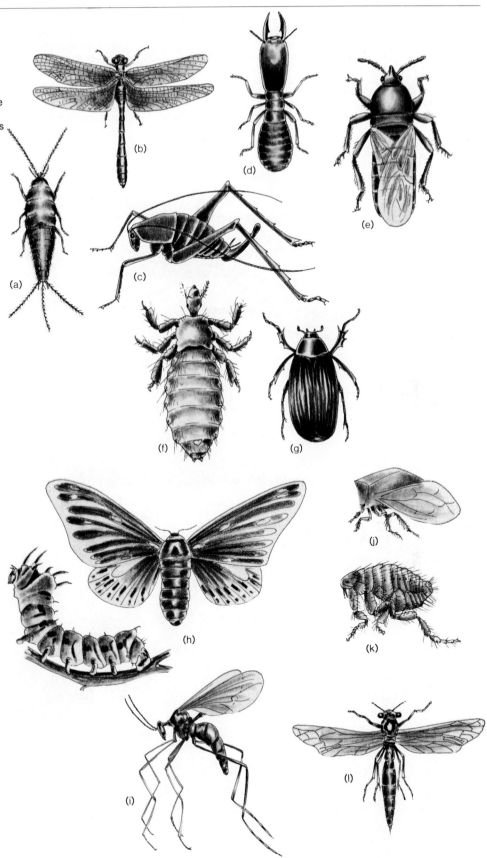

Figure 40.55 Insect structure. (a) A grasshopper. Note that the three thorax segments—prothorax, mesothorax, and metathorax—each have a pair of walking legs. (b) Reproductive structures of a female grasshopper. Sperm are received in the seminal receptacle during mating. The female uses her ovipositor to insert eggs into the ground, where embryos diapause during winter months. (c) Insect (grasshopper) mouthparts in place (*left*) and spread out (*right*). The labrum ("upper lip") does not correspond to mouthparts of arthropods in other classes. The insect labium is a fused structure corresponding to the second pair of maxillae found in other arthropods. (d) Scanning electron micrograph of the compound eye of the fruitfly *Drosophila*. (e) Structure of a compound eye (*above*) with many ommatidia, with details of a single ommatidium (*below*). The corneal lens and the cone focus light on the rhabdom, which is a cylinder consisting of light-sensitive microvilli of receptor cells. Pigment cells form a dark cylinder around the receptor cells that prevents light from passing from one ommatidium to others around it. Each ommatidium receives a separate image, and the insect's nervous system must process information received in multiple images.

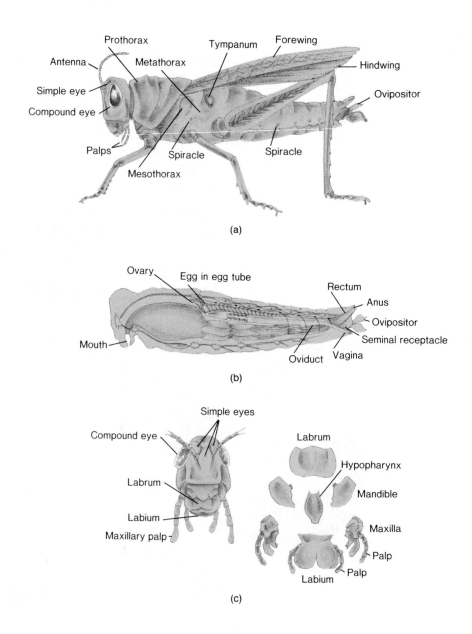

(a)

(b)

(c)

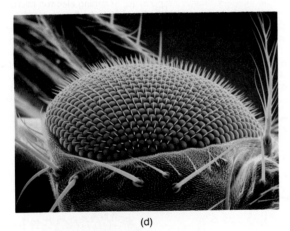

(d)

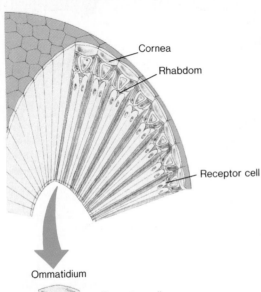

Cornea

Rhabdom

Receptor cell

Ommatidium

Corneal lens

Crystalline cone

Rhabdom

Receptor cell

Pigment cell

Pigment cell

(e)

of kinds of insect mouthpart specializations associated with different feeding habits.

The three segments of the thorax each bear a pair of walking legs, and the presence of six legs in adults is probably the most reliable criterion for identifying a given arthropod as an insect (figure 40.55). Many insects also have wings on the second and third thoracic segments and are excellent fliers.

There are no appendages on the insect abdomen. Many insect abdomens have prominent **spiracles,** the openings leading to the tracheal tubes that function in gas exchange. The abdomen also contains Malpighian tubules, the excretory devices that are attached to the posterior part of the digestive tract. Like other arthropods, insects have open circulatory systems.

Insect Reproduction and Development

Insects' specialized reproductive mechanisms are tremendously varied. Some insects lay eggs on plants that provide a ready food supply for their developing young. Others even use the bodies of different species of insects as "nurseries" for their young. They lay their eggs in or on the bodies of these unfortunate hosts, who eventually serve as a food source for the developing young after they hatch (figure 40.56).

Figure 40.56 A parasitized tobacco hornworm with another insect's eggs attached.

Figure 40.57 Insect development.
(a) Grasshopper development, an example of
incomplete metamorphosis. The nymph that
hatches generally resembles the adult.
Following each molt, the emerging nymph
more closely resembles the adult. (b) Moth
development, an example of complete
metamorphosis. The life history includes
several wormlike larval stages, a pupa, and an
adult. These complex developmental changes
are coordinated by hormonal regulation (see
p. 352).

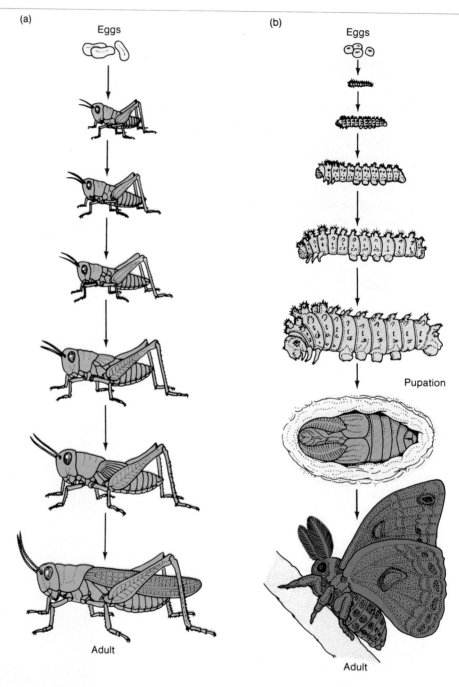

(a) Eggs

Adult

(b) Eggs

Pupation

Adult

Because it has a nonexpandable exoskeleton, an insect's development must involve a series of molts. However, insect development involves much more than simple growth in size; there is also **metamorphosis,** a marked change in body form that converts an immature individual into an adult.

There are two basic patterns of molting and metamorphosis in insects. In **incomplete (gradual) metamorphosis,** the young insect hatches as a **nymph** that generally resembles its parents (figure 40.57a). With each molt, the insect resembles its parents more closely, and it emerges from a final molt as a fully developed adult.

The other major category of insect development involves much more drastic and abrupt changes in form. The familiar developmental stages of butterflies and moths—**egg, larva, pupa,** and **adult (imago)**—illustrate **complete (abrupt) metamorphosis** (figure 40.57b). A wormlike larva hatches from the egg, grows, molts several times (the number of molts depends upon the species), and then **pupates,** producing a hard case around itself.

Dramatic developmental changes occur within the pupal case. Parts of the larval body are broken down and used as raw material for construction of new adult structures, which form by the growth of small clusters of cells called **imaginal discs** (so named because they contribute to the body of the imago, the adult insect). Inactive since early development, the imaginal discs become active during the pupal stage to produce antennae, wings, legs, reproductive organs, and other characteristically adult structures. Finally, the adult emerges as a reproductively mature individual.

Insect Success

The insect exoskeleton is a major factor in insects' evolutionary success, especially in terrestrial environments. The exoskeleton is virtually impermeable and thus provides a barrier to water loss, which is a continuing problem of terrestrial life.

But possibly the greatest factor in insects' overall evolutionary success is their host of marvelous adaptations to almost every imaginable habitat. Some insects are large, flying hunters or conspicuous plant eaters; others are microscopic species that live virtually unnoticed on the bodies of other animals, including humans.

Insects' entire life cycles are adapted to environmental conditions. Often, larvae and adults use different food sources and thus take advantage of seasonal changes in vegetation. Precisely timed periods of **diapause** permit insects to avoid the most adverse climatic conditions. In diapause, an insect has lowered metabolism and increased resistance to environmental stresses, such as the rigors of temperate zone winters. Some species diapause as an embryo within an eggshell (grasshoppers); some diapause as pupae (moths); and others diapause as adults (houseflies). Whatever the diapause form, onset of diapause is timed to precede adverse conditions, while its end coincides precisely with the return of favorable conditions for normal activities.

Some insect reproductive processes show remarkable timing adaptations with cycles of many years' duration. For example, some species of **cicadas** live underground for seventeen years as larvae. Then they pupate and emerge in vast swarms as adults that complete their mating and egg laying within a few days, leaving the next generation underground,

Figure 40.58 Adult seventeen-year cicada (*Magicicada septendecim*).

not to be seen until seventeen years later. Although many cicada adults are eaten by predators when they emerge, the mass emergence of so many individuals in such a short time "saturates" predation and assures adequate reproductive success to start the new generation (figure 40.58).

Insects also show remarkable behavioral adaptations. There are complex individual behavioral responses, such as the evasive maneuvers by which flying moths avoid hunting bats (p. 822). But possibly the social behavioral adaptations of bees, ants, termites, and other colonial insects are even more impressive. Castes of individuals are specialized to perform specific functions in the colony. Both developmental and behavioral differences are based on chemical communication systems in which substances (pheromones) produced by some individuals affect other individuals in the colony. Complex forms of behavioral communication also coordinate activities in the colony. For example, the "dance" of honeybees permits one bee to communicate distance and direction of a food source to other bees in the hive (see p. 817).

Figure 40.90 A mountain lion (order Carnivora) stalking its prey.

Table 40.3
Some Major Orders of Placental Mammals

Insectivora	Primitive, insect-eating mammals. Moles, shrews, and hedgehogs.
Chiroptera	Flying mammals with a broad skin flap extending from elongated fingers to the body and legs. Bats.
Primates	Omnivorous mammals with opposable thumb and fingers, eyes directed forward, well-developed cerebral cortex. Lemurs, monkeys, apes, and humans.
Edentata	Mammals with few or no teeth. Sloths, anteaters, and armadillos.
Rodentia	Mammals with sharp chisellike incisor teeth that grow continuously. Squirrels, beavers, rats, mice, voles, porcupines, hamsters, chinchillas, and guinea pigs.
Lagomorpha	Mammals with chewing teeth, tails reduced or absent, hind limbs longer than forelimbs. Hares, rabbits, and pikas.
Proboscidea	Mammals with long muscular trunk (proboscis); thick, loose skin; incisors elongated as tusks. Elephants.
Cetacea	Marine mammals with fish-shaped bodies, finlike forelimbs, no hind limbs, body insultated with a thick layer of fat (blubber). Whales, dolphins, and porpoises.
Perissodactyla	Herbivorous, odd-toed hoofed mammals. Horses, zebras, tapirs, and rhinoceroses.
Artiodactyla	Herbivorous, even-toed hoofed mammals. Cattle, sheep, pigs, giraffes, deer, antelopes, gazelles, hippopotamuses, camels, bison, and llamas.
Carnivora	Carnivorous mammals with sharp, pointed canine teeth and shearing molars. Cats, dogs, foxes, wolves, hyenas, bears, otters, minks, weasels, skunks, badgers, seals, walruses, and sea lions.

In natural ecosystems, the ungulates are prey to the largest members of the **order Carnivora** (figure 40.90), which includes the great cats—lions, tigers, cheetahs, and leopards. In terms of physical characteristics, these large, fast, ferocious hunters may represent the pinnacle of evolution of placental mammals, but none of them is the single dominant species on earth today.

That distinction is reserved for the human species. Humans are members of the **order Primates.** We humans are not the fastest or the strongest of animals. Our sense organs are not superior to those of all other animals. But we are dominant despite our physical limitations because human evolution has brought one key characteristic to the forefront: our brains are superior to those of other animals. We alone are able to contemplate our existence and the processes that have shaped the living world. Unfortunately, the power generated by our superior brains also has led to alterations of the natural world, many of which threaten our own welfare and that of many other species. It now seems that, for better or worse, the quality of future life on earth will be determined by human activities in the coming years.

Summary

This chapter examines and compares the major animal phyla and the evolutionary relationships among the groups.

Simply organized animals are restricted to relatively stable, aquatic environments or sheltered environments inside other organisms (internal parasites).

Terrestrial animals have more complex body organizations, including impermeable body surfaces that separate a stable internal body environment from the changeable external environment. Exchanges between inside and outside take place only in restricted body areas.

All animals are heterotrophs, but animals display a variety of nutritional specializations. Aquatic animals range from sessile filter feeders to active predators. Terrestrial animals include herbivores and the carnivores that feed on them. All of these animal nutritional strategies are associated with specific structural and functional specializations of the animals' bodies.

Reproductive patterns also are adapted to environmental conditions. This is particulary true of terrestrial animals. Well-adapted terrestrial animals have internal fertilization, which protects gametes from desiccation, and they either lay eggs enclosed in protective shells (for example, insects, arachnids, reptiles, birds) or have reproductive adaptations that permit early development of their young to occur inside the female body (placental mammals).

Questions for Review

1. What are the three basic body layers of animals?
2. Define the term invertebrate.
3. What is extracellular digestion?
4. Explain why, in the United States, the danger of tapeworm infections is significant while the danger of fluke infections is very small.
5. Distinguish between incomplete and complete digestive tracts.
6. Describe some advantages and disadvantages of the arthropod exoskeleton as a body surface covering for terrestrial animals.
7. Name the two major animal phyla that are deuterostomes.
8. List and explain the three primary diagnostic characteristics of the chordates.
9. What characteristics of amphibians limit their capacity to occupy large portions of the terrestrial environment?
10. What reproductive adaptations of reptiles, birds, and mammals contribute to their success as terrestrial animals?

Questions for Analysis and Discussion

1. Many animals that live attached to a substrate or move about very slowly are radially symmetrical. Animals that move about very actively are bilaterally symmetrical. How would you explain this difference?
2. What is the adaptive significance of the very large reproductive potentials of virtually all parasitic animals whose life histories involve several different host animals?
3. Homeothermic animals must obtain more food than poikilothermic animals of comparable size because homeotherms expend considerable energy for heat production. This would seem to be a disadvantage of being homeothermic. Discuss some advantages of being homeothermic.

Suggested Readings

Books

Attenborough, D. 1979. *Life on earth.* Boston: Little, Brown.

Attenborough, D. 1984. *The living planet.* Boston: Little, Brown.

Barnes, R. D. 1980. *Invertebrate zoology.* 4th ed. Philadelphia: Saunders.

Bliss, D. E. 1982. *Shrimps, lobsters, and crabs: Their fascinating life story.* Piscataway, N.J.: New Century Publishers.

Glaessner, M. F. 1984. *The dawn of animal life.* New York: Cambridge University Press.

Hill, J. E., and Smith, J. D. 1984. *Bats: A natural history.* Austin, Texas: University of Texas Press.

Villee, C. A.; Walker, W. F.; and Barnes, R. D. 1984. *General zoology.* 6th ed. Philadelphia: Saunders.

Welty, J. C. 1982. *The life of birds.* 3d ed. Philadelphia: Saunders.

Articles

Alvarez, L. W.; Alvarez, W.; Asora, R.; and Michel, H. V. 1980. Extraterrestrial cause of the Cretaceous-Tertiary extinction. *Science* 208:1095.

Calder, W. A., III. July 1978. The kiwi. *Scientific American.*

Ghiselin, M. T. September 1985. A movable feaster. *Natural History (Peripatus).*

Gilbert, P. W. 1984. Biology and behaviour of sharks. *Endeavour* 8:179.

Goreau, T. F.; Goreau, N. I.; and Goreau, T. J. August 1979. Corals and coral reefs. *Scientific American.*

Gosline, J. M., and DeMont, M. E. January 1985. Jet-propelled swimming in squids. *Scientific American.*

Gould, S. J. September 1986. The archaeopteryx flap. *Natural History.*

Greany, P. D.; Vinson, S. B.; and Lewis, W. J. 1984. Insect parasitoids: Finding new opportunities for biological control. *BioScience* 34:690.

Kolata, G. 1985. Avoiding the schistosome's tricks. *Science* 227:285.

Langston, W., Jr. February 1981. Pterosaurs. *Scientific American.*

Laws, R. 1 September 1983. Antarctica: A convergence of life. *New Scientist.*

Macurda, D. B., Jr., and Meyer, D. L. 1983. Sea lilies and feather stars. *American Scientist* 71:354.

Marx, J. L. 1978. Warm-blooded dinosaurs: Evidence pro and con. *Science* 199:1424.

McLaren, D. 24 November 1983. Impacts that changed the course of evolution. *New Scientist.*

McWhinnie, M. A., and Denys, C. J. March 1980. The high importance of the lowly krill. *Natural History.*

Roper, C. F. E., and Boss, K. J. April 1982. The giant squid. *Scientific American.*

Weinberg, D. February 1986. Decline and fall of the black-footed ferret. *Natural History.*

Würsig, B. March 1979. Dolphins. *Scientific American.*

Appendix 1
Classification of Organisms

This appendix summarizes the major taxonomic groups discussed in the survey of organisms in chapters 37 through 40 and lists some examples of members of many of the groups. The classification system we use here is only one of several systems currently in use by biologists and should not be considered the only possible classification of living things. There is, for example, considerable disagreement among biologists concerning the status of the slime molds and the assignment of various groups of algae to the protist and plant kingdoms.

Botanists use the term "division" for major groups of plants, while zoologists use the term "phylum" for the major groups of animals. We have used these terms in the same way. We also have used the term "division" in connection with groups of protists that formerly were included in the plant kingdom under the old two-kingdom classification system.

Kingdom Monera (Prokaryotae): prokaryotes—bacteria and cyanobacteria (blue-green algae)

[There is no "official" classification of the prokaryotes, but one highly-regarded scheme for classifying bacteria is that proposed by Gibbons and Murray and adopted in *Bergey's Manual of Systematic Bacteriology* (9th edition), the standard reference used in bacterial identification. This scheme includes four divisions: *Division Gracilicutes* (prokaryotes with a complex, Gram-negative type of cell wall); *Division Firmicutes* (prokaryotes that have a Gram-positive cell wall profile and generally react positively with the Gram stain); *Division Tenericutes* (prokaryotes such as mycoplasmas that lack a cell wall and do not synthesize precursors of peptidoglycan); and *Division Mendosicutes* (archaebacteria, prokaryotes that give evidence of quite distant phylogenetic relationships with other prokaryotic organisms because of differences in cell wall construction, protein synthesis mechanisms, and other characteristics).]

Kingdom Protista: eukaryotes, unicellular or colonies without tissue differentiation

[Protozoa—Heterotrophic, Unicellular Protists]

Phylum Mastigophora: flagellated protozoa [*Trichonympha*, *Trypanosoma*]

Phylum Sarcodina: pseudopodial protozoa [*Amoeba*, *Entamoeba*, *Arcella*, foraminiferans, radiolarians]

Phylum Sporozoa: spore-forming protozoa [*Toxoplasma*, *Plasmodium*]

Phylum Ciliophora: ciliated protozoa [*Euplotes*, *Paramecium*, *Stentor*, *Vorticella*, *Didinium*]

Chapter 34

Figure 34.16c: Alice B. Thiede. **Figures 34.5 and 34.6** From Knut Schmidt-Nielsen, *Animal Physiology, 3d ed.* p. 51. © 1970 Prentice-Hall, Inc., Englewood Cliffs, NJ. Adapted by permission. **Figure 34.12** After C. Kabat and D. R. Thompson, 1963, "Wisconsin Quail 1834–1962: Population Dynamics and Habitat Management" in *Technical Bulletin No. 30*, Madison, Wisconsin. Used with permission. **Figure 34.13** Reprinted from "The Rise and Fall of a Reindeer Herd" by V. C. Scheffer, in *Scientific Monthly 73* (December 1951):356–62. Copyright 1951 by the American Association for the Advancement of Science. **Figure 34.15** From: *Population Growth and Land Use* by Colin Clark. © 1967, 1977 by Colin Clark, and reprinted by permission of St. Martin's Press, Inc., and Macmillan, London and Basingstoke. **Figure 34.16a & b** From The Ecology of Man, 2nd Edition, by Robert Leo Smith. Copyright © 1972, 1976 by Robert Leo Smith. Reprinted by permission of Harper & Row Publishers, Inc. **Figure 34.16c** Reprinted by permission from "Life as a Wonderful One-hoss Shay" in *Nature*, 26 August 1982, p. 779. Copyright © 1979 Macmillan Journals, Ltd., London, England. **Figure 34.17** Reprinted with permission of Hafner Press from *The Struggle for Existence* by G. F. Gause. Copyright © 1934. **Figure 34.18** From "Physiological Ecology of Three Codominant Successional Annuals", by N. K. Wieland and F. A. Bazzaz, in *Ecology*, 1975, *56*, 681–688. Copyright © 1975 by Ecological Society of America. Reprinted by permission. **Figure 34.19** Reprinted from "Population Ecology of Some Warblers in Northeastern Coniferous Forests" by R. H. MacArthur, in *Ecology 39*, 599–619. **Figure 34.22** Used by permission of the Division of Agricultural Sciences, University of California.

Chapter 35

Figures 35.1 and 35.15: Susan Strawn, M.S. **Figures 31.9 and 31.10:** Alice B. Thiede. **Figure 35.13** Reprinted with permission of Macmillan Publishing Company from *Communities and Ecosystems* by Robert H. Whittaker. Copyright © 1975 by Robert H. Whittaker. **Figure 35.16** From *Plant Physiology, Second Edition*, by Frank Salisbury and Cleon Ross. © 1978 by Wadsworth Publishing Company, Inc. Reprinted by permission of the publisher. **Figure 35.17** From *The Ecology of Man, 2nd Edition*, by Robert Leo Smith. Copyright © 1972, 1976 by Robert Leo Smith. Reprinted by permission of Harper & Row Publishers, Inc.

Chapter 36

Figures 36.7 and 36.11: Alice B. Thiede. **Figure 36.13:** Susan Strawn, M.S.

Chapter 37

Figure 37.12: Alice B. Thiede. **Figure 37.2** From Whittaker, R. H., "New Concepts of Kingdoms of Organisms" in *Science 163*, 10 January 1969:150–60. © 1969 American Association for the Advancement of Science. Reprinted by permission. **Figure 37.3a & b** From Freeman, *Burrows Textbook of Microbiology, 21st ed.* © 1979 W. B. Saunders Company, Philadelphia, PA. Reprinted by permission. **Figure 37.6** From: *Microbiology, 2/e*, by Eugene W. Nester, C. Evans Roberts, Nancy N. Pearsall & Brian J. McCarthy. Copyright © 1973, 1978 by Holt, Rinehart & Winston. Reprinted by permission of CBS College,

Publishing. **Figure 37.13b** From Cohan-Bazire, G., and J. London, "Organelles of Bacterial Flagella" in Journal of *Bacteriology 94*:458. © 1967 American Society for Bacteriology, Washington, D.C. Reprinted by permission. **Figure 37.20** From Stern, Kingsley R., *Introductory Plant Biology, 3d ed.* © 1979, 1982, 1985 Wm. C. Brown Publishers, Dubuque, Iowa. All Rights Reserved. Reprinted by permission.

Chapter 38

Figures 38.19b and 38.22: Susan Strawn, M.S. **Figures 38.1, 38.6, 38.8, and 38.10** From Vickerman, K. and F. E. G. Cox, *The Protozoa*. © 1967 Houghton Mifflin Company, Boston, MA. Reprinted by permission of the author. **Figure 38.2b & c** Reprinted with permission of Macmillan Publishing Company from *Microbiology: An Introduction to Protista* by Jeanne Stove Poindexter. Copyright © 1971 by Jeanne Stove Poindexter. **Figures 38.3 and 38.7** Reprinted with permission of Macmillan Publishing Company from *The Invertebrates: Function and Form* by Irwin W. Sherman and Vilia G. Sherman. Copyright © 1970 by Macmillan Publishing Company. **Figures 38.15 and 38.19b & c** Reprinted with permission of Macmillan Publishing Company from *Algae and Fungi* by Constantine Alexopoulos and Harold C. Bold. Copyright © 1967 by Macmillan Publishing Company. **Figure 38.17** From Stern, Kingsley R., *Introductory Plant Biology, 3d ed.* © 1979, 1982, 1985 Wm. C. Brown Publishers, Dubuque, Iowa. All Rights Reserved. Reprinted by permission.

Chapter 39

Figures 39.3, 39.4a and b, 39.5, 39.7, 39.8a, 39.12, 39.14, 39.19a, 39.23, and 39.27: Susan Strawn, M.S. **Figure 39.13a** Adapted from *Botany, 5th ed.*, by Carl L. Wilson, Walter E. Loomis, Taylor A. Steeves. Copyright © 1952, 1957, 1962, 1967, 1971 by Holt, Rinehart & Winston, Inc. Used by permission of CBS College Publishing. **Figure 39.27** From: *Botany, 3/e*, by Carl L. Wilson & Walter E. Loomis. Copyright © 1952, 1957, 1962 by Holt, Rinehart & Winston, Inc. Reprinted by permission of CBS College Publishing.

Chapter 40

Figures 40.9 and 40.12: Susan Strawn, M.S. **Figures 40.35a and b, 40.46b, 40.57, and 40.61a:** Martha Blake. **Figure 40.18** Redrawn after World Health Organization, 1959. **Figure 40.20** Used with permission of Kenneth G. Rainis, Ward's Natural Science Establishment, Inc. **Figure 40.22a** Reprinted with permission of Macmillan Publishing Company from *The Invertebrates: Function & Form* by Irwin W. Sherman and Vilia G. Sherman. Copyright © 1970 by Macmillan Publishing Company. **Figures 40.36, 40.38b, and 40.51** Reprinted with permission of Macmillan Publishing Company from *A Life of Invertebrates* by Dr. W. D. Russell-Hunter. Copyright © 1979 by Dr. W. D. Russell-Hunter. **Figure 40.77** From T. W. Torrey and A. Feduccia, *Morphogenesis of the Vertebrates, 4th ed.* © 1979 John Wiley & Sons, Inc. Reprinted by permission. **Figure 40.80b** From *An Introduction to Embryology, 5/e*, by B. I. Balinsky. Copyright © 1981 by CBS College Publishing. Copyright © 1954, 1959, 1965, 1970, and 1975 by W. B. Saunders Company. Reprinted by permission of CBS College Publishing.

Photos

Part Openers

Part One: © E. R. Degginger; **Part Two:** © Leland Johnson; **Part Three:** © R. Hamilton Smith; **Part Four:** © Andrew Bajer; **Part Five:** © James R. Fisher; **Part Six:** © Carl Purcell; **Part Seven:** © Bob Coyle.

Chapter 1

1.1: French Government Tourist Office; **1.2:** Oriental Institute Museum, University of Chicago; **1.4:** © Gerald Corsi/Tom Stack & Associates; **1.5:** Richard J. Feldman/National Institutes of Health; **1.6a:** © Rebecca L. Johnson; **1.6b:** © Pam Hickman/ Valan Photos; **1.10:** VU/© T. E. Adams; **1.11a:** © Bruce Russell/BioMedia Associates; **1.11b:** © Eric Gravé; **1.12:** © George Robbins; **1.13a:** © Bruce Russell/BioMedia Associates; **1.13b:** © S. J. Krasemann/Photo Researchers, Inc.; **1.13c:** © Carolina Biological Supply Company; **1.13d:** © L. West/Bruce Coleman, Inc.; **1.13e:** © Rebecca L. Johnson; **1.13f:** © M. P. L. Fogden/Bruce Coleman, Inc.; **1.14a:** © D. Woodward/Tom Stack & Associates; **1.14b:** © Peter Parks/Oxford Scientific Films; **1.14c:** © Edwin Reschke/Peter Arnold, Inc.; **1.14d:** © David Spier/Tom Stack & Associates; **1.14e:** © Patrice/Tom Stack & Associates; **1.14f:** © Dwight Kuhn; **1.14g:** © John Lidington/Photo Researchers, Inc.; **1.14h:** © Andrew Odum/Peter Arnold, Inc.; **1.14i:** © Gregory G. Dimijian/Photo Researchers, Inc.; **1.15:** © Ed Pacheco.

Chapter 2

2.4: © Manfred Kage/Peter Arnold, Inc.; **2.11:** © John Bova/Photo Researchers, Inc.

Chapter 3

3.1: © John D. Cunningham; **3.5c:** © S. E. Frederick/ courtesy of E. H. Newcomb, University of Wisconsin; **3.6c:** left, © Harold V. Green/Valan Photos; top right, © J. D. Cunningham; bottom right, © Ian Spellerberg; **Box 3.1a:** © Carl Purcell; **Box 3.1b:** © S. J. Krasemann/Valan Photos; **3.12b:** from Stryer, Lubert: BIOCHEMISTRY, 2nd ed. © 1981 W. H. Freeman and Company; **3.14a:** © Jerome Gross; **3.15c:** Richard J. Feldman/National Institutes of Health; **3.15d: both,** Bill Longcore/Photo Researchers, Inc.; **3.16a:** Regents of the University of California, San Francisco; **3.20:** © Gordon F. Leedale/BioPhoto Associates; **3.21b:** © Vincent Marchesi; **3.24a:** © A. M. Glauert and G. M. W. Cook.

Chapter 4

4.1a: Armed Forces Institute of Pathology; **4.1b:** National Library of Medicine; **4.2a:** Historical Pictures Service, Chicago; **4.7:** © J. Herbert Taylor; **4.11:** © Donald Fawcett; **4.12:** © Paul Heidger; **4.14a:** © Keith Porter; **4.14b:** © Barry King, University of California School of Medicine/BPS; **4.15:** © Gordon F. Leedale/BioPhoto Associates; **4.17:** © James Burbach; **4.19:** © Keith Porter; **4.20:** © L. K. Shumway; **4.22:** © Jean-Paul Revel; **4.23a:** © Elias Lazarides; **4.24a:** © L. E. Roth, University of Tennessee, Y. Shigenaka, University of Hiroshima, D. J. Pihlaja, Howe Laboratory of Ophthalmology, Boston/BPS; **4.24b:** VU/© M. Schliwa; **4.25a:** © Keith Porter; **4.27a:** © Jan Löfberg, Institute of Zoology, Uppsala University, Sweden; **4.27b:** © J. André and E. Favret-Fremiet; **4.29a:** VU/© David Phillips; **4.29b:** © D. T. Woodrum and R. W. Linck; **4.32:** © Donald F. Lundgren.

Chapter 5

5.1: © Carl W. May/BPS; **5.5a–c:** © S. J. Singer; **5.11a–d:** M. M. Perry and A. B. Gilbert/*Journal of Cell Science* 39: 257–272, 1979; **5.12:** Fawcett, D. W.: THE CELL: ITS ORGANELLES AND INCLUSIONS, 2nd ed. © 1966 W. B. Saunders; **5.13:** © C. L. Sanders/BPS; **5.16a:** © Leland Johnson; **5.16b:** © E. R. Degginger; **5.17a:** © Carolina Biological Supply Company; **5.17b:** © Nancy Sefton/Photo Researchers, Inc.; **Box 5.2a:** © B. F. King, University of California School of Medicine/BPS; **5.18a:** © Dwight Kuhn; **5.18b:** © Edwin Reschke; **5.18c:** © B. F. King, University of California School of Medicine/BPS; **5.19a–b, 5.20, 5.21a–b, 5.22a–b:** © Edwin Reschke; **5.22c, 5.23:** © Manfred Kage/Peter Arnold, Inc.; **5.24a:** © Dwight Kuhn; **5.24b:** VU/© Fred Hossler; **5.25:** © Harold V. Green/Valan Photos; **5.26, 5.27a:** © Dwight Kuhn; **5.27b:** VU/© R. Moore; **5.28:** © John D. Cunningham.

Chapter 6

6.1a: Mark M. Littler: SCIENCE, Vol. 227, p. 58, fig. 1B, 4 Jan. 1985/© AAAS; **6.1b:** Harbor Branch Foundation, Fort Pierce, FL; **6.2:** © Ray Elliott; **Box 6.1:** © Thomas Eisner; **Box 6.3:** © Joe Munroe/Photo Researchers, Inc.

Chapter 7

7.1: © Stephen J. Krasemann/Valan Photos; **7.4:** © Bob Coyle; **7.6b:** © John Troughton and Leslie Donaldson; **7.7a–b:** © R. Hedrich and W. Pagel; **7.8:** from Salisbury, F. W. and Ross, Cleon: PLANT PHYSIOLOGY, 2nd ed. © 1978 Wadsworth Publishing Company, Inc. Reprinted by permission; **7.9b:** © L. K. Shumway; **7.9c:** © Gordon F. Leedale/BioPhoto Associates; **7.14:** © Rebecca L. Johnson; **Box 7.1a:** © Leland Johnson; **Box 7.1b:** © Dwight Kuhn; **7.24:** © Melvin Calvin; **7.27:** © Doug Wechsler; **7.28a: both,** © Nancy Vander Sluis; **7.28b:** © Dwight Kuhn; **7.30b:** © George H. H. Huey.

Chapter 8

8.1: © Dwight Kuhn; **8.7:** © Barbara Pfeffer/Peter Arnold, Inc.; **8.10:** © Keith Porter; **8.19a:** © H. Fernandez-Morán; **8.25:** © Thomas Kitchin/Valan Photos.

Chapter 9

Box 9.1a: VU/© Randy Moore; **Box 9.1b:** © Dwight Kuhn; **Box 9.1c:** © J. N. A. Lott, McMaster University/BPS; **9.2b:** © John Troughton and Leslie Donaldson; **9.4:** © Bob Coyle; **9.6:** © John Hendrickson; **9.7a–b:** © Walter Hodge/Peter Arnold, Inc.; **9.8:** © John D. Cunningham; **9.9:** Kimball, John W.: BIOLOGY, 2nd ed. © 1968 Addison-Wesley, Reading, MA; **9.11a:** © Carolina Biological Supply Company; **9.13a:** © Dwight Kuhn; **Box 9.2:** © Kjell B. Sanved; **9.20a–c:** © E. J. Hewitt; **9.21a:** © L. E. Roth, University of Tennessee/BPS; **9.22:** © Gordon F. Leedale/BioPhoto Associates; **9.23:** © Robert and Linda Mitchell.

Chapter 10

10.1: © Gregory K. Scott; **10.3:** © J. R. Waaland, University of Washington/BPS; **10.5a:** © Carolina Biological Supply Company; **10.5b:** © Dwight Kuhn; **10.6a:** © Carolina Biological Supply Company; **10.6c:** VU/© P. Manchester; **10.7a:** © Al Bussewitz; **10.9b:** © Thomas Eisner; **10.12:** © John Troughton and Leslie Donaldson; **10.13:** © John D. Cunningham; **Box 10.1:** © Walter Hodge/Peter Arnold; **10.15c:** © J. Cronshaw; **10.16a–b:** Martin Zimmerman/courtesy of Harvard University.

Chapter 11

11.4: Centers for Disease Control, Atlanta; **11.5a:** © Lester Bergman & Associates; **11.5b:** Niilo Hallman; **11.6a–b:** © Gregory Antipa; **11.10a:** © Tom Stack/Tom Stack & Associates; **11.10b:** © Tom McHugh/Photo Researchers, Inc.; **11.13a:** © John Shaw/Tom Stack & Associates; **11.13b:** © E. R. Degginger; **Box 11.1a:** © G. L. Barron, University of Guelph, Ontario; **Box 11.1b:** © Fred and Marian Nickerson; **11.15:** © Carolina Biological Supply Company; **11.17b:** Kessel and Kardon/© W. H. Freeman and Company; **11.19b:** © Omikron/Photo Researchers; **11.24b:** Kessel and Kardon/© W. H. Freeman and Company; **11.26:** © Martin Rotker/Taurus Photos.

Chapter 12

12.1: © Peter J. Bryant, University of California, Irvine/BPS; **12.11b:** © G. M. Hughes; **12.12c:** © Thomas Eisner; **12.13:** © Dwight Kuhn; **12.20c:** © Hans-Rainer Duncker; **12.22a–b:** Kessel and Kardon/© W. H. Freeman and Company; **12.26:** © Wolfgang Kaehler; **12.30a–b:** © Martin Rotker/Taurus Photos.

Chapter 13

13.1: David C. West et al., SCIENCE Vol. 228, fig. 1, p. 1325, 14 June 1985/© AAAS; **13.6b–c:** Kessel and Kardon/© W. H. Freeman and Company; **13.7:** Historical Pictures Service, Inc., Chicago; **13.11:** Carroll Weiss/Camera M. D. Studios; **13.17:** T. Kuwabara, from Bloom, W. and Fawcett, D. W.: A TEXTBOOK OF HISTOLOGY, 10th ed. © 1975 W. B. Saunders Company; **13.18a–c:** © D. W. Fawcett; **13.20:** © Edwin Reschke; **13.23:** © Jean-Paul Revel; **13.25:** Reproduced with permission from ''The Morphology of Human Blood Cells'' by L. W. Diggs, D. Sturm and A. Bell. © Abbott Laboratories; **13.26:** Eila Kairinen and Emil Bernstein, Gillette Research Institute, Rockville, MD, SCIENCE Vol. 173, 27 August 1971/© AAAS; **13.28:** VU/© D. M. Phillips; **13.29a:** Juhl, J. H.: PAUL AND JUHL'S ESSENTIALS OF ROENTGEN INTERPRETATION, 4th ed. © 1981 J. B. Lippincott Company; **13.29c** (1–3): American Heart Association.

Chapter 14

14.1: NASA; **Box 14.1:** © John Crowe; **14.8a:** © David Hughes/Bruce Coleman Ltd.; **14.8b:** © G. E. Schmida/Australasian Nature Transparencies; **14.9a:** © Eric Gravé/Photo Researchers; **14.9c: both,** © Thomas Eisner; **14.15:** © Dennis W. Schmidt/Valan Photos; **14.16:** © Bob Evans/Peter Arnold, Inc.; **14.21b:** © R. B. Wilson; **Box 14.2:** © Dennis W. Schmidt/Valan Photos.

Chapter 15

15.1a: Reprinted by permission from NATURE Vol. 304, July 21, 1983, p. 205. © MacMillan Journals Limited; **15.3:** © Bob Coyle; **15.10:** The Bettmann Archive, Inc.; **15.11a–d:** © Joseph T. Bagnara; **15.14b:** © Carolina Biological Supply Company; **15.15a–b:** © Dwight Kuhn; **15.17:** © Edwin Reschke; **15.18a:** Turner, C. D. and Bagnara, J. T.: GENERAL ENDOCRINOLOGY, 6th ed. © 1976 W. B. Saunders Company. Reprinted by permission of Holt, Rinehart & Winston; **15.30:** © Michael Borque/Valan Photos.

Chapter 16

16.1: Robert Hinckley painting, courtesy of Countway Library of Medicine, Boston; **16.4:** © Carolina Biological Supply Company; **16.9a:** © David Denning/BioMedia Associates; **16.16:** H. Webster, from THE VERTEBRATE PERIPHERAL NERVOUS SYSTEM, John Hubbard, editor. © 1974 Plenum Press; **16.19:** © E. R. Lewis, University of California, Berkeley/BPS; **16.22a:** © John Heuser; **Box 16.1:** Field Museum of Natural History, Chicago.

Chapter 17

17.6a: © Manfred Kage/Peter Arnold, Inc.; **17.11:** © Tom Stack/Tom Stack & Associates.

Chapter 18

18.1: © Jacques Jangoux/Peter Arnold, Inc.; **18.3:** © Edwin Reschke; **18.4: both,** Kessel and Kardon/© W. H. Freeman and Company; **18.5b:** VU/© Michael Gabridge; **18.7a–b:** Carl May, photographer/© Wm. C. Brown Company Publishers; **Box 18.1:** © Allen Ruid; **18.15:** © H. E. Huxley; **18.18a–b:** © Junzo Desaki, Ehime University School of Medicine, Shigenobu, Japan.

Chapter 19

19.1: © John Hendrickson; **19.2a:** © Frank B. Salisbury; **19.9a–b:** © Leland Johnson; **19.11a–d:** Nitsch, J. P.: ''Growth and Morphogenesis of the Strawberry as Related to Auxin,'' © *American Journal of Botany*, 37 (3) March 1950; **19.14:** © Folke Skoog; **19.16:** © Bernard Phinney; **19.17:** © Sylvan Wittwer; **19.20:** from Kendrick, R. E. and Frankland, B.: PHYTOCHROME AND PLANT GROWTH. © Edward Arnold, Ltd.; **19.22:** © Kennon Cooke/Valan Photos; **19.27:** Salisbury, F. B. and Ross, C.: PLANT PHYSIOLOGY, 2nd ed. © 1978 Wadsworth Publishing Company, Inc. reprinted by permission; **Box 19.2a: both,** © E. R. Degginger; **Box 19.2c:** © Alec Duncan/Taurus Photos; **19.32b:** © E. J. Cable/Tom Stack & Associates.

Chapter 20

20.4: E. J. Dupraw; **20.6a–f:** © Andrew Bajer; **20.7:** © Edwin Reschke; **20.9a:** © Robert E. Waterman; **20.10a–b:** © Edwin Reschke; **20.11:** W. P. Wergin and E. H. Newcomb, University of Wisconsin/BPS; **20.12:** © Peter J. Bryant, University of California, Irvine/BPS; **20.14:** © Marta S. Walters; **20.15:** D. von Wettstein; **20.16a:** © James Kezer; **20.17a–b, 20.18a–c:** © Herbert Stern.

Chapter 21

21.1a: © Mary Eleanor Browning/Photo Researchers, Inc.; **21.1b:** courtesy of Eileen Hudak, American Belgian Tervuren Club; **21.2:** American Museum of Natural History; **21.20:** Library of Congress Collection.

Chapter 22

22.1b: VU/© K. G. Murti; **22.1a:** Olins, Donald and Ada: "The Structural Quantum in Chromosomes," AMERICAN SCIENTIST 66(Nov. 1978); **22.3:** courtesy of W. Atlee Burpee Company; **22.4b:** © James Shaffer; **22.6:** Mittwoch, Ursula: SEX CHROMOSOMES. © 1967 Academic Press; **22.7:** Margery Shaw, from Wisniewski, L. P. and Hirschhorn, K.: "A Guide to Human Chromosome Defects," 2nd ed., White Plains: The March of Dimes Birth Defects Foundation, BD: OAS 16(6), 1980; **22.8:** © Digamber S. Borgaonkar; **22.14: all,** © S. D. Sigamani, from *American Journal of Human Genetics* 16(4) 1964, p. 467; **22.15, 22.16:** © Bob Coyle **Box 22.2:** National Jewish Hospital.

Chapter 23

23.1: Eli Lilly Company; **23.6b:** © M. H. F. Wilkins; **23.7c:** from Watson, J. D.: THE DOUBLE HELIX. Atheneum, NY. © 1968 J. D. Watson/A. C. Barrington Brown, photographer; **23.11:** © Ross Inman, University of Wisconsin.

Chapter 24

24.1: © David Micklos/Cold Spring Harbor Laboratory, Research Library Archives; **24.11: both,** © David Greene/Cold Spring Harbor Laboratory, Research Library Archives; **24.14b:** © Lester Bergman and Associates.

Chapter 25

25.4: © Charles C. Brinton, Jr. and Judith Carnahan; **25.11a–c:** Huntington Potter and David Dressler/ LIFE Magazine, © 1980 Time, Inc.; **25.12a:** © Larry W. Moore; **25.13:** Ralph Brinster/NATURE (16 December 1982).

Chapter 26

26.1: © Dan Suzio; **26.3a:** © Jeff Rotman/Peter Arnold, Inc.; **26.3b:** © Robert Ross; **26.5a:** © E. S. Ross; **26.5b:** © Bob Coyle; **Box 26.1:** © M. J. Tyler, University of Adelaide; **26.9:** © Robert Ross; **26.10a:** © William Byrd; **26.10b:** © Gerald Schatten; **26.10c–d:** Tegner, M. J. and Epel, D.: SCIENCE Vol. 179, pp. 685–688, 16 February 1973/© AAAS; **26.12a–b, 26.14a–c:** © Leland Johnson; **26.15a–b:** © Victor Vacquier; **26.15c:** © Leland Johnson; **26.17:** © Lynwood Chase/Photo Researchers, Inc.; **26.18a–c:** © Robert Waterman; **26.19c–d, 26.21a(1–2):** Kessel and Shih: SCANNING ELECTRON MICROSCOPY IN BIOLOGY/© Springer-Verlag, Berlin; **26.21b:** © Kathryn Tosney; **26.26:** © S. Murphree, Middle Tennesee State University/ BPS; **26.28a–c:** © William H. Allen.

Chapter 27

27.9a: © Landrum Shettles; **27.9b:** Carnegie Institution of Washington, Department of Embryology, Davis, CA, Division; **27.9c:** © Landrum Shettles; **27.11a–c:** Carnegie Institution of Washington, Department of Embryology, Davis, CA, Division; **27.16b:** © Landrum Shettles; **27.17:** Carnegie Institution of Washington, Department of Embryology, Davis, CA, Division; **27.19:** © Gregory K. Scott; **Box 27.1:** John Messineo/© AAAS; **27.21a:** © Holly Williamson/Taurus Photos; **27.21b:** Atlanta Centers for Disease Control.

Chapter 28

28.1a–d: © Eberhard Schierenberg; **Box 28.1a:** © Clement L. Markert; **Box 28.1b:** © J. M. Willadsen; **28.10b–c:** © Kenneth B. Raper; **28.10d:** © David Francis; **28.10e:** © David Scharf/Peter Arnold, Inc.; **28.13c:** Johnson, L. G. and Volpe, E. P.: PATTERNS AND EXPERIMENTS IN DEVELOPMENTAL BIOLOGY. © 1973 Wm. C. Brown Co. Publishers; **28.14c:** © Kathryn Tosney; **28.17a(4):** Johnson, L. G. and Volpe, E. P.: PATTERNS AND EXPERIMENTS IN DEVELOPMENTAL BIOLOGY. © 1973 Wm. C. Brown Co. Publishers; **28.18b:** Kathryn Tosney, from N. K. Wessels: TISSUE INTERACTIONS AND DEVELOPMENT. © 1977 Benjamin-Cummings Publishing Company; **28.18c: right,** © J. C. Smith; **28.19a:** © Kathryn Tosney; **28.22:** © Peter J. Bryant, University of California, Irvine/BPS; **28.23b:** © Bo Lambert.

Chapter 29

29.5b: © R. Dourmashkin; **29.9a–c:** © Dorthea Zucker-Franklin, from THE ATLAS OF BLOOD CELLS: FUNCTION AND PATHOLOGY, Vol. 2, Lea & Fibiger, 1981; **29.13b:** Richard J. Feldman, National Institutes of Health; **29.18a:** VU/© T. E. Adams; **29.19:** © Richard J. Goss; **29.22a–b:** © Beth Johnson; **29.23:** R. W. Briggs; **29.29:** © Bob Coyle.

Chapter 30

30.1: © John Moss/Black Star; **30.3a:** © Jerg Kroener; **30.3b:** © Christopher Crowley/Tom Stack & Associates; **30.3c:** © Leonard Lee Rue/Tom Stack & Associates; **30.4c: both,** © Rudolph Freund; **30.9a–b:** © E. R. Degginger; **30.10:** © E. S. Ross; **30.12b:** American Museum of Natural History; **30.14a:** © Don and Pat Valenti; **30.14b:** © Ray Elliott; **30.16a:** © William Ferguson; **30.16b:** © Paul F. Hoffman, Geological Survey of Canada.

Chapter 31

31.3: © Carl Purcell; **31.4:** © BioMedia Associates; **31.7a–b:** © J. A. Bishop and L. M. Cook; **31.9a:** © E. S. Ross; **31.9b:** © Chesher/Photo Researchers, Inc.; **31.10:** © J. A. L. Cooke/Oxford Scientific Films; **31.11:** © Hans Pfletschinger/Peter Arnold, Inc.; **31.12a:** © Perry D. Slocum/Animals, Animals; **31.12b:** © Kenneth Paige; **31.13:** © E. R. Degginger.

Chapter 32

32.1: Bob Anderson, photographer, from "Animal Oddities," © Hubbard Scientific, Northbrook, IL; **32.2a:** © W. H. Müller/Peter Arnold, Inc.; **32.2b:** © Alan Nelson/Root Resources; **32.4a:** © L. E. Gilbert, University of Texas, Austin/BPS; **32.4b:** © Tom McHugh/Photo Researchers, Inc.; **32.5:** © Russ Kinne/Photo Researchers, Inc.; **32.6:** © E. S. Ross; **32.14, 32.15:** © Robert Campbell; **32.17a–b:** American Museum of Natural History; **32.18:** Musee National de Prehistoir des Eyzies, Dordogne/P. Jugie.

Chapter 33

33.1: © Gordon Langsbury/Bruce Coleman, Inc.; **33.2:** © Susan McCartney/Photo Researchers, Inc.; **33.8:** Nine Leen/LIFE magazine, © Time, Inc. **Box 33.1:** © E. R. Degginger; **33.12:** © E. S. Ross; **33.16a–c:** © Lee Miller; **33.21b:** © J. Markham/Bruce Coleman, Inc.; **33.22:** © Michael Borque/Valan Photos; **33.23:** © Tom Stack/Tom Stack & Associates; **33.24:** © T. W. Ransom/BPS; **33.25:** © Anthony Merciera/Photo Researchers, Inc.; **33.26:** © Warren Garst/Tom Stack & Associates.

Chapter 34

34.1: © Larry Brock/Tom Stack & Associates; **34.4, 34.7:** © Robert and Linda Mitchell; **34.8:** © Carl Purcell; **34.20:** © R. Hamilton Smith; **34.23:** © E. R. Degginger; **34.24:** © Ron West; **34.25a:** © Harold V. Green/Valan Photos; **34.25b:** © Dwight Kuhn; **34.26:** © Robert and Linda Mitchell; **34.27:** John Marzluff.

Chapter 35

35.2: © Robert and Linda Mitchell; **35.4a:** © Tom Stack/Tom Stack and Associates; **35.4b:** © Thomas Kitchin/Tom Stack & Associates; **35.5:** © Ray Elliott; **35.6:** © Keith Gunnar/Tom Stack & Associates; **35.7:** © Rick McIntyre/Tom Stack & Associates; **35.8:** © Don and Pat Valenti/Tom Stack & Associates; **Box 35.1b:** © Jack Donnelly, Woods Hole Oceanographic Institution.

Chapter 36

36.1: © Jeff Rotman/Peter Arnold, Inc.; **36.2:** © M. T. Cook/Tom Stack & Associates; **36.3:** © E. R. Degginger; **36.5a:** © J. N. A. Lott, McMaster University/BPS; **36.5b:** © Gary K. Thompson/Tom Stack & Associates; **36.6:** © R. Hamilton Smith; **36.8:** © Tom Walker/Photo Researchers, Inc.; **36.9:** © Douglas Faulkner, Sally Faulkner Collection; **36.10:** © Claudia Mills, University of Washington/BPS; **36.12:** © Keith Gillett/Tom Stack & Associates; **36.14:** © Rod Allin/Tom Stack & Associates; **36.15:** © Bob and Miriam Francis/Tom Stack & Associates; **36.16a:** © E. R. Degginger; **36.16b:** © Dwight Kuhn; **36.17:** © Anne P. Layman/Tom Stack & Associates; **36.18:** © William Ferguson; **36.19a:** © T. M. Smith/EWPIC, Inc.; **36.19b:** © Robert and Linda Mitchell; **36.20:** © Perry Conway/Tom Stack & Associates; **36.21:** © Budd Titlow/Tom Stack & Associates; **36.22:** © Stewart M. Green/Tom Stack & Associates; **36.23:** © Steve Peterson/Tom Stack & Associates; **36.24:** Tim McCabe/USDA Soil Conservation Service.

Chapter 37

37.3c: © Keiichi Namba and Donald L. D. Caspar, Brandeis University, and Gerald Stubbs, Vanderbilt University; **37.4a–c:** © Gordon F. Leedale/BioPhoto Associates; **37.9a–b:** VU/© David M. Phillips; **37.9c:** © Paul W. Johnson and John McN. Sieburth, University of Rhode Island/BPS; **37.10a: both,** Jeanne Poindexter; **37.10b:** © P. L. Grilione and J. Pangborn; **37.11:** © Donald F. Lundgren; **37.13a:** © G. Cohen-Bazire; **37.14:** © T. J. Beveridge, University of Guelph/BPS; **37.15:** Centers for Disease Control, Atlanta, GA/BPS; **37.16:** VU/© Bill Beatty; **37.17:** © Michael Madigan; **37.18:** VU/© Michael Gabridge; **37.19:** © Norma J. Lang/*Journal of Phycology* 1: 127–134, 1965; **37.21:** © J. Robert Waaland, University of Washington/BPS.

Chapter 38

38.1: © Steven Brentano; **38.2a:** © Dwight Kuhn; **38.4a–b:** © J. W. Murray; **38.5:** Kessel and Shih: SCANNING ELECTRON MICROSCOPY IN BIOLOGY/© Springer-Verlag, Berlin; **38.9a–b:** © Bruce Russell/BioMedia Associates; **38.11b:** © Gordon F. Leedale/BioPhoto Associates; **38.12:** © Eric Gravé; **38.14:** © E.S. Ross; **38.16a:** © J. Robert Waaland, University of Washington/BPS; **38.16b:** © C. R. Wittenbach, University of Kansas/BPS; **38.18:** © Carolina Biological Supply Company; **38.19a:** © E. R. Degginger; **38.20:** © 1982 Donald Breneman/Photographic Specialties; **38.21a–c:** © Fred and Marian Nickerson; **38.23a–c:** © E. S. Ross.

Chapter 39

39.2, 39.4c: © Dwight Kuhn; **39.10a–b:** © Carolina Biological Supply Company; **39.11:** © William Ferguson; **39.13b:** © Robert A. Ross; **39.15a:** Field Museum of Natural History; **39.17:** © William Ferguson; **39.18a:** © E. R. Degginger; **39.18b:** © Phil Degginger/E. R. Degginger; **39.19b:** © John D. Cunningham; **39.21a:** © Kennon Cooke/Valan Photos; **39.21b:** © A. Nelson/Tom Stack & Associates; **39.21c:** © R. E. Lyons; **39.21d:** © E. S. Ross; **39.22a:** VU/© C. P. Hickman; **39.22b:** USDA Forest Service; **39.22c:** © Robert and Linda Mitchell; **39.24:** C. Y. Shih; **39.26a:** © Judy Hile; **39.26b:** © E. R. Degginger; **39.26c:** © Bob Coyle; **39.26d:** © Jerg Kroener.

Chapter 40

40.1a: © Nancy Sefton/Photo Researchers, Inc.; **40.2a:** © L. G. Johnson; **40.6c:** © G. B. Chapman; **40.8:** © BioMedia Associates; **40.10:** © Peter Parkes/Oxford Scientific Films; **40.11:** © N. G. Daniel/Tom Stack & Associates; **40.13a:** © Robert Evans/Peter Arnold, Inc.; **40.13b:** © Carl Roessler/Tom Stack & Associates; **40.14:** © M. S. Laverack; **40.15a:** © Bruce Russell/BioMedia Associates; **40.19:** © Ward's Natural Science; **40.21:** © Harvey Blankenspoor; **40.22b:** © John D. Cunningham; **40.25a:** © Dwight Kuhn; **40.26:** © Ed Reschke/Peter Arnold, Inc.; **40.27:** John F. Kessel, photographer, from Markell, E. K. and Voge, M.: MEDICAL PARASITOLOGY, 6th ed. © 1986 W. B. Saunders Company; **40.28:** © E. R. Degginger; **40.29:** © Roland Birke/Peter Arnold, Inc.; **40.30a:** © Leland Johnson; **40.32a:** © Robert A. Ross; **40.32b:** © Doug Faulkner, Sally Faulkner Collection; **40.33:** © David Spier/Tom Stack & Associates; **40.34a:** © Michael DiSpezio; **40.34b:** © William Ferguson; **40.37a:** © Michael DiSpezio; **40.37b:** © Paul L. Janosi/Valan Photos; **40.38c:** © Michael DiSpezio; **40.39:** © E. S. Ross; **40.40:** © Raymond A. Mendez/Animals, Animals; **40.42a–b:** © Michael DiSpezio; **40.43:** © Dwight Kuhn; **40.44a:** © Michael DiSpezio; **40.44b:** © Russ Kinne/Photo Researchers, Inc.; **40.46a:** © Leonard Lee Rue/Photo Researchers, Inc.; **40.47a:** © Tom McHugh/Photo Researchers, Inc.; **40.47b:** © Don and Pat Valenti; **40.48a:** Kessel and Shih/© Springer-Verlag, Berlin; **40.48b:** © Michael DiSpezio; **40.49:**

© Gary Milburn/Tom Stack & Associates; **40.50a:** © BioMedia Associates; **40.50b:** © Zig Lezczynski/Animals, Animals; **40.50c:** © Kent Dannon/Photo Researchers, Inc.; **40.50d:** © Michael DiSpezio; **40.50e:** © Noel R. Kemp/Photo Researchers, Inc.; **40.51b:** © Tom Stack/Tom Stack & Associates; **40.52:** © Tom McHugh/Photo Researchers, Inc.; **40.53:** © Robert and Linda Mitchell; **40.55d:** © Manfred Kage/Peter Arnold, Inc.; **40.56:** © Irene Vandermolen/Photo Researchers, Inc.; **40.58:** © Virginia Weinland/Photo Researchers, Inc.; **40.59:** © Robert and Linda Mitchell; **40.60a:** © Michael DiSpezio; **40.60b:** © Carolina Biological Supply Company; **40.60c:** © Leland Johnson; **40.61b:** © Michael DiSpezio; **40.62a:** © Robert Dunne/Photo Researchers, Inc.; **40.62b:** © H. Wes Pratt/BPS; **40.63a:** © Michael DiSpezio; **40.63b:** VU/© R. F. Myers; **40.63c, 40.64a:** © Michael DiSpezio; **40.64c:** © David L. Meyer; **40.64d:** © Carl Roessler/Tom Stack & Associates; **40.65a:** © C. R. Wyttenbach, University of Kansas/BPS; **40.65b–c:** © Douglas P. Wilson; **40.67a:** © Michael DiSpezio; **40.69:** © Russ Kinne/Photo Researchers, Inc.; **40.70:** Field Museum of Natural History; **40.71a:** © Ed Robinson/Tom Stack & Associates; **40.71b:** © Michael DiSpezio; **40.71c:** © Carl Roessler/Tom Stack & Associates; **40.72a:** © Tom McHugh/Photo Researchers, Inc.; **40.72b:** © David Doubilet; **40.72c:** © Michael DiSpezio; **40.72d:** VU/© R. Wallace; **40.73a–b:** American Museum of Natural History; **40.75:** © L. L. T. Rhodes/Animals, Animals; **40.76a:** © George Porter/Photo Researchers, Inc.; **40.76b:** © Michael DiSpezio; **40.76c:** © Tom McHugh/Photo Researchers, Inc.; **40.78a–c:** © Dwight Kuhn; **40.79:** American Museum of Natural History; **40.80a:** © Cosmos Blank/Photo Researchers, Inc.; **40.81a:** © E. S. Ross; **40.81b:** © John Cancalosi/Tom Stack & Associates; **40.81c:** © E. S. Ross; **40.83a, b (right):** Field Museum of Natural History, Chicago; **40.83b (left):** American Museum of Natural History, Charles R. Knight, artist; **40.83c:** Field Museum of Natural History, Chicago; **40.84, 40.85:** American Museum of Natural History; **40.86a:** © E. R. Degginger; **40.86b:** © Nancy Vander Sluis; **40.87a:** © L. Walker/Photo Researchers, Inc.; **40.87b:** © Helen Williams/Photo Researchers, Inc.; **40.88a–b:** © Tom McHugh/Photo Researchers, Inc.; **40.89a:** © Carl Purcell/Photo Researchers, Inc.; **40.89b:** © Tom McHugh/Photo Researchers, Inc.; **40.90:** © Stephen J. Kraseman/Photo Researchers, Inc.

Renal threshold for glucose, 338
Renal vein, 334, **335**
Renin, 342, **343**
Rennin, 124
Replication fork, **563, 564**
Replication of DNA, **561–66**
  bidirectional, **563, 564**
  in eukaryotic cells, 565, **566**
  experimental proofs of
      semiconservative, 561,
      562, 563
  leading/lagging strands, **565**
  Okazaki fragments, **565**
  in prokaryotic cells, **563–64**
  rolling circle (unidirectional), 564
Replicons, **566**
Repressible operons, 583, **584**
Repressor, 583
Reproduction, **9, 10,** 485, 616, **687**
  asexual, **616–17**
  in green algae, 964, **965**
  human. *See* Human reproduction
  sexual. *See* Sexual reproduction
  See also Genetics
Reproductive cycles, 619–20
  timing of, 620–21
Reproductive isolation, 779
Reptilia, Class, **1050–53**
  eggs, 323–24, **325**
  family tree, **1052**
  fossil, **1050, 1053**
Reservoir (gullet), euglenoid, 943
Residual volume in lungs, 278
Resins, 979
Resolution in microscopy, 65
Resource depletion, 902
Resource partitioning, **853, 854**
Respiration. *See* Breathing; Cell
      respiration;
      Photorespiration
Respiratory center, 279
Respiratory control, 178, **179**
Respiratory pigments, 264, **265,** 291,
      292
Respiratory system, human, **276,**
      277–79, 283
  problems in, 284–85
Respiratory tree, 1039, **1040**
Resting membrane potential, 391, 392,
      393
Restriction endonucleases (enzymes),
      602, **604**
Restriction point, cell cycle, 489
Reticular activating system (RAS), **420**
Reticular formation, **420**
Reticulopodia, 936, **937**
Retina, **37,** 428, **429**
  circulation in, **301**
  light on, **430, 431, 432**
Retinal, **432**
Retroviruses, 721
Reverse mutation, 771
Reverse transcriptase, 721
Reverse transcription, 608
Rh blood types, 308–309
Rheobatrachus silus, 620
Rheumatoid arthritis, 716
Rhizobium and nitrogen fixation, **208,**
      209, 928
Rhizoids, 98, **99, 946,** 950, 968
Rhizomes, 972
Rhizopus stolonifer, **949**
Rhodophyta, 964, 966, **967**
Rhodopsin, **37, 432**
Rhynchocephalia, 1050, **1051**
Riboflavin (vitamin B2), **240, 241**
Ribonuclease, **49, 254**
Ribonucleic acid. *See* RNA

Ribosomal RNA (rRNA), 70, **71,** 579,
      580
Ribosomes, 70, **71, 82, 83,** 920
  and tRNA, 579, **580, 581**
Ribulose bisphosphate. *See* RuBP
Ribulose (mono) phosphate, 150
Rich, Alexander, 560
Rickets, **239**
Riftia pachyptila, 875
Ring dove, **816,** 817
RNA (ribonucleic acid), 556
  -amino acid directory, **577**
  complimentary base pairing
      between DNA and, 561,
      575
  and oncogenic viruses, 721, **722**
  synthesis of, 378
  transcription, 577, **578,** 579
  See also Messenger RNA;
      Ribosomal RNA; Small
      nuclear RNA; Transfer RNA
RNA polymerase, 579
RNA replicase, 915
RNA transcriptase, 915–16
Robertson, J. D., 54
Rocky Mountain spotted fever, **1024**
Rodentia, Order, **1058**
Rods (eye), **431, 432**
Roeder, Kenneth, 822–23
Root(s) and root systems, 190,
      194–201
  cap, 196
  development of, **191**
  hairs, **196,** 197, **198**
  internal structure of, 197, **198–200**
  pressure, 222
  region of elongation, **197**
  region of maturation, **197, 198**
  tip, 191, 196, **197**
  water movement in, 201, **202, 203**
Rosettes, 468
Ross, Cleon, 134
Rotifera, Phylum, **1007,** 1008
Rough (granular) endoplasmic
      reticulum, 70, **71, 82, 83**
Round window (ear), **427**
Roundworm. *See* Nematodes
Rous, Peyton, 721
Rous sarcoma virus, 721
Roux, Wilhelm, 675
rRNA. *See* Ribosomal RNA
r-selected species, 851
r-strategists, 851
RuBP (ribulose bisphosphate), 149, 154
  regeneration, 150, **151**
Rumen, **43, 44,** 45, 248
Rusts, 247
Rutherford, Ernest, 26

Saccinobacculus ambloaxostylus, **81**
St. Martin, Alexis, 234
Salamanders, 1047, **1048, 1049**
  regeneration in, **718**
Salisbury, Frank, 134
Saliva, 251
Salivary amylase, 251, 253, **254**
Salivary glands, 251
Salmon, **333**
Salmonella, 594
Saltatory conduction, 396, **397**
Salt marshes, 890
Salts, body,
  in marine and freshwater fish, 332,
      333
  regulation of, in humans, 340, **341**
  See also Sodium
Salvia plant, **215**

Sampling error, 769
Sand dollars, **1038,** 1039
  sexual reproduction in, 623–30
Sanger, Frederick, 45, 569
San Salvadore Island, 110
Saprophytes, 247, 946
Sapwood, 220
Sarcodina, Phylum, 936, **937, 938**
Sarcolemma, 441, **448,** 450–51
Sarcoma, 720
Sarcomere, **448,** 449
Sarcoplasmic reticulum, 451, **452**
Sargasso Sea, 964
Sargassum, 964
Sautuola, Marcelino de, 4
Savannas, 791, **792,** 897, **898**
Scales, sporophyll, 979
Scallops, 1010
Scanning electron microscope (SEM),
      65
Scar tissue, 717
Schally, Andrew, 360
Schierenberg, Einhard, 672
Schistosoma mansoni, **1000, 1001,**
      1002
Schistosomiasis, 1000–1002
Schizophrenia, 404
Schleiden, Mathias Jakob, 62, 488
Schrödinger, Erwin, 553
Schwann, Theodor, 62, 99, 488
Schwann cells, 396, **397**
Sciatic nerve, 441, **442**
Scientific inquiry, xix–xx
Scientific method, 5, **6**
Sclera, 428, **429**
Sclereids, **106**
Sclerenchyma, **106,** 216
Scolex, 1002, **1003**
Scorpion, 1019, **1023**
Scramble competition, 850
Scurvy, **239**
Scyphozoa, Class, 994, **995,** 996
Sea anemones, 991, 994, **996**
Sea cucumbers, 1039, **1040**
Sea lillies, 1039, **1040**
Seasonal isolation, 779
Sea squirts, **1042,** 1043
Sea stars, 1035–37
Sea urchin, **1038,** 1039
  effect of hypertonic medium on, 91
  sexual reproduction in, 623–30
Seawater,
  human body and, **321**
  ionic composition of, and animal
      body fluid, **33, 320, 327,**
      332
  oxygen dissolved in, **263**
Secondary cell wall, 80, **81**
Secondary consumers, 869
Secondary lysosomes, **73, 74**
Secondary oocyte, 622
Secondary plant tissues, 215, **220**
Secondary production, 874
Secondary protein structure, 47, **48**
Secondary receptors, 423
Secondary response to antigens, **711**
Secondary root, 194
Secondary sex characteristics, 644
Secondary spermatocytes, 621
Secondary succession, **867,** 868
Secretin, 256
Secretory vesicles of Golgi apparatus,
      72
Sedimentary cycles, 877, 878–79, **880**
Seedlings, 639, **640**
Seed plants, 976. *See also*
      Angiosperms;
      Gymnosperms

Seeds, 638, **639,** 976
  coat, 638, **640,** 976
  germination, 191, 470, 471, **472,**
      639, **640**
Segmented bodies, 1013, 1014, 1043
Segregation, law of, 510
Selection pressure, 809
Self-assembly, principle of, 49, 50
Semen, 378
Semicircular canals, 425, **427**
Semilunar valves, 297, **298**
Seminal fluid, 644, **645, 646**
Seminal receptacles, **1016,** 1017
Seminal vesicles, 378, 645, **646**
Seminiferous tubules, 644, **645**
Sendai virus, **56,** 546
Senescence in plants, 482
Senses,
  corresponding areas of the brain
      for, **418, 419**
  ears (hearing), 425, 426, **427**
  receptors in skin, **424**
  smell, 424, **425**
  taste, 425, **426**
  types of, **423**
  vision, 428, **429–32,** 433
Sensitivity periods in fetal
      development, 658
Sensory adaptation, 424
Sensory cells, Hydrozoa, 993
Sepals, 635, **636,** 981
Sequoia tree, **15**
Serine, **46, 185**
Serotonin, **402**
Serum, 312, 321
Sessile, 988
Setae, 969, 1014
Sex attractants, 379
Sex determination, 534, **535,** 536
Sex influence inheritance, 537, **538–40**
Sex linkage inheritance, 535–36, **537**
Sex ratios, 536
Sexual dimorphism, 623
Sexually transmitted disease, 666–68
Sexual reproduction, 488, 618
  in animals, 618–35
  meiosis in, 496–505
  in plants, 635–42
Shaffer, B. M., 680
Sharks, 1044, **1045**
  digestion in, 248, **249**
  excretion in, **332,** 333
Shells, egg, 1050
Shells of electron orbitals, 25
Sherrington, Charles, 397, 400
Shivering, 839, **840**
Shoot apex, 638
Short-day plants, 473, **474, 475**
Shrimp, 1024
SI. *See* International System of Units
Sickle-cell anemia, **51,** 52, 523, **569,**
      609, 772, **773**
Sieve plate, **226**
Sieve tubes, **226**
Signal hypothesis, 71
Signal sequence, 71
Sign stimulus, 827
Silent mutation, 588
Simple fruits, 639, **641**
Simple (passive) diffusion, 89
  vs. carrier-mediated transport, **93**
Simple twitch of muscles, 442, **443**
Simpson, G. G., 766
Singer, Marcus, 718
Singer, S. J., 55, 94
Single-circuit circulatory system, **293**
Single covalent bonds, 28, **29**
Sinks, 225

Nobel Laureates in Physiology/Medicine *continued*

Nobel Laureate	Country*	Year of Award	Accomplishment
Bernardo Alberto Houssay	United States	1947	Discovery of how glycogen is catalytically converted
Carl F. Cori	United States (Hungary)		
Gerty T. Cori	United States (Hungary)		
Paul Müller	Swtz.	1948	Discovery of the insect-killing properties of DDT
Walter Rudolf Hess	Switzerland	1949	Research on brain control of body
Egas Moniz	Portugal		
Edward Calvin Kendall	United States	1950	Discoveries concerning the suprarenal cortex hormones
Philip Showalter Hench	United States		
Tadeus Reichstein	Swtz. (Poland)		
Max Theiler	United States (South Africa)	1951	Development of a vaccine for yellow fever
Selman A. Waksman	United States	1952	Discovery and development of streptomycin
Fritz A. Lipmann	United States (Ger.)	1953	Studies on the metabolism of carbohydrates in cells
Hans Adolph Krebs	England (Germany)		
John F. Enders	United States	1954	Cultivation of polio viruses in cell culture
Thomas H. Weller	United States		
Frederick C. Robbins	United States		
Hugo Theorell	Sweden	1955	Studies on oxidation enzymes

Nobel Laureate	Country*	Year of Award	Accomplishment
Dickinson W. Richards, Jr.	United States	1956	Techniques in treating heart disease
Andre F. Cournand	United States (France)		
Werner Forssmann	Germany		
Daniel Bovet	Italy (Swtz.)	1957	Investigations on antihistamines
Joshua Lederberg	United States	1958	Studies on the biochemistry of microbial genetics
George W. Beadle	United States		
Edward L. Tatum	United States		
Severo Ochoa	United States (Spain)	1959	Discoveries of the mechanisms of synthesis of DNA and RNA
Arthur Kornberg	United States		
Sir R. Macfarlane Burnet	Australia	1960	Studies on immunologic tolerance
Peter Brian Medawar	England (Brazil)		
George von Bekesy	United States (Hungary)	1961	Studies on the function of cochlea
James D. Watson	United States	1962	Determination of the structure of DNA
Francis H. C. Cricks	England		
Maurice H. F. Wilkins	England		
Alan Lloyd Hodgkin	Australia	1963	Research on nerve cells
Andrew Fielding Huxley	England		
John C. Eccles			
Konrad Emil Bloch	Germany	1964	Regulation of cholesterol and fatty acid metabolism
Feodor Lynen	Germany		
Francois Jacob	France	1965	Studies on the regulation of gene activity in the cell
Jacques Monod	France		
Andre Lwoff	France		
Charles Brenton Huggins	United States (Can.)	1966	Treatment of prostate cancer
Francis Peyton Rous	United States		Discovery of tumor-producing viruses

*Country of birth is in parenthesis